# Microbial Technologies for Sustainable Waste Management

# Developments in Microbial Cell Factories: From Design to Commercial Production

Series Editor

**Farshad Darvishi Harzevili**

Series Website

https://www.elsevier.com/books-and-journals/book-series/developments-in-microbial-cell-factories-from-design-to-commercial-production

Developments in Microbial Cell Factories: From Design to Commercial Production

# Microbial Technologies for Sustainable Waste Management


**Edited by**

**Rajneesh Kumar**
School of Biosciences and Technology,
Galgotias University, Uttar Pradesh, India

**Maulin P. Shah**
Department of Research Impact and Outcome,
Research and Development Cell, Lovely Professional University,
Phagwara, Punjab, India

**Dibyajyoti Haldar**
Manipal Institute of Technology, Manipal Academy of
Higher Education, Manipal, India



ELSEVIER

ACADEMIC PRESS
An imprint of Elsevier

Academic Press is an imprint of Elsevier
125 London Wall, London EC2Y 5AS, United Kingdom
50 Hampshire Street, 5th Floor, Cambridge, MA 02139, United States



ISBN: 978-0-443-44093-9

For information on all Academic Press publications visit our website at https://www.elsevier.com/books-and-journals

*Publisher:* Megan Ball
*Acquisitions Editor:* Michelle Fisher
*Editorial Project Manager:* Shreya Jain
*Production Project Manager:* Karthiga Muthukumarappan
*Cover Designer:* Arumugam Kothandan

Typeset by Integra Software Services

# Contents

Contributors ........................................ xvii
About the editors ........................................ xxi

**CHAPTER 1 Microbial remediation of hazardous and toxic waste ........................................ 1**

Poulami Banerjee and Basha Shaik

**1** Introduction ........................................ 1
1.1 Classification and sources of hazardous and toxic waste ........ 3
**2** Impacts on environmental and public health ........................................ 6
**3** Techniques and mechanisms of microbial remediation ........................................ 7
3.1 Microbial remediation of HMs and metalloids ........................................ 9
3.2 Remediation of antibiotics ........................................ 12
3.3 Remediation of HC ........................................ 12
3.4 Remediation of pesticides ........................................ 15
3.5 Remediation of radioactive waste ........................................ 17
3.6 Remediation of dye ........................................ 18
3.7 Microbial remediation of chlorinated solvents ........................................ 20
3.8 Advances and emerging technologies in microbial remediation ........................................ 22
**4** Future perspectives ........................................ 26
**5** Conclusions ........................................ 27
References ........................................ 28

**CHAPTER 2 Microbial consortia for bioremediation: The actinobacteria case in pesticide degradation... 51**

Lucas Ariel Bazán, Enzo Emanuel Raimondo, Claudia Susana Benimeli, and María Soledad Fuentes

**1** Introduction to pesticides pollution ........................................ 51
**2** Pesticides: Definition, classification, mobility, and pollution ....... 51
**3** Eco-friendly solutions to the pesticide pollution issue ........................................ 54
3.1 Bioremediation ........................................ 55
**4** Microbial strategies for bioremediation ........................................ 55
4.1 Natural attenuation ........................................ 55
4.2 Biostimulation ........................................ 56
4.3 Bioaugmentation ........................................ 57
4.4 Combining BS and BA ........................................ 57

**5** Potential microorganisms for environmental bioremediation ...... 58
**6** Phylum Actinomycetota in bioremediation ...... 59
6.1 General characteristics of the phylum Actinomycetota ...... 59
6.2 Bioremediation potential of the phylum Actinomycetota ...... 60
**7** Microbial teamwork: Approaches to mixed cultures and MCs ...... 61
7.1 MCs for bioremediation ...... 62
7.2 Actinomycetota consortia for pesticide bioremediation ...... 64
7.3 BA with free actinobacterial consortia ...... 65
7.4 BA with immobilized actinobacterial consortia ...... 69
**8** Conclusion ...... 70
Acknowledgement ...... 70
References ...... 70

**CHAPTER 3 Hybrid microbial–physical processes for enhanced contaminant removal in water treatment ...... 79**
Rodrigo Gutiérrez-Ramírez, Samira El Omari, Imane Haydari, Zineb Kassab, Mohamed Laabd, Karim Benhabib, and Youness Abdellaoui
**1** Introduction ...... 79
**2** Hybrid microbial systems in waste management ...... 81
2.1 Microbial electrochemical systems ...... 81
2.2 Biophotocatalysis ...... 83
2.3 Bioadsorption systems ...... 85
2.4 Biological membrane filtration ...... 88
2.5 Biological oxidation process ...... 90
2.6 Other emerging combinations ...... 93
**3** Challenges and future perspectives ...... 95
**4** Conclusions ...... 96
References ...... 97

**CHAPTER 4 Microbial encapsulation and immobilization in waste management ...... 105**
Khouloud Barakat and Ahmed El Nemr
**1** Introduction ...... 105
**2** Microbial cell immobilization approaches ...... 107
2.1 Adhesion and adsorption ...... 108
2.2 Entrapment (entanglement) ...... 109
2.3 Cross-linking or covalent binding ...... 110
2.4 Encapsulation ...... 111
2.5 Advantages and disadvantages of microbial immobilization techniques ...... 115
2.6 Major applications of immobilization ...... 115

2.7 Minor applications of immobilization ............................... 118
3 Microbial cells supported or coated by nanomaterials .............. 120
3.1 Nanoencapsulation microbial cells .................................... 121
4 Biotechnology application of nanocell encapsulation ............... 126
5 Conclusions ................................................................................ 128
References ..................................................................................... 129

**CHAPTER 5 Bioprospecting microbes for polyvinyl chloride plastic biodegradation: Potential and challenges ................................................ 143**

Neha Hatwar and Asifa Qureshi

1 Introduction ................................................................................ 143
2 Biodegradation approaches involved in PVC degradation ......... 145
2.1 Pure culture isolation and identification ............................. 145
2.2 Microbial consortia designing ........................................... 145
2.3 Insect gut microbiota ........................................................ 146
2.4 Protoplast fusion and genetic engineering .......................... 146
2.5 Multienvironment pretreatment approach for PVC degradation ........................................................................ 146
2.6 Microbial enzyme application for PVC biodegradation ...... 146
3 Bioprospecting microbes from diverse environmental niches for PVC degradation ...................................................... 146
3.1 Soil ecosystems: Reservoirs of PVC-degrading microbes ............................................................................ 147
3.2 Marine and coastal environments: Microbial communities under extreme conditions .............................. 148
3.3 Landfill sites and industrial waste zones: Microbial adaptation to polymeric waste ............................................ 148
4 Analytical methods used for determining PVC biodegradation ........................................................................... 150
4.1 Molecular weight determination ........................................ 150
4.2 Functional group analysis ................................................. 152
4.3 Surface topography .......................................................... 152
4.4 Mineralization ................................................................... 152
5 Challenges for development of biobased PVC bioremediation process ............................................................ 152
6 Future scope of research on PVC bioremediation in the environment ................................................................... 153
7 Conclusions ................................................................................ 154
References ..................................................................................... 154

**CHAPTER 6 Microbial approaches for electronic waste decomposition and metal recovery ...................... 159**
Ifrah Rashid, Tasaduq Hussain Shah, Sobiya Gul, and Ahali Jahan
1 Introduction ........ 159
1.1 Background on electronic waste ........ 159
1.2 Composition and classification of e-waste ........ 160
1.3 Environmental and economic concerns ........ 160
2 Conventional e-waste management techniques ........ 161
2.1 Physical and chemical methods ........ 161
2.2 Limitations of conventional approaches ........ 162
3 Microbial approaches in e-waste management ........ 162
3.1 Overview of biological techniques ........ 162
3.2 Bioleaching: Microbial metal extraction ........ 163
3.3 Biosorption and bioaccumulation ........ 166
3.4 Enzymatic degradation ........ 166
4 Microbial mechanisms for e-waste decomposition ........ 167
4.1 Microbial strains and e-waste degradation ........ 167
4.2 Metal recovery through microbes ........ 167
5 Advances in microbial technologies for e-waste recycling ........ 168
5.1 Genetic engineering ........ 168
5.2 Nanotechnology integration ........ 169
5.3 Microbial consortia ........ 169
5.4 Bioelectrochemical systems ........ 170
6 Challenges and limitations ........ 170
7 Integration of microbial processes into circular economy models ........ 171
8 Mint innovation: A microbial solution to urban e-waste mining 172
9 Future directions and research perspectives ........ 173
10 Conclusions ........ 174
References ........ 175

**CHAPTER 7 Nanotechnology-enabled microbial approaches in agricultural waste management ........................... 183**
Sevinç Yeşilyurt and Muazzez Gürgan
1 Nanotechnology and its importance in agriculture ........ 183
2 Classification of NPs ........ 184
3 Classification according to chemical composition ........ 185
3.1 Organic NPs ........ 185
3.2 Inorganic NPs ........ 185
3.3 Classification by structural properties ........ 185
3.4 Classification by physical characteristics ........ 186
3.5 Classification by production method ........ 186

3.6 Classification according to usage areas ............................ 186
4 Agricultural use of nanomaterials ............................ 186
5 Effects of nanotechnology in plant cultivation ............................ 187
6 Nanotechnology in managing agricultural waste ............................ 189
7 Nanotechnology and soil health ............................ 189
8 The future of nanotechnology in agriculture ............................ 190
8.1 Nanobiosensors and smart agriculture ............................ 190
8.2 Nanofertilizers and nutrient use efficiency ............................ 191
8.3 Soil health and nanoremediation ............................ 191
8.4 Resilience to climate stress ............................ 192
9 Conclusions ............................ 192
References ............................ 193

**CHAPTER 8 Enzyme-assisted biomass conversion for sustainable waste management ............................ 201**

Yehia A.G. Mahmoud and Aya Ahmed Beltagy

1 Introduction ............................ 201
2 Biomass classification ............................ 202
2.1 Categories of biomass ............................ 202
2.2 Chemical composition of biomass ............................ 205
3 Enzymes involved in biomass conversion ............................ 205
3.1 Cellulolytic enzymes ............................ 205
3.2 Hemicellulases ............................ 208
3.3 Ligninolytic enzymes ............................ 210
4 Microbial production of lignocellulolytic enzymes ............................ 213
4.1 Production of lignocellulolytic enzymes from fungi ............................ 213
4.2 Bacterial lignocellulolytic enzymes ............................ 215
4.3 Lignocellulolytic enzyme production from recombinant microorganisms ............................ 216
4.4 Techniques used for the production of lignocellulolytic enzymes ............................ 217
4.5 Biomass conversion optimization for enhanced enzyme production ............................ 219
5 Recent advances in enzyme technology ............................ 220
6 Case studies and industrial applications of lignocellulolytic enzymes ............................ 221
6.1 Biofuels ............................ 221
6.2 Textile industry ............................ 222
6.3 Pulp and paper industry ............................ 223
6.4 Food and feed ............................ 223

6.5 Bioplastics and biocomposites.....224
**7** Challenges and limitations of enzymatic degradation.....224
**8** Conclusions and future research prospects.....226
References.....227

**CHAPTER 9 Anaerobic digestion: Microbial powerhouses for waste-to-energy conversion.....239**
Rahul Mondal and Subodh Kumar Maiti
**1** Introduction.....239
**2** Feedstock suitable for AD.....239
2.1 Biochemical methane potential.....239
2.2 Feedstocks for AD.....244
2.3 Feedstock characterization.....246
**3** Pretreatment of feedstock.....247
3.1 Need for pretreatment.....247
3.2 Physical pretreatment.....248
3.3 Chemical pretreatment.....249
3.4 Physicochemical pretreatment.....249
3.5 Biological pretreatment.....251
**4** Chemistry of AD.....252
4.1 Process microbiology of AD.....252
4.2 Syntrophy in AD.....260
**5** Process description of AD.....261
5.1 Design considerations and operational parameters.....261
5.2 Inhibitors in AD.....265
**6** Products of AD.....266
6.1 Biogas.....266
6.2 Digestate.....267
**7** Environmental impacts of AD.....268
7.1 Environmental impact assessment.....268
7.2 Life-cycle assessment.....269
**8** Conclusion.....269
References.....270

**CHAPTER 10 Microbial conversion processes for resource recovery and circular economy.....279**
Pradeep Kumar Ramteke and Ajit P. Rathod
**1** Introduction.....279
1.1 Global challenges: Waste, resources, and environment.....280
1.2 Role of microbial processes in converting waste into useful properties.....280
**2** Fundamentals of microbial conversion processes.....282

3 Applications in resource recovery ........ 284
3.1 Bioenergy production ........ 284
3.2 Nutrient recovery ........ 285
3.3 Bioproduct synthesis ........ 286
3.4 Bioremediation ........ 287
4 Case studies: Microbial conversion processes for resource recovery ........ 288
5 Challenges, limitations, and future directions ........ 289
6 Conclusion ........ 290
References ........ 291

**CHAPTER 11 Microbial fuel cells: Harnessing energy from waste ........ 299**

Abid Ullah, Muhammad Waqar, Muhammad Humayun, and Sumayya Shah

1 Introduction ........ 299
2 Energy-generating capacity of MFC ........ 301
2.1 Current density and power density ........ 301
3 Energy harvesting from various wastes by MFC ........ 302
3.1 Domestic wastewater ........ 302
3.2 Food wastes ........ 302
3.3 Landfill leachate ........ 304
3.4 Complex industrial wastes ........ 304
3.5 Sewage sludge ........ 305
3.6 Animal wastes ........ 305
3.7 Plant wastes ........ 306
4 Energy capture and storage with MFC ........ 308
4.1 Energy storage technologies for MFC ........ 309
4.2 Capacitor-based power management system ........ 309
5 Direct power utilization from MFCs ........ 311
5.1 Microbial electrolysis cell ........ 312
5.2 Microbial desalination cell ........ 312
5.3 MFC-assisted photoelectrocatalytic system and biophotoelectrochemical cell ........ 313
6 Challenges and perspectives ........ 314
References ........ 316

**CHAPTER 12 Microbial biotechnologies in industrial applications: Fundamentals, innovations, and future prospects ........ 321**

Deepchandra Joshi, Mustari Abrar, Babu Alappat, and T.R. Sreekrishnan

1 Introduction ........ 321
2 Fundamentals of microbial technologies ........ 323

2.1 Key microbial industrial sectors ........................................ 324
2.2 Advantages over conventional technologies........................ 325
**3** Microbial enzymes.................................................................. 326
**4** Bioelectrochemical systems..................................................... 329
4.1 MFCs scale-up ...................................................................... 331
**5** Challenges and techno-economic barriers ................................ 333
**6** Future prospects........................................................................ 334
References ..................................................................................... 335

**CHAPTER 13 Artificial intelligence integration in microbial waste management ........................................... 341**
Vibha Gajbe, Aarav Gupte, and Archisha Patel
**1** Introduction.............................................................................. 341
**2** AI models in microbial waste management............................... 342
2.1 Artificial neural network ...................................................... 343
2.2 Support vector machine ....................................................... 346
2.3 Random forest...................................................................... 346
2.4 Decision trees....................................................................... 347
**3** Application of AI in microbial waste management ................... 347
3.1 Waste generation prediction................................................ 347
3.2 Waste collection ................................................................... 348
3.3 Waste classification and sorting systems ............................. 348
3.4 Bin detection ........................................................................ 349
3.5 Vehicle and route optimization ........................................... 350
**4** ML for process optimization and performance prediction ........................................................................ 350
4.1 Wastewater treatment efficiency ......................................... 351
4.2 Composting.......................................................................... 352
4.3 Performance evaluation of microbial fuel cells................... 352
**5** Process outcome prediction ..................................................... 353
5.1 Bioreactor performance prediction...................................... 353
5.2 Enhancing microbial biodegradation.................................. 354
**6** Conclusions.............................................................................. 355
**7** Challenges and future scope ..................................................... 356
References ..................................................................................... 356

**CHAPTER 14 Advanced microbial techniques for wastewater treatment ......................................................... 363**
T.G. Ashwini, Anum Haleema Ali, and Niladri Shekhar Samanta
**1** Introduction.............................................................................. 363
**2** Emerging trends in immobilized microbial technology for wastewater treatment........................................................... 364

**3** Selection of immobilization carriers........365
3.1 Natural carrier materials........368
3.2 Nanocarrier materials........368
3.3 Metal–organic frameworks........370
3.4 Biochar carrier materials........370
**4** Novel immobilized microbial technology........373
4.1 Sequential film deposition via LBL for microbial immobilization........373
4.2 Electrostatic spinning........374
**5** Biological nitrogen removal from wastewater by immobilization technology........375
5.1 Gel bed embedding of anammox bacteria for nitrogen removal........375
5.2 Immobilization of AD bacteria for nitrogen removal........376
**6** Immobilized microbial technology for enhanced pollutant removal in refractory wastewater........378
**7** Future scope for enhanced wastewater treatment........378
**8** Conclusions........379
References........379

**CHAPTER 15 Microbial innovations in waste management........385**
Aparna Pandey, Debarati Paul, Samarth Gohel, and Sourish Bhattacharya
**1** Introduction........385
**2** Recycling and circular economy........387
**3** WtE technologies........389
3.1 Thermal conversion technologies........390
3.2 Biological conversion technologies........391
3.3 Bioelectrochemical and electroactive systems........392
3.4 Hybrid/integrated systems........392
**4** An insight into organic municipal waste and its prospects........393
**5** Different pollutants and their toxicity to living organisms........394
5.1 Air pollutants........394
5.2 Water pollutants........394
5.3 Soil pollutants........395
5.4 Radioactive pollutants........396
**6** Types of remediation........396
6.1 Physical remediation........396
6.2 Chemical remediation........397
6.3 Biological remediation (bioremediation)........397
6.4 Thermal remediation........398
6.5 Enzymatic oxidation........398

**7** Different microbes used as bioremediation agents ..................... 399
**8** Mechanisms of microbial bioremediation ................................. 399
8.1 Microbial bioremediation using enzymes ........................... 399
8.2 Enzymatic reduction ........................................................ 400
8.3 Bioaugmentation .............................................................. 400
8.4 Biostimulation .................................................................. 400
8.5 Bioleaching ...................................................................... 401
8.6 Biosorption ...................................................................... 401
8.7 Bioaccumulation ............................................................... 402
8.8 Precipitation ..................................................................... 402
**9** Factors affecting microbial bioremediation ............................... 402
9.1 pH ..................................................................................... 402
9.2 Temperature ...................................................................... 403
9.3 Moisture content ............................................................... 403
9.4 Nutrient availability .......................................................... 403
9.5 Oxygen availability ........................................................... 403
**10** Microbial enzymes used in bioremediation ............................... 403
**11** Molecular approaches for validating microbial remediation ...... 403
11.1 Polymerase chain reaction and qPCR for functional gene tracking ................................................................... 404
11.2 Community profiling (Terminal-Restriction Fragment Length Polymorphism (T-RFLP), Automated Ribosomal Intergenic Spacer Analysis (ARISA)) ............. 405
11.3 SIP and metagenomics ..................................................... 405
11.4 Metatranscriptomics, proteomics, and metabolomics ........ 405
**12** Other bioremediation metabolites produced by microbes .......... 405
**13** Recent advancements in microbial bioremediation .................... 405
**14** Future perspectives and conclusions ......................................... 406
**15** Recommendations ...................................................................... 407
References ....................................................................................... 408

**CHAPTER 16 Sustainable waste management: Challenges, opportunities, and future roadmap ....................... 417**
Barun Kanoo and Shruti Sharma
**1** Introduction ................................................................................. 417
1.1 Overview of India's waste generation trends and its impact ......................................................................... 417
1.2 Relevance of sustainable waste management in the Indian context and its impact on attaining Sustainable Development Goals ......................................... 418
1.3 Scope of the study ............................................................. 419

**2** Current status of MSW management in India ... 419
2.1 Composition of MSW in India ... 419
2.2 Current practices in India ... 419
2.3 Environmental, economic, and social implications of the current practice of MSW management ... 420
**3** Policy frameworks and national initiatives ... 420
3.1 Policy frameworks ... 420
3.2 National initiatives ... 427
**4** Technological interventions and innovations ... 430
4.1 Composting, vermicomposting, and anaerobic digestion ... 430
4.2 Waste-to-energy technologies ... 431
4.3 Thermochemical conversion of waste ... 432
4.4 Digital technologies and IoT for waste management ... 432
**5** Systemic gaps and challenges ... 432
**6** Lessons learned: Success stories around the globe ... 433
6.1 Case Study 1: Teocelo, Veracruz, Mexico ... 433
6.2 Case Study 2: San Fernando, Pampanga, Philippines ... 434
6.3 Case Study 3: Kamikatsu, Japan ... 435
6.4 Case Study 4: Pune, Maharashtra, India ... 436
**7** Conclusions ... 437
References ... 438

**CHAPTER 17 Biotechnological approaches for zero waste agriculture ... 445**
Tapas Ranjan Das, Sundram Singh, Abhishek Singh, and Pooran Chand
**1** Introduction ... 445
**2** Importance of zero-waste agriculture ... 447
**3** Bioplastics from agricultural and dairy waste ... 447
**4** Bioenergy from agricultural and dairy waste ... 448
**5** Genome editing to reduce plant waste ... 448
**6** Biofuel generations and evolution ... 449
6.1 Sustainable biochemical conversion of agricultural waste ... 450
6.2 Advancements in microbial biofuels ... 451
6.3 Vermicomposting: An effective approach to agricultural waste management ... 452
**7** Using agricultural waste for ethanol production ... 454
7.1 Fermentation-based valorization ... 454
7.2 Anaerobic digestion process for biofertilizers and renewable energy ... 454
7.3 Production of biohydrogen ... 455

**8** CRISPR–Cas genome engineering in advancing biofuel production ... 455
8.1 CRISPR for reducing yield and postharvest losses ... 457
8.2 Food waste valorization by CRISPR ... 457
**9** Ethanol pathway optimization by CRISPR–Cas system ... 458
**10** Different strategies for agricultural waste management ... 458
10.1 Traditional approaches ... 459
10.2 Emerging valorization methods ... 460
**11** Transforming agricultural residues using microalgae biotechnology ... 462
**12** Conclusions ... 463
References ... 464

Index ... 471

# Contributors

**Youness Abdellaoui**
Sustainability of Natural Resources and Energy, Environmental Functional Materials, Cinvestav Saltillo; Sustainability of Natural Resources and Energy, SECIHTI-Cinvestav Saltillo, Saltillo, Mexico

**Mustari Abrar**
Department of Biochemical Engineering and Biotechnology; Department of Civil Engineering, Indian Institute of Technology Delhi, New Delhi, Delhi, India

**Babu Alappat**
Civil Engineering, Indian Institute of Technology Delhi, New Delhi, Delhi, India

**Anum Haleema Ali**
Department of Petroleum Engineering, Presidency University, Bengaluru, India

**T.G. Ashwini**
Department of Petroleum Engineering, Presidency University, Bengaluru, India

**Poulami Banerjee**
Hyderabad Zonal Centre, CSIR-National Environmental Engineering Research Institute (CSIR-NEERI), IICT Campus, Tarnaka, Hyderabad, Telangana; Academy of Scientific and Innovative Research (AcSIR), Ghaziabad, Uttar Pradesh, India

**Khouloud Barakat**
National Institute of Oceanography and Fisheries (NIOF), Alexandria, Egypt

**Lucas Ariel Bazán**
Actinobacteria Biotechnology Laboratory, Pilot Plant for Microbiological Industrial Processes (PROIMI-CONICET), San Miguel de Tucumán, Argentina

**Aya Ahmed Beltagy**
Faculty of Science, Botany and Microbiology Department, Tanta University, Tanta, Egypt

**Karim Benhabib**
Eco-Processes, Optimization and Decision Support (EPROAD, EA), University of Picardie Jules Verne, Saint-Quentin, France

**Claudia Susana Benimeli**
Actinobacteria Biotechnology Laboratory, Pilot Plant for Microbiological Industrial Processes (PROIMI-CONICET), San Miguel de Tucumán; Faculty of Exact and Natural Sciences, National University of Catamarca, San Fernando del Valle de Catamarca, Argentina

**Sourish Bhattacharya**
Process Design and Engineering Division, CSIR-Central Salt and Marine Chemicals Research Institute, Bhavnagar, India

**Pooran Chand**
Department of Genetics and Plant Breeding, Sardar Vallabhbhai Patel University of Agriculture and Technology, Meerut, Uttar Pradesh, India

**Tapas Ranjan Das**
Division of Genetics, ICAR-Indian Agricultural Research Institute, New Delhi, India

**Ahmed El Nemr**
National Institute of Oceanography and Fisheries (NIOF), Alexandria, Egypt

**Samira El Omari**
Laboratory of Materials and Environment, Department of Chemistry, Faculty of Science, Ibn Zohr University, Agadir, Morocco; Eco-Processes, Optimization and Decision Support (EPROAD, EA), University of Picardie Jules Verne, Saint-Quentin, France

**María Soledad Fuentes**
Actinobacteria Biotechnology Laboratory, Pilot Plant for Microbiological Industrial Processes (PROIMI-CONICET), San Miguel de Tucumán, Argentina

**Vibha Gajbe**
School of Biotechnology & Bioinformatics, D.Y. Patil Deemed to be University, Navi Mumbai, India

**Samarth Gohel**
Process Design and Engineering Division, CSIR-Central Salt and Marine Chemicals Research Institute, Bhavnagar, India

**Sobiya Gul**
Faculty of Fisheries, Sher-e-Kashmir University of Agricultural Sciences and Technology of Kashmir, Srinagar, India

**Aarav Gupte**
School of Biotechnology & Bioinformatics, D. Y. Patil Deemed to be University, Navi Mumbai, India

**Muazzez Gürgan**
Department of Biology, Tekirdağ Namık Kemal Üniversitesi, Tekirdağ, Turkey

**Rodrigo Gutiérrez-Ramírez**
Sustainability of Natural Resources and Energy, Environmental Functional Materials, Cinvestav Saltillo, Saltillo, Mexico

**Neha Hatwar**
CSIR-National Environmental Engineering Research Institute (NEERI), Bio-prospecting Microbes for PVC Plastic Biodegradation: Potential and Challenges, Nagpur, India

**Imane Haydari**
Materials Science, Energy, and Nano-Engineering (MSN), Mohammed VI Polytechnic University (UM6P), Benguerir, Morocco

**Muhammad Humayun**
Department of Basic Sciences, University of Engineering & Technology, Peshawar, Pakistan

**Ahali Jahan**
Faculty of Fisheries, Sher-e-Kashmir University of Agricultural Sciences and Technology of Kashmir, Srinagar, India

**Deepchandra Joshi**
Department of Biochemical Engineering and Biotechnology, Indian Institute of Technology Delhi, New Delhi, Delhi, India; Institute of Energy and Environment, Pennsylvania State University, State College, University Park, PA, United States

**Barun Kanoo**
Department of Civil Engineering, National Institute of Technology Silchar, Silchar, Assam, India

**Zineb Kassab**
Department of Materials Science, Energy, and Nano-engineering (MSN), College of Chemical Sciences and Engineering (CCSE), Mohammed VI Polytechnic University, Benguerir, Morocco

**Mohamed Laabd**
Laboratory of Materials and Environment, Department of Chemistry, Faculty of Science, Ibn Zohr University, Agadir, Morocco

**Yehia A.G. Mahmoud**
Faculty of Science, Botany and Microbiology Department, Tanta University, Tanta, Egypt

**Subodh Kumar Maiti**
Department of Environmental Science and Engineering, Indian Institute of Technology (Indian School of Mines) Dhanbad, Dhanbad, India

**Rahul Mondal**
Department of Environmental Science and Engineering, Indian Institute of Technology (Indian School of Mines) Dhanbad, Dhanbad, India

**Aparna Pandey**
Process Design and Engineering Division, CSIR-Central Salt and Marine Chemicals Research Institute, Bhavnagar, India

**Archisha Patel**
School of Biotechnology & Bioinformatics, D.Y. Patil Deemed to be University, Navi Mumbai, India

**Debarati Paul**
Process Design and Engineering Division, CSIR-Central Salt and Marine Chemicals Research Institute, Bhavnagar, India

**Asifa Qureshi**
CSIR-National Environmental Engineering Research Institute (NEERI), Bioprospecting Microbes for PVC Plastic Biodegradation: Potential and Challenges, Nagpur, India

**Enzo Emanuel Raimondo**
Actinobacteria Biotechnology Laboratory, Pilot Plant for Microbiological Industrial Processes (PROIMI-CONICET); Faculty of Biochemistry, Chemistry and Pharmacy, National University of Tucumán, San Miguel de Tucumán, Argentina

**Pradeep Kumar Ramteke**
Department of Chemical Engineering, Visvesvaraya National Institute of Technology (VNIT), Nagpur, India

**Ifrah Rashid**
Faculty of Fisheries, Sher-e-Kashmir University of Agricultural Sciences and Technology of Kashmir, Srinagar, India

**Ajit P. Rathod**
Department of Chemical Engineering, Visvesvaraya National Institute of Technology (VNIT), Nagpur, India

**Niladri Shekhar Samanta**
Department of Petroleum Engineering, Presidency University, Bengaluru, India

**Sumayya Shah**
Department of Chemistry, University of Malakand, Malakand, Pakistan

**Tasaduq Hussain Shah**
Faculty of Fisheries, Sher-e-Kashmir University of Agricultural Sciences and Technology of Kashmir, Srinagar, India

**Basha Shaik**
Hyderabad Zonal Centre, CSIR-National Environmental Engineering Research Institute (CSIR-NEERI), IICT Campus, Tarnaka, Hyderabad, Telangana; Academy of Scientific and Innovative Research (AcSIR), Ghaziabad, Uttar Pradesh, India

**Shruti Sharma**
Dillon Consulting Limited, Vancouver, BC, Canada

**Abhishek Singh**
Department of Genetics and Plant Breeding, Sardar Vallabhbhai Patel University of Agriculture and Technology, Meerut, Uttar Pradesh, India

**Sundram Singh**
Animal Biochemistry Division, National Dairy Research Institute, Karnal, Haryana, India

**T.R. Sreekrishnan**
Department of Biochemical Engineering and Biotechnology, Indian Institute of Technology Delhi, New Delhi, Delhi, India

**Abid Ullah**
Advanced Energy and System Engineering, University of Science and Technology South Korea, Daejeon, South Korea

**Muhammad Waqar**
Department of Chemistry, University of Malakand, Malakand, Pakistan

**Sevinç Yeşilyurt**
Department of Soil Science and Plant Nutrition, Tekirdağ Namık Kemal Üniversitesi, Tekirdağ, Turkey

# About the editors

**Dr. Rajneesh Kumar** is a dedicated researcher focusing on wastewater treatment, in which he has made significant contributions. He is currently serving as a Assistant Professor in the School of Biosciences and Technology, Galgotias University, Uttar Pradesh, India. He obtained his Ph.D. from IIT Guwahati, where he worked on developing tools and techniques to assess the recovery potential of aerobic sludge biomass treated with copper, amoxicillin, chlorpyrifos, and piggery wastewater. He has laid a strong interdisciplinary foundation for his research in environmental biotechnology and pollution control. With over four years of research experience, he has contributed significantly to sustainable wastewater treatment, emerging pollutant removal, and microbial process optimization. He has authored and co-authored several peer-reviewed journal articles in reputed international journals and has also contributed to several book chapters. He serves as an Early Career Editorial Board Member of Emerging Contaminants and as a Guest Editor for a special issue in Explora: Environment and Resource. He actively reviews manuscripts for leading journals in the field. His work integrates biotechnological innovation, environmental microbiology, and process engineering toward advancing eco-efficient wastewater treatment technologies.

**Dr. Maulin P. Shah** has emerged as a globally recognized and highly influential researcher in Environmental Microbiology, with an unwavering dedication to scientific inquiry spanning more than two decades. His research excellence is firmly rooted in exploring the intricate and dynamic relationship between environmental pollution and microbial diversity in wastewater ecosystems, employing both cultivation-dependent and advanced cultivation-independent (molecular) approaches. Through this integrated framework, he has significantly enhanced the understanding of microbial structure, function, and resilience in polluted environments. His pioneering research has made substantial and lasting contributions to key domains of environmental microbiology, including biological wastewater treatment, biodegradation, bioremediation, and phytoremediation of environmental pollutants. His scholarly impact is reflected in an extraordinary publication record, comprising over 350 peer-reviewed research papers published in highly reputed national and international scientific journals. These publications have significantly advanced scientific knowledge and continue to serve as critical references for researchers, academicians, and professionals worldwide. Beyond research articles, he holds a distinguished editorial legacy, having served as editor for more than 250 internationally acclaimed scientific books. Further extending his influence, he actively serves on the editorial boards of several top-ranked international journals, where he contributes to maintaining high standards of scientific quality, ethics, and innovation. Through his research, publications, and editorial leadership, he stands as a commanding figure in environmental microbiology, driving scientific progress, fostering global collaboration, and advancing sustainable solutions for environmental protection.

**Dr. Dibyajyoti Haldar** is working as an Associate Professor, Research at Manipal Institute of Technology, Manipal Academy of Higher Education, Manipal, India. His research primarily focuses on biomass valorization and the development of sustainable biobased materials and fuels. His work emphasizes the conversion of lignocellulosic biomass into high-value chemicals, biofuels, and functional biomaterials through integrated biorefinery approaches. Prior to joining MIT Manipal, he served as an Assistant Professor in the Division of Biotechnology at Karunya Institute of Technology and Sciences, Coimbatore, India. He completed his Post-Doc as an Institute Postdoctoral Fellow at the Centre for the Environment of Indian Institute of Technology, Guwahati, India. Regarding his academics, he obtained his Ph.D. in Biochemical Engineering from the National Institute of Technology Agartala, India, and MTech in Environmental Science and Technology from the National Institute of Technology Durgapur, India. He has published more than 65 research and review articles in various SCI-indexed reputed international journals of high impact factors. In addition, he is the co-author of four books published by Elsevier Science. Moreover, he also contributed a few book chapters published in International Publishers. As per the Google Scholar, he has an excellent research profile with an H-index of 29 and an i10 index of 51. He is listed among the World's Top 2% Scientists for 2023–2025 according to the database released by Elsevier.

CHAPTER

# Microbial remediation of hazardous and toxic waste

# 1

**Poulami Banerjee[a,b] and Basha Shaik[a,b]**

[a]*Hyderabad Zonal Centre, CSIR-National Environmental Engineering Research Institute (CSIR-NEERI), IICT Campus, Tarnaka, Hyderabad, Telangana, India*

[b]*Academy of Scientific and Innovative Research (AcSIR), Ghaziabad, Uttar Pradesh, India*

## 1 Introduction

Hazardous and toxic materials are substances, energy, or components that can pollute and damage the environment, endangering health, survival, and other living organisms. Wastes generated as a result of anthropogenic activities contain hazardous materials, which can directly or indirectly harm the environment (Wang et al., 2019c; Gupta and Babu, 1999). The issue of hazardous and toxic waste management, particularly in communities, health facilities, and industrial areas, is often noncompliant with regulations and standard operating procedures. This waste can cause uneven distribution and spread across a wide environmental matrix, and it can be carried by the wind, dumped into rivers, washed away, and eventually end up in the sea or a lake. The Basel, Rotterdam, and Stockholm Conventions are international treaties that address hazardous waste and chemical management. The Basel Convention regulates transboundary waste movement, the Rotterdam Convention encourages shared responsibility, and the Stockholm Convention safeguards human health and the environment from persistent organic pollutants (POPs) (*International trade control measures under the Basel- Rotterdam and Stockholm Conventions*, 2023).

Problematic wastes commonly originate from residential, industrial, agricultural, mining, and recreational activities (Taufan and Purwanto, 2018). While household waste can also include hazardous and toxic substances, apart from industrial sources, it is typically in lesser amounts (Ishchenko et al., 2024). These wastes are produced daily from household activities and products containing toxic materials. Manufacturers often do not list active ingredients in their products (Scudder and Blehm, 1991). As a result, some household products, such as insecticides, porcelain cleaners, glass cleaners, floor cleaners, and anticlogging agents, contain hazardous and toxic substances, which can harm humans by contaminating underground water or plants near the house (Lin and Tsai, 2009).

Hazardous and toxic wastes, including industrial effluents, agricultural runoff, 1
petroleum hydrocarbon (PHC) residues, heavy metals (HMs), POPs, radioactive materials, and other persistent chemical wastes, pose an existential threat to global ecosystems and human health. Several industries, such as dye manufacturing, mining,

Microbial Technologies for Sustainable Waste Management. https://doi.org/10.1016/B978-0-443-44093-9.00005-0

and other industrial activities, have been found to generate environmental pollutants, including various harmful HMs (Laker, 2023; Methneni et al., 2021; Ayilara et al., 2020; Prabagar et al., 2021; Liu et al., 2022; Ayangbenro et al., 2018; Orlović-Leko et al., 2022). POPs, hydrocarbons (HCs), and radioactive wastes (Mishra et al., 2022; Guillemot et al., 2022; Ibrahim et al., 2024) along with HMs cause severe toxicity and adverse impacts on humans and ecosystems. POPs can accumulate in fatty tissues and biomagnify through the trophic chain, affecting both humans and animals. Moreover, POPs and HCs contribute to climate change by depleting the ozone layer and reducing the photosynthetic ability of flora (Sadatshojaie and Rahimpour, 2020; Molina and Segura, 2021; Speight, 2011). Some common examples of radioactive wastes include spent nuclear fuel and radioactive isotopes. These materials emit ionizing radiation, which can cause genetic mutations, lead to cancer, and result in acute health effects in humans. These pollutants can persist in the environment for decades to centuries, rendering the surroundings uninhabitable and disrupting ecological balance (Guillemot et al., 2022; Thakur and Kumar, 2024; Pooja Sharma et al., 2022b).

For remediating these dangerous wastes, conventional methods (physical and chemical) have been used for many years. However, they have their shortcomings. For the chemical method, these include the need for expertise, special equipment, etc. In addition, the physical method of remediation is quite expensive (Mahmoud, 2021). In today's context, conventional remediation methods, such as incineration, chemical stabilization, and landfill containment, are being recognized as economically unsustainable and environmentally inadequate. Moreover, traditional approaches often generate secondary pollutants, relocate contaminants, or fail to address complex waste matrices. To address this problem, microbial remediation emerges as a transformative, eco-friendly strategy that harnesses the metabolic capabilities of microbes such as bacteria, fungi, and archaea. This method decomposes and cleans up hazardous contaminants from the environment (Deka Boruah et al., 2023).

Other promising advances in waste management practices include microbial fuel cells (MFCs), metabolic engineering, multiomics approaches, nanotechnology, etc. MFC technology utilizes microorganisms that are electroactive; this method not only detoxifies hazardous waste, but also generates bioelectricity during the process. Single, double, or triple chambers with air cathode and soil/sediment-based MFC systems have further been developed for specific applications (Vasiliadou and Luu et al., 2021; Logan and Regan, 2006; R. Raj et al., 2023a; Kumar et al., 2018). MFC-based systems have shown remarkable efficiencies in removing industrial dyes, pharmaceuticals, and endocrine-disrupting chemicals (Raj et al., 2024; Rishabh, 2023a; Xie et al., 2022). On the contrary, synthetic biology techniques emerge as a promising tool that can enhance the electron transfer capabilities of microbial communities and increase their chemical stress tolerance. Hence, synthetic biology has the potential to broaden the range of treatable waste streams (Dwivedi et al., 2022). Molecular mechanisms are crucial for understanding microbes' metabolism, genes, nature, diversity, and dynamics in bioremediation. Methods such as metabolic profiling, sequencing, and bioinformatics aid in understanding factors affecting microbes (United States Environmental Protection Agency (USEPA), 2021).

The USEPA document provides criteria for the definition of solid waste and solid and hazardous waste exclusions (https://www.epa.gov/hw/criteria-definition-solid-waste-and-solid-and-hazardous-waste-exclusions) (P. Sharma et al., 2022b). On the contrary, nanotechnology is a thriving method for pollution control worldwide. Nanomaterials can be sourced by different means, including physical and chemical methods (Shanmuganathan et al., 2019).

Considering various aspects and the necessity of microbial remediation, an overview of hazardous and toxic wastes is provided. The sources and impacts of mismanaged and untreated hazardous and toxic wastes are included as the context for the detailed discussion. Advanced integrated methodologies, i.e., MFC and microbiome engineering aspects, which amalgamate microbial remediation with alternative techniques for improved performance and efficiency, followed by future perspectives, were further incorporated into the discussion.

## 1.1 Classification and sources of hazardous and toxic waste

Hazardous wastes are classified based on their characteristics, such as ignitability, corrosivity, reactivity, and toxicity. Several countries follow specific guidelines regarding waste classification for the identification and management of wastes. For example, in the United States, the Resource Conservation and Recovery Act (RCRA) oversees hazardous and nonhazardous waste management programs, with hazardous waste classifications determined by Environmental Protection Agency (EPA) regulations, excluding certain materials with mandates or existing regulations (*USEPA*, 2021). The RCRA and EPA regulate waste disposal.

Based on the risks associated with public and environmental health, hazardous wastes are further categorized into high-, intermediate-, and low-risk compounds. High-risk wastes are extremely toxic, persistent, mobile, ignitable, and bioaccumulative, such as chlorinated solvents, POPs, HMs, and polychlorinated biphenyls (PCBs). Intermediate-risk wastes are mostly insoluble and have low mobility, like metal hydroxide sludges. Low-risk wastes are high-volume, nontoxic, and foul-smelling, like municipal solid waste (MSW) (Adeola, 2011). In addition, liquids that are defined as ignitable or flammable wastes, having a flash point of <140 degrees Fahrenheit (°*F*) or the ability to cause fire through friction or absorption of atmospheric moisture, are classified accordingly. Corrosivity is determined using a pH scale, with corrosive wastes assigned the code D002 and reactive wastes assigned the D003 code. Acute toxicity refers to the lethal dose (LD) of a chemical that takes <50 mg/kg of body weight to kill 50% of the exposed population (<LD50). These toxic wastes are often manufactured and transported across the nation (Lin and Tsai, 2009).

Hazardous materials like PCBs, asbestos, and lead-based paint are considered toxic and pose a significant threat to public safety and ecosystems when uptake occurs in biological systems. Toxic wastes are usually treated as a subcategory of hazardous wastes (Adeola, 2011; Asante-Duah, 2021; LaGrega et al., 2016) and are further differentiated according to their type, such as inorganic and organic compounds, industry, and source of generation.

Wastes in the environment belong to distinct classes such as PCBs, halogenated hydrocarbons, estrogenic equivalents, phthalates, dioxins, perfluorinated compounds, brominated flame retardants, polycyclic aromatic hydrocarbons (PAHs), and HMs (Manzetti et al., 2014).

- Sources of release

Since the 18th century, pollution generation and accumulation have been increasing globally (Hubacek et al., 2007), primarily from mining, petrochemical industries, agricultural and industrial sectors, wastewater treatment plants (WWTPs), effluent treatment plants (ETPs), transportation, incineration, and the deposition of electronic waste (e-waste), along with various other sources (Fig. 1.1). These pollutants have different half-lives, with some persisting in humans for over 10 years (Larsen, 2006) and in the environment for over 250 years (Sinkkonen and Paasivirta, 2000). Estimation of the daily environmental release regarding the total amount of pollutants cannot be determined easily; however, efforts have been made time and again in different continents, for example, the number of pesticide applications in the United States (Manzetti et al., 2014; Grube et al., 2011). Meanwhile, e-waste deposition, primarily in East Asia and Africa, has been found to contribute to the dispersal of HMs (predominantly Cu and Pb), POPs, and PCBs in the environment (Wong et al., 2007; Robinson, 2009).

Industries such as dye production, mining, and industrial waste management generate environmental HMs such as Pb, As, Cd, Cu, Zn, Cr, Ag, Sb, Hg, and As (Methneni et al., 2021; Ayilara et al., 2020; Prabagar et al., 2021). Mining activities release harmful chemicals like cyanide and sulfuric acid (Ayangbenro et al., 2018; Orlović-Leko et al., 2022; Liu et al., 2020), while industrial wastes from cement production, pharmaceutical effluents, and plastics contaminate the top layer of soil and water with these pollutants (Jafari et al., 2019; Kumari and Tripathi, 2020; Zhou et al., 2019; Sun et al., 2019). In addition, waste from hospitals and nuclear power plants contributes to the acute outbreak of severe acute respiratory syndrome and other detrimental health impacts by releasing contaminants into the environment (Atlas and Hazen, 2011; Alloway, 2013).

HCs such as methane ($CH_4$), benzene, toluene, xylene, and chlorofluorocarbons (CFCs) are the major components of oil, natural gas, and pesticides, and they are released from vehicles, industrial activities, and oil spills (Osmani et al., 2026; Yüce et al., 2024). These environmental pollutants, such as crude oil, are primarily due to pipeline destruction, transportation leaks, and unintentional spills (Ogunlaja et al., 2019) and are a significant source of pollution, posing potential human carcinogens, mutagens, and endocrine disruptors.

POPs are a class of pollutants that comprise many substances. Examples include dichlorodiphenyltrichloroethane and dioxins, which are used in agricultural fields as pesticides, pharmaceutical compounds for disease control, and in manufacturing and industrial processes (e.g., PCBs in electrical equipment). POPs are found globally, including in remote regions like the Arctic, where they travel via air and water currents. Their persistence in soil, water, and living organisms leads

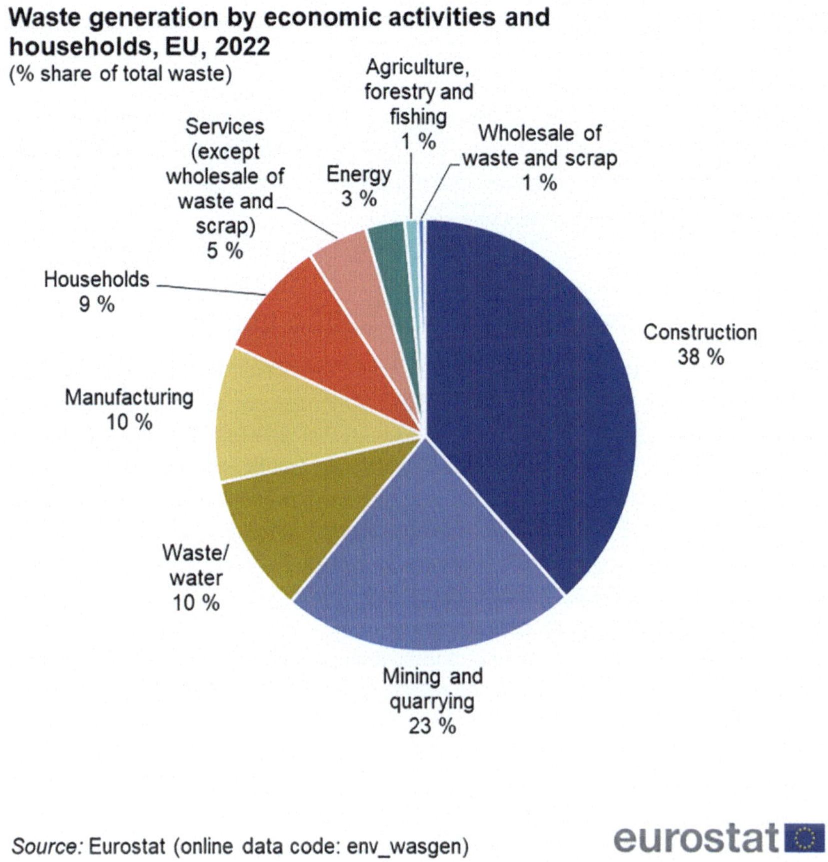


**FIG. 1.1**

Pie chart showing waste generation by economic activities and households as a percentage share of total waste in the EU for the year 2022. A representative image of sector-specific hazardous waste generation. *[Adopted with permission from European Commission (Waste Statistics—Statistics Explained [Internet], 2022) under Creative Commons Attribution 4.0 International licence]. "From Waste Statistics—Statistics Explained. (2022). https://ec.europa.eu/eurostat/statistics-explained/index.php?title=Waste_statistics".*

to long-term ecological damage (*Persistent Organic Pollutants: A Global Issue, A Global Response [Internet]*, 2012).

Radioactive wastes, such as spent nuclear fuel, radioactive isotopes like cesium-137 and strontium-90, and thorium, are generated by nuclear power plants, medical facilities, research institutions, and certain industrial processes (Modupe S. Ayilara and Babalola, 2023).

Other industrial effluents, such as those from the textile industry, contain azo and other complex dyes, phenolic compounds, and phenols. Distilleries discharge a high amount of phenols. Agricultural wastewater contains pesticides, while paper and pulp mill effluents release phenols, chlorinated HCs, furans, and dioxins (Saxena and Bharagava, 2017; Pandey et al., 2021). Phenolic compounds are the most common pollutants present in industrial effluents and are used to produce phenolic resins, alkylphenols, pesticides, dyes, explosives, and cresols. Moreover, untreated pollutants from WWTPs, sewage treatment plants (STPs), and ETPs of the agrifood industries, when disposed into river canals and other water bodies, have harmful effects on the environment (Al-Huqail et al., 2022).

## 2 Impacts on environmental and public health

Organizations like the RCRA, the Toxic Substances Control Act, the Hazardous and Solid (Osmani et al., 2026; Yüce et al., 2024) Waste Amendments, and the Comprehensive Environmental Response, Compensation, and Liability Act have established a scale for characteristic hazard identification of materials and wastes to identify and mitigate their potential leaks. The scale categorizes waste into F-List materials (hazardous waste from nonspecific industrial processes), K-List materials (waste from specific industrial processes like pesticide industries), and P- & U-List materials (commercial sectors, having a higher shelf life and as per *Defining Hazardous Waste: Listed, Characteristic and Mixed Radiological Wastes.| EPA* (2021), that are either 100% pure, technical grade, or the sole active ingredient in a chemical formulation). Their mismanagement or accidental leakage leads to health issues, air pollution, and contamination of groundwater and soil (Marfe and Di Stefano, 2020).

For instance, indoor residential exposure to volatile organic compounds (VOCs) may lead to adverse respiratory health effects (Cakmak et al., 2014). Proper waste management and disposal can prevent the atmospheric release of hazardous gaseous substances, which could otherwise cause increased morbidity and mortality, as they may include hydrogen chloride, carbon monoxide (CO), nitrogen oxides (NOx), sulfur dioxide (SOx), and, in some cases, HMs (Marfe and Di Stefano, 2020). HC compounds significantly contribute to the greenhouse effect and climate change, deplete the ozone layer (especially CFCs), and reduce the photosynthetic ability of flora. Methane oxidizes to carbon dioxide, intensifying global warming. In humans, these compounds increase the risk of cancer and respiratory disorders. Likewise, oil spills, especially in marine environments, cause severe damage to aquatic ecosystems. CFCs break down the ozone layer, increasing exposure to harmful UV radiation for all living organisms (Häder et al., 2014; Voytek, 1990). Behavioral changes, brain and cognitive impairment, neurological and respiratory disorders, dermatitis, and chemical burns are the results of human contact with PHCs released from oil refineries (Saxena et al., 2020; Bharagava et al., 2019).

On the contrary, POPs, which are lipophilic and bioaccumulate in ecosystems for extended periods, have become a growing concern lately (Kahn et al., 2021). These pollutants resist degradation and have extended shelf lives, with low concentrations exhibiting high toxicity and bioaccumulative capability in humans and the environment. They can lead to cancer, as well as neurological, immunological, reproductive, and hormonal disorders in living beings (Varjani et al., 2017; El-Shahawi et al., 2010;

Varjani et al., 2015). PAHs, originating from various sources, are extremely hazardous pollutants. Reportedly, 16 PAHs are listed as priority pollutants in 70 industrial effluents by the USEPA (Varjani et al., 2017), and they are characteristically teratogenic, mutagenic, immunotoxic, and carcinogenic. Among these, Benzo[a]pyrene is documented as the most dangerous carcinogen (Varjani et al., 2017, 2019). These pollutants affect human growth and development as they accumulate in the food chain, making the developmental stage in humans more vulnerable. The production and utilization of this class of pollutants have been banned in many countries due to their adverse effects (El-Shahawi et al., 2010).

Often, acute exposure to compounds like phenols leads to diarrhea and nausea, whereas chronic exposure causes hemolytic anemia, methemoglobinemia, perspiration, dermatological problems, and neurological and cardiovascular disorders (Pandey et al., 2021). For example, nonyl-phenol, an endocrine-disrupting compound, has LD50 of 1475 mg/kg in rats. Other endocrine disruptors include dibutyl phthalate (LD50 7499 mg/kg), benzyl butyl phthalate (LD50 2330 mg/kg), bis(2-ethylhexyl) phthalate (LD50 30,000 mg/kg), hexachlorobenzene (LD50 10,000 mg/kg), benzidine (LD50 309 mg/kg), and 4-aminobiphenyl (LD50 5628 mg/kg) (Saxena et al., 2020; Bharagava et al., 2019; Choudhary and Goldhaber, 1995; Dixit et al., 2015). Pharma compounds such as chlorophenols, like pentachlorophenol, are recalcitrant in nature and cause severe toxicity to humans (Lindholm-Lehto et al., 2015). Organic compounds like azo dyes cause dermatological problems, nausea, and kidney and liver damage. Frequently used agripollutants like pesticides, insecticides, fungicides, herbicides, rodenticides, and fumigants can lead to hormonal and reproductive abnormalities, suppressed immunity, and are even carcinogenic when directly exposed to the substances (Saxena et al., 2020; Pilli et al., 2020). HMs released from various sources are highly toxic to aquatic and terrestrial habitats and their inhabitants. In humans, mercury, cadmium, and lead alter the central nervous system, especially in infants, while lead results in liver and kidney dysfunction, cardiovascular diseases, and compromised activity of the reproductive and immune systems. Cadmium is a potent carcinogen and also causes skeletal disorders, interferes with the normal function of the reproductive system, and is responsible for neurotoxic and nephrotoxic complexities (Zwolak et al., 2019; Omotola Fashola et al., 2020, 2020). Last but not least, the class of radioactive wastes emits ionizing radiation, which can lead to genetic mutations, cancer, and acute health effects in humans. In the environment, they contaminate soil and waterways, affecting plants, animals, and entire ecosystems over extensive periods due to their prolonged half-lives (Modupe S. Ayilara and Babalola, 2023).

## 3 Techniques and mechanisms of microbial remediation

Bioremediation is broadly categorized into in situ and ex situ based on its approaches. In situ bioremediation involves treating the contaminant at the location, while the ex situ process involves the relocation of contaminants to other places for remediation. However, considering the relocation cost, bioremediation is preferred to be conducted in situ (Hu et al., 2023).

Microbes have diverse biochemical mechanisms to neutralize toxic pollutants (Ayilara and Babalola, 2023). Removal of pollutants from the environment by microorganisms occurs through two mechanisms: immobilization and mobilization (Ndeddy Aka and Babalola, 2016; Verma, no date). Mobilization includes: (1) enzymatic oxidation, (2) bioleaching, (3) biostimulation, (4) bioaugmentation, and (5) enzymatic reduction processes. The immobilization approach includes: (1) bioaccumulation, (2) complexation, (3) biosorption, and (4) precipitation (solidification) (Tak et al., 2013; Ayangbenro et al., 2019). Microbes help transform contaminants by generating $CO_2$ and $H_2O$ or other intermediates. Immobilization converts compounds into unavailable forms, such as nitrate-nitrogen ($NO_3^-$) into organic nitrogen (N) (Pratush et al., 2018).

Enzymatic breakdown of complex pollutants (e.g., HCs, pesticides) into simpler compounds like $CO_2$, water, or organic acids occurs through key enzymes, including oxygenases (which insert oxygen into substrates) and hydrolases (which cleave chemical bonds) (Pandey et al., 2021). Partial modification of pollutants commonly happens through processes like oxidation, reduction, or dehalogenation. For example, *Dehalococcoides* bacteria reductively dechlorinate toxic solvents like trichloroethylene (TCE) into ethylene (Ren and Manefield, 2025). In addition, extraction of HMs from contaminated matrices using microbial metabolites (e.g., organic acids) occurs via passive adsorption onto microbial cell walls through ionic or hydrophobic interactions. Active intracellular uptake and sequestration of the contaminants reduce their environmental mobility. Moreover, many microbes can convert these pollutants into stable inorganic forms (e.g., precipitation of metal sulfides by sulfate-reducing bacteria) (Ayilara and Babalola, 2023; Thirumalaivasan et al., 2024). These mechanisms are often complementary. For instance, *Pseudomonas* spp. degrade HCs via oxygenase-initiated biodegradation, while fungi like *Phanerochaete chrysosporium* employ extracellular peroxidases to dismantle recalcitrant PAHs (Thirumalaivasan et al., 2024; Karnwal et al., 2024).

During enzymatic oxidation, pollutants are oxidized from a higher state of oxidation to a lower one. For example, the laccase (oxidoreductase type) enzyme catalyzes the oxidation of aromatic amines and phenols and reduces molecular oxygen ($O_2$) to water ($H_2O$) (Kushwaha et al., 2018; Gangola et al., 2018; Sahay, 2021). *Pycnoporus* sp. and *Leptosphaeria* sp. have been reported to produce laccase (Copete-Pertuz et al., 2018; Tian et al., 2020).

In the enzymatic reduction procedure carried out by microbes, the pollutants are transformed into a lower oxidized state and become insoluble. Obligate and facultative anaerobes are effective for the remediation of compounds such as polychlorinated dibenzo-p-dioxins and dibenzofurans (Zacharia, 2019).

Bioaugmentation is a process where microorganisms are added to the site of contamination to feed on toxic pollutants. This is an effective, speedy, and inexpensive method (Mahmood et al., 2021). The purpose of adding external microbes is to augment resident microbes and cause genetic modification of the native microbes, enhancing their ability to degrade pollutants. The effectiveness of the introduced nonnative strains depends on factors such as their ability to compete with native microbes and adapt to the new environment (Fashola et al., 2016; Ayangbenro and Babalola, 2017; Tribedi et al., 2018). Researchers reported that *Burkholderia* sp. FDS-1 degrades nitrophenolic compounds in polluted soil at a slightly acidic pH and a temperature of 30°C (Tribedi et al., 2018; Ojuederie et al., 2021).

During the biostimulation method, nutrients (N, P, and K), metabolites, electron donors, enzymes, electron acceptors, biosurfactants, etc., are added to the soil, which increases the resident microbes' activity and helps in remediation. It is an inexpensive, greener, and efficient process and is preferable over bioaugmentation due to the competitiveness of indigenous microbes (Tribedi et al., 2018; "Environmental NSIJ of, 2021. Bioremediation of total petroleum hydrocarbons by bioaugmentation and biostimulation in water with floating oil spill containment booms as bioreactor basin," no date), which in turn maintains the natural microbial diversity and balance of the environment.

In the bioaccumulation process, where the absorption rate of a contaminant surpasses its loss rate, it leads to toxic accumulation in the intracellular portions of microbes (Pooja Sharma et al., 2022c). Again, the process of complexation entails the utilization of ligands to form complexes with inorganic metals, which are environmental pollutants, particularly in solid waste contexts (Ayangbenro and Babalola, 2017).

## 3.1 Microbial remediation of HMs and metalloids

Microbial remediation of HMs, especially in highly contaminated environments, is carried out through the immobilization process (Fig. 1.2), using either in situ or ex situ approaches (Pratush et al., 2018). The ex situ process involves transporting polluted soils to be treated by microbial processes, which immobilize the metal ions responsible for the contamination, whereas the in situ procedure involves treatment carried out on-site (Ayangbenro and Babalola, 2017; Cao et al., 2020). Various microbes (Table 1.1) have been found to efficiently degrade HMs. For example, *Enterobacter asburiae* and *Bacillus cereus* are involved in immobilizing HMs, and during the remediation process, microbes shield themselves from the toxic compounds targeted by forming hydrophobic layers or employing solvent efflux pumps (Verma, no date; Fashola et al., 2020).

In the enzymatic oxidation process, HMs lose an electron, and their toxicity is reduced. During this process, the oxidoreductase enzyme released by microbes breaks down radicals into fractions, thereby generating high-molecular-weight compounds (Unuofin et al., 2019). In addition, through enzymatic reduction, hexavalent chromium is reduced to trivalent chromium, and this procedure is catalyzed by chrome reductase (Saxena et al., 2019).

Bioleaching is a method that uses acidophilic microbes to dissolve solid HMs from sediments, especially iron or sulfur pollutants (Sun et al., 2021; Bhandari et al., 2023). For their remediation, iron- or sulfur-oxidizing bacteria, such as *Trioxidanes, Aspergillus, Mucor, Penicillium, Cladosporium,* and *Rhizopus*, are utilized. By creating an acidic environment, these specific microbes solubilize immobilized HMs into an aqueous medium (Medfu Tarekegn et al., 2020).

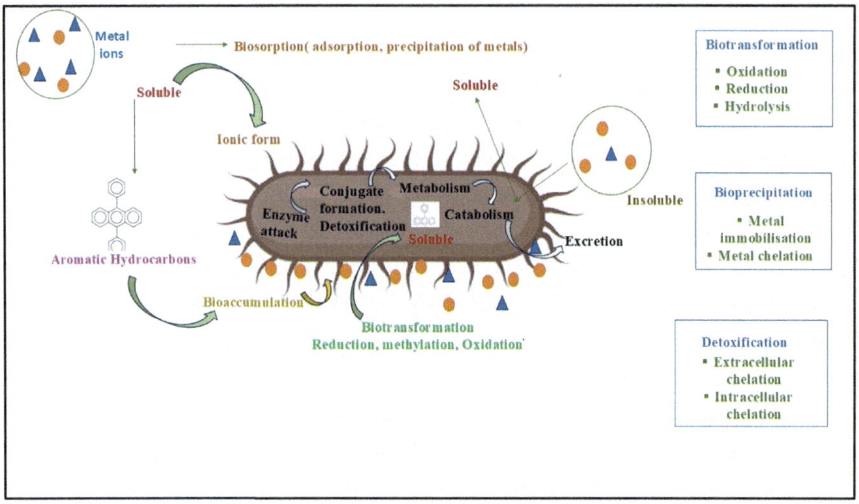


**FIG. 1.2**

Microbial prokaryotic cell involved in various kinds of remediation techniques. Microbial pathways for biosorption, bioaccumulation, biotransformation (oxidation, reduction, methylation), and bioprecipitation are depicted. These processes convert toxic soluble metal ions into less bioavailable or insoluble forms, aiding in the microbial detoxification of heavy metal contaminants. *[Adopted with permission from (Ayub et al., 2025) © 2025 The Author(s). Published by Elsevier B.V.]. "With permission from Ayub, A., Wani, A.K., Malik, S.M., Ayub, M., Chopra, C., Singh, R., Malik, T., 2025. Harnessing microbes and plants for bioremediation of heavy metal contaminants: current paradigms and future perspectives. Environ. Challeng. 20. https://doi.org/10.1016/j.envc.2025.101220".*

During biosorption, HMs are adsorbed from pollutants through the processes of proton and ion displacement, complexation, chelation, and electrostatic forces (Mahmood et al., 2021). Bacteria, fungi, and algae, such as *Rhodococcus erythropolis, Streptomyces* sp. K11, and *Bacillus anthracis*, are bioremediation agents that remove contaminants from solutions (Mathew and Krishnamurthy, 2018; Baltazar et al., 2019; Sedlakova-Kadukova et al., 2019). HM pollutants, such as gold (Au), zinc (Zn), and copper (Cu), are economically important and, therefore, useful in industrial processes. Desorption, as opposed to biosorption, is a method for recovering these compounds (Medfu Tarekegn et al., 2020).

Complexation occurs when microbial surfaces, which contain extracellular polymeric substances (EPSs), interact with HMs, primarily through agents like ligands, siderophores, toxic compounds, and low-molecular-weight organic acids such as alcohols, tricarboxylic acids (TCAs), and citric acids (Pratush et al., 2018; Xie et al., 2020; Xiao et al., 2019). Examples of organisms involved in complexation include *Rhodobacter blasticus* and *Bacillus licheniformis* (Bai et al., 2019; L. Wang et al., 2019b). Siderophores, iron chelators produced by microbes in iron-deficient environments, act as effective agents due to their binding groups, i.e., hydroxamate,

**Table 1.1** Microbial-based remediation of heavy metals and their degradation efficiency. [Adapted with permission from Ayub et al. (2025) © 2025 by Elsevier B.V.]

| Microorganisms | Metal ionic form | Remediation efficiency (%) | Mechanism | References |
|---|---|---|---|---|
| *B. cereus* | $Pb^{2+}$ | 79.26 | Bioaccumulation and biosorption | (Sharma and Shukla, 2021) |
| *Bacillus* sp. | $Pb^{2+}$, $Mn^{2+}$ | 76, 90 | Biosurfactant production | (Ravindran et al., 2020) |
| *R. opacus* | $Ni^{2+}$, $Cr^{6+}$ | – | Adsorption in exopolysaccharides | (Czemierska et al., 2017) |
| *B. cereus* | Iron ($Fe^{2+}$) | 92 | Reduction | (Nayak et al., 2018) |
| *Aspergillus penicillioides* | $Cd^{2+}$, $Cu^{2+}$, $Fe^{2+}$ | 53, 32, 90 | Biosorption | (Raklami et al., 2022) |
| *A. niger* | $Ni^{2+}$ | 70.3 | Biosorption | (Amini et al., 2009) |
| *Phanerochaete chrysosporium* | $Cd^{2+}$, $Ni^{2+}$ | 96, 89 | Bioaccumulation | (Noormohamadi et al., 2019) |
| *Aspergillus restrictus* | $Cd^{2+}$, $Cu^{2+}$, $Fe^{2+}$, $Mn^{2+}$, $Pb^{2+}$, $Zn^{2+}$ | 61, 77, 64, 72, 44, 87 | Biosorption | (Bano et al., 2018) |
| *Sterigmatomyces halophilus* | $Mn^{2+}$, $Pb^{2+}$, $Zn^{2+}$ | 89, 57, 93 | Biosorption | (Bano et al., 2018) |
| *Scenedesmus sp.* | $Cr^{6+}$ | 93 | Biosorption | (Pradhan et al., 2019) |
| *S. platensis* | $Ni^{2+}$, $Zn^{2+}$ | – | Bioaccumulation | (Cepoi et al., 2022) |
| *Scenedesmus sp.* | $Cr^{6+}$ | 93 | Biosorption | (Noormohamadi et al., 2019) |

*With permission from Ayub, A., Wani, A.K., Malik, S.M., Ayub, M., Chopra, C., Singh, R., Malik, T., 2025. Harnessing microbes and plants for bioremediation of heavy metal contaminants: current paradigms and future perspectives. Environ. Challeng. 20. https://doi.org/10.1016/j.envc.2025.101220.*

catecholate, and phenolates, which form complexes with HMs, increasing their solubility (Khan et al., 2018a). Siderophores also produce reactive oxygen species, which enhance their functional capability as bioremediation agents for organic pollutants (Albelda-Berenguer et al., 2019). In addition, cyanobacteria, due to their production of siderophores such as anachelin, synechobactin, and schizokinen, are considered effective bioremediation agents (Årstøl and Hohmann-Marriott, 2019; Sarmah and Rout, 2020).

Microbes that are active in bioaccumulating HMs include *Rhizopus arrhizus, Pseudomonas putida*, and *Aspergillus niger*, which bioremediate mercury, cadmium, and thorium (Pooja Sharma et al., 2022c).

In precipitation, HMs are converted into crystals and precipitates during biogeochemical cycling, reducing their toxicity. This process involves specific enzymes and secondary metabolites, which undergo processes such as galactosis, and includes metals like mineralized manganese (Mn), silver (Ag), and microfossils (Pooja Sharma et al., 2022c). Sulfate-reducing bacteria, *Bacillus subtilis*, and *Oceanobacillus indicireducens* have reportedly been involved in the precipitation of HMs in the environment, affecting the overall health of the ecosystem (Pratush et al., 2018; Maity et al., 2019).

According to Nivetha et al. (2023), biostimulation has been effective in the bioremediation of HMs using microorganisms such as *Bacillus* sp., *Rhodococcus* sp., *Staphylococcus* sp., *Klebsiella* sp., *Pseudomonas* sp., and *Citrobacter* sp. However, it can lead to environmental complications like eutrophication and pollution if the nutrients used are synthetic.

## 3.2 Remediation of antibiotics

Bioremediation of antibiotics is primarily facilitated by bacteria such as *Pseudomonas* spp., *Bacillus* spp., *Acinetobacter* spp., and *Streptomyces* spp. (Kour et al., 2022). Microorganisms modify the structures of antibiotics (AB) through peripheral metabolic pathways like hydrolysis, oxidation, and deamination, while central metabolic pathways, such as the TCA cycle, aid in the mineralization process of antibiotic derivatives, eventually leading to their removal (Akinsemolu and Onyeaka, 2025) (Table 1.2).

PCBs are decomposed by microbes like *Burkholderia xenovorans, P. putida*, and *Rhodococcus* sp. (Furukawa and Fujihara, 2008). The biphenyl degradation pathway oxidizes PCBs into chlorobenzoates (Parnell et al., 2010), which are processed through the β-ketoadipate pathway, known to be a central metabolic pathway. This mechanism reduces toxic effects and prevents PCB accumulation in ecosystems (Akinsemolu and Onyeaka, 2025) (Table 1.2).

On the contrary, through the bioaugmentation procedure, genetically modified microorganisms are developed to target POPs (PAHs and PCBs), thereby enhancing the effectiveness of indigenous microbial populations in degrading these pollutants (Parnell et al., 2010). PAH degradation is primarily driven by *Mycobacterium* (Table 1.2).

## 3.3 Remediation of HC

In the soil, microbes reportedly destroyed 62%–75% of PHCs between 150 and 270 days (Chaîneau et al., 2005; Chaíneau et al., 1995). Free microorganisms destroyed 2.3%–6.8% of petroleum hydrocarbons in 60 days, while with the addition of biochar as a carrier, microbes degraded 7.2%–30.3% of petroleum hydrocarbons in the same duration (Pino et al., 2016). The rate of degradation was reported as 29.8% in sodium alginate–diatomite beads, which act as an immobilized medium, whereas free cells degraded at a lesser rate (21.2%) in 20 days (Wang et al., 2012).

**Table 1.2** Representative microorganisms and metabolic pathways involved in the bioremediation of POPs. [Adapted under a Creative Commons Attribution 4.0 International license from Akinsemolu and Onyeaka (2025) © 2025 by the authors. Licensee MDPI, Basel, Switzerland.]

| Pollutant | Representative microorganisms | Peripheral metabolic pathways | Central metabolic pathways | References |
|---|---|---|---|---|
| Antibiotics | *Pseudomonas* spp., *Bacillus* spp., *Acinetobacter* spp., *Streptomyces* spp. | Hydrolysis, oxidation, deamination | Tricarboxylic acid (TCA) cycle | (Kour et al., 2022) |
| Polychlorinated biphenyls (PCBs) | *Burkholderia xenovorans*, *Pseudomonas putida*, *Rhodococcus* spp. | Biphenyl degradation pathway (dioxygenase-mediated oxidation to chlorobenzoates) | $\beta$-ketoadipate pathway, TCA cycle | (Furukawa and Fujihara, 2008; Parnell et al., 2010) |
| Polycyclic aromatic hydrocarbons (PAHs) | *Mycobacterium vanbaalenii*, *Sphingomonas* spp., *Pseudomonas aeruginosa* | Dioxygenase-mediated oxidation to catechols and quinones | $\beta$-ketoadipate pathway, TCA cycle | (Tiehm and Fritzsche, 1995; Hennessee, 2016; Alvarez, no date) |

*From Akinsemolu, A.A., Onyeaka, H.N., 2025. Harnessing biotechnology for the remediation of organic pollutants in coastal marine ecosystems. Appl. Sci. 15 (12). https://doi.org/10.3390/app15126921.*

Higher temperatures enhance the microbial mineralization of *hexa*-decane. As a result, within the same duration (50 days), 68% $CO_2$ was reported to be generated at 25°C, while 45% $CO_2$ was generated at 4°C and 10°C, indicating better digestion of *hexa*-decane by microbes (Hamamura et al., 2008). In addition, in conditions of elevated soil salinity (>8%) and pH values below 4 or above 9, the activity of *Acinetobacter baylyi* ZJ2 was reportedly modified, leading to decreased production of specific lipopeptide surfactants, which subsequently impaired the degradation of petroleum hydrocarbons by microorganisms (Zou et al., 2014).

Microorganisms readily degrade low-molecular-weight saturated HCs and aromatic HCs, whereas higher-molecular-weight PHCs demonstrate significant resistance to microbial degradation. The sequence of microbial degradation is as follows: N-alkanes > branched-chain alkanes > branched alkenes, particularly in the context of low-molecular-weight compounds. Similarly, the degradation sequence for aromatic compounds is: n-alkyl aromatics > monoaromatics > cyclic alkanes, whereas the degradation of polynuclear aromatics > asphaltenes (Atlas, 1981; Chen et al., 2020). The methylene concentration of asphaltenes declined to 14% and 8% after

45 days of decomposition by *B. subtilis* and *P. aeruginosa*, respectively, while the latter degraded 63.8% of n-*hexa*-decane within a period of 60 days (Zhang et al., 2020; Hajieghrari and Hejazi, 2020).

Bioaugmentation is effective and economically viable in terms of applicability when compared to conventional remediation techniques (Hussain et al., 2019; Tao et al., 2017), as discussed earlier. Bioaugmentation can be accomplished by adding lipophilic bacteria (Bidja Abena et al., 2019) in a wide range of environments contaminated by petroleum, including saline-water, seashores, sludge, and sediment. Oleophilic bacteria are commonly found to thrive only on HCs or carry out the mineralization of harmful and hazardous petroleum contaminants (Bacosa et al., 2012; Kumari et al., 2018; Lee et al., 2018). Different types of bacteria with degradation abilities can be found in various contaminated environmental matrices. Actinomycetes is a common phylum in PAH-contaminated soil. *Acidovorax, Rhodoferax, Hydrogenophaga,* and *Polaromonas* were reportedly found in soil contaminated in the Philippines. *Acidobacteria* exist in soil contaminated with petroleum, phenanthrene, pyrene, and fluoranthene (Bidja Abena et al., 2019).

Several studies have demonstrated that bacteria, including *Rhodococcus* sp., *Pseudomonas* sp., and *Scedosporium boydii*, can degrade HC contaminants (Kumari et al., 2018; Pi et al., 2017; Wang et al., 2021). HCs are primarily degraded by bacteria through aerobic pathways, where oxygen ($O_2$) acts as an electron acceptor, thus speeding up HC catabolism (Yuan et al., 2018; Varjani and Upasani, 2016). In aerobic settings, degradation occurs via oxidation, reduction, hydroxylation, and dehydrogenation reactions. Various enzymes, including monooxygenases, dioxygenases, cytochrome P450, peroxidases, hydroxylases, and dehydrogenases, help microbes break down HC compounds (Liu et al., 2010; Sirajuddin and Rosenzweig, 2015; Margesin et al., 1999; Cao et al., 2009; Wang et al., 2010).

During the reduction of HC contaminants, *Pseudomonas* sp., *Acinetobacter* sp., and *Rhodococcus* sp. are considered the most effective bacteria, with short-chain and medium-chain alkanes oxidized by AlkB or CYP153 enzymes. Certain putative flavin-binding monooxygenases (AlmA) and long-chain alkane monooxygenases (LadA) oxidize long-chain alkanes (Liu et al., 2014). Genes pertaining to the AlkB-type (AlkMa and AlkMb) and the AlmA-type (AlmA) are found in the *Acinetobacter* strain DSM17874 and are responsible for the degradation of alkanes containing different chain lengths (Liu et al., 2021). The degradation of alkanes and PAHs is carried out by *Pseudomonas* sp., which contains nahAc, catechol dioxygenases $C1_2O$ and C23O), AlkB, and cytochrome P450 (Noble et al., 1993; Champion et al., 1978; Varjani, 2017; Hesham et al., 2014; Abouseoud et al., 2008).

Microorganisms use various pathways, including terminal oxidation, subterminal oxidation, ω-oxidation, and β-oxidation, for alkane and PAH metabolism, with terminal oxidation being the most common. This process oxidizes the terminal methyl group into alcohols, followed by aldehydes, fatty acids, and, in the end, $CO_2$ and $H_2O$ which are produced by the β-oxidation pathway (Rojo, 2009; Xia et al., 2005; Oyetibo et al., 2017). PAHs, owing to their structural stability, are predominantly metabolized by a mixed functional oxidase system involving the cytochrome P450 enzyme. This mechanism produces diols and catechol intermediates via ortho- or meta-cleavage routes, which then enter the TCA cycle (Cao et al., 2009; Xia et al., 2005) (Table 1.2).

Based on the type of reductive reaction and the electron acceptors, bacterial anaerobic PAH degradation employs a variety of methods to break and open aromatic rings. Anaerobic PAH degradation is receiving attention in today's world due to anoxic conditions in environmental niches such as phreatic zones, deep aquatic sediments, and water-flooded soils. Aerobic PAH degradation is the usual and preferred method of PAH degradation (Ghosal et al., 2016; Dhar et al., 2020).

Mixed bacterial cultures and consortia can enhance or complete PAH degradation through collaborative catabolic activities and diverse pathways (Vaidya et al., 2018; Haleyur et al., 2019; Patel et al., 2019). Recent studies emphasize these methods, along with the use of immobilized and genetically modified bacteria (Peng et al., 2018). However, the dispersion of inoculum in soil/sediment is a major challenge in degradation, especially in subsurface soil, due to restricted microbial transport and the strong adhesion of cells to soil organic matter.

Bacterial extremophiles, such as halophilic, acidophilic, and thermophilic bacteria, have also been reported for PAH degradation. High temperatures decrease viscosity, thereby increasing PAH diffusion and bioavailability. Hence, employing thermotolerant and thermophilic bacteria is advantageous (Mehetre et al., 2019).

In recent decades, researchers have been involved in bacterial community analysis (Muangchinda et al., 2018; Yang et al., 2020), their PAH degradation pathways (Chen et al., 2016; Vaidya et al., 2017), bacterial genes associated with degradation and related biochemistry (Peng et al., 2019; Sangkharak et al., 2020), bacterial metabolic products (Chen et al., 2016; Wu et al., 2020), and the genes regulating degradation processes (Wu et al., 2020; Kan et al., 2020), all of which have been documented extensively.

## 3.4 Remediation of pesticides

Misuse of biologically active pesticides can alter soil physicochemical characteristics, microbial populations, and metabolic activity, thereby affecting soil fertility and overall metabolic processes. Several studies suggest that nitrogen-fixing and phosphorus-solubilizing bacteria are inactivated (Gao et al., 2015; Vandermaesen et al., 2016). Pesticides affect biochemical reactions such as mineralization, nitrification, denitrification, ammonification, redox reactions, and methanogenesis (Parween et al., 2016). Bacteria mitigate the negative impacts of these pesticides on seed germination and plant growth-promoting (PGP) activities, including phosphate solubilization, EPS, indole acetic acid (IAA) , siderophore, and HCN production, which help plants thrive (Nayak et al., 2020). In addition, pesticides alter bacterial cell growth by transferring pesticides and inorganic ions across the cytoplasmic membrane (Sundaram et al., 2022). The uptake of pesticides disrupts metabolism by binding to amino and sulfide groups, thereby impairing the metabolic activity of bacterial cells (Shahid and Khan, 2017).

*Bradyrhizobium* sp. strain R5 and *Pseudomonas* sp. strain PS6 were found to tolerate elevated herbicide concentrations via the secretion of siderophores, phytohormones, and exopolysaccharides. This discovery may elucidate the increased yield of green gram reported postinoculation, notwithstanding the presence of glyphosate and quizalofop (Shahid and Khan, 2017; Pujar et al., 2022). *Brevibacillus parabrevis* strain BCP-09 was the most efficient degrader, removing 50.48% of beta-cypermethrin (100 mg/L), suggesting it could be used for the bioremediation of environmental and agricultural goods contaminated with it (Tang et al., 2018). An interesting and significant feature of *Azotobacter vinelandii* strain AZ6 suggests that PGP activities were maintained at elevated pesticide levels, even in compromised or damaged cells (Shahid et al., 2019).

Microbes such as *B. cereus* strains BHUJP-P4 and BHUJP-P3 exhibited resistance to potent organophosphorus insecticides. BHUJP-P4 exhibited the greatest resistance to chlorpyrifos at 940 mg/L, while BHUJP-P3 demonstrated significant resistance to monocrotophos at 544 mg/L (Jaiswal et al., 2019). The interaction between Fe and pentachlorophenol demonstrated a synergistic effect in enhancing the production of pyoverdine-type siderophore, which plays a role in pesticide removal. The introduction of iron (Fe) notably elevated the biodegradation of pentachlorophenol by the PGPR strain (Ammeri et al., 2023).

In recent years, there has been a lot of interest in using endophytic bacteria (Fig. 1.3) to clean up pesticides from contaminated sites. It is important to highlight

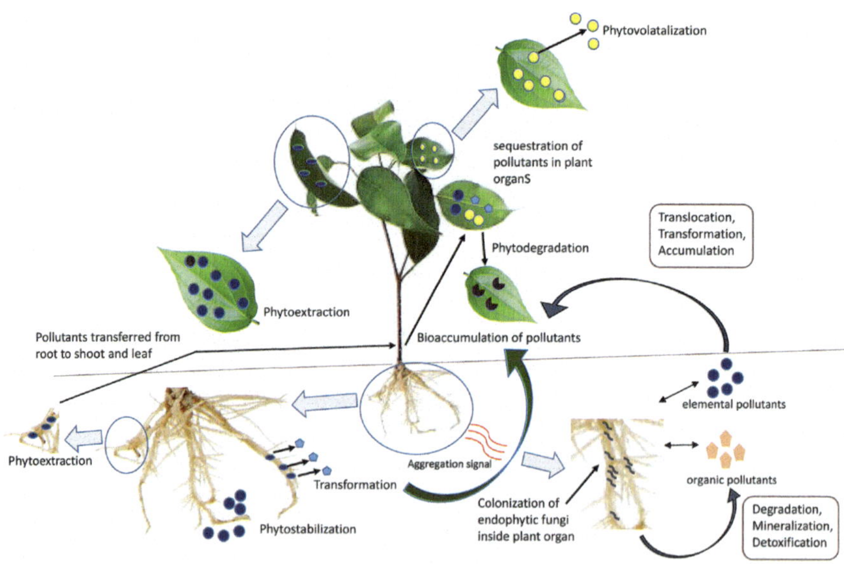


**FIG. 1.3**

Endophytic microorganisms and their mechanisms to carry out remediation. Schematic diagram of phytoremediation and the effect of endophytic fungi-mediated sequestration of organic and elemental pollutants. *[Adapted with permission from Nandy et al. (2020) Creative Commons CC-BY license © 2020 SAAB. Published by Elsevier B.V.]. "From Nandy, S., Das, T., Tudu, C.K., Pandey, D.K., Dey, A., Ray, P., 2020. Fungal endophytes: futuristic tool in recent research area of phytoremediation. South Afr. J. Botany 134, 285–295. https://doi.org/10.1016/j.sajb.2020.02.015".*

that these bacteria are involved in the bioremediation of pesticides in aquatic systems too. *Streptomyces* serves as an example of endophytic bacteria capable of reducing or neutralizing stress caused by specific chemicals, including pesticides and atrazine, in aquatic environments (Mesquini et al., 2015). Microbes have the potential to reduce the harmful impacts of pesticides on aquatic organisms and human populations by transforming pesticides into less harmful forms. Furthermore, they promote the breakdown of pesticides through endophytic microorganisms, increasing the metabolic activity of various microorganisms in both soil and water environments. *Pseudomonas fluorescens*, employed as an endophytic bacterium, accelerates the breakdown of metolachlor and alachlor in soil and aquatic environments (Mesquini et al., 2015). Endophytic bacteria detoxify pesticides in diverse ways. Initially, they disrupt pesticide-degradation enzymes, thereby inhibiting the harmful impacts. In addition, they have the potential to amplify the effects of other biodegrading microorganisms. Ultimately, they enhance the biodegradation of pesticides through the synthesis of certain compounds that inhibit pesticide-degrading microorganisms (Zhang et al., 2018; Gupta et al., 2020; Sharma et al., 2018).

## 3.5 Remediation of radioactive waste

Native subsurface microorganisms (Table 1.3) can efficiently break down organic pollutants like TCE and uranium via natural attenuation mechanisms (Chen et al., 2021). Koarashi et al. (2012) quantified $^{137}Cs$ retained by microbiological and adsorptive processes in forest surface soils from the Fukushima nuclear accident atmospheric

**Table 1.3** The remediation mechanisms of several microorganisms against radionuclides. [Adapted with permission from Thakur and Kumar (2024) © 2024 Elsevier Ltd.]

| Radionuclide | Microorganism | Remediation mechanism | References |
|---|---|---|---|
| U(VI) | *Desulfovibrio* sp. | Ion exchange and bioaccumulation in cells | (Maleke et al., 2015) |
| U(VI) | *Geobacter* sp. | Sorption and precipitation by microbial cells | (Maleke et al., 2015) |
| U(VI) | *Shewanella putrefaciens* | Reduction to less toxic and immobile forms | (Mahadevan, no date) |
| $^{137}Cs$ | *Phytolacca americana Linn* | Reduction and immobilization in sediments | (Song et al., 2012) |
| $^{137}Cs$ | *Amaranthus cruentus* L. | Conversion to less volatile and soluble forms | (Song et al., 2012) |
| $^{226}Ra$ | *Serratia proteamaculans* | Microbial oxidation and immobilization | (Yan and Luo, 2015) |
| $^{4o}K$ | *Serratia proteamaculans* | Metabolism and degradation of tritiated compounds | (Yan and Luo, 2015) |

*With permission from Thakur, A., Kumar, A., 2024. Emerging paradigms into bioremediation approaches for nuclear contaminant removal: from challenge to solution. Chemosphere 352. https://doi.org/10.1016/j.chemosphere.2024.141369.*

fallout. Gerber et al. (2016) found that a gram-negative *Beta-proteobacterium* could remove 130 mg U/g of dry biomass, suggesting its viability for in situ bioremediation of polluted wastewater. Chabalala and Chirwa (2010) found that *Pantoea* sp., *Pseudomonas* sp., and *Enterobacter* sp. converted U(VI) to U(IV) within a certain pH range in the Limpopo uranium mine. *Pseudomonas* sp. showed the most significant preliminary reduction rate, achieving 88% elimination of the contaminant 1 day (24 h) after exposure in batches containing 400 mg/L U(VI).

The bioreduction process of U(VI) to U(IV) effectively removes insoluble uranium from contaminated sites. Microorganisms such as *Geobacter metallireducens* GS-15, *Shewanella oneidensis, Desulfovibrio desulfuricans*, along with indigenous bacterial communities, facilitate the transformation of soluble U(VI) into insoluble U(IV) (Lovley et al., 1993; Lovley and Phillips, 1992). In a packed-bed bioreactor comprising *D. desulfuricans*, more than 99% of U(VI) was removed within 37 hours (Maleke et al., 2015).

Studies suggest that a single-electron reduction system is the most common method for U(VI) reduction to U(IV), primarily reduced by one electron and then to U(IV) through disproportionation in the presence of *Geobacter sulfurreducens* (Banala et al., 2021). Acetate donates electrons to surface soil, improving U-reducing bacterial activity and facilitating soluble U(VI) extraction from polluted areas (Selvakumar et al., 2018). Research shows a proliferation of *Geobacter* sp. and a decrease in U(VI) with the introduction of acetate in uranium-contaminated areas. Distinct functional groups (e.g., hydroxyl, carbonyl, amine, and phosphoryl) belonging to microorganisms, contained on their cell surfaces, facilitate the absorption of U (Anderson et al., 2003). It has been shown that marine yeast *Yarrowia lipolytica* has removed significant U by a mechanism called biphasic binding (Kolhe et al., 2020). Moreover, biosorption and bioreduction are influenced by the pH, temperature, organic matter of the soil substratum, and chemical speciation of U. These factors make them crucial in designing effective microbial remediation strategies (Banala et al., 2021; Lakaniemi et al., 2019; L. Wang et al., 2019c).

## 3.6 Remediation of dye

Textile effluents generate harmful colors such as indigo, carmine, and malachite green (MG). Textile dye categorization involves chemical structures like azo, anthraquinone, nitro, indigoid, and phthalein dyes. Azo dyes are manufactured organic dyes with NR' (amine) functional groups, while anthraquinone dyes have chromophore groups. Meanwhile, natural colors (dyes) come from plants, animals, or minerals, for example, cellulose fiber or protein dyes.

Over the last 15 years, a lot of research has been conducted on the biodegradation process of dyes, and it has been found that a variety of microbes are capable of degrading toxic dyes. For instance, *Pseudomonas* sp., *Bacillus* sp., *Saccharomyces* sp., *Streptomyces* sp., *Candida* sp., etc., can degrade and detoxify dyes due to the presence of certain enzymes. Two processes play a critical role in the decomposition of toxic dyes found in textile industry effluent. There are many enzymes present in different microorganisms that aid in these degradation processes of toxic textile dyes (Alloway, 2013).

Microorganisms like bacteria, fungi, and algae in wastewater and polluted environments aid in the biodegradation and bioremediation of dyes. They can also be isolated and used in laboratory settings for textile effluent treatment and decolonization (Alloway, 2013). Procion Red, a widely used azo dye, is difficult to degrade due to its high molecular weight and strong azo bonds. *Pseudomonas stutzeri* SPM-1, a facultative aerobic Gram-negative, rod-shaped bacterium, can degrade this dye through the activity of azoreductase and NADH-DCIP reductase (Bera and Tank, 2021).

*Pseudomonas* spp.'s efficiency in dye degradation is enhanced by optimal parameters such as pH 6–10 and a temperature of 35°C. The degree of decolonization increases with the availability of carbon (C) and nitrogen (N) sources. *Saccharomyces cerevisiae* (MTCC 463), a unicellular fungus, bioaccumulates azo-reactive textile dyes (Remazol Blue Dye and Remazol Black B Dye) in molasses, decolorizing the hazardous and toxic azo dye methyl red in a static anoxic environment at temperatures between 35°C and 40°C (Jadhav et al., 2010). The dye breakdown process involves enzymes such as laccase, azoreductase, and NADH-DCIP reductase. It has been reported that decolorized textile dye wastewater can be treated with *Saccharomyces* species, which are nonpathogenic, convenient, and affordable biomass. *Pseudomonas* sp. SUK1's internal bacterial peroxidase was isolated and assessed for its potential role in decolonizing dyes and detoxifying carcinogenic anthraquinone dyes (Kalyani et al., 2011). Researchers found that the enzyme's substrate specificity had also been altered. Peroxidase, with its capacity to decolorize various textile dyes, has been proven to be a detoxifying enzyme for dye contaminants.

*Phormidium autumnale*, a soil and freshwater cyanobacterium, decomposes indigo dye through metabolites (anthranilic acid and isatin). Studies show that when Methyl Orange is subjected to symmetrical cleavage by peroxidase, it results in the production of 1,4-benzenediamine, N,N-dimethyl, 4-aminobenzenesulfonic acid, and aniline (Maria Dellamatrice et al., 2017).

*Candida albicans*, a pathogenic yeast, produces reductase, which can break down dyes. Direct Violet 51, a carcinogenic dye, is degraded into nontoxic compounds due to its metabolites (Vitor and Corso, 2008). *C. albicans* and *Neurospora crassa* fungal bodies have reportedly altered industrial effluent dye structures, while *N. crassa* produces oxidative enzymes for phenolic compound biodegradation, inhibiting dye activity (Luke and Burton, 2001). The immobilization of *N. crassa* biofilms allows for the production of oxidase enzymes, which convert hazardous phenol and *p*-cresol wastes into nontoxic compounds. MG, a triphenylmethane dye used in textiles, is degraded by *Streptomyces chrestomyceticus*, a Gram-positive bacterium, reducing its harmful effects (Vignesh et al., 2020). Research shows that the metabolites produced when *Saccharomyces chrestomyceticus* S20 breaks down MG become less harmful to human cells and plants than the parent MG. This biodegradation could be a potential solution for removing MG from various areas. The Rhodamine-B dye, dangerous and mutagenic, especially to human organs, has been extensively used in various industries (Saravanan et al., 2022). According to Goud et al. (2020), endophytes that have strong dye degradation potential may be more advantageous than bioremediation by microbes. This is because endophytes can assist plants in recovering from the stress caused by textile effluents through enzyme-mediated hydrolysis. Eight of the 41 endophytes derived from *Bacillus*, *Microbacterium*, and *Halomonas* demonstrated the highest level of efficiency in the breakdown of dyes and stimulated the growth of plants. According to Mustafa et al. (2021), a biorestorative method was utilized, involving the use of the microbiological tool *Klebsiella pneumoniae* GM-04. *K. pneumoniae,* isolated from industrial effluent, achieved 90% decolonization of DB-284 under ideal conditions. Low dye discoloration was caused by temperature reduction and increased dye concentration. Bacterial enzymes like reductases, laccases, and oxygenases break down azo linkages and convert aromatic amines. *B. subtilis* DS is found to produce an alkali-thermostable laccase (Shamim Hossain et al., 2022).

## 3.7 Microbial remediation of chlorinated solvents

Various industries emit chlorinated volatile organic compounds (Cl-VOCs), which pose significant health risks, and the remediation methods currently in use are not sustainable, as they are very expensive, complex, and generate secondary toxic metabolites. Microbial consortia play an essential role in the degradation of Cl-VOCs from the atmosphere, and they also convert the chloride released into the water into chloride ions (Chheda and Sorial, 2017). Heterotrophic microorganisms are responsible for the formation of biomass through the utilization of VOC molecules. The growth of biomass and the breakdown of contaminants are both dependent on the availability of metabolizable carbon and nutrients in the filter bed (Chaudhary et al., 2023). Aerobic cometabolism and reductive dechlorination are the two processes used to break down Cl-VOCs. Reductive dechlorination is a relatively less successful process that results in the production of potentially dangerous molecules, whereas cometabolism leads to a more rapid and thorough breakdown that does not release any toxic by-products. In the process of cometabolism, for example, monooxygenase is responsible for the breakdown of TCE by producing unstable molecules, which then transform into organic chemicals that are not harmful (Dolinová et al., 2017).

Aerobic bacterial strains such as *Methylorubrum extorquens* CM4, *Leisingera methylohalidivorans* MB2, *Methylobacterium extorquens* DM4, *Pandoraea pnomenusa* LX-1, *Bacillus* sp. D1, *Albibacter methylovorans* sp., *P. aeruginosa*, *Bacillus* sp., *Pseudomonas* sp. strain DCA1, *Ralstonia pickettii* L2, *Pandoraea* sp., *Pandoraea pnomenusa* MCB032, and *Delftia tsuruhatensis* LW26 have all been identified as potential decomposers of Cl-VOCs in the absence of cosubstrates. They also generate a by-product that is less hazardous and does not require a cosubstrate. Research demonstrates a genetically engineered strain of *Escherichia coli* that incorporates six foreign genes from the 1,2-DCA degradation pathway and has the capability to fully mineralize dichloroaniline or DCA (Deng et al., 2024).

It has been documented that anaerobic bacterial strains include *Candidatus dichloromethanomonas elyunquensis* RM, *Dehalobacterium* sp., *Chryseobacterium joostei*, *Dehalococcoides mccartyi*, *Corynebacterium variabile* HRJ4, *Dehalococcoides* strain BAV1, *Pseudomonas guguanensis* NR135725, *Desulfitobacterium* sp. strain Y-51, and *D. mccartyi* strain BTF08, which are involved in the process of reductive dechlorination of Cl-VOCs (Wu et al., 2022).

Microorganisms such as *E. coli, K. pneumoniae*, and *Staphylococcus aureus* can break down TCE in biomass-based fuels generated from bamboo-clay and turkey litter compost. *Rhodococcus opacus* is a strong TCE degrader in airlift bioreactors and two-phase solid–liquid partitioning bioreactors (Baskaran et al., 2020a, 2020b).

The biofilm and microbes within the rotating biological contactor disc are crucial to the biodegradation process. The microbial consortia were identified as *E. coli* and *S. aureus* (Ravi et al., 2015). Recalcitrant drugs present therapeutic difficulties for bacterial strains owing to their limited solubility (Chi et al., 2023). According to Rybarczyk et al. (2021), a biotrickling filter (Fig. 1.4), inoculated with the fungal

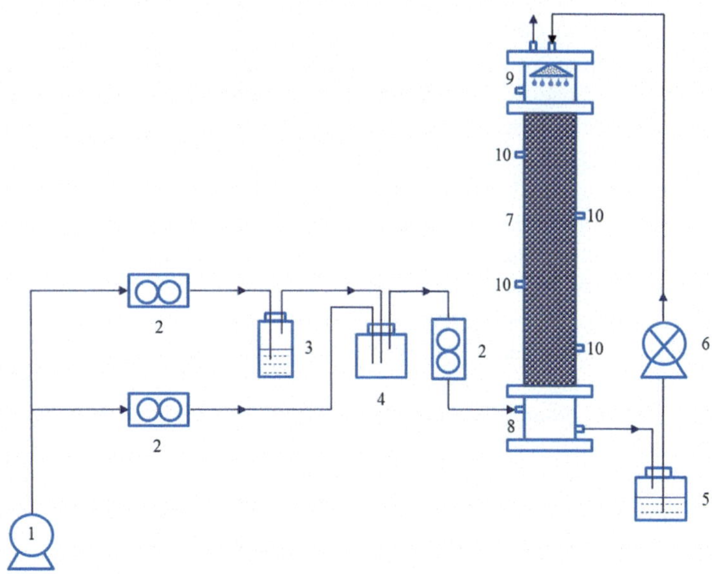


**FIG. 1.4**

Biotrickling filter (BTF) (a bioreactor used in the treatment of volatile organic compounds). Schematic representation of the BTF: (1) air compressor; (2) rotameter; (3) bottle for VOC generation; (4) bottle for gas mixing; (5) reservoir of mineral medium; (6) peristaltic pump; (7) BTF; (8) inlet sampling port; (9) outlet sampling port; (10) sampling port of packing. *[Adapted with permission from Hu et al. (2016) © 2016 Elsevier B.V.]. "With permission from Hu, J., Zhang, L., Chen, J., Luo, Y., Sun, B., Chu, G., 2016. Performance and microbial analysis of a biotrickling filter inoculated by a specific bacteria consortium for removal of a simulated mixture of pharmaceutical volatile organic compounds. Chem. Eng. J. 304, 757–765. https://doi.org/10.1016/j.cej.2016.06.078".*

species *Candida albicans* and *Candida subhashii,* may remove 95%–99% of hydrophobic VOCs from the air. However, pure strains are insufficient for processing significant amounts of chlorinated waste gases on an industrial scale.

Fungal cultures can inhabit interior voids, enhance mass transfer surfaces (Prenafeta-Boldú et al., 2019), and sequester hydrophobic VOCs via their aerial hyphae. Previously, fungal species such as *Phanerochaete chrysosporium*, *Fusarium solani*, *C. subhashii*, *Candida antarctica*, *Candida batistae, Candida bombicola,* and *Torulopsis bombicola* have been employed in the dechlorination of refractory chemicals (Marycz et al., 2023). Nonetheless, specific fungal and yeast strains exhibit limitations in their applicability due to the toxicity of chlorinated compounds. In bioreactors and biofilters, fungal taxa such as *Aspergillus*, *Scedosporium, Trametes, Trichoderma, Paecilomyces, Yarrowia, Gymnopilus, Penicillium, Rhizopus*, and *Kuehneromyces* can break down Cl-VOCs (Marycz et al., 2022).

## 3.8 Advances and emerging technologies in microbial remediation

### 3.8.1 *Microbial fuel cells*

MFCs use microorganisms that can conduct electricity to clean up dangerous waste and generate bioelectricity, thereby acting as an advanced, promising alternative and suitable choice for waste treatment. They aid in getting rid of different kinds of pollutants, such as HMs, organic pollutants, industrial effluents, and leachates from landfills. Combining MFCs with new technologies like engineered wetlands, membrane bioreactors, supercapacitors, bioelectro-Fenton processes, and bioelectrochemical systems makes the system more efficient at breaking down pollutants and producing power (see Fig. 1.5). However, technical problems, such as lower power output and higher material costs, restrict their widespread usage (Chhachhiya et al., 2025). Traditional soil-based and constructed-wetland MFCs are capable of decomposing intricate organic substances, including aromatic insecticides and PAHs (Khan et al., 2018b, Khan et al., 2020). Numerous studies have demonstrated that double-chamber (DC-MFC) designs incorporate photocatalysis, air-cathode systems, Fenton systems, fungal MFCs, and benthic MFCs. Anion exchange membranes in DC-MFCs removed phenol with 100% efficiency, outperforming proton exchange membrane-based systems, which achieved 89% efficiency (Khan et al., 2019; Feng et al., 2015).

#### 3.8.1.1 Air cathode MFCs

Air cathodes are an economical alternative in MFCs that utilize oxygen transfer (in a passive manner) to produce energy from wastewater. They have successfully eliminated transition metals such as vanadium and posttransition metals, including zinc and cadmium (Abourached et al., 2014; Hao et al., 2016). Membrane-less single-chamber MFCs (SC-MFCs) have demonstrated a 77%–97% removal efficiency for hazardous inorganic contaminants. Similarly, in double-chamber MFCs, air cathodes degrade thiosulfate and thiocyanate by means of a cation exchange membrane (CEM) (Yamada et al., 2022).

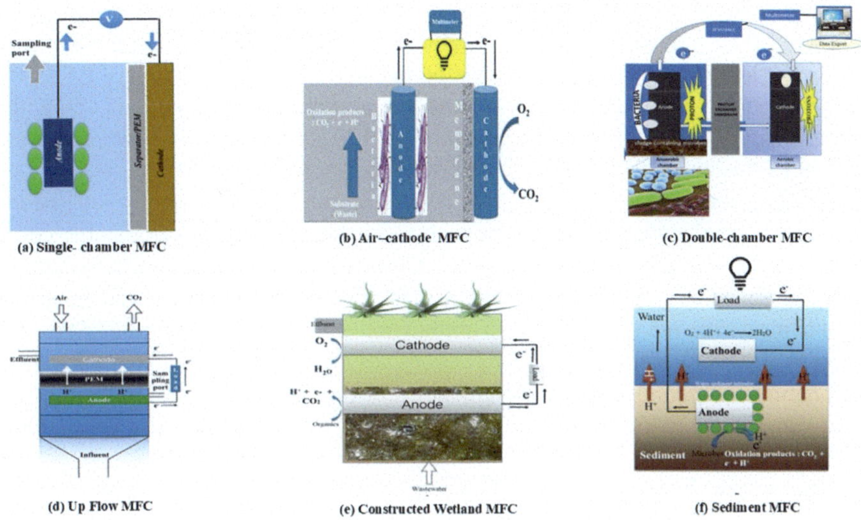


**FIG. 1.5**

Microbial fuel cells (MFCs) in advanced waste treatment procedures. Variations in the design and configurations of MFCs employed for chemical waste treatment: (A) single-chamber, (B) air-cathode, (C) double-chamber, (D) up-flow, (E) constructed-wetland, and (F) sediment *[Adopted with permission from Chhachhiya et al. (2025) © 2024 Elsevier Ltd.]. "With permission from Chhachhiya, N., Tiwari, A., Sharma, R.S., Rai, P.K., Anand, S., Mishra, V., 2025. Transformative potential of optimized microbial fuel cell designs and materials for eco-friendly management of hazardous chemical waste. J. Water Process Eng. 69. https://doi.org/10.1016/j.jwpe.2024.106647".*

#### 3.8.1.2 Soil-based microbial fuel cells

This system utilizes soil as a nutrient-rich anodic medium, successfully eliminating sulfide and arsenic while simultaneously generating electricity (Rabaey et al., 2006; Wang et al., 2022). It enables the efficient remediation of pollutants and the generation of electricity. Approximately 23% of research on SC-MFCs has achieved a maximum power density of 77.5 $MW/m^2$ and elimination efficiencies between 39% and 71% for pesticides. Conversely, the treatment of PAHs, namely through phenanthrene degradation, has achieved a power density of 4.69 $MW/m^2$ (Adelaja et al., 2021; Cao et al., 2017).

#### 3.8.1.3 Constructed wetland microbial fuel cells

This system is highly efficient at getting rid of organic and inorganic chemical waste, especially dangerous transition metals like chromium (Cr). According to a study, it has previously achieved 94% efficiency in the removal of sulfamethoxazole in bioelectrochemical systems (Zhang et al., 2017). Integration of the plant further enhances constructed wetland MFC performance. For instance, *Ipomoea aquatica* in a plant microbial fuel cell increases copper removal efficiency by doubling the rate of the process and further enhances electricity generation (Lin et al., 2022).

#### 3.8.1.4 Sediment microbial fuel cells

Sediment MFCs efficiently process transition metals and gases, making them effective for the treatment of inorganic transition metals and the degradation of PAHs in organic chemical waste. Researchers found that phenanthrene can be removed with 86% efficiency using a CEM, while several other PAHs achieved 41%–76% removal efficiency (Adelaja et al., 2021).

MFCs in chemical waste treatment have been notably enhanced by integrated systems that amalgamate technologies such as supercapacitors, three-chamber configurations, benthic processes, denitrification, photocatalysis, membrane bioreactors, and sludge blanket systems. For example, these systems have demonstrated exceptional efficacy, attaining 87.8% xylene elimination and a power output of 63 $MW/m^2$ (Umar et al., 2021). Some other notable examples are PHCs, phenanthrene, pyrene, and pharma compounds like sulfamethoxazole (Morris and Jin, 2012; Yan et al., 2012; Wang et al., 2016; Mohan and Chandrasekhar, 2011; Marshall and May, 2009; Morris et al., 2009).

A pilot-scale bioelectrochemical system utilizing SC-MFCs achieved a 99% removal rate for 4-chloronitrobenzene (Yuan et al., 2015), while a single-chamber cylindrical reactor combined with a membrane bioreactor achieved 99% phenol removal utilizing polyvinylidene fluoride hollow fiber and ultrafiltration membranes, in conjunction with carbon felt electrodes and a mixed microbial culture (Hou et al., 2021). However, the biodegradation of AB in MFCs is problematic due to their toxicity to microorganisms. To tackle the problem, innovative technologies such as electrosorption MFC and membrane bioreactor MFC have demonstrated immense potential in the treatment of high-concentration antibiotic wastewater (Song et al., 2022).

Depending on the function, MFCs could be mediator-based or mediator-less.

**A.** *Mediator-type MFCs* were developed in the early 1900s and use substances like thionine, methyl viologen, and neutral red to transfer electrons from microorganisms to the anode. Earlier, quinone derivatives were utilized to facilitate the degradation of tetracycline hydrochloride, underscoring the significance of mediators in enhancing MFC performance (Zhang et al., 2022).

**B.** *Mediator-less MFCs* signify a notable innovation as they remove the need for external electron carriers. By utilizing microorganisms with electrochemically active outer membrane proteins, such as cytochromes, this technology directly transfers electrons to the anode. According to Kumar et al. (2015), microorganisms like *Shewanella putrefaciens* and *Aeromonas hydrophila* facilitate this process through conductive pili. This method is safe for the public and the environment and uses metal-reducing bacteria (*S. putrefaciens, Rhodoferax*, and *G. sulfurreducens*) to move electrons. This makes it effective for cleaning up complicated chemical waste, such as PAHs and *p*-nitrophenol (Sherafatmand and Ng, 2015; Mohan et al., 2007; Zhang et al., 2016).

### 3.8.2 *Synthetic biology and omics*

The construction of biosynthetic metabolic pathways enables microbes to degrade previously recalcitrant pollutants or produce valuable by-products (Hassard et al., 2024). Molecular mechanisms play an important role in understanding the metabolism, genetics, nature of diversity, and microbial dynamics associated with bioremediation. Various molecular methods are employed for detailed microbiological analysis, including metabolic and protein profiling, sequencing, and advanced bioinformatics resources. These methods aid in understanding the factors affecting microbes during the bioremediation process (P. Sharma et al., 2022b). During the process of bioremediation, classical and culture-dependent molecular approaches are utilized to monitor the populations of microbial organisms. Several examples of this type of analysis include temperature gradient gel electrophoresis, amplified ribosomal DNA restriction analysis, polymorphism of T-RF length, and automated ribosomal intergenic spacer analysis (Bharagava et al., 2018).

The state-of-the-art multiomics approaches, such as genomics, transcriptomics, proteomics, and metabolomics, along with their integrative applications, are used to identify and optimize microbial communities and metabolic pathways for more efficient remediation (Pooja Sharma et al., 2022b; Khanna et al., 2023). Methodologies that fall within the category of metagenomics include metabolomics, metatranscriptomics, fluxomics, and metabolomics (Mukherjee and Reddy, 2020; R. N. Bharagava et al., 2018; Mukherjee and Reddy, 2020). The investigation of metaproteomics involves the evaluation of all protein samples obtained from environmental samples (Bargiela et al., 2015). The process of identifying and quantifying all metabolites released into an environment is referred to as metabolomics (Liu et al., 2022). In recent years, there has been a rise in the use of next-generation sequencing, which is widely regarded as the most powerful method for gene sequencing (Eisenhofer et al., 2019).

### 3.8.3 *Nanotechnology*

One of the most successful approaches to pollution control on a worldwide scale is nanotechnology. According to Shanmuganathan et al. (2019), nanomaterials can be obtained from both chemical and physical sources. However, the effectiveness of nanoparticles as bioremediation agents is largely determined by a variety of physical factors, including the size, chemical composition, surface coating, and shape of the nanoparticles. In addition, the success of nanoparticles in the bioremediation process depends on other factors, such as the type of contaminants, the type of media, the temperature, and the pH of the environment (Tan et al., 2018). Carbon dot nanoparticles have recently gained popularity in the field of environmental pollution treatment, as they are abundant, have a low toxicity level, and possess unique optical features (Long et al., 2021). As a result, further research needs to be conducted to discover new technologies and mechanisms that can enhance the efficiency of the bioremediation process.

## 4 Future perspectives

In today's world, the establishment of new technologies is providing solutions to conventional challenges in breaking down pollutants, such as slow kinetics, substrate specificity, and microbial exposure to extreme environments. Research on microbial enzymes for the bioremediation of waste materials is essential for a safer and more sustainable environment. Discovering novel microbes for efficient pollutant removal, especially from industrial sources, is imperative in this regard. Furthermore, focusing on genetically modified microbes can enhance biodegrading ability, and a combination of microbial consortia is often more effective. However, field studies are vital to determine optimal degradation temperatures and microbial degradation in conjunction with nanoparticles (Ayilara and Babalola, 2023). The enduring repercussions of fatal effects and bioactive pollutants hinder the permissibility of extended low-dose pesticide exposure. Unfortunately, there is limited information available about pesticide combinations, how they work together, and whether they can be used as antimicrobial agents. Conducting research on the impact of pesticides in natural habitats is challenging. Researchers have emphasized that further investigation is necessary to determine how well nanoparticles and endophytic bacteria work in removing pesticides and the potential concerns this method may pose (Murthy, 2025). In addition, further research is needed to discover a synergistic consortium of fungi for the removal of a variety of Cl-VOCs. The use of bioelectrochemical gas-phase reactors is one way to carry out the process of fungal-based MFCs. This approach could eliminate hydrophobic VOCs and break down pollutants such as methylene blue, acetaminophen, para-aminophenol, and sulfanilamide (Gorin et al., 2024). This technique presents substantial prospects for managing air pollution by ensuring environmental compliance and societal acceptance.

MFCs show incredible efficiencies in removing both inorganic and organic classes of pollutants, while they simultaneously produce energy. Nevertheless, further standardization is necessary due to the performance variability among various categories of waste (Chhachhiya et al., 2025). The authors also emphasized graphene-based electrodes and biochar as examples of material advancements. These advancements could, in turn, enhance electron transfer mechanisms and biofilm development. However, large-scale applications are hindered by difficulties such as high material costs, membrane fouling, and biofilm stability, which need to be addressed. Microbial consortia are essential for the operation of MFCs; yet, the intricate interactions within biofilms hinder the prediction of their behaviors. Synthetic biology and microbial engineering have advanced considerably in recent years, offering promising opportunities for the enhanced optimization of these communities. The incorporation of energy storage technology is essential; yet, economic and operational challenges persist, hindering the industrial use of MFC technologies. In addition, MFCs are considered one of the potential approaches for the execution of sustainable waste management systems, especially within circular economy (CE) frameworks (Chhachhiya et al., 2025). Advancements in materials science, microbial engineering, and system integration are essential for the progress of MFCs. Together, they can help overcome challenges, develop cost-effective and chemically resistant materials, and reduce capital expenditures (Palanisamy et al., 2023; Vidhyeswari et al., 2022).

In the case of metabolic engineering, immobilization must be studied further to eliminate issues regarding mass transfer, enzyme washing out, and physiological variables under modification. Enzymatic biodegradation is distinct; yet molecular and protein engineering, in conjunction with bioreactor design, should collaboratively endorse biocatalysis as a comprehensive biodegradation approach (Pandey et al., 2021). Further study on the mechanisms and commercial applications of these agents is essential to advance the development of next-generation bioproducts that employ these agents for selective pesticide breakdown (Manganyi et al., 2024). New alternative technologies like CRISPR-Cas9 and MFC treatment of complex classes of pollutants represent a new area of research. Replacing or using these technologies in combination with traditional remediation techniques, especially those that are greener, is not only advantageous but also sustainable. For example, CRISPR–Cas9-driven modifications (e.g., engineering *E. coli* to express toluene dioxygenase for aromatic compound breakdown) (Pandey et al., 2021; Ren and Manefield, 2025), synthetic microbial consortia, designer enzymes (Pandey et al., 2021), immobilization of nanomaterials (Karnwal et al., 2024), and waste-to-wealth models are enhancing these processes. However, the purpose of these technologies is to convert detoxified by-products into value-added compounds (Maglione et al., 2024).

Future research must prioritize the real-time monitoring (e.g., biosensors) approach, the design of AI-driven microbial consortia, and policy frameworks that incentivize adoption. With the intensification of industrial pollution, microbial remediation stands as a cornerstone of sustainable environmental management, transforming toxicity into renewal through advanced microbial processes (Pandey et al., 2021). Microbial remediation, when integrated with CE frameworks, is a crucial step for converting waste into bioenergy, biofertilizers, or recovered metals. However, it requires supportive policies, public awareness, and stakeholder engagement. Future research should focus on optimizing the microbial consortia for pollutant degradation and developing biosensors to monitor the remediation process simultaneously.

## 5 Conclusions

Microbial remediation has been explored over the past few years due to its advantages. It is considered a greener and more sustainable approach to hazardous waste management. This method is favored for its cost-effectiveness and ability to degrade even complex mixtures of pollutants in several environmental matrices. Yet, with the advent of new classes of emerging industrial pollutants, microbial consortia and enzymes face performance issues in treating these pollutants. Challenges to be considered include predictability, ecological risks, scalability, regulatory and ethical considerations, and ensuring the effectiveness of engineered strains for remediation in complex environments. Herein, combining microbial remediation with phytoremediation, advanced materials, and real-time monitoring systems could provide more robust solutions. Attention must be given toward the development of enzymes for environmental remediation, with a focus on protein engineering, synthetic biology, and bioreactor design from a CE perspective. Most importantly, assigning the hazardous and toxic category is crucial for prioritizing the remediation process. Moreover, public awareness toward household-generated toxic wastes and legislative procedures for categorization, release, and the polluter-pays principle are imperative in association with the bioremediation process of hazardous substances.

## References

Persistent Organic Pollutants: a Global Issue, A Global Response [Internet]. (2012), Available from: https://www.epa.gov/international-cooperation/persistent-organic-pollutants-global-issue-global-response.

2021 2025 12 12 USEPA https://www.epa.gov/hw/criteria-definition-solid-waste-and-solid-and-hazardous-waste-exclusions.

2021 2025 12 24 Web page Defining Hazardous Waste: Listed, Characteristic and Mixed Radiological Wastes.| EPA. https://www.epa.gov/hw/defining-hazardous-waste-listed-characteristic-and-mixed-radiological-wastes#FandK.

Waste statistics—Statistics Explained [Internet]. (2022), Available from: https://ec.europa.eu/eurostat/statistics-explained/index.php?title=Waste_statistics.

International trade control measures under the Basel- Rotterdam and Stockholm Conventions. (2023),

Abourached, C., Catal, T., Liu, H., 2014. Efficacy of single-chamber microbial fuel cells for removal of cadmium and zinc with simultaneous electricity production. Water Res. 00431354. 51, 228–233. doi:10.1016/j.watres.2013.10.062.

Abouseoud, M., Yataghene, A., Amrane, A., Maachi, R., 2008. Biosurfactant production by free and alginate entrapped cells of Pseudomonas fluorescens. J. Ind. Microbiol. Biotechnol. 14765535. 35 (11), 1303–1308. doi:10.1007/s10295-008-0411-0.

Adelaja, O., Keshavarz, T., Kyazze, G., 2021. Enhanced electrochemical treatment of phenanthrene-polluted soil using microbial fuel cells. Earthline J. Chem. Sci. 2581-9003. 37–63. doi:10.34198/ejcs.6121.3763.

Adeola, F.O., 2011. Taxonomy of Hazardous Wastes. Springer Science and Business Media LLC, pp. 25–53. doi:10.1057/9780230339538_3.

Akinsemolu, A.A., Onyeaka, H.N., 2025. Harnessing biotechnology for the remediation of organic pollutants in coastal marine ecosystems. Appl. Sci. 20763417. 15 (12). Available from: https://www.mdpi.com/journal/applsci/. doi:10.3390/app15126921.

Al-Huqail, A.A., Kumar, P., Eid, E.M., Singh, J., Arya, A.K., Goala, M., Adelodun, B., Fayssal, S.A., Kumar, V., Širić, I., 2022. Risk assessment of heavy metals contamination in soil and two rice (Oryza sativa L.) varieties irrigated with paper mill effluent. Agriculture. 20770472. 12 (11). Available from: www.mdpi.com/journal/agriculture. doi:10.3390/agriculture12111864.

Albelda-Berenguer, M., Monachon, M., Joseph, E., 2019. Siderophores: from natural roles to potential applications. Adv. Appl. Microbiol. 106, 193–225. Available from: http://www.sciencedirect.com/science/journal/00652164. doi:10.1016/bs.aambs.2018.12.001.

Alloway, B.J., 2013. Sources of heavy metals and metalloids in soils. In: Heavy Metals in Soils: Trace Metals and Metalloids in Soils and Their Bioavailability, vol. 22. Springer Netherlands, Dordrecht. pp. 11–50. Available from: https://link.springer.com/book/10.1007/978-94-007-4470-7.

Alvarez, P.J., Vogel, T.M., 1991. Substrate interactions of benzene, toluene, and para-xylene during microbial degradation by pure cultures and mixed culture aquifer slurries. Appl. Environ. Microbiol. 57 (10), 2981–2985. Available from: https://journals.asm.org/doi/abs/10.1128/aem.57.10.2981-2985.1991.

Amini, M., Younesi, H., Bahramifar, N., 2009. Biosorption of nickel(II) from aqueous solution by Aspergillus niger: response surface methodology and isotherm study. Chemosphere. 00456535. 75 (11), 1483–1491. doi:10.1016/j.chemosphere.2009.02.025.

Ammeri, R.W., Kloula, S., Simeone, G.D.R., Mehri, I., Hassan, W., Hassen, A., 2023. Pesticide removal by the bioaugmentation process in secondary treated wastewater. Water Qual. Res. J. 27098052. 58 (2), 111–127. Available from: https://iwaponline.com/wqrj/article-pdf/58/2/111/1227222/wqrjc0580111.pdf. doi:10.2166/wqrj.2023.001.

Anderson, R.T., Vrionis, H.A., Ortiz-Bernad, I., Resch, C.T., Long, P.E., Dayvault, R., Karp, K., Marutzky, S., Metzler, D.R., Peacock, A., White, D.C., Lowe, M., Lovley, D.R., 2003. Stimulating the in situ activity of geobacter species to remove uranium from the groundwater of a uranium-contaminated aquifer. Appl. Environ. Microbiol. 69 (10), 5884–5891. doi:10.1128/AEM.69.10.5884-5891.2003.

Årstøl, E., Hohmann-Marriott, M.F., 2019. Cyanobacterial siderophores—physiology, structure, biosynthesis, and applications. Mar. Drugs. 17 (5) Available from: https://www.mdpi.com/1660-3397/17/5/281. doi:10.3390/md17050281.

Asante-Duah, D.K., 2021. Hazardous Waste Risk Assessment. CRC press. Available from: https://www.taylorfrancis.com/books/mono/10.1201/9781003070009/hazardous-waste-risk-assessment-kofi-asante-duah. doi:10.1201/9781003070009. 9781003070009. (Accessed 22 07 2025).

Atlas, R.M., 1981. Microbial degradation of petroleum hydrocarbons: an environmental perspective. Microbiol. Rev. 0146-0749. 45 (1), 180–209. doi:10.1128/mmbr.45.1.180-209.1981.

Atlas, R.M., Hazen, T.C., 2011. Oil biodegradation and bioremediation: a tale of the two worst spills in U.S. history. Environ. Sci. Technol. 15205851. 45 (16), 6709–6715. doi:10.1021/es2013227.

Ayangbenro, A.S., Babalola, O.O., 2017. A new strategy for heavy metal polluted environments: a review of microbial biosorbents. Int. J. Environ. Res. Public Health. 16604601. 14 (1). Available from: http://www.mdpi.com/1660-4601/14/1/94/pdf. doi:10.3390/ijerph14010094.

Ayangbenro, A.S., Babalola, O.O., Aremu, O.S., 2019. Bioflocculant production and heavy metal sorption by metal resistant bacterial isolates from gold mining soil. Chemosphere. 00456535. 231, 113–120. doi:10.1016/j.chemosphere.2019.05.092.

Ayangbenro, A.S., Olanrewaju, O.S., Babalola, O.O., 2018. Sulfate-reducing bacteria as an effective tool for sustainable acid mine bioremediation. Front. Microbiol. 1664302X. 9. Available from: https://www.frontiersin.org/articles/10.3389/fmicb.2018.01986/full. doi:10.3389/fmicb.2018.01986.

Ayilara, M.S., Babalola, O.O., 2023. Bioremediation of environmental wastes: the role of microorganisms. Front. Agron. 26733218. 5. Available from: www.frontiersin.org/journals/agronomy#. doi:10.3389/fagro.2023.1183691.

Ayilara, M.S., Olanrewaju, O.S., Babalola, O.O., Odeyemi, O., 2020. Waste management through composting: challenges and potentials. Sustainability. 20711050. 12 (11). Available from: http://res.mdpi.com/d_attachment/sustainability/sustainability-12-04456/article_deploy/sustainability-12-04456-v2.pdf. doi:10.3390/su12114456.

Ayub, A., Wani, A.K., Malik, S.M., Ayub, M., Chopra, C., Singh, R., Malik, T., 2025. Harnessing microbes and plants for bioremediation of heavy metal contaminants: current paradigms and future perspectives. Environ. Chall. 26670100. 20. Available from: https://www.sciencedirect.com/science/journal/26670100. doi:10.1016/j.envc.2025.101220.

Bacosa, H.P., Suto, K., Inoue, C., 2012. Bacterial community dynamics during the preferential degradation of aromatic hydrocarbons by a microbial consortium. Int. Biodeterior. Biodegrad. 09648305. 74, 109–115. doi:10.1016/j.ibiod.2012.04.022.

Bai, L., Zhang, Q., Wang, C., Yao, X., Zhang, H., Jiang, H., 2019. Effects of natural dissolved organic matter on the complexation and biodegradation of 17α-ethinylestradiol in freshwater lakes. Environ. Pollut. 02697491. 246, 782–789. doi:10.1016/j.envpol.2018.12.098.

Baltazar, M.D.P.G., Gracioso, L.H., Avanzi, I.R., Karolski, B., Tenório, J.A.S., Nascimento, C.A.O.D., Perpetuo, E.A., 2019. Copper biosorption by Rhodococcus erythropolis isolated from the Sossego Mine—PA—Brazil. J. Mater. Res. Technol. 22387854. 8 (1), 475–483. Available from: http://www.elsevier.com/journals/journal-of-materials-research-and-technology/2238-7854. doi:10.1016/j.jmrt.2018.04.006.

Banala, U.K., Das, N.P.I., Toleti, S.R., 2021. Microbial interactions with uranium: towards an effective bioremediation approach. Environ. Technol. Innov. 23521864. 21. Available from: http://www.journals.elsevier.com/environmental-technology-and-innovation/. doi:10.1016/j.eti.2020.101254.

Bano, A., Hussain, J., Akbar, A., Mehmood, K., Anwar, M., Hasni, M.S., Ullah, S., Sajid, S., Ali, I., 2018. Biosorption of heavy metals by obligate halophilic fungi. Chemosphere. 00456535. 199, 218–222. doi:10.1016/j.chemosphere.2018.02.043.

Bargiela, R., Herbst, F.A., Martínez-Martínez, M., Seifert, J., Rojo, D., Cappello, S., Genovese, M., Crisafi, F., Denaro, R., Chernikova, T.N., Barbas, C., von Bergen, M., Yakimov, M.M., Ferrer, M., Golyshin, P.N., 2015. Metaproteomics and metabolomics analyses of chronically petroleum-polluted sites reveal the importance of general anaerobic processes uncoupled with degradation. Proteomics. 16159861. 15 (20), 3508–3520. Available from: http://onlinelibrary.wiley.com/journal/10.1002/(ISSN)1615-9861. doi:10.1002/pmic.201400614.

Baskaran, D., Rajamanickam, R., Vaidyalingam, B., 2020b. Effect of concentration and gas flow rate on the removal of gas-phase trichloroethylene in a novel packed biofilter. Bioresour. Technol. Rep. 2589014X. 9, 100387. doi:10.1016/j.biteb.2020.100387.

Baskaran, D., Sinharoy, A., Pakshirajan, K., Rajamanickam, R., 2020a. Gas-phase trichloroethylene removal by Rhodococcus opacus using an airlift bioreactor and its modeling by artificial neural network. Chemosphere. 00456535. 247, 125806. doi:10.1016/j.chemosphere.2019.125806.

Bera, S.P., Tank, S.K., 2021. Microbial degradation of Procion Red by Pseudomonas stutzeri. Sci. Rep. 2045-2322. 11 (1). doi:10.1038/s41598-021-82494-9.

Bhandari, G., Gupta, S., Chaudhary, P., Chaudhary, S., Gangola, S., 2023. Bioleaching: A Sustainable Resource Recovery Strategy for Urban Mining of E-waste Microbial Technology for Sustainable E-waste Management. Springer International Publishing, India, pp. 157–175. Available from: https://link.springer.com/book/10.1007/978-3-031-25678-3. doi:10.1007/978-3-031-25678-3_10.

Bharagava, R.N., Saxena, G., Mulla, S.I., 2019. Introduction to Industrial Wastes Containing Organic and Inorganic Pollutants and Bioremediation Approaches for Environmental Management, Vol. i. Springer Singapore, pp. 1–18. Available from: https://link.springer.com/chapter/10.1007/978-981-13-1891-7_1. doi:10.1007/978-981-13-1891-7_1.

Bharagava, R.N., Saxena, G., Purchase, D., Mulla, S.I., 2018. Microbial Diversity in the Genomic EraApplications of Metagenomics in Microbial Bioremediation of Pollutants Microbial Diversity in the Genomic Era. Elsevier, India, pp. 459–477. Available from: https://www.sciencedirect.com/science/article/pii/B9780128148495000265. doi:10.1016/B978-0-12-814849-5.00026-5.

Bidja Abena, M.T., Li, T., Shah, M.N., Zhong, W., 2019. Biodegradation of total petroleum hydrocarbons (TPH) in highly contaminated soils by natural attenuation and bioaugmentation. Chemosphere. 18791298. 234, 864–874. Available from: www.elsevier.com/locate/chemosphere. doi:10.1016/j.chemosphere.2019.06.111.

Boruah, H.P.D., Chauhan, P.S., Acharya, C., 2023. Editorial: trends of microbial technologies in rehabilitation of contaminated environments. Front. Microbiol. 1664-302X. 14. doi:10.3389/fmicb.2023.1268002.

Cakmak, S., Dales, R.E., Liu, L., Kauri, L.M., Lemieux, C.L., Hebbern, C., Zhu, J., 2014. Residential exposure to volatile organic compounds and lung function: results from a population-based cross-sectional survey. Environ. Pollut. 02697491. 194, 145–151. doi:10.1016/j.envpol.2014.07.020.

Cao, B., Nagarajan, K., Loh, K.C., 2009. Biodegradation of aromatic compounds: current status and opportunities for biomolecular approaches. Appl. Microbiol. Biotechnol. 01757598. 85 (2), 207–228. doi:10.1007/s00253-009-2192-4.

Cao, H., Liu, C., Cai, F., Qiao, X., Dichiara, A., CTW Research, 2020. Available from: https://www.sciencedirect.com/science/article/pii/S004313542030419X.

Cao, X., Yu, C., Wang, H., Zhou, F., Li, X., 2017. Simultaneous degradation of refractory organic pesticide and bioelectricity generation in a soil microbial fuel cell with different conditions. Environ. Technol. 1479487X. 38 (8), 1043–1050. Available from: http://www.tandf.co.uk/journals/titles/09593330.asp. doi:10.1080/09593330.2016.1216609.

Cepoi, L., Zinicovscaia, I., Valuta, A., Codreanu, L., Rudi, L., Chiriac, T., Yushin, N., Grozdov, D., Peshkova, A., 2022. Bioremediation capacity of edaphic cyanobacteria Nostoc linckia for chromium in association with other heavy-metals-contaminated soils. Environments. 20763298. 9 (1). Available from: https://www.mdpi.com/2076-3298/9/1/1/pdf. doi:10.3390/environments9010001.

Chabalala, S., Chirwa, E.M.N., 2010. Uranium(VI) reduction and removal by high performing purified anaerobic cultures from mine soil. Chemosphere. 00456535. 78 (1), 52–55. Available from: www.elsevier.com/locate/chemosphere. doi:10.1016/j.chemosphere.2009.10.026.

Chaíneau, C.H., Morel, J.L., Oudot, J., 1995. Microbial degradation in soil microcosms of fuel oil hydrocarbons from drilling cuttings. Environ. Sci. Technol. 15205851. 29 (6), 1615–1621. doi:10.1021/es00006a027.

Chaîneau, C.H., Rougeux, G., Yéprémian, C., Oudot, J., 2005. Effects of nutrient concentration on the biodegradation of crude oil and associated microbial populations in the soil. Soil Biol. Biochem. 37 (8), 1490–1497. doi:10.1016/j.soilbio.2005.01.012.

Champion, P.M., Gunsalus, I.C., Wagner, G.C., 1978. Resonance Raman investigations of cytochrome P450CAM from Pseudomonas putida. J. Am. Chem. Soc. 0002-7863. 100 (12), 3743–3751. doi:10.1021/ja00480a015.

Chaudhary, D.K., Park, J.-H., Kim, P.-G., Ok, Y.S., Hong, Y., 2023. Enrichment cultivation of VOC-degrading bacteria using diffusion bioreactor and development of bacterial-immobilized biochar for VOC bioremediation. Environ. Pollut. 02697491. 320, 121089. doi:10.1016/j.envpol.2023.121089.

Chen, B., Huang, J., Yuan, K., Lin, L., Wang, X., Yang, L., Luan, T., 2016. Direct evidences on bacterial growth pattern regulating pyrene degradation pathway and genotypic dioxygenase expression. Mar. Pollut. Bull. 0025326X. 105 (1), 73–80. doi:10.1016/j.marpolbul.2016.02.054.

Chen, L., Liu, J., Zhang, W., Zhou, J., Luo, D., Li, Z., 2021. Uranium (U) source, speciation, uptake, toxicity and bioremediation strategies in soil-plant system: a review. J. Hazard. Mater. 03043894. 413, 125319. doi:10.1016/j.jhazmat.2021.125319.

Chen, Y.A., Grace Liu, P.W., Whang, L.M., Wu, Y.J., Cheng, S.S., 2020. Effect of soil organic matter on petroleum hydrocarbon degradation in diesel/fuel oil-contaminated soil. J. Biosci. Bioeng. 13474421. 129 (5), 603–612. Available from: http://www.elsevier.com/wps/find/journaldescription.cws_home/505516/description#description. doi:10.1016/j.jbiosc.2019.12.001.

Chhachhiya, N., Tiwari, A., Sharma, R.S., Rai, P.K., Anand, S., Mishra, V., 2025. Transformative potential of optimized microbial fuel cell designs and materials for eco-friendly management of hazardous chemical waste. J. Water Process Eng. 22147144. 69, 106647. doi:10.1016/j.jwpe.2024.106647.

Chheda, D., Sorial, G.A., 2017. Evaluation of co-metabolic removal of trichloroethylene in a biotrickling filter under acidic conditions. J. Environ. Sci. 10010742. 57, 54–61. doi:10.1016/j.jes.2016.12.008.

Chi, Z., Liu, X., Li, H., Liang, S., Luo, Y.-H., Zhou, C., Rittmann, B.E., 2023. Co-metabolic biodegradation of chlorinated ethene in an oxygen- and ethane-based membrane biofilm reactor. Sci. Total Environ. 00489697. 905, 167323. doi:10.1016/j.scitotenv.2023.167323.

Choudhary, G., Goldhaber, S., Toxicological profile for benzidine (update, August 1995). (1995), Available from: https://www.osti.gov/biblio/160909.

Copete-Pertuz, L.S., Plácido, J., Serna-Galvis, E.A., Torres-Palma, R.A., Mora, A., 2018. Elimination of Isoxazolyl-Penicillins antibiotics in waters by the ligninolytic native Colombian strain Leptosphaerulina sp. considerations on biodegradation process and antimicrobial activity removal. Sci. Total Environ. 00489697. 630, 1195–1204. doi:10.1016/j.scitotenv.2018.02.244.

Czemierska, Jarosz-wikołazka, Dobrowolski, Szczes, A., 2017. Bioresource Technology Studies of cadmium (II), lead (II), nickel (II), cobalt (II) and chromium (VI) sorption on extracellular polymeric substances produced by Rhodococcus opacus and Rhodococcus rhodochrous. Bioresour. Technol. 225, 113–120. Available from: https://www.sciencedirect.com/science/article/pii/S0960852416315450.

Dellamatrice, P.M., Silva-Stenico, M.E., Moraes, L.A.B., Fiore, M.F., Monteiro, R.T.R., 2017. Degradation of textile dyes by cyanobacteria. Braz. J. Microbiol. Available from: https://www.scielo.br/j/bjm/a/bQLR75bGGqg4WgKJFnyfC7J/?lang=en.

Deng, Y.-D., Zhang, W.-H., Zuo, Z.-H., Zhang, H., Jing, X., Gao, J.-J., Wang, B., Zhen-Jun, L., Xiao-Yan, F., Wang, L.-J., Wang, Y., Tian, Y.-S., Peng, R.-H., Yao, Q.-H., 2024. The complete degradation of 1,2-dichloroethane in Escherichia coli by metabolic engineering. J. Hazard. Mater. 03043894. 472, 134476. doi:10.1016/j.jhazmat.2024.134476.

Dhar, K., Subashchandrabose, S.R., Venkateswarlu, K., Krishnan, K., Megharaj, M., 2020. Anaerobic microbial degradation of polycyclic aromatic hydrocarbons: a comprehensive review. Rev. Environ. Contam. Toxicol. 251, 25–108. Available from: http://www.springer.com/series/398. doi:10.1007/398_2019_29.

Dixit, S., Yadav, A., Dwivedi, P.D., Das, M., 2015. Toxic hazards of leather industry and technologies to combat threat: a review. J. Clean. Prod. 09596526. 87 (1), 39–49. doi:10.1016/j.jclepro.2014.10.017.

Dolinová, I., Štrojsová, M., Černík, M., Němeček, J., Macháčková, J., Ševců, A., 2017. Microbial degradation of chloroethenes: a review. Environ. Sci. Pollut. Res. 16147499. 24 (15), 13262–13283. Available from: http://www.springerlink.com/content/0944-1344. doi:10.1007/s11356-017-8867-y.

Dwivedi, K.A., Huang, S.-J., Wang, C.-T., Kumar, S., 2022. Fundamental understanding of microbial fuel cell technology: recent development and challenges. Chemosphere. 00456535. 288 (Pt 2), 132446. doi:10.1016/j.chemosphere.2021.132446.

Eisenhofer, R., Minich, J.J., Marotz, C., Cooper, A., Knight, R., Weyrich, L.S., 2019. Contamination in low microbial biomass microbiome studies: issues and recommendations. Trends Microbiol. 0966842X. 27 (2), 105–117. doi:10.1016/j.tim.2018.11.003.

El-Shahawi, M.S., Hamza, A., Bashammakh, A.S., Al-Saggaf, W.T., 2010. An overview on the accumulation, distribution, transformations, toxicity and analytical methods for the monitoring of persistent organic pollutants. Talanta. 00399140. 80 (5), 1587–1597. doi:10.1016/j.talanta.2009.09.055.

Environmental NSIJ of, 2021. Bioremediation of total petroleum hydrocarbons (TPH) by bioaugmentation and biostimulation in water with floating oil spill containment booms as bioreactor basin. Available from: https://www.mdpi.com/1660-4601/18/5/2226.

Fashola, M.O., Ngole-Jeme, V.M., Babalola, O.O., 2016. Heavy metal pollution from gold mines: environmental effects and bacterial strategies for resistance. Int. J. Environ. Res. Public Health. 16604601. 13 (11). Available from: http://www.mdpi.com/1660-4601/13/11/1047/pdf. doi:10.3390/ijerph13111047.

Fashola, M.O., Ngole-Jeme, V.M., Babalola, O.O., 2020. Heavy metal immobilization potential of indigenous bacteria isolated from gold mine tailings. Int. J. Environ. Res. 20082304. 14 (1), 71–86. Available from: www.springer.com/41742. doi:10.1007/s41742-019-00240-6.

Feng, C., Huang, L., Yu, H., Yi, X., Wei, C., 2015. Simultaneous phenol removal, nitrification and denitrification using microbial fuel cell technology. Water Res. 00431354. 76, 160–170. doi:10.1016/j.watres.2015.03.001.

Furukawa, K., Fujihara, H., 2008. Microbial degradation of polychlorinated biphenyls: biochemical and molecular features. J. Biosci. Bioeng. 13891723. 105 (5), 433–449. doi:10.1263/jbb.105.433.

Gangola, S., Sharma, A., Bhatt, P., Khati, P., Chaudhary, P., 2018. Presence of esterase and laccase in Bacillus subtilis facilitates biodegradation and detoxification of cypermethrin. Sci. Rep. 2045-2322. 8 (1). doi:10.1038/s41598-018-31082-5.

Gao, C., Jin, X., Ren, J., Fang, H., Yu, Y., 2015. Bioaugmentation of DDT-contaminated soil by dissemination of the catabolic plasmid pDOD. J. Environ. Sci. (China. 10010742. 27 (C), 42–50. Available from: http://www.journals.elsevier.com/journal-of-environmental-sciences/. doi:10.1016/j.jes.2014.05.045.

Gerber, U., Zirnstein, I., Krawczyk-Bärsch, E., Lünsdorf, H., Arnold, T., Merroun, M.L., 2016. Combined use of flow cytometry and microscopy to study the interactions between the gram-negative betaproteobacterium Acidovorax facilis and uranium(VI). J. Hazard. Mater. 03043894. 317, 127–134. doi:10.1016/j.jhazmat.2016.05.062.

Ghosal, D., Ghosh, S., Dutta, T.K., Ahn, Y., 2016. Current state of knowledge in microbial degradation of polycyclic aromatic hydrocarbons (PAHs): a review. Front. Microbiol. 1664-302X. 7. doi:10.3389/fmicb.2016.01369.

Gorin, M., Shabani, M., Votat, S., Lebrun, L., Mbokou, S.F., Pontié, M., 2024. Application of fungal-based microbial fuel cells for biodegradation of pharmaceuticals: comparative study of individual vs. mixed contaminant solutions. Chemosphere. 00456535. 363, 142849. doi:10.1016/j.chemosphere.2024.142849.

Goud, B.S., Cha, H.L., Koyyada, G., Kim, J.H., 2020. Augmented biodegradation of textile azo dye effluents by plant endophytes: a sustainable, eco-friendly alternative. Curr. Microbiol. 14320991. 77 (11), 3240–3255. Available from: https://link.springer.de/link/service/journals/00284/index.htm. doi:10.1007/s00284-020-02202-0.

Grube, Donaldson, D., Kiely, Wu, L., 2011. USEPA, Washington, DC. https://www.epa.gov/sites/default/files/2015-10/documents/market_estimates2007.pdf.

Guillemot, T., Salazar, G., Rauber, M., Kunz, D., Szidat, S., Wieland, E., 2022. Carbon-14 release and speciation during corrosion of irradiated steel under radioactive waste disposal conditions. Sci. Total Environ. 00489697. 817, 152596. doi:10.1016/j.scitotenv.2021.152596.

Gupta, A., Singh, S.K., Singh, V.K., Singh, M.K., Modi, A., Yeka Zhimo, V., Singh, A.V., Kumar, A., 2020. Endophytic Microbe Approaches in Bioremediation of Organic Pollutants. Elsevier BV, pp. 157–174. doi:10.1016/b978-0-12-818734-0.00007-3.

Gupta, J.P., Babu, B.S., 1999. A new hazardous waste index. J. Hazard. Mater. 03043894. 67 (1), 1–7. Available from: www.elsevier.com/locate/jhazmat. doi:10.1016/S0304-3894(99)00006-0.

Häder, D., Williamson, C.E., Wängberg, S., Rautio, M., Rose, K.C., Gao, K., Helbling, E.W., Sinha, R.P., Worrest, R., 2014. Effects of UV radiation on aquatic ecosystems and interactions with other environmental factors. Photochem. Photobiol. Sci. 14 (1), 108–126. doi:10.1039/c4pp90035a.

Hajieghrari, M., Hejazi, P., 2020. Enhanced biodegradation of n-Hexadecane in solid-phase of soil by employing immobilized Pseudomonas Aeruginosa on size-optimized coconut fibers. J. Hazard. Mater. 18733336. 389. Available from: www.elsevier.com/locate/jhazmat. doi:10.1016/j.jhazmat.2020.122134.

Haleyur, N., Shahsavari, E., Jain, S.S., Koshlaf, E., Ravindran, V.B., Morrison, P.D., Osborn, A.M., Ball, A.S., 2019. Influence of bioaugmentation and biostimulation on PAH degradation in aged contaminated soils: response and dynamics of the bacterial community. J. Environ. Manage. 03014797. 238, 49–58. doi:10.1016/j.jenvman.2019.02.115.

Hamamura, N., Fukui, M., Ward, D.M., Inskeep, W.P., 2008. Assessing soil microbial populations responding to crude-oil amendment at different temperatures using phylogenetic, functional gene (alkB) and physiological analyses. Environ. Sci. Technol. 0013936X. 42 (20), 7580–7586. doi:10.1021/es800030f.

Hao, L., Zhang, B., Cheng, M., Feng, C., 2016. Effects of various organic carbon sources on simultaneous V(V) reduction and bioelectricity generation in single chamber microbial fuel cells. Bioresour. Technol. 09608524. 201, 105–110. doi:10.1016/j.biortech.2015.11.060.

Hassard, F., Curtis, T.P., Dotro, G.C., Golyshin, P., Gutierrez, T., Heaven, S., Horsfall, L., Jefferson, B., Jones, D.L., Krasnogor, N., Kumar, V., Lea-Smith, D.J., Pidou, K.L.C., Liu, Y., Lyu, T., McCarthy, R.R., McKew, B., Smith, C., Yakunin, A., Yang, Z., Zhang, Y., Coulon, F., 2024. Scaling-up engineering biology for enhanced environmental solutions. ACS Synth. Biol. 21615063. 13 (6), 1586–1588. Available from: http://pubs.acs.org/journal/asbcd6. doi:10.1021/acssynbio.4c00292.

Hennessee, Microbiology QLA and E, 2016. Effects of polycyclic aromatic hydrocarbon mixtures on degradation, gene expression, and metabolite production in four Mycobacterium species. Appl. Environ. Microbiol. 82 (11), 3357–3369. Available from: https://journals.asm.org/doi/abs/10.1128/aem.00100-16.

Hesham, A.E.-L., Mawad, A.M.M., Mostafa, Y.M., Shoreit, A., 2014. Study of enhancement and inhibition phenomena and genes relating to degradation of petroleum polycyclic aromatic hydrocarbons in isolated bacteria. Microbiology. 0026-2617. 83 (5), 599–607. doi:10.1134/s0026261714050129.

Hossain, M.S., Paul, G.K., Mahmud, S., Saleh, M.A., Uddin, M.S., Dutta, A.K., Roy, A.K., Saha, A.K., Sheam, M.M., Ahmed, S., Rahman, M.M., Paul, D.K., Biswas, S.K., 2022. Mixed dye degradation by Bacillus pseudomycoides and Acinetobacter haemolyticus isolated from industrial effluents: a combined affirmation with wetlab and in silico studies. Arab. J. Chem. 18785352. 15 (9), 104078. doi:10.1016/j.arabjc.2022.104078.

Hou, B., Zhang, R., Liu, X., Li, Y., Liu, P., Lu, J., 2021. Study of membrane fouling mechanism during the phenol degradation in microbial fuel cell and membrane bioreactor coupling system. Bioresour. Technol. 09608524. 338, 125504. doi:10.1016/j.biortech.2021.125504.

Hu, F., Wang, P., Li, Y., Ling, J., Ruan, Y., Yu, J., Zhang, L., 2023. Bioremediation of environmental organic pollutants by Pseudomonas aeruginosa: mechanisms, methods and challenges. Environ. Res. 10960953. 239. Available from: https://www.sciencedirect.com/journal/environmental-research. doi:10.1016/j.envres.2023.117211.

Hu, J., Zhang, L., Chen, J., Luo, Y., Sun, B., Chu, G., 2016. Performance and microbial analysis of a biotrickling filter inoculated by a specific bacteria consortium for removal of a simulated mixture of pharmaceutical volatile organic compounds. Chem. Eng. J. 13858947. 304, 757–765. Available from: www.elsevier.com/inca/publications/store/6/0/1/2/7/3/index.htt. doi:10.1016/j.cej.2016.06.078.

Hubacek, K., Guan, D., Barua, A., 2007. Changing lifestyles and consumption patterns in developing countries: a scenario analysis for China and India. Futures. 00163287. 39 (9), 1084–1096. doi:10.1016/j.futures.2007.03.010.

Hussain, I., Puschenreiter, M., Gerhard, S., Sani, S.G.A.S., Khan, W.U.D., Reichenauer, T.G., 2019. Differentiation between physical and chemical effects of oil presence in freshly spiked soil during rhizoremediation trial. Environ. Sci. Pollut. Res. 16147499. 26 (18), 18451–18464. Available from: http://www.springerlink.com/content/0944-1344. doi:10.1007/s11356-019-04819-6.

Ibrahim, O.A.A.M., Kadhim, S.A., Hammoodi, K.A., Rashid, F.L., Askar, A.H., 2024. Review of hydrocarbon refrigerants as drop-in alternatives to high-GWP refrigerants in VCR systems: the case of R290. Clean. Eng. Technol. 26667908. 23. Available from: https://www.sciencedirect.com/science/journal/26667908. doi:10.1016/j.clet.2024.100825.

Ishchenko, V., Dworak, S., Fellner, J., 2024. Hazardous household waste management in Ukraine and Austria. J. Mater. Cycles Waste Manag. 1438-4957. 26 (1), 635–641. doi:10.1007/s10163-023-01818-1.

Jadhav, J.P., Kalyani, D.C., Telke, A.A., Phugare, S.S., Govindwar, S.P., 2010. Evaluation of the efficacy of a bacterial consortium for the removal of color, reduction of heavy metals, and toxicity from textile dye effluent. Bioresour. Technol. 09608524. 101 (1), 165–173. doi:10.1016/j.biortech.2009.08.027.

Jafari, A., Ghaderpoori, M., Kamarehi, B., Abdipour, H., 2019. Soil pollution evaluation and health risk assessment of heavy metals around Douroud cement factory, Iran. Environ. Earth Sci. 18666299. 78 (8). Available from: http://www.springerlink.com/content/121380/. doi:10.1007/s12665-019-8220-5.

Jaiswal, D.K., Verma, J.P., Krishna, R., Gaurav, A.K., Yadav, J., 2019. Molecular characterization of monocrotophos and chlorpyrifos tolerant bacterial strain for enhancing seed germination of vegetable crops. Chemosphere. 00456535. 223, 636–650. doi:10.1016/j.chemosphere.2019.02.053.

Kahn, L.G., Harley, K.G., Siegel, E.L., Zhu, Y., Factor-Litva, P., Porucznik, C.A., Klein-Fedyshi, M., Hipwell, A.E., 2021. Persistent organic pollutants and couple fecundability: a systematic review. Hum. Reprod. Update. 14602369. 27 (2), 339–366. Available from: http://humupd.oxfordjournals.org/. doi:10.1093/humupd/dmaa037.

Kalyani, D.C., Phugare, S.S., Shedbalkar, U.U., Jadhav, J.P., 2011. Purification and characterization of a bacterial peroxidase from the isolated strain Pseudomonas sp. SUK1 and its application for textile dye decolorization. Ann. Microbiol. 18692044. 61 (3), 483–491. doi:10.1007/s13213-010-0162-9.

Kan, J., Peng, T., Huang, T., Xiong, G., Hu, Z., 2020. NarL, a novel repressor for CYP108j1 expression during PAHs degradation in Rhodococcus sp. P14. Int. J. Mol. Sci. 14220067. 21 (3). Available from: https://www.mdpi.com/1422-0067/21/3/983/pdf. doi:10.3390/ijms21030983.

Karnwal, A., Martolia, S., Dohroo, A., Al-Tawaha, A.R.M.S., Malik, T., 2024. Exploring bioremediation strategies for heavy metals and POPs pollution: the role of microbes, plants, and nanotechnology. Front. Environ. Sci. 2296665X. 12. Available from: https://journal.frontiersin.org/journal/environmental-science. doi:10.3389/fenvs.2024.1397850.

Khan, A., Singh, P., Srivastava, A., 2018a. Synthesis, nature and utility of universal iron chelator – siderophore: a review. Microbiol. Res. 09445013. 212-213, 103–111. doi:10.1016/j.micres.2017.10.012.

Khan, N., Anwer, A.H., Ahmad, A., Sabir, S., Khan, M.Z., 2020. Investigating microbial fuel cell aided bio-remediation of mixed phenolic contaminants under oxic and anoxic environments. Biochem. Eng. J. 1369703X. 155, 107485. doi:10.1016/j.bej.2019.107485.

Khan, N., Danish Khan, M., Nizami, A.-S., Rehan, M., Shaida, A., Ahmad, A., Khan, M.Z., 2018b. Energy generation through bioelectrochemical degradation of pentachlorophenol in microbial fuel cell. RSC Adv. 2046-2069. 8 (37), 20726–20736. doi:10.1039/c8ra01643g.

Khan, N., Khan, M.D., Ansari, M.Y., Ahmad, A., Khan, M.Z., 2019. Bio-electrodegradation of 2,4,6-Trichlorophenol by mixed microbial culture in dual chambered microbial fuel cells. J. Biosci. Bioeng. 13891723. 127 (3), 353–359. doi:10.1016/j.jbiosc.2018.08.012.

Khanna, N., Chatterji, T., Singh, S., Chaturvedi, P., 2023. Microbial Metabolism in Bioremediation: a Review. In: AIP Conference Proceedings, vol. 1. American Institute of Physics Inc., India, p. 2800. 15517616. http://scitation.aip.org/content/aip/proceeding/aipcp. doi:10.1063/5.0163146.

Koarashi, J., Moriya, K., Atarashi-Andoh, M., Matsunaga, T., Fujita, H., Nagaoka, M., 2012. Retention of potentially mobile radiocesium in forest surface soils affected by the Fukushima nuclear accident. Sci. Rep. 2045-2322. 2 (1). doi:10.1038/srep01005.

Kolhe, N., Zinjarde, S., Acharya, C., 2020. Impact of uranium exposure on marine yeast, Yarrowia lipolytica: insights into the yeast strategies to withstand uranium stress. J. Hazard. Mater. 03043894. 381, 121226. doi:10.1016/j.jhazmat.2019.121226.

Kour, D., Khan, S.S., Kour, H., Kaur, T., Devi, R., Rai, P.K., Judy, C., McQuestion, C., Bianchi, A., Spells, S., Mohan, R., Rai, A.K., Yadav, A.N., 2022. Microbe-mediated bioremediation: current research and future challenges. J. Appl. Biol. Biotechnol. 2347212X. 10, 6–24. Available from: https://jabonline.in/index.php. doi:10.7324/JABB.2022.10s202.

Kumar, R., Singh, L., Wahid, Z.A., Din, M.F.M., 2015. Exoelectrogens in microbial fuel cells toward bioelectricity generation: a review. Int. J. Energy Res. 1099114X. 39 (8), 1048–1067. Available from: http://onlinelibrary.wiley.com/journal/10.1002/(ISSN)1099-114X. doi:10.1002/er.3305.

Kumar, R., Singh, L., Zularisam, A.W., Hai, F.I., 2018. Microbial fuel cell is emerging as a versatile technology: a review on its possible applications, challenges and strategies to improve the performances. Int. J. Energy Res. 1099114X. 42 (2), 369–394. Available from: http://onlinelibrary.wiley.com/journal/10.1002/(ISSN)1099-114X. doi:10.1002/er.3780.

Kumari, S., Regar, R.K., Manickam, N., 2018. Improved polycyclic aromatic hydrocarbon degradation in a crude oil by individual and a consortium of bacteria. Bioresour. Technol. 18732976. 254, 174–179. Available from: www.elsevier.com/locate/biortech. doi:10.1016/j.biortech.2018.01.075.

Kumari, V., Tripathi, A.K., 2020. Remediation of heavy metals in pharmaceutical effluent with the help of Bacillus cereus-based green-synthesized silver nanoparticles supported on alumina. Appl. Nanosci. 21905517. 10 (6), 1709–1719. Available from: https://www.springer.com/materials/nanotechnology/journal/13204. doi:10.1007/s13204-020-01351-9.

Kushwaha, A., Maurya, S., Pathak, R.K., Agarwal, S., Chaurasia, P.K., Singh, M.P., 2018. Laccase from white rot fungi having significant role in food, pharma, and other industries. In: Research Advancements in Pharmaceutical, Nutritional, and Industrial Enzymology. IGI Global, India, pp. 252–277. Available from: https://www.igi-global.com/chapter/laccase-from-white-rot-fungi-having-significant-role-in-food-pharma-and-other-industries/203818. doi:10.4018/978-1-5225-5237-6.ch011.

LaGrega, M.D., Phillip, L., Buckingham, J.C.E., 2016. Hazardous Waste Management, Second ed. Google Books, p.344. Available from: https://books.google.co.in/books?hl=en&lr=&id=NdYYAAAAQBAJ&oi=fnd&pg=PR3&dq=LaGrega+et+al.,+2001&ots=EcQ-K4eV-q&sig=EFQk1bZEDfTb8o2UR4B5qLUcODI&redir_esc=y#v=onepage&q&f=false.

Lakaniemi, A.-M., Douglas, G.B., Kaksonen, A.H., 2019. Engineering and kinetic aspects of bacterial uranium reduction for the remediation of uranium contaminated environments. J. Hazard. Mater. 03043894. 371, 198–212. doi:10.1016/j.jhazmat.2019.02.074.

Laker, M.C., 2023. Environmental impacts of gold mining—with special reference to South Africa. Mining. 26736489. 3 (2), 205–220. Available from: https://www.mdpi.com/journal/mining. doi:10.3390/mining3020012.

Larsen, J.C., 2006. Risk assessments of polychlorinated dibenzo- p -dioxins, polychlorinated dibenzofurans, and dioxin-like polychlorinated biphenyls in food. Mol. Nutr. Food Res. 1613-4125. 50 (10), 885–896. doi:10.1002/mnfr.200500247.

Lee, D.W., Lee, H., Kwon, B.O., Khim, J.S., Yim, U.H., Kim, B.S., Kim, J.J., 2018. Biosurfactant-assisted bioremediation of crude oil by indigenous bacteria isolated from Taean beach sediment. Environ. Pollut. 18736424. 241, 254–264. Available from: www.elsevier.com/inca/publications/store/4/0/5/8/5/6. doi:10.1016/j.envpol.2018.05.070.

Lin, A.Y.C., Tsai, Y.T., 2009. Occurrence of pharmaceuticals in Taiwan's surface waters: impact of waste streams from hospitals and pharmaceutical production facilities. Sci. Total Environ. 00489697. 407 (12), 3793–3802. doi:10.1016/j.scitotenv.2009.03.009.

Lin, C.W., Alfanti, L.K., Cheng, Y.S., Liu, S.H., 2022. Enhancing bioelectricity production and copper remediation in constructed single-medium plant sediment microbial fuel cells. Desalination. 00119164. 542. Available from: https://www.journals.elsevier.com/desalination. doi:10.1016/j.desal.2022.116079.

Lindholm-Lehto, P.C., Knuutinen, J.S., Ahkola, H.S.J., Herve, S.H., 2015. Refractory organic pollutants and toxicity in pulp and paper mill wastewaters. Environ. Sci. Pollut. Res. 16147499. 22 (9), 6473–6499. Available from: http://www.springerlink.com/content/0944-1344. doi:10.1007/s11356-015-4163-x.

Liu, H., Xu, J., Liang, R., Liu, J., 2014. Characterization of the medium- And long-chain n-alkanes degrading pseudomonas aeruginosa strain SJTD-1 and its alkane hydroxylase genes. PLoS ONE. 19326203. 9 (8). Available from: http://www.plosone.org/article/fetchObject.action?uri=info%3Adoi%2F10.1371%2Fjournal.pone.0105506&representation=PDF. doi:10.1371/journal.pone.0105506China.

Liu, J., Zhao, B., Lan, Y., Ma, T., 2021. Enhanced degradation of different crude oils by defined engineered consortia of Acinetobacter venetianus RAG-1 mutants based on their alkane metabolism. Bioresour. Technol. 18732976. 327. Available from: www.elsevier.com/locate/biortech. doi:10.1016/j.biortech.2021.124787.

Liu, L., Wu, Q., Miao, X., Fan, T., Meng, Z., Chen, X., Zhu, W., 2022. Study on toxicity effects of environmental pollutants based on metabolomics: a review. Chemosphere. 00456535. 286 (Pt 2), 131815. doi:10.1016/j.chemosphere.2021.131815.

Liu, X., Shi, H., Bai, Z., Zhou, W., Liu, K., Wang, M., Yujie, H., 2020. Heavy metal concentrations of soils near the large opencast coal mine pits in China. Chemosphere. 00456535. 244, 125360. doi:10.1016/j.chemosphere.2019.125360.

Liu, Y.C., Li, L.Z., Wu, Y., Tian, W., Zhang, L.P., Xu, L., Shen, Q.R., Shen, B., 2010. Isolation of an alkane-degrading Alcanivorax sp. strain 2B5 and cloning of the alkB gene. Bioresour. Technol. 09608524. 101 (1), 310–316. doi:10.1016/j.biortech.2009.08.028.

Logan, B.E., Regan, J.M., 2006. Microbial fuel cells—challenges and applications. Environ. Sci. Technol. 0013936X. 40 (17), 5172–5180. Available from: http://pubs.acs.org/journal/esthag. doi:10.1021/es0627592.

Long, C., Jiang, Z., Shangguan, J., Qing, T., Zhang, P., Feng, B., 2021. Applications of carbon dots in environmental pollution control: a review. Chem. Eng. J. 13858947. 406, 126848. doi:10.1016/j.cej.2020.126848.

Lovley, D.R., Phillips, E.J., 1992. Reduction of uranium by Desulfovibrio desulfuricans. Appl. Environ. Microbiol. 58 (3), 850–856. doi:10.1128/aem.58.3.850-856.1992.

Lovley, D.R., Widman, P.K., Woodward, J.C., Phillips, E.J., 1993. Reduction of uranium by cytochrome c3 of Desulfovibrio vulgaris. Appl. Environ. Microbiol. 59 (11), 3572–3576. doi:10.1128/aem.59.11.3572-3576.1993.

Luke, A.K., Burton, S.G., 2001. A novel application for Neurospora crassa: progress from batch culture to a membrane bioreactor for the bioremediation of phenols. Enzyme Microb. Technol. 01410229. 29 (6-7), 348–356. doi:10.1016/s0141-0229(01)00390-8.

Maglione, G., Zinno, P., Tropea, A., Mussagy, C.U., Dufossé, L., Giuffrida, D., Mondello, A., 2024. Microbes' role in environmental pollution and remediation: a bioeconomy focus approach. AIMS Microbiol. 24711888. 10 (3), 723–755. Available from: https://www.aimspress.com/aimspress-data/aimsmicro/2024/3/PDF/microbiol-10-03-033.pdf. doi:10.3934/microbiol.2024033.

Mahadevan, G.D., Zhao, F., 2017. A concise review on microbial remediation cells (MRCs) in soil and groundwater radionuclides remediation. J. Radioanal. Nucl. Chem. 314 (3), 1477–85. Available from: https://link.springer.com/article/10.1007/s10967-017-5612-4.

Mahmood, A., Bilal, B., Naeem, Z., Iram, S., 2021. Physical, Chemical, and Biological Remediation Techniques for Textile Effluents in Context with Developed and Developing Countries Rhizobiont in Bioremediation of Hazardous Waste. Springer Singapore, Pakistan, pp. 409–441. Available from: https://link.springer.com/book/10.1007/978-981-16-0602-1. doi:10.1007/978-981-16-0602-1_18.

Mahmoud, G.A.E., 2021. Microbial Scavenging of Heavy Metals Using Bioremediation Strategies Rhizobiont in Bioremediation of Hazardous Waste. Springer Singapore, Egypt, pp. 265–289. Available from: https://link.springer.com/book/10.1007/978-981-16-0602-1. doi:10.1007/978-981-16-0602-1_12.

Maity, J.P., Chen, G.S., Huang, Y.H., Sun, A.C., Chen, C.Y., 2019. Ecofriendly heavy metal stabilization: microbial induced mineral precipitation (MIMP) and biomineralization for heavy metals within the contaminated soil by indigenous bacteria. Geomicrobiol. J. 15210529. Available from: http://www.tandf.co.uk/journals/titles/01490451.asp. doi:10.1080/01490451.2019.1597216.

Maleke, M., Williams, P., Castillo, J., Botes, E., Ojo, A., DeFlaun, M., van Heerden, E., 2015. Optimization of a bioremediation system of soluble uranium based on the biostimulation of an indigenous bacterial community. Environ. Sci. Pollut. Res. 16147499. 22 (11), 8442–8450. Available from: http://www.springerlink.com/content/0944-1344. doi:10.1007/s11356-014-3980-7.

Manganyi, M.C., Dikobe, T.B., Maseme, M.R., 2024. Exploring the potential of endophytic microorganisms and nanoparticles for enhanced water remediation. Molecules. 1420-3049. 29 (12), 2858. doi:10.3390/molecules29122858.

Manzetti, S., Van Der Spoel, E.R., Van Der Spoel, D., 2014. Chemical properties, environmental fate, and degradation of seven classes of pollutants. Chem. Res. Toxicol. 15205010. 27 (5), 713–737. Available from: http://pubs.acs.org/journal/crtoec. doi:10.1021/tx500014w.

Marfe, G., Di Stefano, C., 2020. Hazardous Waste Management and Health Risks Hazardous Waste Management and Health Risks. Bentham Science Publishers, Italy, pp. 1–205. Available from: https://benthambooks.com/book/9789811454745/. doi:10.2174/97898114547451200101.

Margesin, R., Zimmerbauer, A., Schinner, F., 1999. Soil lipase activity - A useful indicator of oil biodegradation. Biotechnol. Tech. 0951208X. 13 (12), 859–863. doi:10.1023/A:1008928308695.

Marshall, C.W., May, H.D., 2009. Electrochemical evidence of direct electrode reduction by a thermophilic Gram-positive bacterium, Thermincola ferriacetica. Energy Environ. Sci. 1754-5692. 2 (6), 699. doi:10.1039/b823237g.

Marycz, M., Brillowska-Dąbrowska, A., Cantera, S., Gębicki, J., Muñoz, R., 2023. Fungal co-culture improves the biodegradation of hydrophobic VOCs gas mixtures in conventional biofilters and biotrickling filters. Chemosphere. 00456535. 313, 137609. doi:10.1016/j.chemosphere.2022.137609.

Marycz, M., Brillowska-Dąbrowska, A., Muñoz, R., Gębicki, J., 2022. A state of the art review on the use of fungi in biofiltration to remove volatile hydrophobic pollutants. Rev. Environ. Sci. Biotechnol. 15729826. 21 (1), 225–246. Available from: http://www.kluweronline.com/issn/1569-1705. doi:10.1007/s11157-021-09608-7.

Mathew, B.B., Krishnamurthy, N.B., 2018. Screening and identification of bacteria isolated from industrial area groundwater to study lead sorption: kinetics and statistical optimization of biosorption parameters. Groundw. Sustain. Dev. 2352801X. 7, 313–327. Available from: http://www.journals.elsevier.com/groundwater-for-sustainable-development/. doi:10.1016/j.gsd.2018.07.007.

Medfu Tarekegn, M., Zewdu Salilih, F., Ishetu, A.I., 2020. Microbes used as a tool for bioremediation of heavy metal from the environment. Cogent Food Agric. 23311932. 6 (1). Available from: www.tandfonline.com/action/journalInformation?journalCode=oafa20. doi:10.1080/23311932.2020.1783174.

Mehetre, G.T., Dastager, S.G., Dharne, M.S., 2019. Biodegradation of mixed polycyclic aromatic hydrocarbons by pure and mixed cultures of biosurfactant producing thermophilic and thermo-tolerant bacteria. Sci. Total Environ. 00489697. 679, 52–60. doi:10.1016/j.scitotenv.2019.04.376.

Mesquini, J.A., Sawaya, A.C.H.F., López, B.G.C., Oliveira, V.M., Miyasaka, N.R.S., 2015. Detoxification of atrazine by endophytic Streptomyces sp. isolated from sugarcane and

detection of nontoxic metabolite. Bull. Environ. Contam. Toxicol. 14320800. 95 (6), 803–809. Available from: http://www.springerlink.com. doi:10.1007/s00128-015-1673-7.
Methneni, N., Antonio Morales-González, J., Jaziri, A., Mansour, H.B., Fernandez-Serrano, M., 2021. Persistent organic and inorganic pollutants in the effluents from the textile dyeing industries: ecotoxicology appraisal via a battery of biotests. Environ. Res. 00139351. 196, 110956. doi:10.1016/j.envres.2021.110956.
Mishra, A., Kumari, M., Swati, R.K., Iqbal, K., Thakur, I.S., 2022. Persistent organic pollutants in the environment: risk assessment, hazards, and mitigation strategies. Bioresour. Technol. Rep. 2589014X. 19, 101143. doi:10.1016/j.biteb.2022.101143.
Mohan, S.V., Chandrasekhar, K., 2011. Self-induced bio-potential and graphite electron accepting conditions enhances petroleum sludge degradation in bio-electrochemical system with simultaneous power generation. Bioresour. Technol. 09608524. 102 (20), 9532–9541. doi:10.1016/j.biortech.2011.07.038.
Mohan, S.V., Raghavulu, S.V., Srikanth, S., Sarma, P.N., 2007. Bioelectricity production by mediatorless microbial fuel cell under acidophilic condition using wastewater as substrate: influence of substrate loading rate. Curr. Sci. 92 (12). Available from: https://www.jstor.org/stable/24107621.
Molina, L., Segura, A., 2021. Biochemical and metabolic plant responses toward polycyclic aromatic hydrocarbons and heavy metals present in atmospheric pollution. Plants. 22237747. 10 (11). Available from: https://www.mdpi.com/2223-7747/10/11/2305/pdf. doi:10.3390/plants10112305.
Morris, J.M., Jin, S., 2012. Enhanced biodegradation of hydrocarbon-contaminated sediments using microbial fuel cells. J. Hazard. Mater. 03043894. 213-214, 474–477. doi:10.1016/j.jhazmat.2012.02.029.
Morris, J.M., Jin, S., Crimi, B., Pruden, A., 2009. Microbial fuel cell in enhancing anaerobic biodegradation of diesel. Chem. Eng. J. 13858947. 146 (2), 161–167. doi:10.1016/j.cej.2008.05.028.
Muangchinda, C., Rungsihiranrut, A., Prombutara, P., Soonglerdsongpha, S., Pinyakong, O., 2018. 16S metagenomic analysis reveals adaptability of a mixed-PAH-degrading consortium isolated from crude oil-contaminated seawater to changing environmental conditions. J. Hazard. Mater. 03043894. 357, 119–127. doi:10.1016/j.jhazmat.2018.05.062.
Mukherjee, A., Reddy, M.S., 2020. Metatranscriptomics: an approach for retrieving novel eukaryotic genes from polluted and related environments. 3 Biotech. 21905738. 10 (2). Available from: http://www.springerlink.com/content/2190-572x/. doi:10.1007/s13205-020-2057-1.
Murthy, M.K., 2025. Environmental dynamics of pesticides: sources, impacts on amphibians, nanoparticles, and endophytic microorganism remediation. Environ. Sci. Pollut. Res. 16147499. 32 (13), 7860–7893. Available from: https://www.springer.com/journal/11356. doi:10.1007/s11356-025-36216-7.
Mustafa, G., Zahid, M.T., Ali, S., Abbas, S.Z., Rafatullah, M., 2021. Biodegradation and discoloration of disperse blue-284 textile dye by Klebsiella pneumoniae GM-04 bacterial isolate. J. King Saud Univ. Sci. 10183647. 33 (4), 101442. doi:10.1016/j.jksus.2021.101442.
Nandy, S., Das, T., Tudu, C.K., Pandey, D.K., Dey, A., Ray, P., 2020. Fungal endophytes: futuristic tool in recent research area of phytoremediation. S. Afr. J. Bot. 02546299. 134, 285–295. Available from: http://www.elsevier.com. doi:10.1016/j.sajb.2020.02.015.
Nayak, A.K., Panda, S.S., Basu, A., Dhal, N.K., 2018. Enhancement of toxic Cr (VI), Fe, and other heavy metals phytoremediation by the synergistic combination of native Bacillus

cereus strain and Vetiveria zizanioides L. Int. J. Phytoremediation. 15497879. 20 (7), 682–691. Available from: http://www.tandf.co.uk/journals/titles/15226514.asp. doi:10.1080/15226514.2017.1413332.

Nayak, Adhya, T.K., Panda, A.N., Das, Raina, V., 2020. Bioremediation of chlorpyrifos and 3,5,6-trichloro-2-pyridinol (TCP) by a paddy field bacterial isolate: insights from genome analysis for possible biodegradation pathway. bioRxiv. Available from: https://scholar.archive.org/work/iht6mtpjrnghbk4m43cx3vjnsy/access/wayback/https://www.biorxiv.org/content/biorxiv/early/2020/09/02/2019.12.12.866210.full.pdf.

Ndeddy Aka, R.J., Babalola, O.O., 2016. Effect of bacterial inoculation of strains of pseudomonas aeruginosa, alcaligenes feacalis and bacillus subtilis on germination, growth and heavy metal (cd, cr, and ni) uptake of brassica juncea. Int. J. Phytoremediation. 15497879. 18 (2), 200–209. Available from: http://www.tandf.co.uk/journals/titles/15226514.asp. doi:10.1080/15226514.2015.1073671.

Nivetha, N., Srivarshine, B., Sowmya, B., Rajendiran, M., Panchamoorthy Saravanan, R.R., Rajasimman, M., Pham, T.H.T., Shanmugam, V., Dragoi, E.-N., 2023. A comprehensive review on bio-stimulation and bio-enhancement towards remediation of heavy metals degeneration. Chemosphere. 00456535. 312 (Pt 1), 137099. doi:10.1016/j.chemosphere.2022.137099.

Noble, M.E.M., Cleasby, A., Johnson, L.N., Egmond, M.R., Frenken, L.G.J., 1993. The crystal structure of triacylglycerol lipase from Pseudomonas glumae reveals a partially redundant catalytic aspartate. FEBS Lett. 00145793. 331 (1-2), 123–128. doi:10.1016/0014-5793(93)80310-Q.

Noormohamadi, H.R., Fat'hi, M.R., Ghaedi, M., Ghezelbash, G.R., 2019. Potentiality of white-rot fungi in biosorption of nickel and cadmium: modeling optimization and kinetics study. Chemosphere. 00456535. 216, 124–130. doi:10.1016/j.chemosphere.2018.10.113.

Ogunlaja, A., Ogunlaja, O.O., Okewole, D.M., Morenikeji, O.A., 2019. Risk assessment and source identification of heavy metal contamination by multivariate and hazard index analyses of a pipeline vandalised area in Lagos State, Nigeria. Sci. Total Environ. 00489697. 651 (Pt 2), 2943–2952. doi:10.1016/j.scitotenv.2018.09.386.

Ojuederie, O.B., Chukwuneme, C.F., Samuel, O., Ayilara, O.M., Adegboyega, T.T., Babalola, O.O., 2021. Contribution of Microbial Inoculants in Sustainable Maintenance of Human Health, Including Test Methods and Evaluation of Safety of Microbial Pesticide Microorganisms Biopesticides: Botanicals and Microorganisms for Improving Agriculture and Human Health. Logos Verlag Berlin GmbH, South Africa, pp. 215–248. Available from: https://www.logos-verlag.de/cgi-bin/engbuchmid?isbn=5264andlng=deuandid=.

Omotola Fashola, M., Mpode Ngole-Jeme, V., Oluranti Babalola, O., 2020. Physicochemical properties, heavy metals, and metal-tolerant bacteria profiles of abandoned gold mine tailings in Krugersdorp, South Africa. Can. J. Soil Sci. 19181841. 100 (3), 217–233. Available from: http://pubs.aic.ca/toc/cjss/current. doi:10.1139/cjss-2018-0161.

Orlović-Leko, P., Farkaš, B., Galić, I., 2022. A short review of environmental and health impacts of gold mining. Reliab. Theory Appl. 19322321. 17, 242–248. Available from: https://cyberleninka.ru/article/n/a-short-review-of-environmental-and-health-impacts-of-gold-mining. doi:10.24412/1932-2321-2022-470-242-248.

Osmani, I., Haziri, A., Murtaj, B., Nuro, A., 2026. Organic Pollutants in the Water Ecosystem of Batllava Lake, Kosovo: OCPs, PCBs, PAHs, and BTEX Compounds. Water Air Soil. Pollut. 237 (11), doi:10.1007/s11270-026-09335-1.

Oyetibo, G.O., Chien, M.F., Ikeda-Ohtsubo, W., Suzuki, H., Obayori, O.S., Adebusoye, S.A., Ilori, M.O., Amund, O.O., Endo, G., 2017. Biodegradation of crude oil and phenanthrene by heavy metal resistant Bacillus subtilis isolated from a multi-polluted industrial wastewater creek. Int. Biodeterior. Biodegrad. 09648305. 120, 143–151. Available from: www.elsevier.com/inca/publications/store/4/0/5/8/9/9. doi:10.1016/j.ibiod.2017.02.021.

Palanisamy, G., Thangarasu, S., Dharman, R.K., Patil, C.S., Negi, T.P.P.S., Kurkuri, M.D., Pai, R.K., Oh, T.H., 2023. The growth of biopolymers and natural earthen sources as membrane/separator materials for microbial fuel cells: a comprehensive review. J. Energy Chem. 20954956. 80, 402–431. Available from: https://elsevier.com/journals/journal-of-energy-chemistry/2095-4956. doi:10.1016/j.jechem.2023.01.018.

Pandey, A.K., Gaur, V.K., Udayan, A., Varjani, S., Kim, S.H., Wong, J.W.C., 2021. Biocatalytic remediation of industrial pollutants for environmental sustainability: research needs and opportunities. Chemosphere. 18791298. 272. Available from: www.elsevier.com/locate/chemosphere. doi:10.1016/j.chemosphere.2021.129936.

Parnell, J.J., Denef, V.J., Park, J., Tsoi, T., Tiedje, J.M., 2010. Environmentally relevant parameters affecting PCB degradation: carbon source- and growth phase-mitigated effects of the expression of the biphenyl pathway and associated genes in Burkholderia xenovorans LB400. Biodegradation. 09239820. 21 (1), 147–156. Available from: www.wkap.nl/journalhome.htm/0923-9820. doi:10.1007/s10532-009-9289-4.

Parween, T., Bhandari, P., Jan, S., Raza, S.K., 2016. Interaction between Pesticide and Soil Microorganisms and Their Degradation: A Molecular Approach Plant, Soil and Microbes: Volume 2: Mechanisms and Molecular Interactions. Springer International Publishing, India, pp. 23–43. doi:10.1007/978-3-319-29573-2_2.

Patel, A.B., Singh, S., Patel, A., Jain, K., Amin, S., Madamwar, D., 2019. Synergistic biodegradation of phenanthrene and fluoranthene by mixed bacterial cultures. Bioresour. Technol. 09608524. 284, 115–120. doi:10.1016/j.biortech.2019.03.097.

Peng, T., Luo, A., Kan, J., Liang, L., Huang, T., Zhong, H., 2019. Identification of A ring-hydroxylating dioxygenases capable of anthracene and benz[a]anthracene oxidization from Rhodococcus sp. P14. Microb. Physiol. 2673-1665. 28 (4), 183–189. doi:10.1159/000494384.

Peng, X., Xu, P.F., Du, H., Tang, Y., Meng, Y., Yuan, L., Sheng, L.P., 2018. Degradation of polycyclic aromatic hydrocarbons: a review. Appl. Ecol. Environ. Res. 17850037. 16 (5), 6419–6440. Available from: http://aloki.hu/pdf/1605_64196440.pdf. doi:10.15666/aeer/1605_64196440.

Pi, Y., Chen, B., Bao, M., Fan, F., Cai, Q., Ze, L., Zhang, B., 2017. Microbial degradation of four crude oil by biosurfactant producing strain Rhodococcus sp. Bioresour. Technol. 09608524. 232, 263–269. doi:10.1016/j.biortech.2017.02.007.

Pilli, S., Sellamuthu, B., Pandey, A.K., Tyagi, R.D., 2020. Treatment of wastewater containing pharmaceuticals: biological treatment. In: Current Developments in Biotechnology and Bioengineering: Environmental and Health Impact of Hospital Wastewater. Elsevier, India, pp. 463–520. Available from: https://www.sciencedirect.com/book/9780128197226. doi:10.1016/B978-0-12-819722-6.00013-4.

Pino, N.J., Muñera, L.M., Peñuela, G.A., 2016. Bioaugmentation with immobilized microorganisms to enhance phytoremediation of PCB-contaminated soil. Soil Sediment Contam. 15497887. 25 (4), 419–430. Available from: http://www.tandfonline.com/toc/bssc20/current. doi:10.1080/15320383.2016.1148010.

Prabagar, S., Dharmadasa, R.M., Lintha, A., Thuraisingam, S., Prabagar, J., 2021. Accumulation of heavy metals in grape fruit, leaves, soil and water: a study of influential factors and evaluating ecological risks in Jaffna, Sri Lanka. Environ. Sustain. Indic. 26659727. 12, 100147. doi:10.1016/j.indic.2021.100147.

Pradhan, D., Sukla, L.B., Mishra, B.B., Devi, N., 2019. Biosorption for removal of hexavalent chromium using microalgae Scenedesmus sp. J. Clean. Prod. 09596526. 209, 617–629. Available from: https://www.journals.elsevier.com/journal-of-cleaner-production. doi:10.1016/j.jclepro.2018.10.288.

Pratush, A., Kumar, A., Hu, Z., 2018. Adverse effect of heavy metals (As, Pb, Hg, and Cr) on health and their bioremediation strategies: a review. Int. Microbiol. 16181905. 21 (3), 97–106. Available from: https://rd.springer.com/journal/10123. doi:10.1007/s10123-018-0012-3.

Prenafeta-Boldú, F.X., Roca, N., Villatoro, C., Vera, L., Sybren de Hoog, G., 2019. Prospective application of melanized fungi for the biofiltration of indoor air in closed bioregenerative systems. J. Hazard. Mater. 03043894. 361, 1–9. doi:10.1016/j.jhazmat.2018.08.059.

Pujar, N.K., Premakshi, H.G., Ganeshkar, M.P., Kamanavalli, C.M., 2022. Biodegradation of pesticides used in agriculture by soil microorganisms. Microorganisms Sustain. 30, 213–235. Available from: https://www.springer.com/series/14379. doi:10.1007/978-981-16-4574-7_11.

Rabaey, K., De Sompel, K.V., Maignien, L., Boon, N., Aelterman, P., Clauwaert, P., De Schamphelaire, L., Pham, H.T., Vermeulen, J., Verhaege, M., Lens, P., Verstraete, W., 2006. Microbial fuel cells for sulfide removal. Environ. Sci. Technol. 0013936X. 40 (17), 5218–5224. doi:10.1021/es060382u.

Raj, R., Das, S., Behera, M., Ghangrekar, M.M., 2023a. Bio-Electrochemical Systems for Micropollutant Removal Microconstituents in the Environment: Occurrence, Fate, Removal, and Management. Wiley, India, pp. 455–490. Available from: https://onlinelibrary.wiley.com/doi/book/10.1002/9781119825289. doi:10.1002/9781119825289.ch18.

Raj, R., Dhanda, A., Das, S., Ghangrekar, M.M., 2024. Green-activated charcoal-anchored iron oxide-driven microbial electro-fenton system for sustainable mitigation of refractory contaminants. ACS ES&T Water. 26900637. 4 (9), 4129–4139. Available from: https://pubs.acs.org/page/aewcaa/about.html. doi:10.1021/acsestwater.4c00402.

Raj, R., Sathe, S.M., Das, S., Ghangrekar, M.M., 2023b. Nickel–iron-driven heterogenous bio-electro-fenton process for the degradation of methylparaben. Chemosphere. 00456535. 341, 139989. doi:10.1016/j.chemosphere.2023.139989.

Raklami, A., Meddich, A., Oufdou, K., Baslam, M., 2022. Plants—microorganisms-based bioremediation for heavy metal cleanup: recent developments, phytoremediation techniques, regulation mechanisms, and molecular responses. Int. J. Mol. Sci. 14220067. 23 (9). Available from: https://www.mdpi.com/1422-0067/23/9/5031/pdf. doi:10.3390/ijms23095031.

Ravi, R., Philip, L., Swaminathan, T., 2015. Modified rotating biological contactor for removal of dichloromethane vapours. Environ. Technol. 1479487X. 36 (5), 566–572. Available from: http://www.tandf.co.uk/journals/titles/09593330.asp. doi:10.1080/09593330.2014.953553.

Ravindran, A., Sajayan, A., Priyadharshini, G.B., Selvin, J., Kiran, G.S., 2020. Revealing the efficacy of thermostable biosurfactant in heavy metal bioremediation and surface treatment in vegetables. Front. Microbiol. 1664302X. 11. Available from: https://www.frontiersin.org/journals/microbiology#. doi:10.3389/fmicb.2020.00222.

Ren, Y., Manefield, M., 2025. Evolution of pollutant biodegradation. Appl. Microbiol. Biotechnol. 14320614. 109 (1). Available from: https://www.springer.com/journal/253. doi:10.1007/s00253-025-13418-0.

Robinson, B.H., 2009. E-waste: an assessment of global production and environmental impacts. Sci. Total Environ. 00489697. 408 (2), 183–191. doi:10.1016/j.scitotenv.2009.09.044.

Rojo, F., 2009. Degradation of alkanes by bacteria: minireview. Environ. Microbiol. 14622920. 11 (10), 2477–2490. Available from: http://onlinelibrary.wiley.com/journal/10.1111/(ISSN)1462-2920. doi:10.1111/j.1462-2920.2009.01948.x.

Rybarczyk, P., Marycz, M., Szulczyński, B., Brillowska-Dąbrowska, A., Rybarczyk, A., Gębicki, J., 2021. Removal of cyclohexane and ethanol from air in biotrickling filters inoculated with Candida albicans and Candida subhashii. Arch. Environ. Prot. 20834810. 47 (1), 26–34. Available from: https://journals.pan.pl/dlibra/journal/96938. doi:10.24425/aep.2021.136445.

Sadatshojaie, A., Rahimpour, M.R., 2020. CO2 Emission and Air Pollution (Volatile Organic Compounds, Etc.)–related Problems Causing Climate Change. Elsevier BV, pp. 1–30. doi:10.1016/b978-0-12-816778-6.00001-1.

Sahay, R., 2021. Synthetic applications of Laccase and its catalytic potentials. Int. J. Adv. Eng. Res. Sci. 23496495. 8 (6), 112–120. doi:10.22161/ijaers.86.12.

Sangkharak, K., Choonut, A., Rakkan, T., Prasertsan, P., 2020. The degradation of phenanthrene, pyrene, and fluoranthene and its conversion into medium-chain-length Polyhydroxyalkanoate by novel polycyclic aromatic hydrocarbon-degrading bacteria. Curr. Microbiol. 14320991. 77 (6), 897–909. Available from: https://link.springer.de/link/service/journals/00284/index.htm. doi:10.1007/s00284-020-01883-x.

Saravanan, S., Femina Carolin, C., Senthil Kumar, P., Chitra, B., Rangasamy, G., 2022. Biodegradation of textile dye Rhodamine-B by Brevundimonas diminuta and screening of their breakdown metabolites. Chemosphere. 00456535. 308 (Pt 1), 136266. doi:10.1016/j.chemosphere.2022.136266.

Sarmah, P., Rout, J., 2020. Role of algae and cyanobacteria in bioremediation: prospects in polyethylene biodegradation. In: Advances in Cyanobacterial Biology. Elsevier, India, pp. 333–349. Available from: https://www.sciencedirect.com/book/9780128193112. doi:10.1016/B978-0-12-819311-2.00022-X.

Saxena, G., Bharagava, R.N., 2017. Organic and inorganic pollutants in industrial wastes: ecotoxicological effects, health hazards, and bioremediation approaches. In: Environmental Pollutants and Their Bioremediation Approaches. CRC Press, India, pp. 23–56. Available from: http://www.tandfebooks.com/doi/book/10.1201/b22171. doi:10.1201/b22171.

Saxena, G., Kishor, R., Bharagava, R.N., 2019. Application of Microbial Enzymes in Degradation and Detoxification of Organic and Inorganic Pollutants. Springer Science and Business Media LLC, pp. 41–51. doi:10.1007/978-981-13-1891-7_3.

Saxena, G., Purchase, D., Bharagava, R.N., 2020. Environmental hazards and toxicity profile of organic and inorganic pollutants of tannery wastewater and bioremediation approaches. Bioremediation Ind. Waste Environ. Saf. i. Available from: https://link.springer.com/chapter/10.1007/978-981-13-1891-7_17. doi:10.1007/978-981-13-1891-7_17.

Scudder, K., Blehm, K.D., 1991. Household hazardous waste: assessing public attitudes and awareness. J. Environ. Health. 00220892. 53 (6), 18–20.

Sedlakova-Kadukova, J., Kopcakova, A., Gresakova, L., Godany, A., Pristas, P., 2019. Bioaccumulation and biosorption of zinc by a novel Streptomyces K11 strain isolated from highly alkaline aluminium brown mud disposal site. Ecotoxicol. Environ. Saf. 01476513. 167, 204–211. doi:10.1016/j.ecoenv.2018.09.123.

Selvakumar, R., Ramadoss, G., Mridula, P.M., Rajendran, K., Thavamani, P., Ravi Naidu, M.M., 2018. Challenges and complexities in remediation of uranium contaminated soils: a review. J. Environ. Radioact. 18791700. 192, 592–603. Available from: www.elsevier.com/locate/jenvrad. doi:10.1016/j.jenvrad.2018.02.018.

Shahid, M., Khan, M.S., 2017. Assessment of glyphosate and quizalofop mediated toxicity to greengram [Vigna radiata (L.) Wilczek], stress abatement and growth promotion by herbicide tolerant Bradyrhizobium and Pseudomonas species. Int. J. Curr. Microbiol. Appl. Sci. 23197692. 6 (12), 3001–3016. doi:10.20546/ijcmas.2017.612.351.

Shahid, M., Zaidi, A., Ehtram, A., Khan, M.S., 2019. In vitro investigation to explore the toxicity of different groups of pesticides for an agronomically important rhizosphere isolate Azotobacter vinelandii. Pestic. Biochem. Physiol. 00483575. 157, 33–44. doi:10.1016/j.pestbp.2019.03.006.

Shanmuganathan, R., Karuppusamy, I., Saravanan, M., Muthukumar, H., Ponnuchamy, K., Ramkumar, V.S., Pugazhendhi, A., 2019. Synthesis of silver nanoparticles and their biomedical applications—a comprehensive review. Curr. Pharm. Des. 13816128. 25 (24), 2650–2660. doi:10.2174/1381612825666190708185506.

Sharma, A., Kumar, V., Handa, N., Bali, S., Kaur, R., Khanna, K., Thukral, A.K., Bhardwaj, R., 2018. Potential of endophytic bacteria in heavy metal and pesticide detoxification. Microorganisms Sustain. 5, 307–336. Available from: https://www.springer.com/series/14379. doi:10.1007/978-981-10-5514-0_14.

Sharma, B., Shukla, P., 2021. Lead bioaccumulation mediated by Bacillus cereus BPS-9 from an industrial waste contaminated site encoding heavy metal resistant genes and their transporters. J. Hazard. Mater. 03043894. 401, 123285. doi:10.1016/j.jhazmat.2020.123285.

Sharma, P., Bano, A., Singh, S.P., Dubey, N.K., Chandra, R., Iqbal, H.M.N., 2022b. Recent advancements in microbial-assisted remediation strategies for toxic contaminants. Clean. Chem. Eng. 27727823. 2, 100020. doi:10.1016/j.clce.2022.100020.

Sharma, P., Dutta, D., Udayan, A., Nadda, A.K., Lam, S.S., Kumar, S., 2022c. Role of microbes in bioaccumulation of heavy metals in municipal solid waste: impacts on plant and human being. Environ. Pollut. 02697491. 305, 119248. doi:10.1016/j.envpol.2022.119248.

Sharma, P., Singh, S.P., Iqbal, H.M.N., Tong, Y.W., 2022a. Omics approaches in bioremediation of environmental contaminants: an integrated approach for environmental safety and sustainability. Environ. Res. 10960953. 211. Available from: http://www.elsevier.com/inca/publications/store/6/2/2/8/2/1/index.htt. doi:10.1016/j.envres.2022.113102.

Sherafatmand, M., Ng, H.Y., 2015. Using sediment microbial fuel cells (SMFCs) for bioremediation of polycyclic aromatic hydrocarbons (PAHs). Bioresour. Technol. 09608524. 195, 122–130. Available from: http://www.journals.elsevier.com/bioresource-technology/. doi:10.1016/j.biortech.2015.06.002.

Sinkkonen, S., Paasivirta, J., 2000. Degradation half-life times of PCDDs, PCDFs and PCBs for environmental fate modeling. Chemosphere. 00456535. 40 (9-11), 943–949. doi:10.1016/S0045-6535(99)00337-9.

Sirajuddin, S., Rosenzweig, A.C., 2015. Enzymatic oxidation of methane. Biochemistry. 15204995. 54 (14), 2283–2294. Available from: http://pubs.acs.org/journal/bichaw. doi:10.1021/acs.biochem.5b00198.

Song, N., Zhang, X., Wang, F., Zhang, C., Tang, S., 2012. Elevated CO2 increases Cs uptake and alters microbial communities and biomass in the rhizosphere of Phytolacca americana Linn (pokeweed) and Amaranthus cruentus L. (purple amaranth) grown on soils spiked with various levels of Cs. J. Environ. Radioact. 0265931X. 112, 29–37. doi:10.1016/j.jenvrad.2012.03.002.

Song, X., Jo, C., Han, L., Zhou, M., 2022. Recent advance in microbial fuel cell reactor configuration and coupling technologies for removal of antibiotic pollutants. Curr. Opin. Electrochem. 24519103. 31, 100833. doi:10.1016/j.coelec.2021.100833.

Speight, J.G., 2011. Environmental Effects of Hydrocarbons. Elsevier BV 539–576. doi:10.1016/b978-0-7506-8632-7.10015-5.

Sun, L., Guo, D., Liu, K., Meng, H., Zheng, Y., Yuan, F., Zhu, G., 2019. Levels, sources, and spatial distribution of heavy metals in soils from a typical coal industrial city of Tangshan, China. CATENA. 03418162. 175, 101–109. doi:10.1016/j.catena.2018.12.014.

Sun, W., Cheng, K., Sun, K.Y., Ma, X., 2021. Microbially mediated remediation of contaminated sediments by heavy metals: a critical review. Curr. Pollut. Rep. 21986592. 7 (2), 201–212. Available from: https://springer.com/environment/pollution+and+remediation/journal/40726. doi:10.1007/s40726-021-00175-7.

Sundaram, S.K., Taj, N.A., Menaka, R., Ajitha, C., 2022. Biodegradation of pesticides through metagenomic approaches. J. Adv. Sci. Res. 0976-9595. 13 (2), 29–37. doi:10.55218/jasr.202213205.

Tak, H.I., Ahmad, F., Babalola, O.O., 2013. Advances in the application of plant growth-promoting rhizobacteria in phytoremediation of heavy metals. Rev. Environ. Contam. Toxicol. 01795953. 223, 33–52. Available from: http://www.springer.com/series/398. doi:10.1007/978-1-4614-5577-6_2.

Tan, W., Peralta-Videa, J.R., Gardea-Torresdey, J.L., 2018. Interaction of titanium dioxide nanoparticles with soil components and plants: current knowledge and future research needs-a critical review. Environ. Sci. Nano. 20518161. 5 (2), 257–278. Available from: http://pubs.rsc.org/en/journals/journal/en. doi:10.1039/c7en00985b.

Tang, J., Liu, B., Chen, T.T., Yao, K., Zeng, L., Zeng, C.Y., Zhang, Q., 2018. Screening of a beta-cypermethrin-degrading bacterial strain Brevibacillus parabrevis BCP-09 and its biochemical degradation pathway. Biodegradation. 15729729. 29 (6), 525–541. Available from: www.wkap.nl/journalhome.htm/0923-9820. doi:10.1007/s10532-018-9850-0.

Tao, K., Liu, X., Chen, X., Hu, X., Cao, L., Yuan, X., 2017. Biodegradation of crude oil by a defined co-culture of indigenous bacterial consortium and exogenous Bacillus subtilis. Bioresour. Technol. 18732976. 224, 327–332. Available from: www.elsevier.com/locate/biortech. doi:10.1016/j.biortech.2016.10.073.

Taufan, H.S., Purwanto, P., 2018. The management of toxic and hazardous waste materials in the food industry. E3S Web Conf. 22671242, 73. www.e3s-conferences.org/. doi:10.1051/e3sconf/20187307020.

Thakur, A., Kumar, A., 2024. Emerging paradigms into bioremediation approaches for nuclear contaminant removal: from challenge to solution. Chemosphere. 00456535. 352, 141369. doi:10.1016/j.chemosphere.2024.141369.

Thirumalaivasan, N., Gnanasekaran, L., Kumar, S., Durvasulu, R., Sundaram, T., Rajendran, S., Nangan, S., Kanagaraj, K., 2024. Utilization of fungal and bacterial bioremediation techniques for the treatment of toxic waste and biowaste. Front. Mater. 2296-8016. 11. doi:10.3389/fmats.2024.1416445.

Tian, Q., Dou, X., Huang, L., Wang, L., Meng, D., Zhai, L., Shen, Y., You, C., Guan, Z., Liao, X., 2020. Characterization of a robust cold-adapted and thermostable laccase from Pycnoporus sp. SYBC-L10 with a strong ability for the degradation of tetracycline and oxytetracycline by laccase-mediated oxidation. J. Hazard. Mater. 03043894. 382, 121084. doi:10.1016/j.jhazmat.2019.121084.

Tiehm, A., Fritzsche, C., 1995. Utilization of solubilized and crystalline mixtures of polycyclic aromatic hydrocarbons by a Mycobacterium sp.Appl. Microbiol. Biotechnol. 14320614. 42 (6), 964–968. doi:10.1007/BF00191198.

Tribedi, P., Goswami, M., Chakraborty, P., Mukherjee, K., Mitra, G., Bhattacharyya, P., Dey, S., 2018. Bioaugmentation and biostimulation: a potential strategy for environmental remediation. J. Microbiol. Exp. 2373437X. 6 (5). doi:10.15406/jmen.2018.06.00219.

Umar, M.F., Rafatullah, M., Abbas, S.Z., Ibrahim, M.N.M., Ismail, N., 2021. Bioelectricity production and xylene biodegradation through double chamber benthic microbial fuel cells fed with sugarcane waste as a substrate. J. Hazard. Mater. 18733336. 419. Available from: www.elsevier.com/locate/jhazmat. doi:10.1016/j.jhazmat.2021.126469.

United States Environmental Protection Agency (USEPA), 2021. Criteria for the Definition of Solid Waste and Solid and Hazardous Waste Exclusions. https://www.epa.gov/hw/criteria-definition-solid-waste-and-solid-and-hazardous-waste-exclusions#tablesw.

Unuofin, J.O., Okoh, A.I., Nwodo, U.U., 2019. Aptitude of oxidative enzymes for treatment of wastewater pollutants: a laccase perspective. Molecules. 14203049. 24 (11). Available from: https://www.mdpi.com/1420-3049/24/11/2064/pdf. doi:10.3390/molecules24112064.

USEPA, 2024. Hazardous Air Pollutants: Sources and Exposure. US EPA, pp. 6–20. https://www.epa.gov/haps/hazardous-air-pollutants-sources-and-exposure.

Vaidya, S., Devpura, N., Jain, K., Madamwar, D., 2018. Degradation of chrysene by enriched bacterial consortium. Front. Microbiol. 1664302X. 9. Available from: https://www.frontiersin.org/articles/10.3389/fmicb.2018.01333/full. doi:10.3389/fmicb.2018.01333.

Vaidya, S., Jain, K., Madamwar, D., 2017. Metabolism of pyrene through phthalic acid pathway by enriched bacterial consortium composed of Pseudomonas, Burkholderia, and Rhodococcus (PBR). 3 Biotech. 21905738. 7 (1). Available from: http://www.springerlink.com/content/2190-572x/. doi:10.1007/s13205-017-0598-8.

Vandermaesen, J., Horemans, B., Bers, K., Vandermeeren, P., Herrmann, S., Sekhar, A., Seuntjens, P., Springael, D., 2016. Application of biodegradation in mitigating and remediating pesticide contamination of freshwater resources: state of the art and challenges for optimization. Appl. Microbiol. Biotechnol. 14320614. 100 (17), 7361–7376. Available from: https://link.springer.de/link/service/journals/00253/index.htm. doi:10.1007/s00253-016-7709-z.

Varjani, S.J., 2017. Microbial degradation of petroleum hydrocarbons. Bioresour. Technol. 18732976. 223, 277–286. Available from: www.elsevier.com/locate/biortech. doi:10.1016/j.biortech.2016.10.037.

Varjani, S.J., Gnansounou, E., Pandey, A., 2017. Comprehensive review on toxicity of persistent organic pollutants from petroleum refinery waste and their degradation by microorganisms. Chemosphere. 00456535. 188, 280–291. doi:10.1016/j.chemosphere.2017.09.005.

Varjani, S.J., Rana, D.P., Jain, A.K., Bateja, S., Upasani, V.N., 2015. Synergistic ex-situ biodegradation of crude oil by halotolerant bacterial consortium of indigenous strains isolated from on shore sites of Gujarat, India. Int. Biodeterior. Biodegrad. 09648305. 103, 116–124. Available from: www.elsevier.com/inca/publications/store/4/0/5/8/9/9. doi:10.1016/j.ibiod.2015.03.030.

Varjani, S.J., Upasani, V.N., 2016. Biodegradation of petroleum hydrocarbons by oleophilic strain of Pseudomonas aeruginosa NCIM 5514. Bioresour. Technol. 18732976. 222, 195–201. Available from: www.elsevier.com/locate/biortech. doi:10.1016/j.biortech.2016.10.006.

Varjani, S., Kumar, G., Rene, E.R., 2019. Developments in biochar application for pesticide remediation: current knowledge and future research directions. J. Environ. Manage. 03014797. 232, 505–513. doi:10.1016/j.jenvman.2018.11.043.

Vasiliadou, P.L., Luu, L., et al, 2021. A sustainable approach for the production of green energy with the holistic treatment of wastewater through microbial electrochemical technologies: a review. Front. Sustain. 2, 792028. Available from: https://www.frontiersin.org/articles/10.3389/frsus.2021.792028/full.

Verma, S., Kuila, A., 2019. Bioremediation of heavy metals by microbial process. Environ. Technol. Innov. 14, 100369. Available from: https://www.sciencedirect.com/science/article/pii/S2352186418305911.

Vidhyeswari, D., Surendhar, A., Bhuvaneshwari, S., 2022. General aspects and novel PEMss in microbial fuel cell technology: a review. Chemosphere. 00456535. 309 (Pt 1), 136454. doi:10.1016/j.chemosphere.2022.136454.

Vignesh, A., Manigundan, K., Santhoshkumar, J., Shanmugasundaram, T., Gopikrishnan, V., Radhakrishnan, M., Joseph, J., Ayyasamy, P.M., Kumar, G.D., Meganathan, R., Balagurunathan, R., 2020. Microbial degradation, spectral analysis and toxicological assessment of malachite green by Streptomyces chrestomyceticus S20. Bioprocess Biosyst. Eng. 16157605. 43 (8), 1457–1468. Available from: https://rd.springer.com/journal/449. doi:10.1007/s00449-020-02339-z.

Voytek, M.A., 1990. Addressing the biological effects of decreased ozone on the antarctic environment. AMBIO 19 (2), https://www.jstor.org/stable/4313660. (Accessed 14 06 2026).

Wang, B., Teng, Y., Yao, H., Christie, P., 2021. Detection of functional microorganisms in benzene [a] pyrene-contaminated soils using DNA-SIP technology. J. Hazard. Mater. 18733336. 407. Available from: www.elsevier.com/locate/jhazmat. doi:10.1016/j.jhazmat.2020.124788.

Wang, H., Liu, J., Gui, C., Yan, Q., Wang, L., Wang, S., Juanjuan, L., 2022. Synergistic remediation of Cr(VI) contaminated soil by iron-loaded activated carbon in two-chamber microbial fuel cells. Environ. Res. 00139351. 208, 112707. doi:10.1016/j.envres.2022.112707.

Wang, L., Liu, Y., Jun, M., Zhao, F., 2016. Rapid degradation of sulphamethoxazole and the further transformation of 3-amino-5-methylisoxazole in a microbial fuel cell. Water Res. 00431354. 88, 322–328. doi:10.1016/j.watres.2015.10.030.

Wang, L., Liu, Y., Shu, X., Lu, S., Xie, X., Shi, Q., 2019b. Complexation and conformation of lead ion with poly-γ-glutamic acid in soluble state. PLoS ONE. 19326203. 14 (9). Available from: https://journals.plos.org/plosone/article/file?id=10.1371/journal.pone.0218742&type=printable. doi:10.1371/journal.pone.0218742.

Wang, L., Wang, W., Lai, Q., Shao, Z., 2010. Gene diversity of CYP153A and AlkB alkane hydroxylases in oil-degrading bacteria isolated from the Atlantic Ocean. Environ. Microbiol. 14622920. 12 (5), 1230–1242. doi:10.1111/j.1462-2920.2010.02165.x.

Wang, L., Xiao, S., Chen, X., Chen, S., Wang, S., Wang, C., Tang, Y., Dong, F., 2019c. YtiB and ythA genes reduce the uranium removal capacity of Bacillus atrophaeus. Int. J. Mol. Sci. 14220067. 20 (7). Available from: https://www.mdpi.com/1422-0067/20/7/1766/pdf. doi:10.3390/ijms20071766.

Wang, Z.Y., Xu, Y., Wang, H.Y., Zhao, J., Gao, D.M., LI, F.M., Xing, B., 2012. Biodegradation of Crude Oil in Contaminated Soils by Free and Immobilized Microorganisms. Pedosphere. 10020160. 22 (5), 717–725. Available from: http://pedosphere.issas.ac.cn. doi:10.1016/S1002-0160(12)60057-5.

Wang, Li, Z.Y., Liu, Y., Xiong, Jiang, X., Irawan, S., et al, 2019a. The Routing of Hazardous and Toxic (B3) Medical Waste Transportation Using Network Analysis (Case Study: primary Health Care Services, Depok, Indonesia). J. Phys. Conf. Ser. 1364 (1), 012046. Available from: https://iopscience.iop.org/article/10.1088/1742-6596/1364/1/012046.

Wong, C.S.C., Wu, S.C., Duzgoren-Aydin, N.S., Aydin, A., Wong, M.H., 2007. Trace metal contamination of sediments in an e-waste processing village in China. Environ. Pollut. 02697491. 145 (2), 434–442. doi:10.1016/j.envpol.2006.05.017.

Wu, Y., Xu, Y., Zhou, N., 2020. A newly defined dioxygenase system from Mycobacterium vanbaalenii PYR-1 endowed with an enhanced activity of dihydroxylation of high-molecular-weight. Front. Environ. Sci. Eng. 14 (1) Available from: https://link.springer.com/article/10.1007/s11783-019-1193-5. doi:10.1007/s11783-019-1193-5.

Wu, Z., Man, Q., Niu, H., Lyu, H., Song, H., Li, R., Ren, G., Zhu, F., Peng, C., Benhang, L., Xiaodong, M., 2022. Recent advances and trends of trichloroethylene biodegradation: a critical review. Front. Microbiol. 1664-302X. 13. doi:10.3389/fmicb.2022.1053169.

Xia, Y., Min, H., Rao, G., Lv, Z.M., Liu, J., Ye, Y.F., Duan, X.J., 2005. Isolation and characterization of phenanthrene-degrading Sphingomonas paucimobilis strain ZX4. Biodegradation 16 (5), 393–402. doi:10.1007/s10532-004-2412-7.

Xiao, F., Cheng, J., Cao, W., Yang, C., Chen, J., Luo, Z., 2019. Removal of heavy metals from aqueous solution using chitosan-combined magnetic biochars. J. Colloid Interface Sci. 00219797. 540, 579–584. doi:10.1016/j.jcis.2019.01.068.

Xie, J., Li, H., Wang, S., Chen, H., Jiang, W., 2022, Available from: http://eeer.org/upload/eer-2021-119.pdf.

Xie, Q., Liu, N., Lin, D., Ruohua, Q., Zhou, Q., Fei, G., 2020. The complexation with proteins in extracellular polymeric substances alleviates the toxicity of Cd (II) to Chlorella vulgaris. Environ. Pollut. 02697491. 263 (Pt A), 114102. doi:10.1016/j.envpol.2020.114102.

Yamada, K., Fukushima, T., Freguia, S., 2022. Coke-oven wastewater treatment in a dual-chamber microbial fuel cell with thiocyanatedegrading biofilm enriched at the air cathode. Water Sci. Technol. 19969732. 85 (7), 2254–2264. Available from: https://iwaponline.com/wst/article-abstract/85/7/2254/87594. doi:10.2166/wst.2022.087.

Yan, X., Luo, X., 2015. Radionuclides distribution, properties, and microbial diversity of soils in uranium mill tailings from southeastern China. J. Environ. Radioact. 0265931X. 139, 85–90. doi:10.1016/j.jenvrad.2014.09.019.

Yan, Z., Song, N., Cai, H., Tay, J.-H., Jiang, H., 2012. Enhanced degradation of phenanthrene and pyrene in freshwater sediments by combined employment of sediment microbial fuel cell and amorphous ferric hydroxide. J. Hazard. Mater. 03043894. 199-200, 217–225. doi:10.1016/j.jhazmat.2011.10.087.

Yang, X., Li, E., Liu, F., Xu, M., 2020. Interactions of PAH-degradation and nitrate-/sulfate-reducing assemblages in anaerobic sediment microbial community. J. Hazard. Mater. 03043894. 388, 122068. doi:10.1016/j.jhazmat.2020.122068.

Yuan, X., Zhang, X., Chen, X., Kong, D., Liu, X., Shen, S., 2018. Synergistic degradation of crude oil by indigenous bacterial consortium and exogenous fungus Scedosporium boydii. Bioresour. Technol. 18732976. 264, 190–197. Available from: www.elsevier.com/locate/biortech. doi:10.1016/j.biortech.2018.05.072.

Yuan, Y., You, S.J., Zhang, J.N., Gong, X.B., Wang, X.H., Ren, N.Q., 2015. Pilot-scale bioelectrochemical system for efficient conversion of 4-chloronitrobenzene. Environ. Technol. 1479487X. 36 (14), 1847–1854. Available from: http://www.tandf.co.uk/journals/titles/09593330.asp. doi:10.1080/09593330.2015.1013572.

Yüce, B., Güzel, B., Canlı, O., Onat, B., 2024. Characterization, source identification and risk assessment of POPs (PAHs, PCBs, and OCPs) in indoor dusts from two metropolitan cities (Istanbul and Kocaeli) of Türkiye with intensive industrial and human activities. Stoch. Environ. Res. Risk. Assess. 38 (4), 1383–1401. doi:10.1007/s00477-023-02631-w.

Zacharia, 2019. Degradation Pathways of Persistent Organic Pollutants (POPs) in the Environment. Persistent Organic Pollutants. Available from: https://books.google.com/

books?hl=en&lr=&id=Swj8DwAAQBAJ&oi=fnd&pg=PA17&ots=aBZsW_D8dn&sig=nDDhMIeu5NMddAaIVuWMFnEDGIo.

Zhang, G., Liang, D., Zhao, Z., Qi, J., Huang, L., 2022. Enhanced performance of microbial fuel cell with electron mediators from tetracycline hydrochloride degradation. Environ. Res. 00139351. 206, 112605. doi:10.1016/j.envres.2021.112605.

Zhang, H., Hu, Z., Hou, S., Xu, T., 2020. Effects of microbial degradation on morphology, chemical compositions and microstructures of bitumen. Constr. Build. Mater. 09500618. 248. Available from: https://www.journals.elsevier.com/construction-and-building-materials. doi:10.1016/j.conbuildmat.2020.118569.

Zhang, H., Yang, L., Ma, Y.Y., Zhu, C., Lin, S., Liao, R.Z., 2018. Theoretical studies on catalysis mechanisms of serum paraoxonase 1 and phosphotriesterase diisopropyl fluorophosphatase suggest the alteration of substrate preference from paraoxonase to DFP. Molecules. 14203049. 23 (7). Available from: http://www.mdpi.com/1420-3049/23/7/1660/pdf. doi:10.3390/molecules23071660.

Zhang, L., Jiang, X., Shen, J., Xu, K., Li, J., Sun, X., Han, W., Wang, L., 2016. Enhanced bioelectrochemical reduction of p-nitrophenols in the cathode of self-driven microbial fuel cells. RSC Adv. 2046-2069. 6 (35), 29072–29079. doi:10.1039/C6RA04293G.

Zhang, S., Yang, X.-L., Li, H., Song, H.-L., Wang, R.-C., Dai, Z.-Q., 2017. Degradation of sulfamethoxazole in bioelectrochemical system with power supplied by constructed wetland-coupled microbial fuel cells. Bioresour. Technol. 09608524. 244 (Pt 1), 345–352. doi:10.1016/j.biortech.2017.07.143.

Zhou, Y., Liu, X., Wang, J., 2019. Characterization of microplastics and the association of heavy metals with microplastics in suburban soil of central China. Sci. Total Environ. 00489697. 694, 133798. doi:10.1016/j.scitotenv.2019.133798.

Zou, C., Wang, M., Xing, Y., Lan, G., Ge, T., Yan, X., Gu, T., 2014. Characterization and optimization of biosurfactants produced by Acinetobacter baylyi ZJ2 isolated from crude oil-contaminated soil sample toward microbial enhanced oil recovery applications. Biochem. Eng. J. 1873295X. (90), 49–58. Available from: www.elsevier.com/locate/bej. doi:10.1016/j.bej.2014.05.007.

Zwolak, A., Sarzyńska, M., Szpyrka, E., Stawarczyk, K., 2019. Sources of soil pollution by heavy metals and their accumulation in vegetables: a review. Water Air Soil Pollut. 15732932. 230 (7). Available from: http://www.kluweronline.com/issn/0049-6979/. doi:10.1007/s11270-019-4221-y.

CHAPTER

# 2

# Microbial consortia for bioremediation: The actinobacteria case in pesticide degradation

**Lucas Ariel Bazán[a], Enzo Emanuel Raimondo[a,b], Claudia Susana Benimeli[a,c] and María Soledad Fuentes[a]**

[a]*Actinobacteria Biotechnology Laboratory, Pilot Plant for Microbiological Industrial Processes (PROIMI-CONICET), San Miguel de Tucumán, Argentina*

[b]*Faculty of Biochemistry, Chemistry and Pharmacy, National University of Tucumán, San Miguel de Tucumán, Argentina*

[c]*Faculty of Exact and Natural Sciences, National University of Catamarca, San Fernando del Valle de Catamarca, Argentina*

## 1 Introduction to pesticides pollution

As the global human population continues to rise, enhancing agricultural productivity remains a critical priority (Fuentes et al., 2023). Food production systems are increasingly challenged by pest outbreaks and climate-related stresses, both of which can severely compromise crop quality and yield. Schmitz et al. (2022) reported that pest diseases contribute to approximately 45% of global food production losses each year. This highlights the importance of adopting agrochemical strategies that are both effective and environmentally responsible to balance food security with sustainable agricultural practices.

Pesticides are among the agrochemicals most widely used worldwide, applied to enhance agricultural productivity. Their overutilization has led to widespread contamination, not only in treated agricultural areas but also in surrounding soils, surface and groundwater systems, and the atmosphere (Medfu Tarekegn et al., 2020). The presence of pesticide residues has been reported across all these environmental compartments, posing health risks to both terrestrial and marine ecosystems (Makarani and Kaushal, 2025).

## 2 Pesticides: Definition, classification, mobility, and pollution

According to the Food and Agriculture Organization (FAO), pesticides are defined as any substance or mixture of substances of chemical or biological origin intended

Microbial Technologies for Sustainable Waste Management. https://doi.org/10.1016/B978-0-443-44093-9.00004-9

for repelling, destroying, or controlling any pest, or for regulating plant growth. The term encompasses insecticides, herbicides, fungicides, rodenticides, molluscicides, wood preservatives, and various other agents used in pest control. It also includes plant growth regulators, defoliants, and desiccants (Srivastava et al., 2019). Their widespread application began in the 1940s, and currently, more than one thousand pesticides are available on the market, including chemical, microbial, semichemical, and botanical formulations (FAO, 2021).

The most common criteria for classifying pesticides are their chemical class, functional group, mode of action, and toxicity (Raffa and Chiampo, 2021). Based on their chemical composition, several categories are typically recognized, such as organochlorines, organophosphates, carbamates, and pyrethrins/pyrethroids, among others (Raffa and Chiampo, 2021). Among the organochlorine pesticides (OPs) that can be mentioned are dichlorodiphenyltrichloroethane (DDT), dichlorodiphenyldichloroethane, dieldrin, and aldrin. The organophosphate group includes hexaethyl tetraphosphate, parathion, and chlorpyrifos (CP). The carbamate group contains compounds such as furadan, sevin, and fenobucarb. Within the pyrethrin and pyrethroid category, common examples are cyfluthrin, bifenthrin, and allethrin. Another pesticide classification that could be mentioned is based on target pests, which allows categorizing them into insecticides, herbicides, rodenticides, bactericides, fungicides, and larvicides (Raffa and Chiampo, 2021).

It is also important to mention that not all pesticides are synthetic; some are of biological origin. These biopesticides are derived from natural sources such as animals, plants, or microorganisms. Based on the nature of their active ingredients, biopesticides are typically categorized into biochemical pesticides, plant-incorporated protectants (PIPs), and microbial pesticides (Hassaan and El Nemr, 2020). The mechanisms of biochemical pesticides do not rely on toxicity, but rather on natural substances that interfere with insect pest mating, such as sex pheromones, or on the use of traps that attract and capture insects through aromatic plant extracts. Pesticides in the PIPs group are derived from plants. Through genetic engineering, genes that encode pesticidal compounds are inserted into the plant genome. Both the inserted genetic material and the pesticidal substance synthesized by the plant are referred to as PIPs. In contrast, microbial pesticides contain active ingredients that consist of microorganisms or their byproducts. Their mode of action is based on toxins produced by the microorganisms or by infection processes. Compared to biochemical pesticides, microbial pesticides offer the advantage of greater specificity in targeting pests (Yadav and Devi, 2017).

Although pesticides provide numerous agronomic and economic benefits, they are hazardous chemical compounds associated with significant risks to both human health and the environment (Kunene and Mahlambi, 2020; Silva et al., 2019). Prolonged and intensive use promotes their accumulation in environmental compartments, such as groundwater, surface water, soil, and the atmosphere, especially in the case of recalcitrant compounds that resist degradation processes. The risk level of pesticide use changes depending on factors such as the chemical properties of the active ingredients, for instance, toxicity and persistence, and the usage patterns. The latter includes application method, timing, dosage, the type of crop, and the physical and chemical characteristics of the soil (Coscollà and Yusà, 2016). The persistence of pesticides in the environment is also influenced by environmental factors, as degradation may occur through physical, chemical, or, more commonly, biological processes driven by climatic and microbial conditions.

Soil contamination resulting from pesticide use raises major concerns regarding soil health, particularly its functional capacity, biodiversity, and implications for food safety. In addition, these compounds can be transported by wind or water erosion, also generating concern due to the exposure of nontarget organisms and human populations to residual contamination (Pérez and Eugenio, 2018; ITPS FAO, 2017). Indeed, diffuse pollution caused by agrochemical inputs is now recognized as one of the most serious threats to soil systems worldwide (Mulhern, 2015).

It is important to note that while a certain level of persistence is required for pesticides to be effective in controlling pests within the target area, this same property facilitates their accumulation across diverse environmental compartments over time. Residues are not confined to the application site; instead, they may be redistributed to nearby or even remote regions via multiple physical and chemical pathways. Prolonged persistence enhances the likelihood of volatilization and atmospheric dispersion, thereby extending the environmental footprint of the applied compounds far beyond their intended use (Van Dijk and Guicherit, 1999; Galon et al., 2021).

Of the applied pesticides, $<1\%$ effectively reach their targets, while the remaining fraction is dispersed into surrounding ecosystems, creating a risk to nontarget organisms and human health (Elhamalawy et al., 2024).

Pesticide pollution occurs through two primary pathways: point and nonpoint (diffuse) sources. Point source pollution has an identifiable origin, for instance, the discharge of industrial effluent into water bodies. In contrast, nonpoint source pollution originates from multiple and diffuse activities, such as large-scale agricultural practices (Crini and Lichtfouse, 2019).

One of the principal processes driving pesticide drift is volatilization (Majewski and Capel, 1996; Gil and Sinfort, 2005). Additional pathways include soil erosion caused by windborne or water-driven forces and agricultural runoff. Moreover, another contributor to diffuse agricultural pollution is livestock farming (Campbell et al., 2005; Galon et al., 2021).

Meteorological factors significantly influence the fate of pesticides that fail to reach their target. The transport and behavior of these compounds in the atmosphere can be affected by elements such as rain, snow, wind, humidity, temperature, sunlight, and thermal inversions (Herzfeld and Sargent, 2011). Understanding these environmental influences is essential for evaluating pesticide dispersion patterns and potential impacts.

Based on the above, different studies aimed at comprehending pesticide dispersion mechanisms in soil, water, and air were carried out. These studies focused on evaluating the risks that pesticides pose to the environment and human health, as well as on the development of innovative treatments to mitigate their negative effects on biodiversity and ecosystem functions (Crini and Lichtfouse, 2019; Silva et al., 2019; Galon et al., 2021; Beaumelle et al., 2023; Barathi et al., 2024; Shekhar et al., 2024).

## 3 Eco-friendly solutions to the pesticide pollution issue

In response to the environmental problems associated with the pesticide pollution, several countries, particularly those with advanced regulatory frameworks such as members of the European Union, have implemented measures aimed at reducing the ecological impact of their application (European Commission, 2021). However, persistent pesticide residues from earlier decades remain in the environment, and the continuous introduction of new active ingredients into the market further compounds this issue (Wang et al., 2020). Therefore pesticide accumulation directly threatens the biodiversity of the different ecosystems and undermines their essential functions (Bernhardt et al., 2017).

Once released into the environment, pesticides can follow multiple pathways, including transformation or degradation, sorption–desorption processes, volatilization, uptake by plants, surface runoff, and leaching into groundwater (Chowdhury et al., 2008). Among these, transformation, particularly degradation, is a key process that determines the environmental fate and mobility of pesticides. This transformation includes mechanisms such as oxidation, hydrolysis, photolysis, and biodegradation. Through these processes, pesticides are converted into transformation products or fully mineralized into inorganic components (Karpouzas and Walker, 2000; Singh et al., 2006). Although abiotic degradation can contribute to pesticide breakdown, the most dominant and ecologically relevant pathway is represented by microbial biodegradation (Karpouzas and Walker, 2000; Chen et al., 2012; Silva et al., 2015; Cycoń et al., 2014).

In response to the persistence and toxicity of pesticides, numerous strategies have been developed to reduce their environmental impact. These include a range of physical, chemical, biological, and combined treatment approaches (Srivastava et al., 2014; Crini and Lichtfouse, 2019). Conventional treatments, such as physicochemical methods, are widely used, yet they often lack sustainability and long-term effectiveness. In contrast, biological methods, especially those that harness microbial and plant-based processes, are increasingly favored for being more environmentally friendly, less invasive, cost-effective, and restorative of natural soil functions (Kidd et al., 2009; Alvarez et al., 2017). Each treatment strategy offers distinct advantages and limitations regarding efficiency, cost, feasibility, and ecological consequences. Due to the complex nature of pesticide-contaminated environments, no single method has proven fully adequate on its own. As a result, integrated approaches are often employed to achieve effective and economically viable remediation (Crini and Lichtfouse, 2019).

One such promising strategy is bioremediation, which utilizes living organisms, mainly plants and microorganisms, to remove, degrade, or immobilize environmental contaminants (Srivastava et al., 2014).

## 3.1 Bioremediation

Bioremediation is a biotechnological process that utilizes living organisms, such as bacteria, fungi, algae, animals, and plants, or enzymatic systems, and even combinations of them, to degrade, transform, or stabilize environmental contaminants in soil, sediments, and water. These biological systems metabolize pollutants, transforming them into less harmful substances such as carbon dioxide, water, and less toxic byproducts (Massot et al., 2022). Bioremediation strategies offer a cost-effective, efficient, and environmentally sustainable approach to cleaning up contaminated sites, minimizing the need for harsh chemicals and energy-intensive methods, reducing environmental impact, and encouraging ecosystem restoration (Aparicio et al., 2022).

Bioremediation technologies are broadly categorized based on the location of their application into two main classes: in situ and ex situ (Romantschuk et al., 2023). In situ bioremediation treats contaminated matrices at the site of localization, minimizing disturbance and often being more cost-effective. However, it may be limited by factors such as soil permeability and the ability to control environmental conditions. Ex situ bioremediation entails the removal of contaminated material to be treated elsewhere, allowing for greater control over treatment conditions. While this approach can be more efficient and thorough, it is typically more expensive due to the costs associated with the withdrawal and transportation of materials.

# 4 Microbial strategies for bioremediation

In bioremediation processes, microorganisms are key players because of their ability to metabolize a wide variety of harmful compounds. Their diverse metabolic capabilities enable them to transform or neutralize both organic and inorganic pollutants, making them essential for cleaning up contaminated environments (Mokrani et al., 2024). Microorganisms break down organic contaminants as a source of carbon and energy or utilize them in their metabolic processes, ultimately reducing the concentration of harmful substances (Priya et al., 2024).

Outlined below are diverse bioremediation approaches that utilize microorganisms.

## 4.1 Natural attenuation

Natural attenuation (NA), also called intrinsic bioremediation, refers to naturally occurring processes in soil and groundwater, such as dispersion, dilution, sorption, volatilization, and biotic and abiotic degradation, that reduce the concentration, toxicity, mobility, or volume of environmental contaminants without human intervention (Fig. 2.1A). One of the main advantages of NA is its cost-effectiveness, as it requires less equipment and labor than other remediation technologies, resulting in lower operational costs and minimal waste generation (Aparicio et al., 2022). However, NA is a very slow process, as only about 10% of the total soil microbial population is composed of degrading microorganisms. In addition, there are many factors that

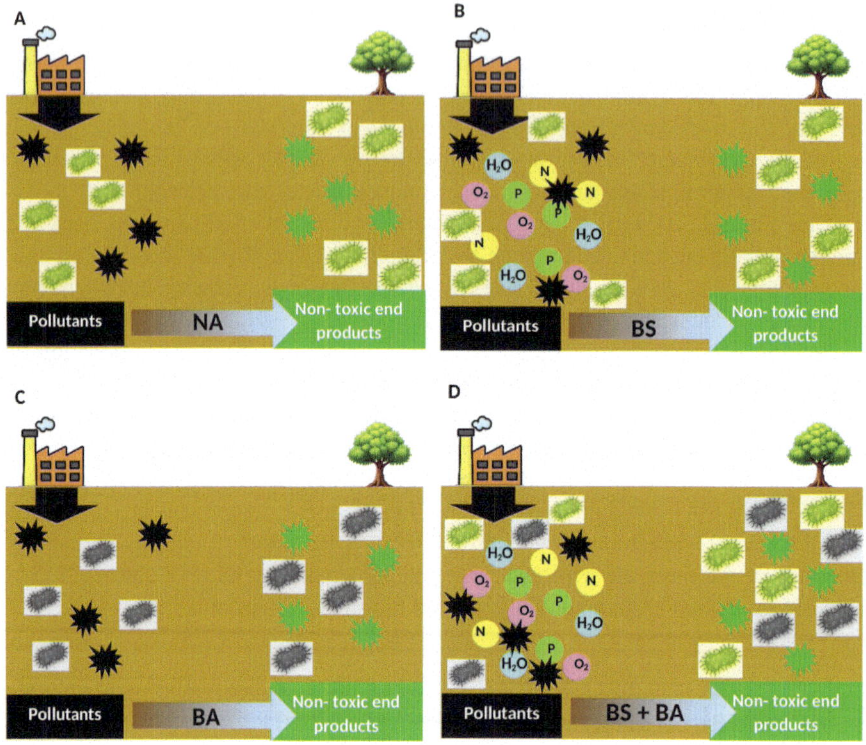


**FIG. 2.1**

Different strategies for microbial bioremediation: (A) natural attenuation (NA), (B) biostimulation (BS), (C) bioaugmentation (BA), and (D) combination of biostimulation and bioaugmentation (BS + BA). Green microbes represent autochthonous microbiota, while gray microbes represent exogenous microbiota.

limit NA effectiveness, such as a shortage of nutrients, pollutant type and concentration, the occurrence of cocontamination, the size and depth of the contaminated area, and the soil type (Raimondo et al., 2019).

This technique requires pollutants to be readily degradable, and its success can be affected by factors such as subsurface geology, hydrology, and microbiology (Raimondo et al., 2019).

## 4.2 Biostimulation

Biostimulation (BS) is a bioremediation technique that involves altering matrix conditions, such as physical and chemical properties, to enhance the activity of native microorganisms. By adjusting environmental factors like nutrient levels, oxygen availability, and moisture, this approach stimulates the indigenous microbial community to degrade hazardous and toxic pollutants (Fig. 2.1B) (Raimondo et al., 2020b).

The addition of rate-limiting nutrients (carbon, nitrogen, or phosphorus), trace minerals, organic amendments, electron acceptors or donors, water, or oxygen improves the degradative activity of native microbiota or promotes cometabolism. Even surfactants can be added with the aim of increasing contaminant bioavailability.

Numerous reports describe the use of organic materials, such as sugarcane bagasse, among others, as biostimulant agents for bioremediation (Aparicio et al., 2022). For instance, Raimondo et al. (2020b, 2020) demonstrated the removal of the pesticide lindane (LIN) in soils of different textural classes by means of BS with by products of the sugar industry, such as bagasse and sugarcane filter cake. Both by products are inexpensive alternatives for the BS of soils contaminated with LIN, significantly enhancing pesticide removal compared to NA strategies.

## 4.3 Bioaugmentation

Bioaugmentation (BA) is a bioremediation strategy that involves the inoculation of individual strains or microbial consortia (MCs) into contaminated environments with the necessary catalytic capacities to degrade the target pollutants (Fig. 2.1C). The addition of exogenous microorganisms complements the metabolic deficiencies of the indigenous microbial communities, favoring the elimination of the toxic compound or its conversion into less toxic or easily biodegradable products and improving the bioremediation process efficiency (Aparicio et al., 2022).

The efficiency of BA in polluted environment treatment is markedly influenced by the selection of the microbial inoculum. The introduced microorganisms must possess the metabolic capabilities to degrade specific pollutants and demonstrate resilience to challenging environmental conditions prevalent in contaminated systems. Microorganisms that can adapt to such stressful conditions are more likely to survive, proliferate, and perform the desired biodegradation processes efficiently (Chettri et al., 2024). In addition, the success of BA depends on the chemical structure, concentration, and availability of the pollutants; the soil physicochemical properties; and the interactions among indigenous and inoculated microorganisms. Consequently, careful selection and monitoring are essential to ensure effective pollutant degradation and minimal ecological disruption (Raimondo et al., 2020).

Several authors have improved bioremediation using BA in pesticide-contaminated soils, such as phenylurea herbicides (Lara-Moreno et al., 2022), OPs (Raimondo et al., 2020b, 2020), 2-Methyl-4-chlorophenoxyacetic acid (Pan et al., 2022), carbamates (Mishra et al., 2021), or dinitroanilines (Alba et al., 2022).

## 4.4 Combining BS and BA

BA involves introducing highly concentrated and specialized microbial populations into environments contaminated with recalcitrant and toxic compounds. BA is particularly advantageous in scenarios where (1) the native microbial populations are insufficient or (2) the indigenous microorganisms do not possess the specific metabolic pathways required to break down the pollutants present (Chettri et al., 2024). On the contrary, BS focuses on enhancing the natural biodegradation capabilities of indigenous microorganisms by modifying environmental conditions (Aparicio et al., 2022). The implementation of either or both of these bioremediation techniques is primarily determined by two key factors: the degrading ability of the indigenous microbes and the extent of contamination at the site to be treated. However, a common approach to the bioremediation of polluted matrices that has gained attention in recent decades is the combination of two strategies, BS and BA (Fig. 2.1D). For instance, Nayak et al. (2021) demonstrated that CP degradation was enhanced in soil microcosm assays by a combined approach. This approach included BS with N, P, and K in the presence of compost as a carrier material, and BA with *Ochrobactrum* sp. CPD-03, achieving a pesticide degradation efficiency of 76% and 74% in nonsterile and sterile microcosms, respectively, over 56 days. In another comparative study, bioremediation of LIN-contaminated soil mesocosms by BA with an actinobacterial consortium coupled with BS using sugarcane filter cake increased pesticide removal and reduced the half-life of the contaminants compared to nonbioremediated controls. This was observed across three types of soils under environmental conditions with minimal intervention (Raimondo et al., 2020a).

These findings highlight the versatility and efficacy of BS plus BA in addressing a wide range of environmental contaminants. The choice between these methods or their combination should be guided by site-specific factors, including the nature of the pollutants, soil characteristics, and the existing microbial community.

## 5 Potential microorganisms for environmental bioremediation

Bioremediation is an attractive procedure that involves the utilization of different microorganisms to detoxify contaminated environmental matrices. Microorganisms stand out as exceptional systems in the scope of pollutant detoxification, offering an economical, simple, and environmentally friendly approach to the reclamation of contaminated sites (Aparicio et al., 2022).

Researchers worldwide are focusing on finding a wide range of microorganisms to clean-up polluted environments, adapting their approaches to diverse ecological and geographical contexts (Bhowmick et al., 2024). For example, yeasts are increasingly recognized for their ability to break down and neutralize various environmental pollutants. Species such as *Candida tropicalis* have been identified as potent degraders of petroleum hydrocarbons in marine environments, showcasing their adaptability and efficiency in oil-contaminated sites (Hegazy et al., 2024).

Among fungi, several species have been employed in aquatic ecosystems for bioremediation of different toxic and recalcitrant compounds, for example, converting chemically suspended matter, microbial, nutritional, and oxygen-depleting aquatic contaminants into ecologically less hazardous products using multiple modes of action (AbuQamar et al., 2023). In this context, species of *Aspergillus* sp., *Phanerochaete* sp., *Trichoderma* sp., and *Penicillium* sp. have been reported as potential contributors to the remediation of heavy metal contamination. In addition, a plethora of fungal species stands out as a suitable approach for effectively eliminating textile dyes that have accumulated in the environment (Bhowmick et al., 2024).

Algal bioremediation is increasingly recognized as a superior alternative to traditional pollution treatment methods, owing to its multifaceted mechanisms and broad-spectrum applicability. This algal bioremediation employs a variety of processes to remove contaminants, including biosorption, bioaccumulation, biouptake, biodegradation, and photodegradation. These combined capabilities enable algae to effectively address a wide range of both traditional and emerging pollutants, encompassing inorganic substances like heavy metals and organic compounds such as pharmaceuticals and pesticides. Among the microalgal genera frequently used in environmental cleanup processes are *Spirulina, Scenedesmus, Phormidium, Oscillatoria, Nodularia, Desmodesmus, Cyanothece, Chlorella, Botryococcus*, and *Arthrospira* (Karnwal et al., 2024).

Bacteria, in particular, have developed a variety of strategies to survive in environments contaminated with toxic metals. These adaptive mechanisms include metallothionein and siderophore production, the presence of active efflux systems, morphological alterations, and the synthesis of exopolymeric substances, among others. These mechanisms not only enable bacteria to thrive in metal-contaminated environments but also make them valuable agents in bioremediation efforts aimed at detoxifying polluted ecosystems (Karnwal et al., 2024).

Bacteria are also well-known for their ability to degrade or mineralize organic pollutants through enzyme-catalyzed catabolic action. During bioremediation studies, numerous catabolic genes that participate in such degradation processes and are remarkably conserved have been identified. These genes are present in both gram-positive and gram-negative bacterial species (Ali et al., 2022). Among the genera with such capabilities are *Alcaligenes*, *Bacillus*, *Flavobacterium*, *Nocardia*, *Pseudomonas*, *Streptomyces*, *Rhodococcus*, and *Arthrobacter*, among others. These microorganisms have demonstrated effectiveness in degrading and/or removing a variety of contaminants, such as pesticides, azo dyes, emerging contaminants like pharmaceuticals, heavy metals, hydrocarbons, and effluents from diverse industries (Kour et al., 2022).

# 6 Phylum Actinomycetota in bioremediation

## 6.1 General characteristics of the phylum Actinomycetota

Microorganisms belonging to the phylum *Actinomycetota*, commonly called actinobacteria, are recognized for their vast metabolic diversity, their ability to adapt to hazardous environments, and their capacity to colonize different substrates and niches. They are important producers of a wide variety of metabolites, which have numerous applications in medicine, veterinary science, and nutrition, such as antibiotics, anticancer agents, antivirals, enzyme inhibitors, and other biologically active compounds (Saez et al., 2025). Regarding environmental applications, *Actinomycetota* can degrade a broad range of organic compounds, including pesticides, hydrocarbons, dyes, and other xenobiotics, which makes them excellent candidates for bioremediation processes (Espinosa-Ortiz et al., 2022).

*Actinomycetota* are gram-positive or gram-variable bacteria with a rigid cell wall composed of complex compounds such as peptidoglycan, polysaccharides, teichoic acids, and teichuronic acids, which allow them to maintain the shape of the cell and prevent bursting under high osmotic pressure (Amin et al., 2020). These microorganisms represent one of the most diverse bacterial groups, exhibiting a wide array of morphological characteristics, including rod-shaped forms (*Acidiferrimicrobium*), coccoid forms (*Micrococcus*), rod-coccoid forms (*Arthrobacter*), and bent rods (*Sinomonas*). They also include fragmented hyphae (*Nocardia*) and differentiated branched mycelia (*Streptomyces* and *Frankia*); some develop elongated filaments but lack true mycelium (*Rhodococcus*), while others do not produce mycelia at all (*Corynebacterium*). In addition, certain members are distinguished by producing branched substrate hyphae that break up into flagellated motile elements (*Oerskovia*) (Rao et al., 2022). Moreover, members of the phylum *Actinomycetota* can form complex structures, including spores, spore chains, sporangia, and sporangiospores. They are generally aerobic, although some genera are facultative or obligate anaerobes. These bacteria are chemoheterotrophic, utilizing diverse energy sources and complex polymers (Amin et al., 2020).

*Actinomycetota* have been identified across a wide array of ecosystems, spanning virtually the entire planet. As free-living microorganisms, they abound in soil environments, as well as in marine and freshwater ecosystems. However, they can also establish associations with eukaryotic hosts in a wide range of environments, including the exoskeletons of ants, the respiratory and dermal systems of mammals, and the root systems and inner tissues of plants. The diverse ecological niches colonized by *Actinomycetota* highlight their remarkable physiological adaptability, enabling them to thrive across a wide spectrum of environmental conditions (Lewin et al., 2016).

Members of the phylum *Actinomycetota* exhibit relatively large genomes compared to other bacterial phyla. Nonetheless, there is considerable variation in genome sizes, ranging from under 1 Mb to >12 Mb. Typically, their genomes have an elevated guanine + cytosine (G + C) content, with certain genera like *Streptomyces* and *Pseudonocardia* exceeding the 70% threshold in G + C content. Nevertheless, there are cases like *Gardnerella vaginalis* and *Tropheryma whipplei* in which the G + C content is <50% (Barka et al., 2016).

## 6.2 Bioremediation potential of the phylum Actinomycetota

*Actinomycetota* have attracted significant attention as promising candidates for the bioremediation of various environmental matrices contaminated with persistent inorganic and organic pollutants, such as heavy metals, hydrocarbons, pesticides, and aliphatic and aromatic compounds (Romano-Armada et al., 2020; Alvarez et al., 2017).

Within the microbial world, *Actinomycetota* are exceptional because of their flexible metabolism and capacity to tolerate severe environments. They produce secondary metabolites, such as enzymes that help break down a variety of pollutants, siderophores, EPS, biosurfactants, and bionanoparticles, which assist in trapping organic contaminants and immobilizing heavy metals (Makarani and Kaushal, 2025).

Many studies have demonstrated the capacity of certain *Actinomycetota* genera, including *Arthrobacter*, *Frankia*, *Gordonia*, *Kocuria*, *Nocardioides*, *Streptomyces*, and *Rhodococcus*, to break down pesticides. These microorganisms have shown the ability to metabolize a wide range of pesticide chemical structures, including organophosphates, organochlorines, triazines, synthetic pyrethroids, and benzimidazoles (Makarani and Kaushal, 2025; Alvarez et al., 2017). For instance, *Streptomyces* sp. are promising for the bioremediation of pesticide-contaminated matrices polluted with LIN, CP, cypermethrin, carbofuran, diuron, alachlor, and neonicotinoids (Briceño et al., 2024, 2018a). Similarly, *Rhodococcus* strains have been reported for their ability to metabolize atrazine and carbendazim, among others (Mawang et al., 2021).

Several *Actinomycetota* species have been proven capable of biodegrading petroleum hydrocarbons. These microorganisms include representatives of the genera *Streptomyces*, *Actinomyces*, *Rhodococcus*, *Nocardia*, *Micrococcus*, and *Gordonia* (Mawang et al., 2021).

Besides, numerous studies have highlighted the capacity of various *Actinomycetota* genera to remediate environments contaminated with heavy metals, demonstrating their bioremediation potential. Among them, species belonging to the genera *Streptomyces*, *Amycolatopsis*, *Arthrobacter*, *Rhodococcus*, *Nocardia*, and *Micrococcus* can be mentioned. The mechanisms involved in these processes are reduction, bioaccumulation, biosorption, and enzymatic reduction (Alvarez et al., 2017).

## 7 Microbial teamwork: Approaches to mixed cultures and MCs

Microorganisms do not exist in isolation; all are part of complex communities composed of many different individuals with distinct functions that form an active, dynamic, and ever-changing community. Cells within a mixed community interact, communicate, influence, and alter the biochemical and physiological processes among the members. This vast potential among microorganisms provides significant opportunities for various environmental, industrial, and agricultural applications (Sivasubramaniam and Franks, 2016). For this reason, the simultaneous use of different microorganisms represents a promising biotechnological strategy. It is important to note that the utilization of consortia or mixed cultures has become a research trend in applied microbiology, bioprocess engineering, and biotechnology (Ghosh et al., 2016). Despite this, it is not clear whether these terms can be used as synonyms, as there are differing opinions on the use of these concepts.

By definition, a pure culture is a microbial culture that contains a single type of microorganism (Clark et al., 2009). On this basis, the term "mixed culture" could be proposed as a microbial culture that contains different types of microorganisms, where the existence of interactions is unknown. On the contrary, considering that the term "microbial consortia" indicates not only living together but also division of labor, it encompasses all conditions of composition by single, two, and multiple species (Jia et al., 2016). Accordingly, the term "microbial consortium" could be used to describe a stable mixture of diverse microorganisms that interact with each other. The prevalence of positive interactions (commensalism or mutualism) among microbial members of a consortium can significantly enhance the product outcome of the bioprocess, ensuring their industrial application and long-term stability. In contrast, negative interactions (parasitism, predation, or amensalism) lead to the elimination of microbial members from the consortium, resulting in the destruction of community structure and disruption of cumulative performance. Therefore knowledge of the type of interaction among microorganisms is essential for optimizing the performance of designed consortia (Ghosh et al., 2016).

Microbial associations, in general, arise as useful tools in several biotechnological processes, such as the bioremediation of contaminated environments. In this context, previous works have reported the use of a variety of terms to describe these associations, highlighting a lack of consensus among specialists in the field. In this regard, Massot et al. (2022) analyzed the variety of microbial associations referred to as MCs in the specific context of pollutant biodegradation and bioremediation. They identified that the origin of microbes and the level of human intervention are usually considered as criteria to classify them into natural microbial consortia (NMC), artificial microbial consortia (AMC), and synthetic microbial consortia (SMC). The first category (NMC) includes associations of microorganisms obtained from a single source, while the second category (AMC) involves members that come from different sources. The last category (SMC) represents a class of AMC in which the microbial composition is specifically defined to accomplish a certain task. It is important to highlight that cross-kingdom combinations are possible in all cases (Massot et al., 2022; Saez et al., 2025).

## 7.1 MCs for bioremediation

The success of bioremediation technologies applied to polluted environments highly depends on the biodegrading capabilities of native microbial populations or exogenous microorganisms used as inoculants (Venosa and Zhu, 2003). Native populations (as undefined MCs) present in contaminated sites are adapted to the climatic, physicochemical, and nutrient conditions but frequently do not include species with the enzymatic abilities required to allow bioremediation to proceed at increased rates, resulting in long-term processes. As mentioned above, an alternative for this is BS; thus by creating a favorable environment, BS can increase the rate of pesticide degradation, making the process more efficient and applicable to large agricultural areas (Lopes et al., 2022). In this sense, Prada-Vásquez et al. (2017) studied the BS of native microbial populations to bioremediate DDT in soil, in both laboratory-scale and pilot-scale experiments, under anaerobic conditions. Their studies showed that the BS process, using a combination of an electron donor substance (molasses) and an electron acceptor ($KNO_3$), enhanced the insecticide biodegradation rates. On the contrary, Qiu et al. (2009) confirmed that P was a limiting nutrient during the degradation of coexisting dichlobenil and atrazine by bacterial degraders in soil. Dichlobenil was completely degraded in 60 h in the P-supplemented soil extract, in comparison to <40% degradation without P supplementation. Notably, the increase in the degree of atrazine degradation was even greater, as it was completely degraded in the P-supplemented extract within 40 h, compared to <10% herbicide degradation in the absence of P supplementation.

The BA with exogenous microorganisms possessing specific metabolic capabilities to enhance biodegradation processes in contaminated environments has been widely reported (Omenna et al., 2024). For this purpose, two types of MCs can be used: a defined consortium or a functional consortium. The use of a defined consortium facilitates the examination of general characteristics of all the members in pure culture and the monitoring of the dynamics of these members in the community throughout a given cultivation period. Furthermore, by constructing a "knockout community" in which one of the members is eliminated, the roles played by the eliminated member in situ and its impact on the community can be evaluated (Kato et al., 2005). In this context, a cellulose-degrading defined mixed culture (designated SF356) consisting of five bacterial strains (*Clostridium straminisolvens* CSK1, *Clostridium* sp. strain FG4, *Pseudoxanthomonas* sp. strain M1-3, *Brevibacillus* sp. strain M1-5, and *Bordetella* sp. strain M1-6) exhibited both functional and structural stability. In this case, various intertwined relationships were identified, not only through the detection of synergistic or commensalistic interactions but also by detecting inhibitory effects (e.g., inhibition of cellulose degradation due to the production of an excess amount of acetic acid) and competition (e.g., between *Brevibacillus* sp. strain M1-5 and *Bordetella* sp. strain M1-6). The evidence suggests that achieving a balance of various types of relationships, both positive and negative, is essential for the stable coexistence of the members in SF356. Similarly, Horváthová et al. (2018) evaluated the biotreatment of sediment historically contaminated with PCBs using BA with binomial and trinomial consortia constructed from four bacterial strains (*Achromobacter xylosoxidans*, *Stenotrophomonas maltophilia*, *Ochrobactrum anthropi*, and *Rhodococcus ruber*) isolated from the contaminated sediment. The objective of the research was to determine the biodegradation ability of the individual strains and artificially prepared consortia. Under the experimental conditions, it was observed that PCB biodegradation by consortia action was more effective than biodegradation by the addition of individual strains. In addition, it was noted that the naturally occurring microflora contributed to PCB biodegradation. This result indicates that exogenous and indigenous microorganisms were in a synergistic relationship. The findings demonstrate that the BA approach represents an effective method to increase the amount of pollutant degraded compared to the degradation achieved by the indigenous microflora alone.

At this point, it is important to mention that, unlike NMC, in SMC, it is feasible to reconfigure metabolic pathways and program interactions to obtain the desired function. The different tools from synthetic biology allow microbial interactions to be designed and tightly controlled, creating valuable model systems (Wintermute and Silver, 2010). In other words, synthetic biology can reduce the microorganism to a programmable chassis whose functions can be predicted using standardized DNA building blocks (Leonard et al., 2008).

In contrast to a defined consortium, a functional consortium can be used for the BA process. Undefined but functional consortia exhibit desired attributes for specific functions. There are various approaches to the isolation and enrichment of microbial functional consortia in several biological processes (Adav et al., 2009), and many of them have been reported. For example, in the conversion of crude glycerol, an anaerobic MC was obtained from anaerobic activated sludge through 19 serial transfers. It adapted well to high glycerol concentrations (120 g $L^{-1}$) and wide substrate concentration fluctuations (from 15 to 80 g $L^{-1}$), producing 60.61 and 82.66 g $L^{-1}$ of 1,3-PDO in batch and fed-batch fermentation, respectively (Zhou et al., 2017). In another study, Ho et al. (2010) utilized a method called the toxicity-to-extinction approach to obtain the minimal functional consortium capable of effectively degrading meta- (m-), para- (p-), and ortho- (o-) cresols. In this case, microbial diversity decreased as the cresol concentration increased until the desired microbial functional consortium was achieved.

The above exposure demonstrates some successful cases of MCs applied in BA individual processes. However, studies on pollutant bioremediation in the field have been directed toward the utilization of both BS and BA techniques together, as the best alternative biodegradation strategy. For instance, when studying the combination of BA and BS for the bioremediation of pentachlorophenol (PCP)-contaminated forest soil, Ammeri et al. (2022) observed 71% PCP removal after 30 days in forest soil that was contaminated, inoculated with a bacterial consortium B1, and amended with forest compost (FC). This strategy showed the best performance among different assayed treatments. The FC or B1 single application showed less PCP depletion, with 52% and 41% removal, respectively. The results demonstrate the advantage of using a combination of different bioremediation techniques when MCs are applied as BA tools.

## 7.2 Actinomycetota consortia for pesticide bioremediation

The distinctive metabolic and physiological characteristics of *Actinomycetota* position them as promising candidates for the remediation of environments contaminated with organic pollutants, particularly pesticides (Antezana et al., 2023). Furthermore, the remarkable metabolic versatility encoded in their genomes enables them to produce a wide variety of catalytic enzymes that contribute to the breakdown of numerous environmental pollutants (Briceño et al., 2018b). Moreover, strains isolated from contaminated environments are known to secrete extracellular enzymes like monooxygenases and dioxygenases, which enhance their ability to degrade various pesticides (Ravi and Gupta, 2022). On the contrary, *Actinomycetota* have the ability to produce amphipathic compounds such as bioemulsifiers, which act at interfaces (Fuentes et al., 2016). These compounds are key for the bioremediation of poorly soluble pollutants as they promote their desorption from soil particles, increasing their bioavailability for microbial degradation (D'Almeida et al., 2024).

There are studies that demonstrate the abilities of actinobacterial consortia to enhance pollutant removal, individually or in toxic mixtures, due to the cooperation among consortium members, which allows a biodegradation process more efficient than that achieved by a pure culture (Fuentes et al., 2011, 2014, 2016). The significance of using defined actinobacterial consortia lies in their ability to serve as reproducible technologies that can provide valuable insights for a comprehensive understanding of the interactions among the microorganisms within the consortium (Antezana et al., 2023).

Within mixed microbial populations, antagonistic phenomena are common, especially when considering the ability of *Actinomycetota* to produce a wide array of secondary metabolites that may inhibit the growth or metabolic functions of other consortium members. Therefore in the design of defined actinobacterial consortia for bioremediation purposes, the evaluation of antagonistic interactions represents a critical step (Antezana et al., 2023).

The compatibility among *Actinomycetota* can be assessed using a modified version of the method described by Bell et al. (1980). In this test, each strain is inoculated at the center of a Petri dish containing solid medium and arranged perpendicularly to the other strains under evaluation, allowing all possible pairwise combinations. The appearance of growth inhibition zones after incubation reveals antagonistic interactions. Conversely, the absence of such effects suggests that the strains are compatible and may be cocultured as a defined MC (Fuentes et al., 2014, 2016). In this context, several actinobacterial strains did not exhibit growth inhibition in antagonism assays, supporting their compatibility and subsequent application as consortia for bioremediation purposes (Fuentes et al., 2011, 2014, 2016).

## 7.3 BA with free actinobacterial consortia

OPs are synthetic pesticides characterized by their persistence, bioaccumulative potential, and adverse effects on the environment and human health. Although they were historically used for pest control, their long-term environmental impact and toxicological concerns have led to restrictions or bans in many regions (Tzanetou and Karasali, 2022). As a result, various studies have investigated the bioremediation of OP-contaminated matrices by applying actinobacterial consortia. For instance, Fuentes et al. (2011) evaluated LIN degradation in liquid culture using both pure and mixed cultures of *Streptomyces* sp., testing consortia composed of two to six strains to assess pesticide degradation and associated dechlorinase activity. Notably, combinations of two, three, and four strains achieved the highest removal efficiencies, while consortia formed by five and six strains were less effective in eliminating the pesticide. Moreover, in mixed cultures, the specific dechlorinase activity increased up to twelvefold compared to that achieved in individual strains. Among the tested consortia, the combination of *Streptomyces* sp. A2, A5, A11, and M7 showed great potential as a biotechnological tool for LIN biodegradation (Fuentes et al., 2011).

Besides, the consortium *Streptomyces* sp. A2–A5–A11-M7 was tested for the bioremediation of sterile soil microcosms artificially contaminated with LIN (1.66 mg $kg^{-1}$), incubated at 30°C for 28 days. This consortium demonstrated robust bacterial growth in the LIN-polluted soil, and no growth inhibition was observed. This may be attributed to the fact that the strains were originally isolated from an environment heavily contaminated with OPs (Fuentes et al., 2010). It is important to mention that the LIN removal efficiency was 31.4% after the incubation period.

Similarly, Fuentes et al. (2014) evaluated the bioremediation potential of methoxychlor (MTX) by *Streptomyces* strains A3, A6, A12, A14, and M7, which were previously isolated from sediments and soils contaminated with various OPs, along with *Streptomyces coelicolor* A3 obtained from the German Collection of Microorganisms and Cell Cultures (DSMZ). The study aimed to develop BA strategies for the removal of this pesticide from different systems. To this end, the strains were tested in 39 combinations, and the consortium of *Streptomyces* sp. A6, A12, A14, and M7 was selected due to its high MTX removal capacity and elevated specific dechlorinase activity in liquid culture. The BA of soil microcosms with this consortium achieved significant cell proliferation ($1.0 \times 10^{11}$ CFU $g^{-1}$) and enabled 56.4% MTX removal, demonstrating that the selected strains were not only naturally adapted to soil conditions but also to the presence of this type of contaminant. In parallel, and considering that soil-slurry bioreactors represent a widely used ex situ technique for soil bioremediation, the authors also evaluated the performance of this consortium under such conditions. These systems are often used as a preliminary approach to assess the intrinsic restoration potential of contaminated soils (Robles-González et al., 2008). Moreover, bioreactor inoculation enables the implementation of a controlled bioremediation process, allowing for the optimization of operational parameters to enhance both microbial degradation efficiency and pollutant bioavailability. In this context, MTX removal was assessed in two types of soil-slurry bioreactors: one supplemented with distilled water (SB-water) and the other with tryptic soy broth (SB-TSB), a nutrient-rich medium designed to stimulate microbial activity. The highest microbial growth was observed on the third day in SB-TSB systems spiked with MTX. Notably, the consortium was able to remove MTX in both setups, with removal efficiencies of 56.2% in SB-TSB and 45.6% in SB-water by the end of the experiment. These results highlight the versatility of the selected *Streptomyces* consortium, supporting its potential application in both in situ (soil-based) and ex situ (bioreactor-based) bioremediation strategies for MTX-contaminated environments.

More recently, Fuentes et al. (2023) demonstrated that combining soil-slurry bioreactors operated in batch mode with BA using a defined actinobacterial consortium represents a promising strategy for the remediation of soils contaminated with chlordane (CLD). In this study, a consortium composed of *Streptomyces* sp. A2, A5, and A13, previously characterized for their CLD-degrading capacity (Fuentes et al., 2016), was inoculated at 2 g $L^{-1}$ into soil-slurry bioreactors containing 1.66 mg $L^{-1}$ of the pesticide. Again, the two previously mentioned conditions were evaluated:

SB-TSB and SB-water. In both cases, the presence of CLD did not inhibit microbial development; however, higher biomass production was observed in SB-TSB, as indicated by a shorter duplication time (7.27 h) compared to SB-water (10.88 h). Notably, the highest CLD removal efficiency (34.8%) was achieved in SB-water bioaugmented with the actinobacterial consortium, accompanied by a corresponding increase in the release of chloride ions, suggesting active dechlorination.

Agricultural practices aimed at meeting global food demands have inevitably led to the widespread application of various pesticides, often used simultaneously or sequentially. As a consequence, mixtures of these xenobiotic compounds may coexist in the environment, contributing to the contamination of soil, water, and air, and posing risks to ecosystems and trophic chains. These concerns are particularly critical when OPs are involved (Raffa and Chiampo, 2021). Therefore MCs designed for bioremediation must be capable of simultaneously degrading pesticide mixtures across different systems and soil types.

In this context, the MC, constituted by *Streptomyces* sp. A2, A5, A11, and M7, was evaluated for its ability to degrade a mixture of three previously studied OPs (Fuentes et al., 2017). This consortium exhibited satisfactory growth after its inoculation at a biomass concentration of 2 g $L^{-1}$ into a minimal liquid medium supplemented with 1.66 mg $L^{-1}$ of each pesticide, followed by incubation for 16 days at 30°C and 200 rpm. Under the tested conditions, significant removal efficiencies were observed for MTX (99.5%), CLD (99.8%), and LIN (40.4%). When tested in microcosms formulated with sterile soils of different textures, the consortium exhibited variable pesticide degradation rates and efficiencies. The highest removal percentages were observed in clay silty loam, followed by sandy and loam soils. In nonsterile microcosms of clay silty loam bioaugmented with the actinobacterial consortium, CLD removal was limited (~5%), whereas higher removal rates were recorded for LIN (25%) and MTX (21%). Moreover, in sterile slurry systems formulated with the same soil and inoculated with the actinobacterial consortium, distinct degradation patterns were observed among the pesticides: MTX showed the highest dissipation (36%), followed by CLD (15%) and LIN (13.5%). These findings underscore the influence of soil texture and experimental conditions on the performance of MCs in pesticide removal. These observations were subsequently corroborated by Raimondo et al. (2019), who evaluated the effectiveness of this consortium in nonsterile soils of different textures contaminated with LIN (2 mg $kg^{-1}$). They reported higher removal rates in bioaugmented microcosms compared to the nonbioaugmented controls, with LIN degradation values of 70.3%, 36.3%, and 30.7% in sandy soil, silty loam soil, and clayey soil, respectively. Moreover, the pesticide half-life was reduced by 77.3, 50.3, and 10.7 days, respectively, across these soil types. These studies are highly relevant, as they provide valuable insights into microbial behavior and growth requirements, which are essential to consider prior to any in situ decontamination intervention.

The BS with organic amendments demonstrated its effectiveness in enhancing the bioremediation performance of *Actinomycetota* consortia in pesticide-contaminated soils, likely due to the provision of essential macro- and micronutrients. Furthermore, these materials improve the structure of the soil by reducing bulk density and increasing porosity, aeration, and water retention, thereby fostering actinobacterial growth and a more efficient degradation of the contaminant (Raimondo et al., 2020b, 2020).

Previous studies demonstrated that the *Streptomyces* A2–A5–A11–M7 consortium was capable of degrading LIN more efficiently in soils of different textures when combined with BS using sugarcane bagasse or sugarcane filter cake, compared to soils (BA) without these organic amendments (Raimondo et al., 2020b, 2020). The combination of sugarcane filter cake with the actinobacterial consortium was the most effective strategy to clean-up LIN-contaminated soils. Based on these results, the process was scaled up from the microcosm to the mesocosm level and monitored over a 63-day period under conditions with minimal environmental intervention (Raimondo et al., 2020a). At the end of the assay, the BA and BS integrated approach improved LIN removal, increased microbial abundance and enzymatic activity, and significantly reduced the pesticide half-life compared to the nontreated controls. Among the evaluated soil types, the highest dissipation rate (82.6%) was observed in BA–BS sandy soil, highlighting the influence of soil texture on bioremediation efficiency.

Likewise, *Actinomycetota* strains have demonstrated great versatility in the formulation of interkingdom consortia with fungi, aimed at the treatment of biomixtures contaminated with LIN. Biomixtures are the main component of bioprophylactic systems designed to minimize point-source environmental contamination resulting from pesticide handling, accidental spills, and improper disposal of wastewater (Beltrán-Flores et al., 2024). In this regard, Saez et al. (2018) demonstrated that the consortium constituted by LIN-degrading actinobacteria and the white-rot fungus *Trametes versicolor* SNGG1 enhanced insecticide dissipation in biomixtures formulated with soils of different textural classes, peat, and sugarcane bagasse. Inoculation of the biomixtures with the fungus–actinobacteria coculture achieved 81–87% LIN removal after 66 days of incubation, reducing the pesticide's half-life by sixfold compared to its natural degradation in noninoculated soil. The application of interkingdom consortia is considered a promising strategy to better approximate natural conditions and maximize their biotechnological potential (Saez et al., 2025).

On the contrary, organophosphorus pesticides (OPP) constitute the most widely used group of pesticides in the global market. Their extensive application has resulted in the widespread contamination of soils, water bodies, the atmosphere, and food supplies worldwide. Moreover, OPP are recognized as major contaminants in the food chain and have been associated with a range of human metabolic health issues, including type 2 diabetes (Briceño et al., 2016). For these reasons, the removal of OPP by actinobacterial consortia has been actively investigated. In this context, Briceño et al. (2016) demonstrated that a four-strain actinobacterial consortium composed of *Streptomyces* sp. AC5, AC9, GA11, and ISP13 efficiently removed CP and its primary metabolite, 3,5,6-trichloro-2-pyridinol (TCP), from a minimal liquid medium, with 78% cell viability following CP and TCP exposure. Notably, glucose supplementation reduced the biomass doubling time and accelerated pesticide degradation. Moreover, the consortium was able to simultaneously remove four OPP, including CP, diazinon (DZ), azinphos-methyl, and methidathion, achieving maximum removal efficiencies of 69.3% for CP and 58.9% for DZ. Later, Briceño et al. (2018a) reported that the same MC successfully removed a CP and DZ mixture from liquid medium, soil, and a biobed biomixture. In liquid medium, the microorganisms increased their biomass and degraded the pesticide mixture, achieving removal rates of 0.036 $h^{-1}$ for CP and 0.015 $h^{-1}$ for DZ, with corresponding half-lives of 19 and 46 h, respectively. In sterilized soil and biobed biomixtures inoculated with the quad-consortium, actinobacteria successfully colonized the substrates, as evidenced by increased population levels determined via quantitative polymerase chain reaction, along with elevated dehydrogenase and acid phosphatase enzymatic activities. In both soil and biomixture systems, CP removal was limited (6–14%), whereas DZ degradation was more efficient, with removal rates of 0.024 and 0.060 $day^{-1}$ and corresponding half-lives of 29 and 11 days, respectively.

## 7.4 BA with immobilized actinobacterial consortia

The use of immobilized cells offers several advantages compared to free cells, including improved microbial retention and cell viability, as well as protection against toxic effects. Furthermore, immobilized cultures typically exhibit greater metabolic activity and higher tolerance to environmental fluctuations. An additional benefit is their potential for reuse, which contributes to reducing the operational costs of the biotechnological process (Saez et al., 2014).

An example of the application of immobilized actinobacterial consortia for bioremediation is provided by Saez et al. (2012), who evaluated four different matrices, polyvinyl alcohol, cloth sachets, silicone tubes, and agar, for immobilizing *Streptomyces* strains, either as pure cultures or as part of a consortium, to assess LIN (1.66 mg $L^{-1}$) removal in minimal medium. Pesticide degradation by immobilized pure cultures was significantly higher than that obtained by free cells. Notably, cells immobilized in cloth sachets achieved approximately 25% greater LIN removal than the abiotic control. Although consortia immobilized in cloth sachets did not significantly outperform immobilized pure cultures, a maximum removal efficiency of 81% was recorded after 96 h using the consortium composed of *Streptomyces* strains A2, A5, A11, and M7. Moreover, the immobilized consortium remained effective for two additional 96-hour reuse cycles. Subsequently, Saez et al. (2014) demonstrated that the same consortium, immobilized in cloth sachets and applied at an inoculum size of $10^7$ CFU $g^{-1}$, efficiently removed LIN from a concentrated soil-slurry system contaminated with 50 mg $kg^{-1}$ of the pesticide, achieving a reduction of 35.3 mg $kg^{-1}$ after a week. In a diluted soil-slurry system, the removal efficiency was slightly lower, reaching 28.7 mg $kg^{-1}$.

Likewise, immobilized actinobacterial consortia represent a promising tool for treating environments contaminated with mixtures of xenobiotics, whether through in situ or ex situremediation techniques. In this context, Fuentes et al. (2013) compared the removal of a pesticide mixture composed of CP and PCP by a six-strain actinobacterial consortium, either as free cells or immobilized in alginate beads.

The results showed that CP removal reached 40.17% with free cells and 71.05% with immobilized cells, while PCP removal was 5.24% and 14.74%, respectively. The improved removal efficiencies detected with entrapped actinobacteria were likely due to the optimal diffusion of pollutants through the alginate matrix, as well as the protective effect of the support against adverse environmental conditions. In addition, Briceño et al. (2020) reported that the inoculation of a continuous stirred-tank bioreactor with a mixed culture of pesticide-degrading actinobacterial strains immobilized on polyurethane foam effectively reduced the concentrations of the insecticides CP and DZ. The bioreactor, operated at flow rates between 10 and 40 mL $h^{-1}$, achieved high removal efficiencies for both compounds; however, a decline in efficiency was observed at a flow rate of 50 mL $h^{-1}$. The system showed the capacity to handle proportionally high pesticide loadings, demonstrating its potential as a viable strategy for the treatment of wastewater polluted with OPP.

## 8 Conclusion

Environmental remediation is crucial for restoring and preserving the ecosystems of our planet. In response to the persistence and toxicity of pesticides, widely used around the world, numerous strategies have been developed to reduce their environmental impact. Among these, bioremediation strategies are increasingly chosen worldwide because they offer a cost-effective, efficient, and environmentally sustainable approach to cleaning polluted sites. Actinobacterial consortia-based bioremediation, in particular, represents a promising alternative for restoring matrices impacted by pesticides due to the metabolic versatility of their constituents that, together, can degrade a broad range of pollutants, including pesticides. Based on the information described, actinobacterial consortia could be proposed as efficient biotechnological tools with great potential to remediate environments impacted by pesticides.

## Acknowledgement

The authors gratefully acknowledge the financial support of the Agencia Nacional de Promoción de la Investigación, el Desarrollo Tecnológico y la Innovación (PICT 2021-0983, PICT 2019-2825), and the Consejo Nacional de Investigaciones Científicas y Técnicas (PIP 0030, PIP 670).

## References

AbuQamar, S.F., El-Fattah, H.I.A., Nader, M.M., Zaghloul, R.A., El-Mageed, T.A.A., Selim, S., Omar, B.A., Mosa, W.F., Saad, A.M., El-Tarabily, K.A., El-Saadony, M.T., 2023. Exploiting fungi in bioremediation for cleaning-up emerging pollutants in aquatic ecosystems. Mar. Environ. Res. 18790291. 190. Available from: www.elsevier.com/locate/marenvrev. doi:10.1016/j.marenvres.2023.106068.

Adav, S.S., Lee, D.J., Lai, J.Y., 2009. Functional consortium from aerobic granules under high organic loading rates. Bioresour. Technol. 09608524. 100 (14), 3465–3470. doi:10.1016/j.biortech.2009.03.015.

Alba, L.-M., Esmeralda, M., Jaime, V., 2022. Enhanced Biodegradation of Phenylurea Herbicides by Ochrobactrum anthrophi CD3 Assessment of Its Feasibility in Diuron-Contaminated Soils. Int. J. Environ. Res. Public Health. 1660-4601. 19 (3), 1365. doi:10.3390/ijerph19031365.

Ali, M., Song, X., Ding, D., Wang, Q., Zhang, Z., Tang, Z., 2022. Bioremediation of PAHs and heavy metals co-contaminated soils: challenges and enhancement strategies. Environ. Pollut. 02697491. 295, 118686. doi:10.1016/j.envpol.2021.118686.

Alvarez, A., Saez, J.M., Costa, J.S.D., Colin, V.L., Fuentes, M.S., Cuozzo, S.A., Benimeli, C.S., Polti, M.A., Amoroso, M.J., 2017. Actinobacteria: current research and perspectives for bioremediation of pesticides and heavy metals. Chemosphere. 18791298. 166, 41–62. Available from: www.elsevier.com/locate/chemosphere. doi:10.1016/j.chemosphere.2016.09.070.

Amin, D.H., Abdallah, N.A., Abolmaaty, A., Tolba, S., Wellington, E.M.H., 2020. Microbiological and molecular insights on rare Actinobacteria harboring bioactive prospective. Bull. Natl. Res. Cent. 2522-8307. 44 (1). doi:10.1186/s42269-019-0266-8.

Ammeri, R.W., Simeone, G.D.R., Hidri, Y., Abassi, M.S., Mehri, I., Costa, S., Hassen, A., Rao, M.A., 2022. Combined bioaugmentation and biostimulation techniques in bioremediation of pentachlorophenol contaminated forest soil. Chemosphere. 18791298. 290. Available from: www.elsevier.com/locate/chemosphere. doi:10.1016/j.chemosphere.2021.133359.

Antezana, P.E., Colin, V.L., Bourguignon, N., Benimeli, C.S., Fuentes, M.S., 2023. Applied of actinobacteria consortia-based bioremediation to restore co-contaminated systems. Res. Microbiol. 174 (4). doi:10.1016/j.resmic.2023.104028.

Aparicio, J.D., Raimondo, E.E., Saez, J.M., Costa-Gutierrez, S.B., Álvarez, A., Benimeli, C.S., Polti, M.A., 2022. The current approach to soil remediation: a review of physicochemical and biological technologies, and the potential of their strategic combination. J. Environ. Chem. Eng. 22133437. 10 (2). Available from: http://www.journals.elsevier.com/journal-of-environmental-chemical-engineering/. doi:10.1016/j.jece.2022.107141.

Barathi, S., Sabapathi, N., Kandasamy, S., Lee, J., 2024. Present status of insecticide impacts and eco-friendly approaches for remediation-a review. Environ. Res. 00139351. 240 (Pt 1), 117432. doi:10.1016/j.envres.2023.117432.

Barka, E.A., Vatsa, P., Sanchez, L., Gaveau-Vaillant, N., Jacquard, C., Meier-Kolthoff, J.P., Klenk, H.-P., Clément, C., Ouhdouch, Y., van Wezel, G.P., 2016. Correction for Barka et al., Taxonomy, Physiology, and Natural Products of Actinobacteria. Microbiol. Mol. Biol. Rev. 1092-2172. 80 (4). doi:10.1128/mmbr.00044-16.

Beaumelle, L., Tison, L., Eisenhauer, N., Hines, J., Malladi, S., Pelosi, C., Thouvenot, L., Phillips, H.R.P., 2023. Data and analysis code from: pesticide effects on soil fauna communities: a meta-analysis. Zenodo Open Access Repository. doi:10.5281/zenodo.7937860.

Bell, D.K., Wells, H.D., Markham, C.R., 1980. In vitro antagonism of Trichoderma species against six fungal plant pathogens. Phytopathology 72, 379–382.

Beltrán-Flores, E., Sarrà, M., Blánquez, P., 2024. A review on the management of rinse wastewater in the agricultural sector. Chemosphere. 00456535. 352, 141283. doi:10.1016/j.chemosphere.2024.141283.

Bernhardt, E.S., Rosi, E.J., Gessner, M.O., 2017. Synthetic chemicals as agents of global change. Front. Ecol. Environ. 15409309. 15 (2), 84–90. Available from: http://www.esajournals.org/loi/fron. doi:10.1002/fee.1450.

Bhowmick, K., Roy, D., Rana, D., Ghosh, A., Sadhukhan, S., Chakraborty, M., Chattopadhyay, D., Ghosh, T.K., 2024. Potential microbes in bioremediation: a review. Mater. Today Sustain. 25892347. 28, 101032. doi:10.1016/j.mtsust.2024.101032.

Briceño, G., Diez, M.C., Palma, G., Jorquera, M., Schalchli, H., María Saez, J., Benimeli, C.S., 2024. Neonicotinoid Effects on Soil Microorganisms: responses and Mitigation Strategies. Sustainability. 2071-1050. 16 (9), 3769. doi:10.3390/su16093769.

Briceño, G., Fuentes, M.S., Saez, J.M., Diez, M.C., Benimeli, C.S., 2018b. Streptomyces genus as biotechnological tool for pesticide degradation in polluted systems. Crit. Rev. Environ. Sci. Technol. 15476537. 48 (10-12), 773–805. Available from: www.tandf.co.uk/journals/titles/10643389.asp. doi:10.1080/10643389.2018.1476958.

Briceño, G., Levio, M., González, M.E., Saez, J.M., Palma, G., Schalchli, H., Diez, M.C., 2020. Performance of a continuous stirred tank bioreactor employing an immobilized actinobacteria mixed culture for the removal of organophosphorus pesticides. 3 Biotech. 21905738. 10 (6). Available from: http://www.springerlink.com/content/2190-572x/. doi:10.1007/s13205-020-02239-9.

Briceño, G., Schalchli, H., Mutis, A., Benimeli, C.S., Palma, G., Tortella, G.R., Diez, M.C., 2016. Use of pure and mixed culture of diazinon-degrading Streptomyces to remove other organophosphorus pesticides. Int. Biodeterior. Biodegrad. 09648305. 114, 193–201. doi:10.1016/j.ibiod.2016.06.018.

Briceño, G., Vergara, K., Schalchli, H., Palma, G., Tortella, G., Fuentes, M.S., Diez, M.C., 2018a. Organophosphorus pesticide mixture removal from environmental matrices by a soil Streptomyces mixed culture. Environ. Sci. Pollut. Res. 16147499. 25 (22), 21296–21307. Available from: http://www.springerlink.com/content/0944-1344. doi:10.1007/s11356-017-9790-y.

Campbell, N., D'Arcy, B., Frost, A., Novotny, V., Sansom, A., 2005. Diffuse Pollution: An Introduction to the Problems and Solutions. IWA Publishing, pp. 1–900222.

Chen, S., Geng, P., Xiao, Y., Meiying, H., 2012. Bioremediation of β-cypermethrin and 3-phenoxybenzaldehyde contaminated soils using Streptomyces aureus HP-S-01. Appl. Microbiol. Biotechnol. 0175-7598. 94 (2), 505–515. doi:10.1007/s00253-011-3640-5.

Chettri, D., Verma, A.K., Verma, A.K., 2024. Bioaugmentation: an approach to biological treatment of pollutants. Biodegradation. 15729729. 35 (2), 117–135. Available from: https://www.springer.com/journal/10532. doi:10.1007/s10532-023-10050-5.

Chowdhury, A., Pradhan, S., Saha, M., Sanyal, N., 2008. Impact of pesticides on soil microbiological parameters and possible bioremediation strategies. Indian J. Microbiol. 0046-8991. 48 (1), 114–127. doi:10.1007/s12088-008-0011-8.

Clark, D.P., Martinko, J.M., Madigan, M.T., Dunlap, P.V., 2009. Biología de los microorganismos. Pearson.

Coscollà, C., Yusà, V., 2016. Pesticides and Agricultural Air Quality. Compr. Anal. Chem. 0166526X. 73, 423–490. Available from: http://www.elsevier.com/wps/find/bookseriesdescription.cws_home/BS_CAC/description#description. doi:10.1016/bs.coac.2016.04.012.

Crini, G., Lichtfouse, E., 2019. Advantages and disadvantages of techniques used for wastewater treatment. Environ. Chem. Lett. 16103661. 17 (1), 145–155. Available from: https://www.springer.com/journal/10311. doi:10.1007/s10311-018-0785-9.

Cycoń, M., Żmijowska, A., Piotrowska-Seget, Z., 2014. Enhancement of deltamethrin degradation by soil bioaugmentation with two different strains of Serratia marcescens. Int. J. Environ. Sci. Technol. 1735-1472. 11 (5), 1305–1316. doi:10.1007/s13762-013-0322-0.

D'Almeida, A.P., de Azevedo, D.C.S., Melo, V.M.M., de Albuquerque, T.L., Rocha, M.V.P., 2024. Bioemulsifier production by Acinetobacter venetianus AMO1502: potential for bioremediation and environmentally friendly applications. Mar. Pollut. Bull. 18793363. 203. Available from: https://www.sciencedirect.com/science/journal/0025326X. doi:10.1016/j.marpolbul.2024.116436.

Elhamalawy, O., Bakr, A., Eissa, F., 2024. Impact of pesticides on non-target invertebrates in agricultural ecosystems. Pestic. Biochem. Physiol. 10959939. 202. Available from: https://www.sciencedirect.com/science/journal/00483575. doi:10.1016/j.pestbp.2024.105974.

Espinosa-Ortiz, E.J., Rene, E.R., Gerlach, R., 2022. Potential use of fungal-bacterial co-cultures for the removal of organic pollutants. Crit. Rev. Biotechnol. 15497801. 42 (3), 361–383. Available from: http://www.tandfonline.com/loi/ibty20. doi:10.1080/07388551.2021.1940831.

European Commission, 2021. Proposal for a regulation of the European Parliament and of the council on the sustainable use of plant protection products and amending Regulation (EU). COM.

FAO, Q&A on Pests and Pesticide Management. (2021).

Fuentes, M.S., Álvarez, A., Cuozzo, S.A., Benimeli, C.S., 2023. Combination of slurry-bioreactors and actinobacteria consortia as strategy to bioremediate chlordane-contaminated soils. Chemosphere. 00456535. 337, 139270. doi:10.1016/j.chemosphere.2023.139270.

Fuentes, M.S., Alvarez, A., Saez, J.M., Benimeli, C.S., Amoroso, M.J., 2014. Methoxychlor bioremediation by defined consortium of environmental Streptomyces strains. Int. J. Environ. Sci. Technol. 1735-1472. 11 (4), 1147–1156. doi:10.1007/s13762-013-0314-0.

Fuentes, M.S., Benimeli, C.S., Cuozzo, S.A., Amoroso, M.J., 2010. Isolation of pesticide-degrading actinomycetes from a contaminated site: bacterial growth, removal and dechlorination of organochlorine pesticides. Int. Biodeterior. Biodegrad. 09648305. 64 (6), 434–441. doi:10.1016/j.ibiod.2010.05.001.

Fuentes, M.S., Briceño, G.E., Saez, J.M., Benimeli, C.S., Diez, M.C., Amoroso, M.J., 2013. Enhanced removal of a pesticides mixture by single cultures and consortia of free and immobilized streptomyces strains. Biomed. Res. Int. 2013. 10.1155/2013/392573

Fuentes, M.S., Colin, V.L., Amoroso, M.J., Benimeli, C.S., 2016. Selection of an actinobacteria mixed culture for chlordane remediation. Pesticide effects on microbial morphology and bioemulsifier production. J. Basic Microbiol. 15214028. 56 (2), 127–137. Available from: http://onlinelibrary.wiley.com/journal/10.1002/(ISSN)1521-4028. doi:10.1002/jobm.201500514.

Fuentes, M.S., Raimondo, E.E., Amoroso, M.J., Benimeli, C.S., 2017. Removal of a mixture of pesticides by a Streptomyces consortium: influence of different soil systems. Chemosphere. 18791298. 173, 359–367. Available from: www.elsevier.com/locate/chemosphere. doi:10.1016/j.chemosphere.2017.01.044.

Fuentes, M.S., Sáez, J.M., Benimeli, C.S., Amoroso, M.J., 2011. Lindane biodegradation by defined consortia of indigenous Streptomyces strains. Water, Air, and Soil Pollution. 15732932. 222 (1-4), 217–231. doi:10.1007/s11270-011-0818-5.

Galon, L., Bragagnolo, L., Korf, E.P., Dos Santos, J.B., Barroso, G.M., Ribeiro, V.H.V., 2021. Mobility and environmental monitoring of pesticides in the atmosphere — a review. Environ. Sci. Pollut. Res. 16147499. 28 (25), 32236–32255. Available from: https://link.springer.com/journal/11356. doi:10.1007/s11356-021-14258-x.

Ghosh, S., Chowdhury, R., Bhattacharya, P., 2016. Mixed consortia in bioprocesses: role of microbial interactions. Appl. Microbiol. Biotechnol. 0175-7598. 100 (10), 4283–4295. doi:10.1007/s00253-016-7448-1.

Gil, Y., Sinfort, C., 2005. Emission of pesticides to the air during sprayer application: a bibliographic review. Atmos. Environ. 13522310. 39 (28), 5183–5193. doi:10.1016/j.atmosenv.2005.05.019.

Hassaan, M.A., El Nemr, A., 2020. Pesticides pollution: classifications, human health impact, extraction and treatment techniques. Egypt. J. Aquat. Res. 20903278. 46 (3), 207–220. Available from: http://www.elsevier.com/journals/egyptian-journal-of-aquatic-research/1687-4285. doi:10.1016/j.ejar.2020.08.007.

Hegazy, G.E., Soliman, N.A., Farag, S., El-Helow, E.R., Hassan, H.Y., Abdel-Fattah, Y.R., 2024. Isolation and characterization of Candida tropicalis B: a promising yeast strain for biodegradation of petroleum oil in marine environments. Microb. Cell Fact. 14752859. 23 (1). Available from: https://microbialcellfactories.biomedcentral.com/. doi:10.1186/s12934-023-02292-y.

Herzfeld, D., Sargent, K., 2011. Private Pesticide Applicator Safety Education Manual. Protecting the Environment, ed. 19. University of Minessota Extension.

Ho, K.L., Chen, Y.Y., Lee, D.J., 2010. Functional consortia for cresol-degrading activated sludges: toxicity-to-extinction approach. Bioresour. Technol. 09608524. 101 (23), 9000–9005. doi:10.1016/j.biortech.2010.06.148.

Horváthová, H., Lászlová, K., Dercová, K., 2018. Bioremediation of PCB-contaminated shallow river sediments: the efficacy of biodegradation using individual bacterial strains and their consortia. Chemosphere. 00456535. 193, 270–277. doi:10.1016/j.chemosphere.2017.11.012.

ITPS FAO, 2017. Global Assessment of the Impact of Plant Protection Products on Soil Functions and Soil Ecosystems. FAO.

Jia, X., Liu, C., Song, H., Ding, M., Jin, D., Qian, M., Yuan, Y., 2016. Design, analysis and application of synthetic microbial consortia. Synth. Syst. Biotechnol. 2405805X. 1 (2), 109–117. doi:10.1016/j.synbio.2016.02.001.

Karnwal, A., Martolia, S., Dohroo, A., Al-Tawaha, A.R.M.S., Malik, T., 2024. Exploring bioremediation strategies for heavy metals and POPs pollution: the role of microbes, plants, and nanotechnology. Front. Environ. Sci. 2296665X. 12. Available from: https://journal.frontiersin.org/journal/environmental-science. doi:10.3389/fenvs.2024.1397850.

Karpouzas, D.G., Walker, A., 2000. Factors influencing the ability of Pseudomonas putida epI to degrade ethoprophos in soil. Soil Biol. Biochem. 00380717. 32 (11-12), 1753–1762. doi:10.1016/S0038-0717(00)00093-6.

Kato, S., Haruta, S., Cui, Z.J., Ishii, M., Igarashi, Y., 2005. Stable coexistence of five bacterial strains as a cellulose-degrading community. Appl. Environ. Microbiol. 00992240. 71 (11), 7099–7106. doi:10.1128/AEM.71.11.7099-7106.2005.

Kidd, P., Barceló, J., Bernal, M.P., Navari-Izzo, F., Poschenrieder, C., Shilev, S., Clemente, R., Monterroso, C., 2009. Trace element behaviour at the root-soil interface: implications in phytoremediation. Environ. Exp. Bot. 00988472. 67 (1), 243–259. doi:10.1016/j.envexpbot.2009.06.013.

Kour, D., Khan, S.S., Kour, H., Kaur, T., Devi, R., Rai, P.K., Judy, C., McQuestion, C., Bianchi, A., Spells, S., Mohan, R., Rai, A.K., Yadav, A.N., 2022. Microbe-mediated bioremediation: current research and future challenges. J. Appl. Biol. Biotechnol. 2347212X. 10, 6–24. Available from: https://jabonline.in/index.php. doi:10.7324/JABB.2022.10s202.

Kunene, P.N., Mahlambi, P.N., 2020. Optimization and application of ultrasonic extraction and Soxhlet extraction followed by solid phase extraction for the determination of triazine pesticides in soil and sediment. J. Environ. Chem. Eng. 22133437. 8 (2), 103665. doi:10.1016/j.jece.2020.103665.

Lara-Moreno, A., Morillo, E., Merchán, F., Madrid, F., Villaverde, J., 2022. Bioremediation of a trifluralin contaminated soil using bioaugmentation with novel isolated bacterial strains and cyclodextrin. Sci. Total Environ. 00489697. 840, 156695. doi:10.1016/j.scitotenv.2022.156695.

Leonard, E., Nielsen, D., Solomon, K., Prather, K.J., 2008. Engineering microbes with synthetic biology frameworks. Trends Biotechnol. 01677799. 26 (12), 674–681. doi:10.1016/j.tibtech.2008.08.003.

Lewin, G.R., Carlos, C., Chevrette, M.G., Horn, H.A., McDonald, B.R., Stankey, R.J., Fox, B.G., Currie, C.R., 2016. Evolution and Ecology of Actinobacteria and Their Bioenergy Applications. Annu. Rev. Microbiol. 15453251. 70, 235–254. Available from: http://arjournals.annualreviews.org/loi/micro. doi:10.1146/annurev-micro-102215-095748.

Lopes, P.R.M., Cruz, V.H., de Menezes, A.B., Gadanhoto, B.P., Moreira, B.R.D.A., Mendes, C.R., Mazzeo, D.E.C., Dilarri, G., Montagnolli, R.N., 2022. Microbial bioremediation of pesticides in agricultural soils: an integrative review on natural attenuation, bioaugmentation and biostimulation. Rev. Environ. Sci. Biotechnol. 15729826. 21 (4), 851–876. Available from: https://www.springer.com/journal/11157. doi:10.1007/s11157-022-09637-w.

Majewski, M.S., Capel, P.D., 1996. Pesticides in the Atmosphere: Distribution, Trends, and Governing Factors, Ann Arbor Press, Inc.

Makarani, N., Kaushal, R.S., 2025. Advances in actinobacteria-based bioremediation: mechanistic insights, genetic regulation, and emerging technologies. Biodegradation. 15729729. 36 (2). Available from: https://www.springer.com/journal/10532. doi:10.1007/s10532-025-10118-4.

Massot, F., Bernard, N., Alvarez, L.M.M., Martorell, M.M., Cormack, W.P.M., Ruberto, L.A.M., 2022. Microbial associations for bioremediation. What does "microbial consortia" mean? Appl. Microbiol. Biotechnol. 14320614. 106 (7), 2283–2297. Available from: https://link.springer.de/link/service/journals/00253/index.htm. doi:10.1007/s00253-022-11864-8.

Mawang, C.I., Azman, A.S., Fuad, A.S.M., Ahamad, M., 2021. Actinobacteria: an eco-friendly and promising technology for the bioaugmentation of contaminants. Biotechnol. Rep. 2215017X. 32. Available from: http://www.journals.elsevier.com/biotechnology-reports/. doi:10.1016/j.btre.2021.e00679.

Medfu Tarekegn, M., Salilih, F.Z., Iniyehu Ishetu, A., Yildiz, F., 2020. Microbes used as a tool for bioremediation of heavy metal from the environment. Cogent Food Agric. 2331-1932. 6 (1), 1783174. doi:10.1080/23311932.2020.1783174.

Mishra, S., Pang, S., Zhang, W., Lin, Z., Bhatt, P., Chen, S., 2021. Insights into the microbial degradation and biochemical mechanisms of carbamates. Chemosphere. 00456535. 279, 130500. doi:10.1016/j.chemosphere.2021.130500.

Mokrani, S., Houali, K., Yadav, K.K., Arabi, A.I.A., Eltayeb, L.B., AwjanAlreshidi, M., Benguerba, Y., Cabral-Pinto, M.M.S., Nabti, E.H., 2024. Bioremediation techniques for soil organic pollution: mechanisms, microorganisms, and technologies - A comprehensive review. Ecol. Eng. 09258574. 207. Available from: https://www.sciencedirect.com/science/journal/09258574. doi:10.1016/j.ecoleng.2024.107338.

Mulhern, G., 2015. Soil Threats in Europe: status, Methods, Drivers, and Effects on Ecosystem Services. European Asylum Support Office. Malta.

Nayak, T., Adhya, T.K., Rakshit, M., Raina, V., 2021. Combined biostimulation and bioaugmentation for chlorpyrifos degradation in laboratory microcosms. 3 Biotech. 21905738. 11 (10). Available from: https://www.springer.com/journal/13205. doi:10.1007/s13205-021-02980-9.

Omenna, E.C., Omage, K., Ezaka, E., Azeke, M.A., 2024. Bio-augmentation and bio-stimulation with kenaf core enhanced bacterial enzyme activities during bio-degradation of petroleum hydrocarbon in polluted soil. Sci. Rep. 20452322. 14 (1). Available from: https://www.nature.com/srep/. doi:10.1038/s41598-023-50882-y.

Pan, D., Yue, X., Yaxin, N., Zhang, H., Hua, R., Xiangwei, W., 2022. The efficient persistence and migration of Cupriavidus gilardii T1 contribute to the removal of MCPA in laboratory and field soils. Environ. Pollut. 02697491. 304, 119220. doi:10.1016/j.envpol.2022.119220.

Pérez, A.P., Eugenio, N.R. 2018 2018 EUR 29124 EN 10.2760/093804 Publications Office of the European Union Status of Local Soil Contamination in Europe: revision of the Indicator “Progress in the Management Contaminated Sites in Europe”.

Prada-Vásquez, M.A., Cardona-Gallo, S.A., Loaiza-Usuga, J.C., 2017. Anaerobic biodegradation of DDT in contaminated soil by biostimulation: laboratory and pilot-scale studies. AUC GEOGRAPHICA. 2336-1980. 52 (2), 199–207. doi:10.14712/23361980.2017.16.

Priya, A.K., Muruganandam, M., Anuj Kumar, N.S., Shkir, M., Pandit, B., Imran, M., Prakash, C., Ubaidullah, M., 2024. Recent advances in microbial-assisted degradation and remediation of xenobiotic contaminants; challenges and future prospects. J. Water Process Eng. 22147144. 60, 105106. doi:10.1016/j.jwpe.2024.105106.

Qiu, Y., Pang, H., Zhou, Z., Zhang, P., Feng, Y., Sheng, G.D., 2009. Competitive biodegradation of dichlobenil and atrazine coexisting in soil amended with a char and citrate. Environ. Pollut. 02697491. 157 (11), 2964–2969. doi:10.1016/j.envpol.2009.06.003.

Raffa, C.M., Chiampo, F., 2021. Bioremediation of agricultural soils polluted with pesticides: a review. Bioengineering. 23065354. 8 (7). Available from: https://www.mdpi.com/2306-5354/8/7/92/pdf. doi:10.3390/bioengineering8070092.

Raimondo, E.E., Aparicio, J.D., Bigliardo, A.L., Fuentes, M.S., Benimeli, C.S., 2020. Enhanced bioremediation of lindane-contaminated soils through microbial bioaugmentation assisted by biostimulation with sugarcane filter cake. Ecotoxicol. Environ. Saf. 10902414. 190. Available from: http://www.elsevier.com/inca/publications/store/6/2/2/8/1/9/index.htt. doi:10.1016/j.ecoenv.2019.110143.

Raimondo, E.E., Aparicio, J.D., Briceño, G.E., Fuentes, M.S., Benimeli, C.S., 2019. Lindane Bioremediation in Soils of Different Textural Classes by an Actinobacteria Consortium. J. Soil Sci. Plant Nutr. 07189516. 19 (1), 29–41. Available from: https://www.springer.com/life+sciences/plant+sciences/journal/42729. doi:10.1007/s42729-018-0003-7.

Raimondo, E.E., Saez, J.M., Aparicio, J.D., Fuentes, M.S., Benimeli, C.S., 2020a. Bioremediation of lindane-contaminated soils by combining of bioaugmentation and biostimulation: effective scaling-up from microcosms to mesocosms. J. Environ. Manage. 10958630. 276. Available from: https://www.sciencedirect.com/journal/journal-of-environmental-management. doi:10.1016/j.jenvman.2020.111309.

Raimondo, E.E., Saez, J.M., Aparicio, J.D., Fuentes, M.S., Benimeli, C.S., 2020b. Coupling of bioaugmentation and biostimulation to improve lindane removal from different soil types. Chemosphere. 18791298. 238. Available from: www.elsevier.com/locate/chemosphere. doi:10.1016/j.chemosphere.2019.124512.

Rao, M.P.N., Lohmaneeratana, K., Bunyoo, C., Thamchaipenet, A., 2022. Actinobacteria–Plant Interactions in Alleviating Abiotic Stress. Plants. 22237747. 11 (21). Available from: www.mdpi.com/journal/plants. doi:10.3390/plants11212976.

Ravi, R.K., Gupta, S., 2022. Bioremediation of Pesticides Using Actinobacteria. Informa UK Limited, pp. 273–301. doi:10.1201/9781003281177-15.

Robles-González, I.V., Fava, F., Poggi-Varaldo, H.M., 2008. A review on slurry bioreactors for bioremediation of soils and sediments. Microb. Cell Fact. 14752859. 7. doi:10.1186/1475-2859-7-5.

Romano-Armada, N., Yañez-yazlle, M.F., Irazusta, V.P., Rajal, V.B., Moraga, N.B., 2020. Potential of bioremediation and PGP traits in streptomyces as strategies for bio-reclamation of salt-affected soils for agriculture. Pathogens. 20760817. 9 (2). Available from: https://www.mdpi.com/2076-0817/9/2/117/pdf. doi:10.3390/pathogens9020117.

Romantschuk, M., Lahti-Leikas, K., Kontro, M., Galitskaya, P., Talvenmäki, H., Simpanen, S., Allen, J.A., Sinkkonen, A., 2023. Bioremediation of contaminated soil and groundwater by in situ biostimulation. Front. Microbiol. 1664302X. 14. Available from: https://www.frontiersin.org/journals/microbiology#. doi:10.3389/fmicb.2023.1258148.

Saez, J.M., Álvarez, A., Benimeli, C.S., Amoroso, M.J., 2014. Enhanced lindane removal from soil slurry by immobilized Streptomyces consortium. Int. Biodeterior. Biodegrad. 09648305. 93, 63–69. Available from: www.elsevier.com/inca/publications/store/4/0/5/8/9/9. doi:10.1016/j.ibiod.2014.05.013.

Saez, J.M., Benimeli, C.S., Amoroso, M.J., 2012. Lindane removal by pure and mixed cultures of immobilized actinobacteria. Chemosphere. 18791298. 89 (8), 982–987. Available from: www.elsevier.com/locate/chemosphere. doi:10.1016/j.chemosphere.2012.06.057.

Saez, J.M., Bigliardo, A.L., Raimondo, E.E., Briceño, G.E., Polti, M.A., Benimeli, C.S., 2018. Lindane dissipation in a biomixture: effect of soil properties and bioaugmentation. Ecotoxicol. Environ. Saf. 10902414. 156, 97–105. Available from: http://www.elsevier.com/inca/publications/store/6/2/2/8/1/9/index.htt. doi:10.1016/j.ecoenv.2018.03.011.

Saez, J.M., Raimondo, E.E., Costa-Gutierrez, S.B., Aparicio, J.D., Angelucci, D.M., Donati, E., Polti, M.A., Tomei, M.C., Benimeli, C.S., 2025. Enhancing environmental decontamination and sustainable production through synergistic and complementary interactions of actinobacteria and fungi. Heliyon. 24058440. 11 (3). Available from: https://www.sciencedirect.com/science/journal/24058440. doi:10.1016/j.heliyon.2025.e42135.

Schmitz, A., Badgujar, C., Mansur, H., Flippo, D., McCornack, B., Sharda, A., 2022. Design of a Reconfigurable Crop Scouting Vehicle for Row Crop Navigation: a Proof-of-Concept Study. Sensors. 1424-8220. 22 (16), 6203. doi:10.3390/s22166203.

Shekhar, C., Khosya, R., Thakur, K., Mahajan, D., Kumar, R., Kumar, S., Sharma, A.K., 2024. A systematic review of pesticide exposure, associated risks, and long-term human health impacts. Toxicol. Rep. 22147500. 13, 101840. doi:10.1016/j.toxrep.2024.101840.

Silva, V., Mol, H.G.J., Zomer, P., Tienstra, M., Ritsema, C.J., Geissen, V., 2019. Pesticide residues in European agricultural soils – a hidden reality unfolded. Sci. Total Environ. 18791026. 653, 1532–1545. Available from: www.elsevier.com/locate/scitotenv. doi:10.1016/j.scitotenv.2018.10.441.

Silva, V.P., Moreira-Santos, M., Mateus, C., Teixeira, T., Ribeiro, R., Viegas, C.A., Arora, P.K., 2015. Evaluation of Arthrobacter aurescens Strain TC1 as Bioaugmentation Bacterium in Soils Contaminated with the Herbicidal Substance Terbuthylazine. PLoS ONE. 1932-6203. 10 (12), e0144978. doi:10.1371/journal.pone.0144978.

Singh, B.K., Walker, A., Wright, D.J., 2006. Bioremedial potential of fenamiphos and chlorpyrifos degrading isolates: influence of different environmental conditions. Soil Biol. Biochem. 00380717. 38 (9), 2682–2693. Available from: www.elsevier.com/inca/publications/store/3/3/2. doi:10.1016/j.soilbio.2006.04.019.

Sivasubramaniam, D., Franks, A.E., 2016. Bioengineering microbial communities: their potential to help, hinder and disgust. Bioengineered. 21655987. 7 (3), 137–144. Available from: http://www.tandfonline.com/loi/kbie20. doi:10.1080/21655979.2016.1187346.

Srivastava Anamika Jangid Nirmala Kumari Srivastava Manish Rawat Varun 2019 Pesticides as Water Pollutants Handbook of Research on The Adverse Effects of Pesticide Pollution in Aquatic Ecosystems Khursheed Mamta IGI Global doi:10.4018/978-1-5225-6111-8.ch001 English

Srivastava, J., Naraian, R., Kalra, S.J.S., Chandra, H., 2014. Advances in microbial bioremediation and the factors influencing the process. Int. J. Environ. Sci. Technol. 1735-1472. 11 (6), 1787–1800. doi:10.1007/s13762-013-0412-z.

Tzanetou, E.N., Karasali, H., 2022. A Comprehensive Review of Organochlorine Pesticide Monitoring in Agricultural Soils: the Silent Threat of a Conventional Agricultural Past. Agriculture. 12 (5). doi:10.3390/agriculture12050728.

Van Dijk, H.F.G., Guicherit, R., 1999. Atmospheric dispersion of current-use pesticides: a review of the evidence from monitoring studies. Water, Air, and Soil Pollution 00496979. 115, 21–70. http://www.kluweronline.com/issn/0049-6979/. doi:10.1023/A:1005293020536.

Venosa, A.D., Zhu, X., 2003. Biodegradation of crude oil contaminating marine shorelines and freshwater wetlands. Spill Sci. Technol. Bull. 13532561. 8 (2), 163–178. Available from: http://www.elsevier.com/inca/publications/store/3/0/3/0/6/index.htt. doi:10.1016/S1353-2561(03)00019-7.

Wang, Z., Walker, G.W., Muir, D.C.G., Nagatani-Yoshida, K., 2020. Toward a Global Understanding of Chemical Pollution: a First Comprehensive Analysis of National and Regional Chemical Inventories. Environ. Sci. Technol. 15205851. 54 (5), 2575–2584. Available from: http://pubs.acs.org/journal/esthag. doi:10.1021/acs.est.9b06379.

Wintermute, E.H., Silver, P.A., 2010. Dynamics in the mixed microbial concourse. Genes Dev. 15495477. 24 (23), 2603–2614. Available from: http://genesdev.cshlp.org/content/24/23/2603.full.pdf+html. doi:10.1101/gad.1985210.

Yadav, I., Devi, N., 2017. Pesticides Classification and Its Impact on Human and Environment. Environ. Sci. Eng. 6 (7), 140–158.

Zhou, J.J., Shen, J.T., Jiang, L.L., Sun, Y.Q., Mu, Y., Xiu, Z.L., 2017. Selection and characterization of an anaerobic microbial consortium with high adaptation to crude glycerol for 1,3-propanediol production. Appl. Microbiol. Biotechnol. 14320614. 101 (15), 5985–5996. Available from: https://link.springer.de/link/service/journals/00253/index.htm. doi:10.1007/s00253-017-8311-8.

CHAPTER

# Hybrid microbial–physical processes for enhanced contaminant removal in water treatment

3

**Rodrigo Gutiérrez-Ramírez[a], Samira El Omari[b,c], Imane Haydari[d], Zineb Kassab[e], Mohamed Laabd[b], Karim Benhabib[c], and Youness Abdellaoui[a,f]**

[a]*Sustainability of Natural Resources and Energy, Environmental Functional Materials, Cinvestav Saltillo, Saltillo, Mexico*

[b]*Laboratory of Materials and Environment, Department of Chemistry, Faculty of Science, Ibn Zohr University, Agadir, Morocco*

[c]*Eco-Processes, Optimization and Decision Support (EPROAD, EA), University of Picardie Jules Verne, Saint-Quentin, France*

[d]*Materials Science, Energy, and Nano-Engineering (MSN), Mohammed VI Polytechnic University (UM6P), Benguerir, Morocco*

[e]*Department of Materials Science, Energy, and Nano-engineering (MSN), College of Chemical Sciences and Engineering (CCSE), Mohammed VI Polytechnic University, Benguerir, Morocco*

[f]*Sustainability of Natural Resources and Energy, SECIHTI-Cinvestav Saltillo, Saltillo, Mexico*

## 1 Introduction

The drastic industrial growth and demographic expansion, coupled with climate change, are the main factors behind the imminent scarcity of freshwater resources. Addressing the increasing water demand necessitates implementing an eco-friendlier strategy for sustainably managing water resources. The situation becomes even more critical when wastewater contaminates groundwater or freshwater sources (Ortiz-Prado et al., 2022). According to the UNESCO report published in 2023, approximately 26% of the global population lacks access to safe drinking water. This crisis is expected to intensify soon due to the continuous discharge of hazardous pollutants and micropollutants into aquatic ecosystems. It is estimated that approximately 80% of the contaminated effluents globally reenter the natural environment without being treated (Dhamorikar et al., 2024). The discharge of untreated wastewater has a detrimental effect on the environment by causing severe harm to aquatic organisms, polluting water supplies, and emitting harmful substances into the atmosphere. Moreover, wastewater can cause acute and long-term threats to public health, contributing to the deaths of millions of people each year from waterborne illnesses. The consumption of contaminated drinking water causes about 80% of diseases and

Microbial Technologies for Sustainable Waste Management. https://doi.org/10.1016/B978-0-443-44093-9.00003-7

nearly half of pediatric deaths worldwide. To sum up, ensuring life's continuity on Earth requires the urgent development of effective and sustainable water treatment technologies (Ren et al., 2021).

Effluents discharged from several industrial sectors (e.g., petrochemical, textile, leather, pharmaceutical, pulp, plastics, fertilizers, etc.) contain a large variety of harmful chemical compounds (Hasan et al., 2025). Many of these compounds have complex and stable molecular structures, making them hardly or nonbiodegradable (Hasan et al., 2025). Thus conventional microbial bioremediation (e.g., activated sludge) is insufficiently effective in treating these contaminants, which has resulted in the development of a more promising alternative by combining physical wastewater remediation processes with microbial ones (Ensano et al., 2016; Ezugbe and Rathilal, 2020). Numerous advanced wastewater purification methods (e.g., adsorption, ion exchange, flocculation–coagulation, membrane filtration, and catalysis) are currently employed for treating such contaminants (Tran et al., 2023). Excluding advanced oxidation and microbial processes, the other wastewater processes simply concentrate and transfer pollution from one phase to another, thereby contributing to secondary environmental burdens (Jabbar and Esmail Ebrahim, 2022). In contrast, advanced oxidation and microbial treatments, which involve degradation or complete mineralization of contaminants, are more environmentally sustainable and effective in treating wastewater. However, their widespread application is limited by high fabrication and maintenance costs, making them less economically viable.

The aim of any sustainable water treatment approach is the complete degradation of pollutants, consequently avoiding the release of poisonous by-products (El Omari et al., 2025). The integration/hybridization of water remediation processes is a novel and powerful strategy for achieving this objective (Bhanot et al., 2020). This strategy synergistically combines two or more water treatment processes to build a reliable and sophisticated water treatment system for achieving superior treatment performance and minimizing process leakages (Vimonses et al., 2010). The hybridization approach is based on the sequential arrangement of several individual water treatment processes. In the resulting hybrid systems, the contaminated effluent passes through each treatment unit, where each stage contributes separately to the whole decontamination process. The effluent obtained from the last unit is considered the treated water. On the contrary, the integration approach is a combination of several individual treatment processes into a unique package with a smaller footprint. In this approach, the multiple treatment units operate simultaneously in a merged package without distinctions between individual units (Madaeni and Salehi, 2016).

Recently, the hybrid/integrated systems that combine microbial and physical processes have garnered prominent research interest because of their excellent water treatment performance, which results from the synergistic interplay of individual processes (El Omari et al., 2025). Based on the literature survey, the hybrid/integrated microbial–physical systems can remove a broader spectrum of water contaminants more efficiently than individual processes (Ensano et al., 2016; Ezugbe and Rathilal, 2020; Pervez et al., 2020). The current chapter provides an overview of the fundamental working principles and recent progress in hybrid microbial–physical processes for wastewater remediation. Furthermore, the underlying mechanisms, relevant case studies, merits, and shortcomings of different hybrid microbial–physical systems are outlined. Finally, the chapter discusses the potential challenges and future research prospects for improving the effectiveness and practical implementation of the hybrid microbial–physical processes.

## 2 Hybrid microbial systems in waste management

### 2.1 Microbial electrochemical systems

Microbial electrochemical systems (MESs) represent a rapidly advancing technology that integrates microbial systems, electrochemistry, and materials science to develop applications in energy production, environmental remediation, and sensing devices (Hassan et al., 2021). MESs harness the biocatalytic activity of living microbes to extract electrons from biodegradable organic matter, facilitating interactions between microbial cells (acting as electron donors) and electrode surfaces (serving as electron acceptors). A key challenge in MES development is ensuring efficient integration and enhancing electron transfer between microorganisms—capable of donating or accepting electrons—and the electrode surface, which is essential for optimizing system performance (Zhao et al., 2021).

#### *2.1.1 Principles and mechanisms of MESs*

MESs harness the metabolic activity of electroactive microorganisms to drive oxidation–reduction reactions at electrode surfaces. These systems offer a sustainable approach for contaminant degradation, resource recovery, and energy production in water treatment. The efficiency of MES depends on three key principles: electron transfer mechanisms, redox reactions, and resource recovery (Logan et al., 2019). Electroactive microorganisms, such as *Geobacter* and *Shewanella* species, facilitate electron transfer between organic or inorganic contaminants and the electrode through two primary mechanisms: direct and mediated electron transfer (MET). In direct electron transfer, bacteria utilize membrane-bound cytochromes, MESs, conductive pili (nanowires), or redox proteins to establish direct contact with the electrode, ensuring efficient electron transfer. In contrast, MET involves the production or utilization of soluble electron shuttles, such as flavins, quinones, and humic substances, which transport electrons between microbial cells and electrodes, enhancing charge transfer over longer distances. These mechanisms play a crucial role in the efficiency of MES for contaminant degradation and energy recovery (Hassan et al., 2021).

MES facilitate contaminant degradation through electrochemical redox processes occurring at both the anode and cathode. In anodic reactions (oxidation), microorganisms break down organic pollutants such as phenols, hydrocarbons, and wastewater organics, transferring electrons to the anode. This process not only degrades contaminants but also leads to the formation of $CO_2$ or intermediate metabolites. Conversely, in cathodic reactions (reduction), the electrons flow to the cathode, where they drive reduction processes such as metal ion precipitation (e.g., $Cr^{6+}$ to $Cr^{3+}$), hydrogen gas production via proton reduction, or nitrate reduction to nitrogen gas. These redox reactions play a crucial role in contaminant removal, resource recovery, and sustainable water treatment (Hernandez and Osma, 2020). MESs not only degrade pollutants but also serve as efficient resource recovery systems. They can generate bioelectricity by functioning as microbial fuel cells, where electrons flow through an external circuit to produce electrical energy. In addition, in microbial electrolysis cells, cathodic reactions drive hydrogen gas production, providing a renewable energy source. MESs also play a role in metal recovery and desalination, facilitating the extraction of valuable metals from wastewater while simultaneously removing ionic contaminants through electrodialysis-like processes. These multifunctional capabilities make MES a promising technology for sustainable water treatment and resource recovery (Kumar et al., 2021).

### 2.1.2 Construction and configuration of MESs

The construction and configuration of MESs are critical for their efficiency in contaminant removal and resource recovery. A typical MES consists of key components, including an anode, where electroactive bacteria form a biofilm and facilitate oxidation reactions; a cathode, where reduction processes occur; and, in some designs, a proton exchange membrane to separate the two electrodes while allowing ion transport (Katuri et al., 2020). Depending on the application, MES can be configured into different designs. In single-chamber MESs, both the anode and cathode are placed within the same reactor, enabling a more compact and cost-effective design, though it may lead to oxygen diffusion issues affecting microbial activity. A dual-chamber MES features separate compartments for anodic oxidation and cathodic reduction, improving system efficiency and selectivity by preventing unwanted interactions between oxidation and reduction processes (Li et al., 2020). In addition, hybrid MESs integrate physical or chemical treatment processes, such as adsorption, advanced oxidation, or membrane filtration, to enhance contaminant removal efficiency and broaden the applicability of MESs in complex wastewater treatment scenarios. The choice of configuration depends on factors such as pollutant type, desired end products, and operational feasibility, highlighting the flexibility of MES technology in environmental applications (Hassan et al., 2021).

### 2.1.3 Applications of MESs for waste management

MESs offer innovative solutions for water treatment by coupling microbial activity with electrochemical reactions. They effectively degrade organic pollutants commonly found in industrial wastewater, breaking down complex compounds into simpler intermediates or carbon dioxide (Yi et al., 2022). These systems effectively degrade organic contaminants by leveraging electroactive microorganisms that break down complex pollutants, such as hydrocarbons, pesticides, and phenolic compounds, into simpler molecules or carbon dioxide ($CO_2$), significantly reducing pollutant loads before discharge or reuse (Karbelkar et al., 2021). In addition, MESs play a crucial role in the removal of heavy metals and inorganic pollutants through reduction reactions at the cathode, facilitating the precipitation and recovery of toxic metals like lead ($Pb^{2+}$), cadmium ($Cd^{2+}$), and hexavalent chromium ($Cr^{6+}$), thereby minimizing environmental impact and enabling metal valorization (Kumar et al., 2021). Their electrochemical capabilities extend to disinfection and pathogen removal, where MES generate powerful disinfectants such as hydrogen peroxide ($H_2O_2$) and hydroxyl radicals ($\cdot OH$), making them particularly useful for treating drinking water and ensuring safe wastewater reuse. Moreover, MES can be integrated with electrodialysis systems to achieve desalination, efficiently reducing water salinity by transporting ions across ion exchange membranes, providing an energy-efficient alternative to traditional desalination techniques like reverse osmosis (Ramírez-Moreno et al., 2021; Ren et al., 2021). Beyond pollution control, MES contribute to bioelectricity production by functioning as microbial fuel cells, converting the chemical energy of pollutants into electrical energy that can power low-energy devices such as water quality sensors, promoting energy self-sufficiency in treatment facilities. Furthermore, these systems facilitate resource recovery by enabling the production of valuable by-products, including hydrogen or methane gas, ammonia, and phosphate-based fertilizers like struvite, transforming wastewater from a pollutant source into a resource stream (Jadhav et al., 2017). Through their ability to simultaneously treat wastewater, generate energy, and recover valuable materials, MESs provide a sustainable and eco-friendly alternative to conventional water treatment technologies, paving the way for more efficient and resilient wastewater management strategies.

## 2.2 Biophotocatalysis

Biophotocatalysis represents an innovative approach that merges the advantages of biological processes with photocatalytic technology, offering an eco-friendly and efficient method for the removal of contaminants from water. This hybrid technique harnesses the power of biocatalysts, such as microorganisms, alongside photocatalysts to enhance pollutant degradation under light irradiation. Biophotocatalysis takes advantage of light energy to drive the generation of reactive oxygen species, such as hydroxyl radicals, which are highly effective in breaking down organic pollutants, including pharmaceuticals, pesticides, and industrial chemicals (Kumar et al., 2020).

### *2.2.1 Understanding the mechanism of biophotocatalysis*

The photocatalytic reaction is a multifaceted process that begins when a significant amount of visible light is absorbed by the material's surface. When the energy of the absorbed photon is equal to or greater than the semiconductor's band gap energy, electrons in the valence band are excited into the conduction band, leaving behind holes in the valence band. Taking $TiO_2$ as an example, the generation of electrons and holes can follow three potential pathways: (1) the electrons and holes separate and move to the surface, where they may participate in redox reactions; (2) they are trapped at defect sites; or (3) they recombine, releasing energy (Zhao et al., 2022). However, only the first process supports the photocatalytic reaction, driving reduction and oxidation, while the second and third processes do not contribute. To initiate the redox reaction, charges must undergo separation, thermalization, trapping, recombination, and transport. Volatile organic compounds are oxidized or generate hydroxyl radicals and superoxide anions, as shown in Eqs. (1)–(8) (Zhao et al., 2022).

$$TiO_2 + hv -> e^-(TiO_2) + h^+(TiO_2) \tag{1}$$

$$h^+(TiO_2) + H_2O -> TiO_2 + H^+OH^· \tag{2}$$

$$h^+(TiO_2) + OH^- -> TiO_2 + OH^· \tag{3}$$

$$e^-(TiO_2) + O_2 -> TiO_2 + O_2^{·-} \tag{4}$$

$$O_2^{·-} + H^+ -> H_2^· \tag{5}$$

$$HO_2^· + HO_2^· -> H_2O_2 + O_2 \tag{6}$$

$$e^-(TiO_2) + H_2O_2 -> OH^- + OH^· \tag{7}$$

$$VOC + O_2 + OH^· -> H_2O + CO_2 + \text{other products} \tag{8}$$

In a biophotocatalytic system, microorganisms assist in degrading organic contaminants using their metabolic pathways (Rajput et al., 2022). This process can be described as follows:

$$C_nH_mO_p + O_2 -> CO_2 + H_2O + \text{Energy} \tag{9}$$

The combined photocatalytic and biocatalytic processes work in synergy, enabling the efficient removal and mineralization of a wide range of organic pollutants in water treatment systems, as represented by the overall Eq. (10). This integrated system offers a promising approach for sustainable and effective water purification.

$$\text{Contaminants} + TiO_2 + hv + \text{Microorganisms} -> CO_2 + H_2O + \text{Nontoxic products} \tag{10}$$

### 2.2.2 Applications of biophotocatalysis in waste management

Biophotocatalysis, the integration of photocatalytic processes with biocatalysis, has gained significant attention in recent years due to its enhanced efficiency in contaminant removal from water (Baldassarre and Ciccarella, 2021). This process leverages the combined strengths of photocatalysts, such as titanium dioxide ($TiO_2$), and microorganisms to degrade a wide range of organic pollutants in water. The applications of biophotocatalysis in water treatment are diverse and offer a promising approach to addressing water contamination challenges (Ding et al., 2018).

One of the most prominent applications of biophotocatalysis is in the degradation of organic pollutants, such as dyes, pharmaceuticals, and industrial chemicals. The use of biophotocatalysis for the degradation of pharmaceutical compounds, such as tetracycline and acetaminophen, which are commonly found in wastewater. The researchers demonstrated that the addition of microbial biomass to a $TiO_2$ photocatalytic system significantly enhanced the degradation rate of these pollutants compared to photocatalysis alone. The microorganisms acted as secondary degraders, mineralizing intermediate by-products into nontoxic end products, further demonstrating the synergistic effect of combining photocatalysis with biocatalysis (Li et al., 2021). Biophotocatalysis has also been applied to the treatment of heavy metals and other inorganic pollutants. Investigation of the potential of biophotocatalysis for the removal of heavy metals, such as lead (Pb) and arsenic (As), from contaminated water. The addition of photocatalysts like $TiO_2$ under ultraviolet (UV) light enhanced the ability of microorganisms to adsorb and detoxify heavy metals, demonstrating the potential of this combined approach in tackling both organic and inorganic contaminants in water treatment systems (Zuo et al., 2021). Moreover, biophotocatalysis has found application in the removal of emerging contaminants, such as endocrine-disrupting chemicals. These chemicals, which include certain pesticides, plasticizers, and personal care products, are difficult to remove through conventional water treatment methods. Biophotocatalysis was shown to effectively degrade endocrine-disrupting chemicals in water. The photocatalytic process under UV light generated ROS ( $OH^{\cdot}$ and $O_2^{\cdot-}$ ) that broke down these pollutants, and the microbes then mineralized the intermediate products, leading to complete degradation (Baldassarre and Ciccarella, 2021). Overall, biophotocatalysis is a promising technology that integrates photocatalytic and biocatalytic processes for enhanced water treatment. By combining the high reactivity of photocatalysts with the ability of microorganisms to further degrade pollutants, this technology offers a sustainable and efficient solution for tackling a wide variety of contaminants in water. The continuous advancements in biophotocatalysis are expected to expand its applications in both industrial and environmental water treatment, providing a viable strategy for improving water quality and ensuring sustainable water resources (Kumar et al., 2020).

## 2.3 Bioadsorption systems

Adsorption, a process that enables the removal of soluble substances using materials with high specific surface area, offers multiple advantages for wastewater treatment, such as the ability to remove micropollutants, applicability in batch or continuous processes, low environmental impact, and low cost (Bonilla-Petriciolet et al., 2017). These advantages allow adsorption processes to be easily integrated with other wastewater treatment methods, enhancing contaminant removal efficiency. A biological–adsorption system aims to harness the synergistic effects of physical adsorption and biodegradation for enhanced contaminant removal. In addition to capturing pollutants, the adsorbent material can serve as a carrier or support matrix for microorganisms, facilitating their growth and immobilization (Ahmad et al., 2017). Adsorbents can be directly introduced into the system as suspended solids, arranged as fixed or fluidized beds, or continuously supplied in the form of dry particles or colloidal suspensions. This flexibility enhances the interaction between the adsorbent surface, active biomass, and contaminants, thereby improving both removal efficiency and system stability. Integrating adsorption with biological processes can also promote the degradation of recalcitrant compounds, protect microbial communities by reducing exposure to toxic substances, and increase the solids concentration in activated sludge systems, which ultimately contributes to reduced operational costs and more robust wastewater treatment performance (Ahmed et al., 2021).

### *2.3.1 Applications of bioadsorption systems for waste management*

García et al. (2021) highlighted the use of activated carbon as an adsorbent material, both in granular activated carbon (GAC) and powdered activated carbon (PAC) forms, within various hybrid configurations. These include: (1) powdered activated carbon membrane bioreactor (PAC-MBR) (Fig. 3.1A), (2) conventional activated sludge system with granular activated carbon (CAS-GAC) (Fig. 3.1B), (3) biofilm or biological activated carbon (BAC) (Fig. 3.1C), and (4) conventional activated sludge reactor with powdered activated carbon (PACT).

Asif et al. (2020) studied the effect of PAC concentration on the microbial community and the removal of various micropollutants. The PAC-MBR hybrid system achieved higher removal efficiency (98%) for five micropollutants (caffeine, naproxen, ofloxacin, azithromycin, and theophylline) compared to the control MBR process (60%–80%). The increased removal efficiency was associated with the growth of various microbial species (*Methylotenera, Haliangium, Comamonas*, and *Methanomethylovorans*) capable of biodegrading emerging contaminants, following enrichment with PAC (20 g $L^{-1}$). The pores of PAC acted as microenvironments that promoted the development of these slow-growing species. Campinas et al. (2022) conducted a full-scale study using PACT for the treatment of municipal wastewater containing pharmaceutical compounds. The addition of PAC (10–25 g $L^{-1}$) improved effluent quality by increasing the removal of pharmaceutical compounds such as diclofenac (69%) and carbamazepine (83%), among others, as well as reducing parameters such as total phosphorus, color, and organic matter, and improving

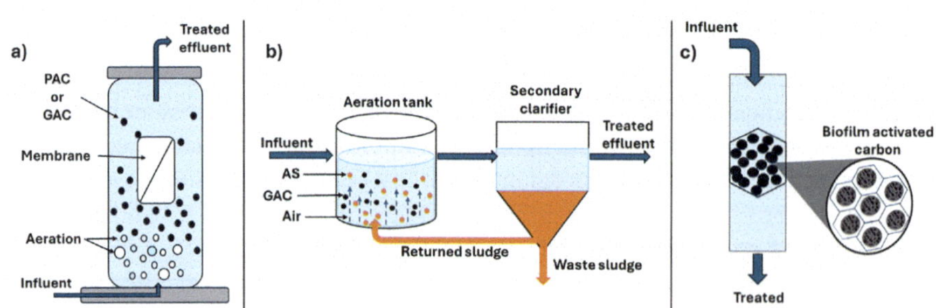


**FIG. 3.1**

Schematic illustration of (A) PAC-MBR, (B) CAS-GAC, and (C) BAC.

transmittance. They also demonstrated that PAC addition did not significantly affect nitrification processes, redox potential, or dissolved oxygen in the oxidation ditch, allowing for stable energy consumption. Alvarino et al. (2016) removed a mixture of organic micropollutants (carbamazepine, diclofenac, diazepam, erythromycin, roxithromycin, among others) using the SeMPAC system, a sequencing batch membrane reactor with direct addition of PAC. The removal of these contaminants was attributed to both adsorption on PAC and biotransformation carried out by microorganisms. In the case of carbamazepine and diazepam, PAC addition contributed to an increase in the biotransformation percentage (99%), as it served as a support for biofilm growth, resulting in sludge enriched with microorganisms effective in biotransformation. Verma et al. (2023) used a granular activated carbon sequencing batch reactor (GAC-SBR) for the treatment of landfill leachate and municipal wastewater. The GAC served both as support for biofilm growth and as an adsorbent material, showing improvements in removal efficiency compared to conventional treatment. The GAC-SBR system effectively removed 83% of chemical oxygen demand (COD) from stabilized landfill leachate after 10 days of hydraulic retention time (HRT), and the removal efficiency for turbidity, phosphate, ammonium, and nitrate exceeded 80%.

As previously discussed, adsorbent materials can serve as support for microbial growth and in the retention and degradation of contaminants through microbial activity. This latter function has been extensively studied as a strategy for the bioregeneration of the adsorbent, in which microbial metabolism facilitates the desorption and degradation of contaminants without the use of chemical compounds (El Gamal et al., 2018). This enables the reusability and extended lifespan of the adsorbent, minimizing operational costs and environmental impact.

### 2.3.2 Considerations and limitations

There are several factors to consider in the operation of hybrid adsorption–biological treatment systems. These include the adsorption capacity of the adsorbent, the HRT, as well as influent characteristics such as pH, temperature, and the presence of competing ions. The physicochemical properties of the adsorbent, such as surface area, pore size, and the presence of functional groups, play a crucial role in the removal efficiency of the target contaminant (Mailler et al., 2016). HRT is a key operational parameter that directly influences the efficiency of hybrid systems. In adsorption processes, HRT is typically short due to the rapid uptake of contaminants, whereas in biological treatments, it tends to be longer due to microbial metabolic activity. This mismatch may lead to operational desynchronization. To ensure that both processes operate compatibly, careful system design and thorough study of operational conditions (e.g., adsorption and biodegradation kinetics, flow dynamics, mass transfer, system stability) are necessary (Jiang et al., 2018). A comprehensive understanding of these factors is crucial for process optimization and large-scale application.

On the contrary, there are still compounds that cannot be removed or biodegraded due to their high chemical stability and resistance. In response, the development and application of advanced nanomaterials (e.g., functionalized and hierarchically structured nanomaterials) as adsorbents, as well as the use of genetic engineering to modify strains capable of degrading such compounds, represent promising strategies for improving the capture and removal of these persistent pollutants (García et al., 2021).

## 2.4 Biological membrane filtration

Membrane technology is based on the physical separation of suspended solids and high-molecular-weight solutes from water by filtration through semipermeable membranes. These systems are classified according to the membrane pore size into microfiltration (between 1 μm and 100 nm), ultrafiltration (UF) (<100 nm), nanofiltration (10–1 nm), and reverse osmosis (<1 nm) (Ezugbe and Rathilal, 2020). Membrane systems have been integrated with biological processes to combine their operational advantages and improve contaminant removal from water, as well as to recover valuable compounds (Racar et al., 2020). According to Ensano et al. (2016), some of the benefits of using this hybrid treatment include reduced energy consumption, lower sludge production, smaller operational volume, and decreased membrane fouling, among others.

### *2.4.1 Applications of biological membrane filtration systems for waste management*

One of the most prominent hybrid technologies for contaminant removal is the membrane bioreactor (MBR), which combines an activated sludge system with membrane filtration. This reactor enables the efficient retention of contaminants and suspended solids, followed by their degradation through microbial activity (Anyango et al., 2024). However, this technology has limitations when it comes to treating persistent contaminants. In this context, Wang et al. (2018) demonstrated that integrating a reverse osmosis (RO) or nanofiltration (NF) process downstream of an MBR can significantly enhance the removal efficiency of emerging contaminants. The proposed MBR-RO and MBR-NF treatment systems achieved removal rates above 90% for a wide range of emerging contaminants found in municipal wastewater, mainly beta-blockers, pesticides, and pharmaceuticals. Jiang et al. (2018) combined an MBR system with a moving bed biofilm reactor and studied the effect of different HRTs on the removal of 22 micropollutants. The combination of biofilm and suspended biomass increased the quantity and diversity of microorganisms, which benefited from longer HRTs (18 and 24 h) due to enhanced degradation of the present contaminants. This hybrid system showed higher efficiency in the removal of pharmaceuticals and personal care products, especially those more resistant to conventional processes, highlighting the potential of this configuration. Furthermore, the combination of technologies contributed to more stable operation and reduced membrane fouling, both key aspects for long-term process performance.

On the contrary, one limitation of MBRs is membrane fouling caused by microbial growth and biofilm formation on the membrane surface. To address this, Waqas et al. (2021) integrated an UF system after a rotating biological contactor (RBC). The

RBC-UF hybrid system achieved a removal efficiency of 87.9% for COD, 98.9% for ammonium, 45.2% for total nitrogen, and 97.9% for turbidity. The high removal performance is attributed to the RBC's promotion of greater microbial density and activity through biofilm development on the rotating discs, which reduces suspended solids and organic load in the effluent. This, in turn, increases membrane permeability and extends membrane lifespan by reducing fouling. The configuration presented by Waqas et al. (2021) offers a significant advantage over the MBR system, as it provides greater operational stability, longer HRTs, lower operational costs, and higher removal efficiency.

Another hybrid technology employed is the use of BAC filters coupled with membranes. While BAC performs adsorption and biodegradation functions through microbial growth and accumulation on its surface, membrane integration complements remediation by physically retaining suspended solids, soluble microbial products, extracellular polymeric substances, and microorganisms, thus improving effluent quality (Jin et al., 2013). The coupling of BAC serves as a pretreatment step, reducing the contaminant load and thereby lowering membrane fouling. A clear example is the study by Sbardella et al. (2018), in which they evaluated a biofilter coupled with an UF membrane for the removal of pharmaceutically active compounds. The BAC-UF system showed removal efficiencies of 78%, 89%, 83%, and 79% for antibiotics, beta-blockers, psychiatric drugs, and a mixture of other therapeutic groups, respectively. While most of the removal was due to adsorption, biodegradation processes also contributed to the overall removal. The biofilm was able to degrade 22% of ciprofloxacin, 30% of ofloxacin, and 35% of sulfamethoxazole, among others.

### 2.4.2 Considerations and limitations

Among the main considerations for the design and operation of these hybrid systems, the system configuration and the type of membrane to be used stand out. The selection of a specific configuration (Fig. 3.2) depends on several factors, such Anyango et al. (2024). On the contrary, the choice of membrane type will depend on the compounds targeted for retention or removal, as well as the characteristics and composition of the water to be treated, due to its potential impact on membrane fouling. To achieve optimal operational performance, it is essential to thoroughly study the

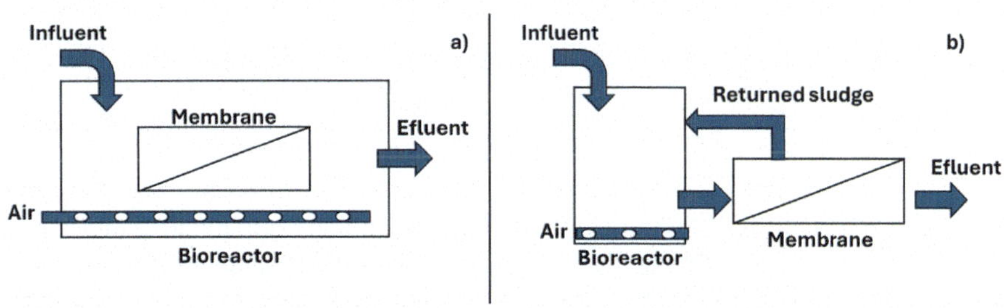


**FIG. 3.2**

Schematic illustration of (A) submerged membrane reactor and (B) external membrane reactor.

interaction between the components and properties of the water and the membrane material to ensure proper system performance and prolonged usability.

Although the described hybrid technology has proven effective in contaminant removal due to its versatility both in reactor configuration (to address membrane fouling and reduce operational costs) and in membrane substitution (to manage different compositions of permeate or bottom fluids caused by by-products of biodegradation), it still faces limitations when dealing with persistent compounds (Dhangar and Kumar, 2020). The presence of hard-to-degrade by-products generated during biological processes requires further investigation to achieve complete removal. In this context, it becomes necessary to couple advanced oxidation processes (AOPs), such as ozonation, UV photolysis, photo-Fenton, electro-Fenton, ultrasound, among others, with the biological treatments mentioned (MBR and BAC) (Alfonso-Muniozguren et al., 2021). The combination of these technologies represents a robust strategy for achieving advanced water purification, especially in scenarios requiring high effluent quality for reuse or safe environmental discharge.

## 2.5 Biological oxidation process

AOPs are a group of technologies used for the generation and application of hydroxyl radicals (·OH), which are highly reactive species capable of degrading pollutants. AOPs include the use of UV radiation, hydrogen peroxide ($H_2O_2$), ozone ($O_3$), plasma, Fenton processes, among others. Their application has gained importance in wastewater treatment, especially due to their ability to degrade and mineralize contaminants (Deng and Zhao, 2015). Given the limitations of biological treatment systems in degrading organic micropollutants owing to their recalcitrant molecular structure, the integration of AOPs with biological systems is increasingly relevant to enhance contaminant removal efficiency and broaden the range of treatable pollutants (Nidheesh et al., 2022).

This hybrid system exploits the synergy between microbial biodegradation and the oxidative capacity of reactive species, with the aim of achieving more efficient and complete elimination of contaminants in wastewater (Thakur et al., 2023). This approach is especially useful for the treatment of persistent, toxic, or recalcitrant organic compounds, which can hardly be removed by conventional methods (Babu Ponnusami et al., 2023). It is also useful for reducing organic load and COD/biochemical oxygen demand (BOD), as well as decreasing sludge production.

### 2.5.1 Applications of biological oxidation process for waste management

AOPs can be employed as a pretreatment to a biological system, where the generation of free radicals breaks down complex compounds into smaller molecules (Fig. 3.3A), which can subsequently be removed by microorganisms through various processes such as biodegradation, bioremediation, cometabolism, and bioaccumulation (Xiong et al., 2018). Ge and Deng (2015) reported that a treatment based on the Fenton process combined with microalgae achieved removal rates of 78% for enrofloxacin (2 mg $L^{-1}$) and 72% for ciprofloxacin (10 mg $L^{-1}$). These results

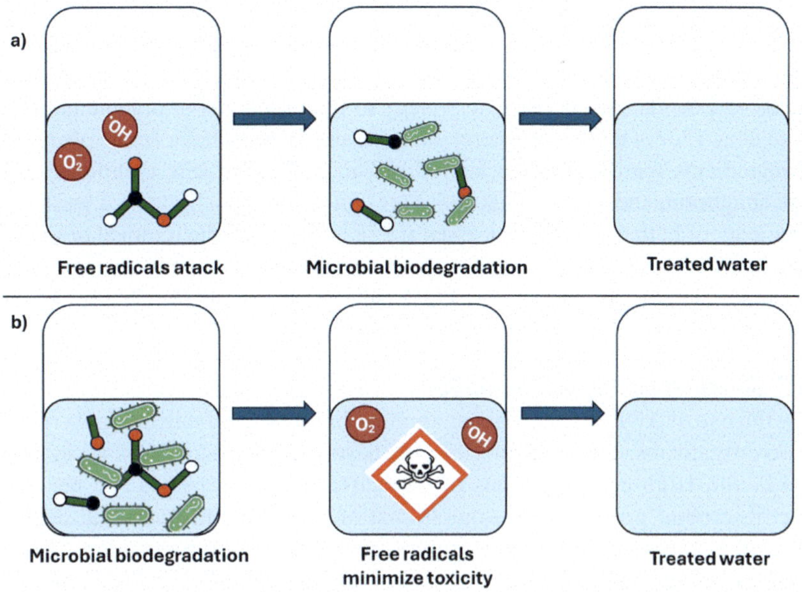


**FIG. 3.3**

Schematic illustration of (A) AOPs as pretreatment and (B) AOPs as a posttreatment to a biological system.

outperformed those obtained by other treatments, which included a control (photolysis only), $Fe^{3+}$ with photolysis, algae with photolysis, and algae in dark reaction conditions. This study highlights the synergy between $Fe^{3+}$ and microalgae for the photoinduced degradation of antibiotics in aqueous solution.

In some cases, the degradation of pollutants by AOPs can generate residues that are even more toxic than the original compound. In such scenarios, the integration of biological treatments is also useful, as microbial activity can reduce the residual toxicity and complete the removal of the generated by-products, thereby improving both the efficiency and safety of the treatment process. Du et al. (2015) achieved the degradation of cefradine using UV irradiation in a short period of time (0.237 h). However, they found it necessary to apply an algae-based treatment to reduce the toxicity of the resulting by-products. They reported that using a hybrid UV–algae treatment reduced effluent toxicity from 93% to 55%, while the combination of UV–activated sludge–algae reduced it from 40% to 23%. Both results highlight the importance of incorporating algae-based treatment to mitigate the residual toxicity produced by the UV irradiation process.

On the contrary, AOPs can also be used as a posttreatment to biological processes to improve effluent quality by reducing COD and BOD, while also minimizing toxicity and eliminating recalcitrant contaminants still present in the effluent (Fig. 3.3B). Moussavi et al. (2019) removed 93.6% of total organic carbon from hospital effluent that had already undergone biological treatment by applying $H_2O_2$ in a vacuum UV photoreactor. This treatment also enabled the degradation of most residual substances, detoxifying the effluent to a level safe for discharge into the environment.

The simultaneous application of different AOPs enhances the oxidation rates of organic compounds and serves as a strategy to treat mixtures of organic pollutants. Krakkó et al. (2021) used high-energy photons and $O_3$ to remove active pharmaceutical ingredients (carbamazepine, ciprofloxacin, clarithromycin, diclofenac, metoprolol, sitagliptin, and sulfamethoxazole) present in biologically treated wastewater. The synergy of both oxidative processes at 254 nm enabled the removal of five contaminants in the 80%–100% range and two others in the 40%–80% range, through the action of $H_2O_2$ and the generated OH radicals. After this (V)UV/$O_3$ treatment, the wastewater showed no genotoxicity or acute toxicity.

### 2.5.2 Considerations and limitations

While the use of AOPs enhances contaminant breakdown, facilitating its removal by microorganisms, the concentration of these oxidizing agents is a key operational factor. High concentrations of oxidant-generating species can negatively impact microbial growth, cause mutations, or even lead to microbial death. Li et al. (2015) reported continuous growth inhibition of the green alga *Chlorella pyrenoidosa* over a 48-h hybrid treatment when using $H_2O_2/Fe^{2+}$ at molar ratios greater than 4:1, due to the residual impact of the involved species. The radicals generated can induce oxidative stress or interfere with algal photosynthesis, thereby reducing growth. Therefore studying the biocompatibility between AOPs and the employed microorganisms is essential. Yang et al. (2017) investigated the optimal application of UV irradiation and its compatibility with different algae species (*Chlorella pyrenoidosa, Scenedesmus obliquus,* and *Selenastrum capricornutum*) for the degradation of cefradine (100 mg $L^{-1}$) and amoxicillin (100 mg $L^{-1}$). UV light at 185 nm inhibited algal growth, whereas the combination of 365 nm UV light with the green alga *S. obliquus* achieved 99.84% biodegradation of the antibiotics after 24 h of exposure. UV light at 365 nm acted as a growth trigger for the algae, which were responsible for biodegrading the antibiotics.

On the contrary, HRT is also a critical factor in process optimization, as it affects reagent consumption, microbial stability, process efficiency, and more. Su et al. (2019) reported that HRT significantly influences the removal of catechol, COD, and $NH_3$-N in a baffled anaerobic–bioelectrical–Fenton coupled reactor. Maximum catechol removal (99.07%) was achieved at 20 h, while COD removal was 77.65%, 80%, and 93.7% at 18 h, 24 h, and 5 days, respectively. These results indicate that microorganisms played a key role in contaminant degradation at longer retention times, achieving efficient removal of organic matter in hospital wastewater.

Although the implementation of AOPs has proven useful for degrading pollutants, facilitating microbial activity, reducing operational costs, and extending the lifespan of biological systems, further research is needed beyond pilot-scale studies. It is necessary to assess their performance under real-world, industrial-scale conditions, considering factors such as influent variability, contaminant degradation kinetics, energy consumption, capital expenditure, and the long-term sustainability of the process (Babu Ponnusami et al., 2023).

## 2.6 Other emerging combinations

In addition to the hybrid microbial systems mentioned in the previous sections, several other hybrid microbial processes have been implemented to increase treatment efficiency and system stability. These include hybrid ion exchange and biological processes. These systems combine ion exchange materials, such as zeolites and resins, with microbial technology to create a hybrid system capable of limiting toxicity resulting from high ammonia concentrations, enhancing performance, and removing heavy metals and organic substances, as reported by Zhang et al. (2024). The purpose of ion exchange is to protect microorganisms in the first stage before applying conventional biological treatment. The same principle was adopted by Qian et al. (2017) using ion exchange resins followed by biodegradation in a dual physical–biological hybrid system. Physical water treatment aims to retain organic matter, while a biofilm is developed on the resins to degrade the retained matter. Many key factors control the performance of these hybrid microbial processes. These include the activity of the microbial community, which means that the effective decomposition of organic substances is directly related to the health and vitality of microorganisms. In addition, it has also been found that operational parameters, such as pH, temperature, reaction time, water properties, and the availability of nutrients essential for bacterial growth, have a significant effect on microbial degradation effectiveness (Lee et al., 2023).

The hybridization of the ion exchange and microbial degradation processes not only reduces wastewater treatment costs by using abundant waste materials (instead of expensive synthetic resins) but is also an environmentally friendly solution (Fig. 3.4). The hybrid ion exchange/microbial system is also a promising sustainable option that contributes further to waste management (Jokar et al., 2019). Despite its advantages, the ion exchange/microbial system exhibits certain shortcomings. In particular, the ion exchange capacity of waste materials is limited compared to that of

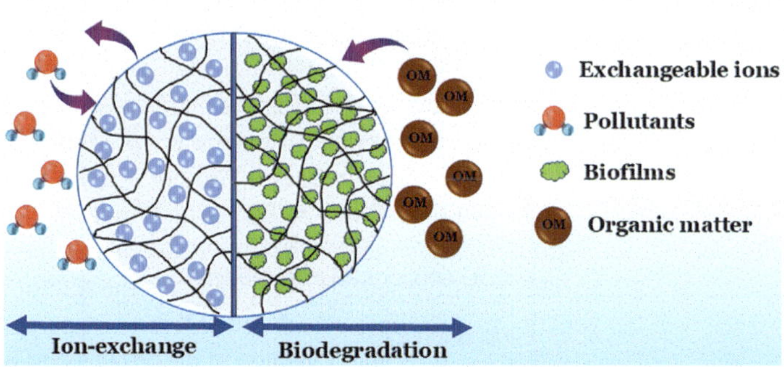


**FIG. 3.4**

Schematic illustration of the ion exchange/microbial hybrid system.

synthetic resins. Moreover, the growth of bacterial biofilm over time on the surface of the ion exchange material can lead to pore blockage, which can negatively affect the filtration process and consequently reduce water treatment performance.

It is also worth noting that bioremediation techniques are hybridized with the electrocoagulation process, which is applied as a secondary step to improve the quality of treated water. In this context, several forms of combination have been considered, such as biological treatment followed by electrocoagulation (electrocoagulation as posttreatment), electrocoagulation followed by biological treatment (electrocoagulation as pretreatment), electrocoagulation/MBR systems, electrocoagulation/biofilm systems, and electrocoagulation combined with advanced biological processes (Table 3.1). An iron or aluminum electrode is generally used during the coagulation phase to generate metal ions capable of coagulating pollutant particles and promoting the development of microorganisms. Electrocoagulation is mainly based on electric current, which enhances microorganisms' growth even at low temperatures. As a result, it can be used for water treatment even in the winter or in cold countries. This process is very interesting in terms of reduced energy consumption, ease of implementation, and efficient treatment of wastewater even under severe conditions

**Table 3.1** Other hybrid microbial–physical systems and their performance in pollutant removal.

| Hybrid system | Treatment efficiency | References |
|---|---|---|
| Biological ion exchange and biological activated carbon hybrid system | Dissolved organic carbon (DOC) removal rate: 59.9%<br>Assimilable organic carbon (AOC) removal rate: 61.2% | (Lee et al., 2023) |
| Ion exchange filter with biological activity | 60% DOC removal | (Schulz et al., 2017) |
| Bioelectrocoagulation reactor with Fe–C electrodes | Achieved the highest ammonia nitrogen removal efficiency (~98.23% at 20°C; ~ 76.26% at 10°C) | (Qian et al., 2017) |
| Electrocoagulation + microfiltration | Removal efficiencies for $NH_4$-N, $PO_4$-P, and COD were 99%, 99%, and 92%, respectively | (Ensano et al., 2019) |
| Rotating hanging media bioreactor, submerged membrane bioreactor, and electrocoagulation | Total nitrogen (T–N) removal: 100%<br>Total phosphorus (T–P) removal: 99.33%<br>Ammonium ($NH_4^+$-N) removal: 100% (completely removed)<br>COD concentration in the final effluent: 11.46 mg/L (indicating good effluent quality)<br>$BOD_5$ (biochemical oxygen demand) concentration in the final effluent: 0.26 mg/L (very low) | (Nguyen et al., 2014) |

with low carbon-to-nitrogen ratios (Qian et al., 2017). The type of microorganisms (e.g., algae, bacteria, or a mixed community), electrode type, and operating conditions are critical factors in controlling the overall performance of the hybrid system. It is, therefore, necessary to optimize this combination to achieve a good synergy between the two technologies and to maximize the removal of pollutants (Al-Qodah et al., 2020).

The application of the bioelectrocoagulation hybrid system is restricted by certain technical challenges that need to be thoroughly analyzed and evaluated. Over time, corrosion or passivation can affect electrodes, reducing the treatment performance of the electrocoagulation process. Furthermore, the additional costs of using electrochemical equipment and sludge treatment represent another financial barrier to large-scale application.

## 3 Challenges and future perspectives

Wastewater treatment is a major environmental challenge, especially in developing countries. The 2030 Agenda for Sustainable Development includes water quality among its 17 goals. That's why serious efforts are needed to confront this issue. The water situation in the world is becoming more and more serious because of population growth and industrialization. Every day, aquatic systems receive enormous amounts of hazardous compounds (e.g., heavy metals, textile dyes, organic and inorganic compounds, pharmaceuticals, pesticides, and so on). The presence of these contaminants, even at low concentrations, can have harmful effects on human health and aquatic ecosystems. The conventional treatment of these contaminants in water has become less effective. In this context, research has recently focused on the application of hybrid systems to improve wastewater treatment efficiency. Microbial-based hybrid systems commonly use microorganisms and biofilm-based processes for water treatment through the degradation of organic pollutants. They are known to be economical and highly effective for treating water (Asheghmoalla and Mehrvar, 2024). However, these systems also present several challenges that need to be overcome. First, laboratory studies often use synthetic wastewater, so the composition of the wastewater is very complicated, which can influence the results. Therefore it is imperative to evaluate these hybrid systems in the treatment of real wastewater. Moreover, the implementation of these systems on an industrial scale remains a challenge due to insufficient technical–economic analysis and limited data on their effectiveness against micropollutants, especially persistent and toxic compounds (Zakaria and Dhar, 2022). Numerous biological remediation processes may produce more toxic and nonbiodegradable by-products, which must be considered in the design of biological-based hybrid systems. In systems that use MBRs, additional maintenance costs may be incurred owing to the membrane fouling caused by the presence of microorganisms. The microorganisms in bioremediation systems are generally unstable and sensitive to changes in environmental conditions. This leads to an imbalance in the bacterial community, which affects the removal of pollutants from wastewater. Finally, combining biological processes with other advanced technologies presents technical and economic challenges. On the technical side, since the operating mechanisms are different, achieving optimum interaction is very complicated. On the contrary, the design and operation of these coupled technologies can significantly increase costs, limiting large-scale implementation (Adjei et al., 2025).

The prospects for wastewater treatment using hybrid microbial–physical systems aim to increase treatment performance and reduce operating costs. Applying artificial intelligence or machine learning is a solution to optimize water treatment performance, which improves system durability and reduces costs through automatic control of operating parameters and real-time monitoring. An in-depth study of the interactions between micropollutants and microorganisms is one of the most important aspects of improving hybrid systems to ensure smooth, homogeneous operation and thus a robust and effective treatment (Lin et al., 2013). The development of global kinetic models capable of describing the action of the entire microbial–physical hybrid system is strongly recommended for a thorough understanding of the interaction between the different processing phases, thus facilitating the transition to an industrial scale. The influence of the various operating parameters of these systems must also be assessed. This is essential because the chemical and biological composition of wastewater changes, and each type of wastewater requires specific operating conditions for optimal treatment performance.

## 4 Conclusions

Hybrid microbial–physical processes represent a strategic response to complex aquatic pollution, combining microbial metabolic versatility with physicochemical operations—electrochemical, photocatalytic, adsorptive, membrane-based, oxidative, and ion exchange/coagulation—to create multifunctional treatment systems capable of detoxifying, transforming, and valorizing resistant contaminants. The reviewed case studies demonstrate consistent synergy patterns. MESs enable simultaneous organic mineralization, metal precipitation, desalination, and energy harvesting through electron transfer; biophotocatalytic reactors use light-driven radicals to break down recalcitrant molecules for microbial processing; adsorption–biofilm hybrids preconcentrate micropollutants while supporting in situ bioregeneration; MBRs, often enhanced with activated carbon or electrochemical polishing, combine high biomass retention with comprehensive solids and pathogen removal; and emerging pairings like ion exchange/biodegradation extend treatment capabilities while reducing chemical demand. Optimally configured hybrids routinely achieve >90% removal efficiencies for pharmaceuticals, dyes, nutrients, and heavy metals while reducing sludge generation and enabling resource recovery. Successful design requires harmonizing key parameters: electron or radical flux, adsorbent capacity, membrane specification, oxidant dose, hydraulic retention, pH/temperature, and nutrient balance. Parameter misalignment causes biofilm inhibition, membrane fouling, electrode passivation, or toxic intermediate accumulation, highlighting the need for integrated kinetic models and adaptive control. While laboratory demonstrations abound, full-scale translation remains limited by insufficient pilot testing with real wastewaters, incomplete techno-economic assessments, and limited long-term stability data.

Future progress depends on three converging areas. First, materials innovation—hierarchically structured sorbents, conductive antifouling membranes, biocompatible photocatalysts, and enhanced microbial consortia—must expand pollutant range and treatment resilience. Second, digitalization through machine-learning-guided monitoring promises stable performance under variable loads while minimizing operational costs. Third, rigorous field demonstrations with comprehensive sustainability metrics are essential to demonstrate practical viability to utilities and regulators. In advancing UN Sustainable Development Goal 6, hybrid microbial–physical technologies offer a more efficient and circular approach, transforming wastewater from an environmental liability to a source of clean water, energy, and recoverable resources. Continued interdisciplinary collaboration across microbiology, materials science, reactor engineering, modeling and data analytics will be crucial for advancing these systems from promising prototypes to standard practice in next-generation water treatment.

## References


Adjei, K.Y., Aliyu, A., Owhe-Ureghe, E., Oriakhi, E.A., Bakare-Abidola, T., Olaoye, J., Adepoju, Y.O., 2025. Biological treatment of emerging organic micropollutants in wastewater: recent advances and perspectives. World J. Biol. Pharm. Health Sci. 2582-5542. 21 (3), 488–505. doi:10.30574/wjbphs.2025.21.3.0181.

Ahmad, M., Liu, S., Mahmood, N., Mahmood, A., Ali, M., Zheng, M., Ni, J., 2017. Synergic adsorption–biodegradation by an advanced carrier for enhanced removal of high-strength nitrogen and refractory organics. ACS Appl. Mater. Interfaces. 1944-8244. 9 (15), 13188–13200. doi:10.1021/acsami.7b01251.

Ahmed, S.F., Mofijur, M., Nuzhat, S., Chowdhury, A.T., Rafa, N., Uddin, M.A., Inayat, A., Mahlia, T.M.I., Ong, H.C., Chia, W.Y., Show, P.L., 2021. Recent developments in physical, biological, chemical, and hybrid treatment techniques for removing emerging contaminants from wastewater. J. Hazard. Mater. 1873-3336. 416. Available from: www.elsevier.com/locate/jhazmat. doi:10.1016/j.jhazmat.2021.125912.

Al-Qodah, Z., Al-Qudah, Y., Assirey, E., 2020. Combined biological wastewater treatment with electrocoagulation as a post-polishing process: a review. Sep. Sci. Technol. 1520-5754. 55 (13), 2334–2352. Available from: www.tandf.co.uk/journals/titles/01496395.asp. doi:10.1080/01496395.2019.1626891.

Alfonso-Muniozguren, P., Serna-Galvis, E.A., Bussemaker, M., Torres-Palma, R.A., Lee, J., 2021. A review on pharmaceuticals removal from waters by single and combined biological, membrane filtration and ultrasound systems. Ultrason. Sonochem. 1873-2828. 76. Available from: www.elsevier.com/inca/publications/store/5/2/5/4/5/1. doi:10.1016/j.ultsonch.2021.105656.

Alvarino, T., Komesli, O., Suarez, S., Lema, J.M., Omil, F., 2016. The potential of the innovative SeMPAC process for enhancing the removal of recalcitrant organic micropollutants. J. Hazard. Mater. 0304-3894. 308, 29–36. doi:10.1016/j.jhazmat.2016.01.040.

Anyango, G.W., Bhowmick, G.D., Rajasekar, V., 2024. Membrane bioreactors and other emerging membrane technologies. In: Ghangrekar, M.M., Yadav, S., Yadava, R.N. (Eds.), Biological and Hybrid Wastewater Treatment Technology. Springer, pp. 91–107. doi:10.1007/978-3-031-63046-0_4.

Asheghmoalla, M., Mehrvar, M., 2024. Integrated and hybrid processes for the treatment of actual wastewaters containing micropollutants: a review on recent advances. Processes. 2227-9717. 12 (2), 339. doi:10.3390/pr12020339.

Asif, M.B., Ren, B., Li, C., Maqbool, T., Zhang, X., Zhang, Z., 2020. Powdered activated carbon—membrane bioreactor (PAC-MBR): impacts of high PAC concentration on micropollutant removal and microbial communities. Sci. Total Environ. 1879-1026. 745. Available from: www.elsevier.com/locate/scitotenv. doi:10.1016/j.scitotenv.2020.141090.

Babu Ponnusami, A., Sinha, S., Ashokan, H., Paul, M.V., Hariharan, S.P., Arun, J., Gopinath, K.P., Hoang Le, Q., Pugazhendhi, A., 2023. Advanced oxidation process (AOP) combined biological process for wastewater treatment: a review on advancements, feasibility and practicability of combined techniques. Environ. Res. 1096-0953. 237. Available from: https://www.sciencedirect.com/journal/environmental-research. doi:10.1016/j.envres.2023.116944.

Baldassarre, F., Ciccarella, G., 2021. Biology-inspired photocatalysis: recent advances in biomimetic photocatalytic nanosystems synthesis and applications. In: García-López, E.I., Palmisano, L. (Eds.), Materials Science in Photocatalysis. Elsevier, Italy, pp. 603–648. Available from: https://www.sciencedirect.com/book/9780128218594. doi:10.1016/B978-0-12-821859-4.00022-2.

Bhanot, P., Celin, S.M., Sreekrishnan, T.R., Kalsi, A., Sahai, S.K., Sharma, P., 2020. Application of integrated treatment strategies for explosive industry wastewater—a critical review. J. Water Process Eng. 2214-7144. 35, 101232. doi:10.1016/j.jwpe.2020.101232.

Bonilla-Petriciolet, A., Mendoza-Castillo D.I. Reynel-Ávila H.E., (Eds.) 2017. In: Adsorption Processes for Water Treatment and Purification. Springer International Publishing, Mexico, pp. 1–256. http://www.springer.com/in/book/9783319581354. doi:10.1007/978-3-319-58136-1.

Campinas, M., Viegas, R.M.C., Almeida, C.M.M., Martins, A., Silva, C., Mesquita, E., Coelho, M.R., Silva, S., Cardoso, V.V., Benoliel, M.J., Rosa, M.J., 2022. Powdered activated carbon full-scale addition to the activated sludge reactor of a municipal wastewater treatment plant: pharmaceutical compounds control and overall impact on the process. J. Water Process Eng. 2214-7144. 49. Available from: http://www.journals.elsevier.com/journal-of-water-process-engineering/. doi:10.1016/j.jwpe.2022.102975.

Deng, Y., Zhao, R., 2015. Advanced oxidation processes (AOPs) in wastewater treatment. Curr. Pollut. Rep. 2198-6592. 1 (3), 167–176. doi:10.1007/s40726-015-0015-z.

Dhamorikar, R.S., Lade, V.G., Kewalramani, P.V., Bindwal, A.B., 2024. Review on integrated advanced oxidation processes for water and wastewater treatment. J. Ind. Eng. Chem. 1226-086X. 138, 104–122. Available from: https://www.sciencedirect.com/science/journal/1226086X. doi:10.1016/j.jiec.2024.04.037.

Dhangar, K., Kumar, M., 2020. Tricks and tracks in removal of emerging contaminants from the wastewater through hybrid treatment systems: a review. Sci. Total Environ. 0048-9697. 738, 140320. doi:10.1016/j.scitotenv.2020.140320.

Ding, R., Yan, W., Wu, Y., Xiao, Y., Gang, H., Wang, S., Chen, L., Zhao, F., 2018. Light-excited photoelectrons coupled with bio-photocatalysis enhanced the degradation efficiency of oxytetracycline. Water Res. 0043-1354. 143, 589–598. doi:10.1016/j.watres.2018.06.068.

Du, Y., Zhang, S., Guo, R., Chen, J., 2015. Understanding the algal contribution in combined UV-algae treatment to remove antibiotic cefradine. RSC Adv. 2046-2069. 5 (74), 59953–59959. doi:10.1039/c5ra10806c.

El Gamal, M., Mousa, H.A., El-Naas, M.H., Zacharia, R., Judd, S., 2018. Bio-regeneration of activated carbon: a comprehensive review. Emirates Sep. Purif. Technol. 1873-3794. 197, 345–359. Available from: http://www.journals.elsevier.com/separation-and-purification-technology/. doi:10.1016/j.seppur.2018.01.015.

El Omari, S., Gutiérrez-Ramírez, R., Azaryouh, L., El Messaoudi, N., Kassab, Z., Albourine, A., Laabd, M., Abdellaoui, Y., 2025. Hybrid process technologies for sustainable water decontamination. In: Gautam, S., Joshi, S.K., Ambade, B. (Eds.), Blue Sky, Blue Water. Springer Science and Business Media LLC, pp. 155–176. doi:10.1007/978-3-031-82559-0_8.

Ensano, B.M.B., Borea, L., Naddeo, V., Belgiorno, V., de Luna, M.D.G., Ballesteros, F.C., 2016. Combination of electrochemical processes with membrane bioreactors for wastewater treatment and fouling control: a review. Front. Environ. Sci. 2296-665X. 4. Available from: http://journal.frontiersin.org/article/10.3389/fenvs.2016.00057/full. doi:10.3389/fenvs.2016.00057.

Ensano, B.M.B., Borea, L., Naddeo, V., de Luna, M.D.G., Belgiorno, V., 2019. Control of emerging contaminants by the combination of electrochemical processes and membrane bioreactors. Environ. Sci. Pollut. Res. 1614-7499. 26 (2), 1103–1112. Available from: http://www.springerlink.com/content/0944-1344. doi:10.1007/s11356-017-9097-z.

Ezugbe, E.O., Rathilal, S., 2020. Membrane technologies in wastewater treatment: a review. Membranes. 2077-0375. 10 (5). Available from: https://www.mdpi.com/2077-0375/10/5/89/pdf. doi:10.3390/membranes10050089.

García, L., Leyva-Díaz, J.C., Díaz, E., Ordóñez, S., 2021. A review of the adsorption-biological hybrid processes for the abatement of emerging pollutants: removal efficiencies, physicochemical analysis, and economic evaluation. Sci. Total Environ. 1879-1026. 780. Available from: www.elsevier.com/locate/scitotenv. doi:10.1016/j.scitotenv.2021.146554.

Ge, L., Deng, H., 2015. Degradation of two fluoroquinolone antibiotics photoinduced by Fe(iii)-microalgae suspension in an aqueous solution. Photochem. Photobiol. Sci. 1474-905X. 14 (4), 693–699. doi:10.1039/c3pp50149c.

Hasan, A., Tabassum, B., Hashim, M., Khan, N., 2025. The negative impacts of industrial water pollution on human health and benefits of zero liquid discharge system. Springer Water. 236, 61–74. Available from: https://www.springer.com/series/13419. doi:10.1007/978-3-031-84909-1_4.

Hassan, R.Y.A., Febbraio, F., Andreescu, S., 2021. Microbial electrochemical systems: principles, construction and biosensing applications. Sensors. 1424-8220. 21 (4), 1279. doi:10.3390/s21041279.

Hernandez, C.A., Osma, J.F., 2020. Microbial electrochemical systems: deriving future trends from historical perspectives and characterization strategies. Front. Environ. Sci. 2296-665X. 8. Available from: http://journal.frontiersin.org/journal/environmental-science. doi:10.3389/fenvs.2020.00044.

Jabbar, Z.H., Esmail Ebrahim, S., 2022. Recent advances in nano-semiconductors photocatalysis for degrading organic contaminants and microbial disinfection in wastewater: a comprehensive review. Environ. Nanotechnol. Monit. Manag. 2215-1532. 17, 100666. doi:10.1016/j.enmm.2022.100666.

Jadhav, D.A., Ghosh Ray, S., Ghangrekar, M.M., 2017. Third generation in bio-electrochemical system research—a systematic review on mechanisms for recovery of valuable by-products from wastewater. Renew. Sustain. Energy Rev. 1879-0690. 76, 1022–1031. Available from: https://www.journals.elsevier.com/renewable-and-sustainable-energy-reviews. doi:10.1016/j.rser.2017.03.096.

Jiang, Q., Ngo, H.H., Nghiem, L.D., Hai, F.I., Price, W.E., Zhang, J., Liang, S., Deng, L., Guo, W., 2018. Effect of hydraulic retention time on the performance of a hybrid moving bed biofilm reactor-membrane bioreactor system for micropollutants removal from municipal wastewater. Bioresour. Technol. 1873-2976. 247, 1228–1232. Available from: www.elsevier.com/locate/biortech. doi:10.1016/j.biortech.2017.09.114.

Jin, P., Jin, X., Wang, X., Feng, Y., Wang, X.C., 2013. Biological activated carbon treatment process for advanced water and wastewater treatment. In: Matovic, M.D. (Ed.), Biomass Now—Cultivation and Utilization. InTech. doi:10.5772/52021.

Jokar, M., Mirghaffari, N., Soleimani, M., Jabbari, M., 2019. Preparation and characterization of novel bio ion exchanger from medicinal herb waste (chicory) for the removal of $Pb^{2+}$ and $Cd^{2+}$ from aqueous solutions. J. Water Process Eng. 2214-7144. 28, 88–99. doi:10.1016/j.jwpe.2019.01.007.

Karbelkar, A.A., Reynolds, E.E., Ahlmark, R., Furst, A.L., 2021. A microbial electrochemical technology to detect and degrade organophosphate pesticides. ACS Cent. Sci. 2374-7951. 7 (10), 1718–1727. Available from: http://pubs.acs.org/journal/acscii. doi:10.1021/acscentsci.1c00931.

Katuri, K.P., Kamireddy, S., Kavanagh, P., Muhammad, A., Conghaile, P., Kumar, A., Saikaly, P.E., Leech, D., 2020. Electroactive biofilms on surface functionalized anodes: the anode respiring behavior of a novel electroactive bacterium, desulfuromonas acetexigens. Water Res. 1879-2448. 185. Available from: www.elsevier.com/locate/watres. doi:10.1016/j.watres.2020.116284.

Krakkó, D., Illés, Á., Licul-Kucera, V., Dávid, B., Dobosy, P., Pogonyi, A., Demeter, A., Mihucz, V.G., Dóbé, S., Záray, G., 2021. Application of (V)UV/O3 technology for post-treatment of biologically treated wastewater: a pilot-scale study. Chemosphere. 1879-1298. 275. Available from: www.elsevier.com/locate/chemosphere. doi:10.1016/j.chemosphere.2021.130080.

Kumar, A., Sharma, G., Naushad, M., Al-Muhtaseb, A.H., García-Peñas, A., Mola, G.T., Si, C., Stadler, F.J., 2020. Bio-inspired and biomaterials-based hybrid photocatalysts for environmental detoxification: a review. Chem. Eng. J. 1385-8947. 382. Available from: www.elsevier.com/inca/publications/store/6/0/1/2/7/3/index.htt. doi:10.1016/j.cej.2019.122937.

Kumar, B., Agrawal, K., Verma, P., 2021. Microbial electrochemical system: a sustainable approach for mitigation of toxic dyes and heavy metals from wastewater. J. Hazard. Toxic Radioact. Waste. 2153-5493. 25 (2). doi:10.1061/(asce)hz.2153-5515.0000590.

Lee, Y., Noh, J.H., Park, J.W., Yoon, S.W., Kim, S.Y., Son, H.J., Lee, W., Maeng, S.K., 2023. Integrating biological ion exchange with biological activated carbon treatment for drinking water: a novel approach for NOM removal, trihalomethane formation potential, and biological stability. Water Res. 1879-2448. 245. Available from: www.elsevier.com/locate/watres. doi:10.1016/j.watres.2023.120598.

Li, H., Pan, Y., Wang, Z., Chen, S., Guo, R., Chen, J., 2015. An algal process treatment combined with the Fenton reaction for high concentrations of amoxicillin and cefradine. RSC Adv. 2046-2069. 5 (122), 100775–100782. doi:10.1039/c5ra21508k.

Li, T., Yang, X.L., Chen, Q.L., Song, H.L., He, Z., Yang, Y.L., 2020. Enhanced performance of microbial fuel cells with electron mediators from anthraquinone/polyphenol-abundant herbal plants. ACS Sustain. Chem. Eng. 2168-0485. 8 (30), 11263–11275. Available from: http://pubs.acs.org/journal/ascecg. doi:10.1021/acssuschemeng.0c03058.

Li, Y., Chen, L., Tian, X., Lin, L., Ding, R., Yan, W., Zhao, F., 2021. Functional role of mixed-culture microbe in photocatalysis coupled with biodegradation: total organic carbon removal of ciprofloxacin. Sci. Total Environ. 0048-9697. 784, 147049. doi:10.1016/j.scitotenv.2021.147049.

Lin, H., Peng, W., Zhang, M., Chen, J., Hong, H., Zhang, Y., 2013. A review on anaerobic membrane bioreactors: applications, membrane fouling and future perspectives. Desalination. 0011-9164. 314, 169–188. doi:10.1016/j.desal.2013.01.019.

Logan, B.E., Rossi, R., Ragab, A., Saikaly, P.E., 2019. Electroactive microorganisms in bioelectrochemical systems. Nat. Rev. Microbiol. 1740-1534. 17 (5), 307–319. Available from: http://www.nature.com/nrmicro/index.html. doi:10.1038/s41579-019-0173-x.

Madaeni, S.S., Salehi, E., 2016. Membrane-adsorption integrated systems/processes. In: Basile, A., Charcosset, C. (Eds.), Integrated Membrane Systems and Processes. Wiley, pp. 343–373. doi:10.1002/9781118739167.ch13.

Mailler, R., Gasperi, J., Coquet, Y., Derome, C., Buleté, A., Vulliet, E., Bressy, A., Varrault, G., Chebbo, G., Rocher, V., 2016. Removal of emerging micropollutants from wastewater by activated carbon adsorption: experimental study of different activated carbons and factors influencing the adsorption of micropollutants in wastewater. J. Environ. Chem. Eng. 2213-3437. 4 (1), 1102–1109. doi:10.1016/j.jece.2016.01.018.

Moussavi, G., Fathi, E., Moradi, M., 2019. Advanced disinfecting and post-treating the biologically treated hospital wastewater in the UVC/$H_2O_2$ and VUV/$H_2O_2$ processes: performance comparison and detoxification efficiency. Process Saf. Environ. Prot. 0957-5820. 126, 259–268. doi:10.1016/j.psep.2019.04.016.

Nguyen, D.D., Ngo, H.H., Yoon, Y.S., 2014. A new hybrid treatment system of bioreactors and electrocoagulation for superior removal of organic and nutrient pollutants from municipal wastewater. Bioresour. Technol. 1873-2976. 153, 116–125. Available from: www.elsevier.com/locate/biortech. doi:10.1016/j.biortech.2013.11.048.

Nidheesh, P.V., Couras, C., Karim, A.V., Nadais, H., 2022. A review of integrated advanced oxidation processes and biological processes for organic pollutant removal. Chem. Eng. Commun. 1563-5201. 209 (3), 390–432. Available from: www.tandf.co.uk/journals/titles/00986445.asp. doi:10.1080/00986445.2020.1864626.

Ortiz-Prado, E., Simbaña-Rivera, K., Cevallos, G., Gómez-Barreno, L., Cevallos, D., Lister, A., Fernandez-Naranjo, R., Ríos-Touma, B., Vásconez-González, J., Izquierdo-Condoy, J.S., 2022. Waterborne diseases and ethnic-related disparities: a 10 years nationwide mortality and burden of disease analysis from Ecuador. Front. Public Health 2296-2565. 10. Available from: http://journal.frontiersin.org/journal/public-health/section/public-health-education-and-promotion#archive. doi:10.3389/fpubh.2022.1029375.

Pervez, M.N., Balakrishnan, M., Hasan, S.W., Choo, K.H., Zhao, Y., Cai, Y., Zarra, T., Belgiorno, V., Naddeo, V., 2020. A critical review on nanomaterials membrane bioreactor (NMS-MBR) for wastewater treatment. Npj Clean Water 2059-7037. 3 (1). Available from: https://www.nature.com/npjcleanwater/. doi:10.1038/s41545-020-00090-2.

Qian, G., Hu, X., Li, L., Ye, L., Lv, W., 2017. Effect of iron ions and electric field on nitrification process in the periodic reversal bio-electrocoagulation system. Bioresour. Technol. 0960-8524. 244 (Pt 1), 382–390. doi:10.1016/j.biortech.2017.07.155.

Racar, M., Dolar, D., Karadakić, K., Čavarović, N., Glumac, N., Ašperger, D., Košutić, K., 2020. Challenges of municipal wastewater reclamation for irrigation by MBR and NF/RO: physico-chemical and microbiological parameters, and emerging contaminants. Sci. Total Environ. 0048-9697. 722, 137959. doi:10.1016/j.scitotenv.2020.137959.

Rajput, M., Brar, A., Vivekanand, V., Pareek, N., 2022. Recent advancements in biohydrogen production: thermochemical and biological conversion routes. In: Nanda, S., Sarangi, P.K. (Eds.), Biohydrogen. Informa UK Limited, pp. 1–31. doi:10.1201/9781003277156-1.

Ramírez-Moreno, M., Esteve-Núñez, A., Ortiz, J.M., 2021. Desalination of brackish water using a microbial desalination cell: analysis of the electrochemical behaviour. Electrochim. Acta. 0013-4686. 388, 138570. doi:10.1016/j.electacta.2021.138570.

Ren, G., Han, H., Wang, Y., Liu, S., Zhao, J., Meng, X., Li, Z., 2021. Recent advances of photocatalytic application in water treatment: a review. Nanomaterials. 2079-4991. 11 (7), 1804. doi:10.3390/nano11071804.

Sbardella, L., Comas, J., Fenu, A., Rodriguez-Roda, I., Weemaes, M., 2018. Advanced biological activated carbon filter for removing pharmaceutically active compounds from treated wastewater. Sci. Total Environ. 0048-9697. 636, 519–529. doi:10.1016/j.scitotenv.2018.04.214.

Schulz, M., Winter, J., Wray, H., Barbeau, B., Bérubé, P., 2017. Biologically active ion exchange (BIEX) for NOM removal and membrane fouling prevention. Water Supply. 1606-9749. 17 (4), 1178–1184. doi:10.2166/ws.2017.016.

Su, C., Deng, Q., Lu, Y., Qin, R., Chen, S., Wei, J., Chen, M., Huang, Z., 2019. Effects of hydraulic retention time on the performance and microbial community of an anaerobic baffled reactor-bioelectricity Fenton coupling reactor for treatment of traditional Chinese medicine wastewater. Bioresour. Technol. 0960-8524. 288, 121508. doi:10.1016/j.biortech.2019.121508.

Thakur, A.K., Kumar, R., Kumar, A., Shankar, R., Khan, N.A., Gupta, K.N., Ram, M., Arya, R.K., 2023. Pharmaceutical waste-water treatment via advanced oxidation based integrated processes: an engineering and economic perspective. J. Water Process Eng. 2214-7144. 54. Available from: http://www.journals.elsevier.com/journal-of-water-process-engineering/. doi:10.1016/j.jwpe.2023.103977.

Tran, N.N., Escribà-Gelonch, M., Sarafraz, M.M., Pho, Q.H., Sagadevan, S., Hessel, V., 2023. Process technology and sustainability assessment of wastewater treatment. Ind. Eng. Chem. Res. 1520-5045. 62 (3), 1195–1214. Available from: http://pubs.acs.org/journal/iecred. doi:10.1021/acs.iecr.2c03471.

Verma, M., Chakraborty, S., Kumari, S., Gupta, A., Kumar, D., Iqbal, J., Banu, J.R., Pugazhendi, A., Kumar, R.N., 2023. Co-treatment of stabilized landfill leachate and municipal wastewater in a granular activated carbon-sequencing batch reactor (GAC-SBR). Process Saf. Environ. Prot. 0957-5820. 174, 424–432. Available from: http://www.elsevier.com/wps/find/journaldescription.cws_home/713889/description#description. doi:10.1016/j.psep.2023.04.015.

Vimonses, V., Jin, B., Chow, C.W.K., Saint, C., 2010. An adsorption-photocatalysis hybrid process using multi-functional-nanoporous materials for wastewater reclamation. Water Res. 0043-1354. 44 (18), 5385–5397. Available from: https://www.sciencedirect.com/science/journal/00431354. doi:10.1016/j.watres.2010.06.033.

Wang, Y., Wang, X., Li, M., Dong, J., Sun, C., Chen, G., 2018. Removal of pharmaceutical and personal care products (PPCPs) from municipalwaste water with integrated membrane systems, MBR-RO/NF. Int. J. Environ. Res. Public Health. 1660-4601. 15 (2). Available from: http://www.mdpi.com/1660-4601/15/2/269/pdf. doi:10.3390/ijerph15020269.

Waqas, S., Bilad, M.R., Huda, N., Harun, N.Y., Md Nordin, N.A.H., Shamsuddin, N., Wibisono, Y., Khan, A.L., Roslan, J., 2021. Membrane filtration as post-treatment of rotating biological contactor for wastewater treatment. Sustainability. 2071-1050. 13 (13). Available from: https://www.mdpi.com/2071-1050/13/13/7287/pdf. doi:10.3390/su13137287.

Xiong, J.Q., Kurade, M.B., Jeon, B.H., 2018. Can microalgae remove pharmaceutical contaminants from water? Trends Biotechnol. 1879-3096. 36 (1), 30–44. Available from: www.elsevier.com/locate/tibtech. doi:10.1016/j.tibtech.2017.09.003.

Yang, K., Lu, J., Jiang, W., Jiang, C., Chen, J., Wang, Z., Guo, R., 2017. An integrated view of the intimate coupling UV irradiation and algal treatment on antibiotic: compatibility, efficiency and microbic impact assessment. J. Environ. Chem. Eng. 2213-3437. 5 (5), 4262–4268. doi:10.1016/j.jece.2017.08.028.

Yi, G., Wang, B., Feng, Y., Fang, D., Yang, L., Liu, W., Zhang, Y., Shao, P., Pavlostathis, S.G., Luo, S., Luo, X., Wang, A., 2022. The ins and outs of photo-assisted microbial electrochemical systems for synchronous wastewater treatment and bioenergy recovery. Resour. Conserv. Recycl. 0921-3449. 181, 106230. doi:10.1016/j.resconrec.2022.106230.

Zakaria, B.S., Dhar, B.R., 2022. A review of stand-alone and hybrid microbial electrochemical systems for antibiotics removal from wastewater. Processes. 2227-9717. 10 (4). Available from: https://www.mdpi.com/2227-9717/10/4/714/pdf. doi:10.3390/pr10040714.

Zhang, T., Payne, K., Zhang, J., Purswani, P., Karpyn, Z., Wang, M., 2024. Hybrid ion exchange and biological processes for water and wastewater treatment: a comprehensive review of process applications and mathematical modeling. Rev. Environ. Sci. Bio/Technol. 1569-1705. 23 (1), 163–188. doi:10.1007/s11157-023-09677-w.

Zhao, J., Li, F., Cao, Y., Zhang, X., Chen, T., Song, H., Wang, Z., 2021. Microbial extracellular electron transfer and strategies for engineering electroactive microorganisms. Biotechnol. Adv. 0734-9750. 53, 107682. doi:10.1016/j.biotechadv.2020.107682.

Zhao, W., Adeel, M., Zhang, P., Zhou, P., Huang, L., Zhao, Y., Ahmad, M.A., Shakoor, N., Lou, B., Jiang, Y., Lynch, I., Rui, Y., 2022. A critical review on surface-modified nano-catalyst application for the photocatalytic degradation of volatile organic compounds. Environ. Sci. Nano. 2051-8161. 9 (1), 61–80. Available from: http://pubs.rsc.org/en/journals/journal/en. doi:10.1039/d1en00955a.

Zuo, W., Yu, Y., Huang, H., 2021. Making waves: microbe-photocatalyst hybrids may provide new opportunities for treating heavy metal polluted wastewater. Water Res. 0043-1354. 195, 116984. doi:10.1016/j.watres.2021.116984.

CHAPTER

# Microbial encapsulation and immobilization in waste management

# 4

**Khouloud Barakat and Ahmed El Nemr**
*National Institute of Oceanography and Fisheries (NIOF), Alexandria, Egypt*

## 1 Introduction

Human health suffers significantly due to air, water, and soil pollution. Concern over the deterioration of the environment in many parts of the world is growing (Manisalidis et al., 2020). These contaminants are nonbiodegradable chemical substances commonly released into ecosystems through mining, fungicides, industrial discharge, and petroleum extraction (Ali et al., 2019). Pollutants are the cause of eutrophication and the pervasive organic and toxic pollution brought on by industrialization and rapid economic growth. Furthermore, many people are in danger of contracting diseases linked to cancer as a direct result of the environment's pollution. This has led the industry to search for appropriate treatment methods and to implement and tighten strict regulations (Lu et al., 2015; Teh et al., 2016).

Among the various treatment methods—membrane processes, osmosis, foam flotation, coagulation, dialysis, adsorption, and photocatalytic degradation—these traditional physicochemical processes are commonly used to reduce the toxicity of effluents (Al-Rubaiey, 2022). Some disadvantages of these conventional processes include the production of toxic sludge, which is not always capable of removing the full contaminant load, in addition to the high cost of operation and maintenance due to engineering expertise, energy requirements, economic implications, processing efficiency, and infrastructure (Khader et al., 2022).

This is due to the fact that biological processes are used in conventional waste treatment. Using various methods, including biosorption, bioaccumulation, and biodegradation, bioremediation enhances naturally occurring remedial activities that depend on microorganisms to degrade, change, or eliminate hazardous organic pollutants (Gül, 2018). The slow buildup of contaminants in microbiological living tissue is known as bioaccumulation (Curtean-Bănăduc et al., 2020). Microorganisms' breakdown of organic materials is known as biodegradation (Azeez and Al-Zuhairi, 2020). Biosorption is an economical and ecologically acceptable technique for removing pollutants from effluents, especially those that are not biodegradable, using a range of dead and live cells and agricultural waste as an efficient biosorbent (Torres, 2020).

Microbial Technologies for Sustainable Waste Management. https://doi.org/10.1016/B978-0-443-44093-9.00008-6

In waste management, the immobilization of microbial cells as a treatment technique has gained global attention. The cell immobilization method provides a substantial advantage over more traditional systems. Microbial cells are localized to a specific area as part of the cell immobilization process. This keeps the cell's catalytic activities intact for repeated and continuous use, restricts their freedom of migration, and allows them to display hydrodynamic characteristics different from those of the surrounding environment. Immobilized cells have been explored as a possible substitute technology for environmental research applications and have gained popularity in the biodegradation of soil pollutants and contaminated groundwater, as well as in industrial, agricultural, biocontrol, and pesticide applications (Ogundolie et al., 2024).

Microbial cell immobilization has been studied utilizing a variety of methods, including entrapment or encapsulation using polymers, self-aggregation by flocculation or the application of cross-linking agents, and adsorption or attachment of cells to an inert substrate (Giese et al., 2020) (Fig. 4.1).

Microbial cell adsorption makes use of the cells' innate capacity to stick to solid surfaces to create biofilms, which might be a single layer or many millimeters thick (Saha and Rao, 2021). Techniques such as metal oxide chelation, silanization onto a silica substrate, and chemical cross-linking via glutaraldehyde (GA) can be applied to microbial cells that do not adhere spontaneously. In addition, microbial cell immobilization has been used for various support materials, including activated charcoal, porous glass, and inorganic ceramics. The ease of use and affordability of the cell immobilization procedure are two benefits of this approach. It is severely limited, nevertheless, by widespread cell leakage from the support, making it challenging to extract cell-free effluent for later processing (Khatoon and Rai, 2018). Cell immobilization via aggregation has been achieved by using the ability of some cells, such as yeast and fungal mycelium, to flocculate, forming cell aggregates. It is also

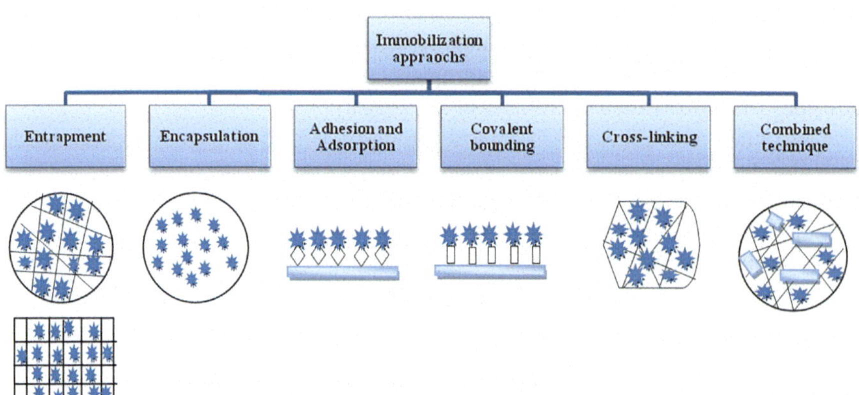


**FIG. 4.1**

Different immobilization approaches used for waste treatment.

possible to induce flocculation in nonflocculating microorganisms under specific circumstances (Vallejo et al., 2012).

Microbial cells are encapsulated or trapped within a polymeric substance to produce beads that are permeable to nutrients, gases, and metabolites while maintaining cell viability within the beads. The two types of conventional microbial immobilization materials are natural organic polymers (alginate, carrageenan, agar, and agarose) and synthetic organic polymers (acrylamide, polyurethane, and polyvinyl) (Berillo et al., 2021). However, these carriers have certain drawbacks, such as the restricted ability to carry substrates and the mechanical brittleness of materials in reaction to environmental fluctuations. According to Jiang (2015), the primary issues that lower the degrading efficiency of immobilized cells are mass transfer constraints and steric hindrance for substrate diffusion.

Using the right support materials is essential for successful microbial cell immobilization (Rusu et al., 2021). For waste treatment, from environmental contaminants to support materials, it is necessary to have properties such as being nonbiodegradable, insoluble, porous, nontoxic, lightweight, high in biomass retention, chemically stable, and permitting a simple immobilization procedure at a low cost (Karm et al., 2020). The two primary characteristics of inorganic carriers are their resistance to microbial destruction and thermo-stability. However, when compared to inorganic materials, biomass materials—which come from plants, animals, and microbes—show superior performance in terms of their unique physical properties and perfect biocompatibility, as well as their good mass transfer performance, affordability, nontoxicity, and ease of immobilization techniques (Cottet et al., 2020).

A new generation of environmental treatment methods that can be used to eliminate harmful pollutants has recently been demonstrated using nanotechnology (Nadi et al., 2019; Zaki et al., 2022). Because of their unique properties, nanoparticles (NPs) provide living cells with flexible and adaptable support (Giese et al., 2020). The mass transfer constraint of immobilized cells has been addressed in two ways by NPs due to their large specific surface area and high surface activity: (1) a conventional carrier, usually an organic one, spreads or embeds magnetic nanoparticles (MNPs) (Ghorbannezhad et al., 2018; Qiao et al., 2020; Jiang, 2015; Shi et al., 2014), and (2) to create an NP–bacteria assembly, NPs are applied to the surface of the target bacteria (Wang et al., 2020; Bestawy et al., 2020). Furthermore, coated cells demonstrated good reusability, simple separation by magnetic force, and high operational stability.

The methods used for microbial cell immobilization, particularly encapsulation, are described in this chapter, along with their advantages and disadvantages. Furthermore, significant applications of these techniques in nanotechnology are also discussed.

## 2 Microbial cell immobilization approaches

As shown in Fig. 4.1, there are four major methods for immobilizing cells. Crosslinking or covalent bonding is the most costly, whereas physical adsorption is the most inexpensive and cost-effective (Najim et al., 2024). Encapsulation and entrapment are in the middle, with encapsulation possibly costing more in the long run because of its complexity (Han et al., 2023). The dangers of secondary contamination from physical adsorption and encapsulation are typically low to high. Covalent bonding or cross-linking presents little risk; however, entrapment may present modest concerns if the matrix deteriorates. The best methods for stability and confinement are encapsulation and covalent bonding or cross-linking connections. Although they are also efficient, physical adsorption and entrapment may be constrained by saturation and degradation (Wang et al., 2021). The following sections provide more details on these techniques.

## 2.1 Adhesion and adsorption

Van der Waals forces, hydrophobic interactions, dipole–dipole interactions, and hydrogen bonding can all occur during the adsorption process for the immobilization technique (Fig. 4.2). According to (Blanco et al., 2023), adsorption is the most straightforward technique for immobilizing microbial cells. The cells are first drawn to the surface of carriers and subsequently adsorbed onto the matrix. This method involves immobilized microorganisms and a carrier's surface physically binding (Bouabidi et al., 2019). For immobilization, the interaction used depends on the characteristics of the adsorbate and the substrate. Due to its ease of cell desorption, which improves reusability, minimal interference with microbial growth, flexibility, cost-effectiveness (economical for large-scale wastewater treatment), and utilization of a variety of solid supports or carriers (activated carbon, silica, and resins) chosen according to the particular waste treatment conditions, physical adsorption is preferred (Suzana et al., 2013). The taxonomic and metabolic condition of the cells influences cell sorption, which in turn affects the immobilization efficiency and restricts the affinity between the cell and the carrier. As Kubota et al. (2008) point out, different microorganisms exhibit varied adsorption tendencies. The proper sorbents can be chosen to protect microorganisms from harmful substances, maintain their metabolic function, and catalyze microbial metabolism. Support materials with high surface

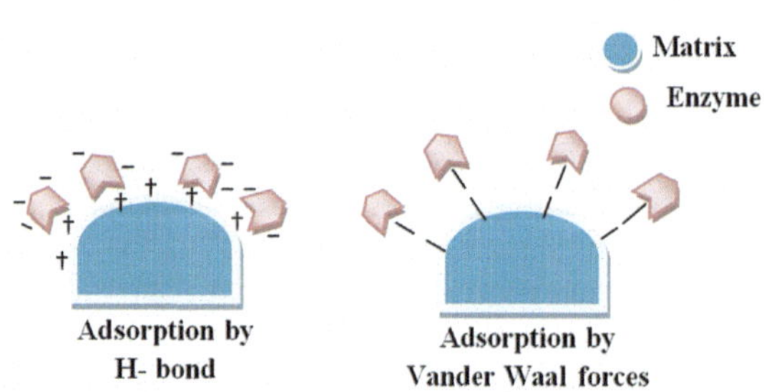


**FIG. 4.2**

Enzyme immobilization by adsorption using van der Waals forces (A) and hydrogen bonding (B).

area, void volume, permeability, and durability are advised for adsorption studies (Monsan et al., 1987). However, a significant disadvantage of this approach is that it uses weak forces for cell attachment, which may not be sufficient in dynamic situations with vigorous agitation and flow rate, leading to cell desorption (Saini et al., 2023). Additional difficulties include the possibility of nonspecific binding, a shorter operational life, and sensitivity to abiotic variables (temperature, ionic strength, and pH), all of which lead to a decrease in immobilization efficiency and a loss of biocatalytic activity (Tripathi and Melo, 2021).

## 2.2 Entrapment (entanglement)

The most often-used method of immobilizing microorganisms is entrapment. Microbes must be enclosed within a stiff network (Trelles and Rivero, 2020). The trap is sufficiently airtight to keep cells from escaping while permitting substrates and products to diffuse. Collagen, alginate, agar, carrageenan, cellulose, gelatin, polyacrylamide, polystyrene, epoxy resin, polyester, polyurethane, and/or photocross-linkable resins are examples of naturally occurring gel polymers (Fig. 4.3) that are used in microbial entrapment, making it one of the least disruptive methods for preventing microbial movement during this irreversible process (Berillo et al., 2021). In addition, immobilized cells stay inside the matrix during bioremediation because entrapment reduces cell leakage (Tong and Derek, 2023; Saberi Riseh et al., 2021). Its main advantages are its strong cell protection, great adaptability to complex industrial settings or toxicants and drastic environmental changes, and mechanical and operational stability for long-term treatment. To improve the penetration of particular substrates and the removal of contaminants, matrices can also be specially designed to be selectively permeable (Mohidem et al., 2023). Because of its straightforward approach and gentle conditions, microbial entrapment in alginate gel is commonly employed (El-Naas et al., 2017). It has been widely used in fermentation and biotransformation processes to produce organic acids, enzymes, antibiotics, and alcohols (Taghizadeh et al., 2020). However, its main drawback is that the entrapment matrix prevents the

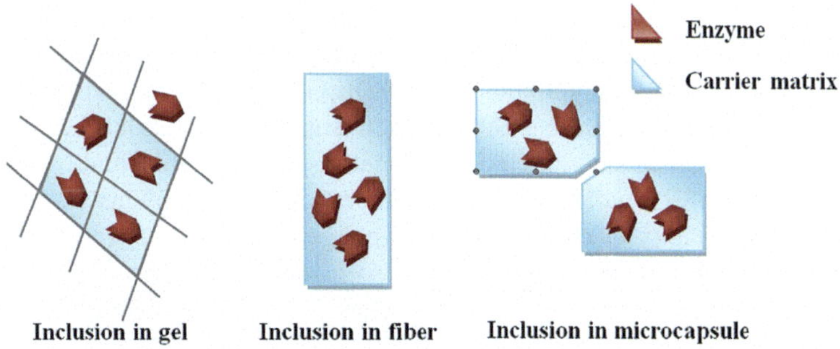


**FIG. 4.3**

Schematic representation of enzyme entrapment in gels, fibers, and microcapsules.

flow of nitrogen, gas (oxygen), and substrates to immobilized cells, which lowers the effectiveness of bioremediation (Dzionek et al., 2022). Because matrix integrity and uniformity in cell distribution must be preserved, scaling it up could be difficult. Sustaining these two essential elements raises operating expenses and makes the product unsuitable for use on an industrial scale (Han et al., 2023). Potential biofilm formation on matrix surfaces, retarded degradation over time (lower reusability), process blockage, and decreased efficiency are further disadvantages (Thomas et al., 2018). A study has demonstrated that even when bovine serum albumin, 3-aminopropyltriethoxysilane, GA (cross-linker), and ammonium sulfate (precipitant) are added, the formation of magnetic cross-linked enzyme aggregates (CLEAs) or their entrapment in particles with improved mechanical characteristics improves enzyme immobilization. The authors demonstrated that increasing the porosity of CLEAs might prevent substrate diffusion issues. Despite these developments, diffusion restrictions were still a problem because of the thick structure of CLEAs, which may hinder substrate mobility and reduce efficiency (Sampaio et al., 2022).

## 2.3 Cross-linking or covalent binding

One frequently employed cell immobilization method is immobilization with covalent binding. It is a cost-effective method that uses a binding agent (cross-linker) to create a covalent linkage between functional groups on the carrier material (solid support) and in the walls of microorganisms (Zhang et al., 2021). Microbial cell walls contain functional compounds such as amide, carboxyl, and phosphoryl (Fortman et al., 2018). In immobilization, carbodiimide hydrochloride (N-Ethyl-N'-(3-dimethylaminopropyl)carbodiimide: EDC) and GA are binding agents. Covalent binding agents typically inhibit cells and are hazardous. Thus the covalent binding approach is primarily used for enzyme immobilization instead of whole-cell immobilization (Nguyen and Kim, 2017; Mahmoudi et al., 2024).

Nevertheless, the technique's disadvantages include the use of hazardous cross-linkers (carbodiimide and GA), which may lower cell viability and metabolism, negatively affecting the remediation process and increasing operational expenses. Cell recovery is challenging since cross-linking is irreversible (Alkayyali et al., 2019). In addition, wastewater types with significant concentrations of organic matter or substances that disrupt the bonding process may not be suitable for the procedures. With polyethylene glycol (PEG) serving as a modifying agent, the ability of multipoint covalent attachment to support surfaces has been investigated using common carriers, including cyanates, tosyl chloride, and N-hydroxysuccinimide esters. By avoiding structural alterations caused by heat and organic cosolvents, its capacity to create multiple linkages between the enzyme and the support helps stabilize the enzyme. However, achieving reproducible results remains a significant obstacle, especially when it comes to maintaining the intended bonding parameters (Guisan et al., 2022). Therefore further optimization research is needed to improve the reliability of findings, cell reusability, and activity retention.

Navarro and Durand (1977) reported that *Saccharomyces carlsbergensis* is successfully bound to porous silica beads covalently, even though successful covalent immobilization in yeast is uncommon. Long-term maintenance of biocatalytic activity and reduced cell leakage (due to the strength of the connections created) are further advantages of this technology. These elements are essential for continuous waste treatment and increased operational efficiency. Górecka and Jastrzębska (2011) and Bouabidi et al. (2019) claimed that covalent bonds increase the resistance of immobilized cells to harmful substances and severe abiotic conditions. Rasheed et al. (2020) immobilized *Pseudomonas aeruginosa* on the surface of eggshell powder, a cheap carrier, and then used GA as a binding agent to extract Cr(VI) from an aqueous solution. In covalent binding, carboxylic and amino groups form a covalent bond (Fig. 4.4), whereas in cross-linking: single biomolecules are assembled into three-dimensional aggregates using a multifunctional linker on the transducer surface (Martinkova et al., 2017).

## 2.4 Encapsulation

Another nonreversible immobilization technique that functions similarly to entrapment is encapsulation, which encloses the target. The so-called coating, or encapsulation, is a unique microbial immobilizing agent for the trap. The encapsulated substance is trapped by a mechanical or physicochemical process (Krishnamoorthi et al., 2015). There are semipermeable membranes around the biomass. Although the membrane walls restrict the biomass, the membrane's permeability allows nutrients and free substrates to enter. Limiting access to the capsule's interior shields the biomass from the elements and prevents it from expanding its metabolic activity and

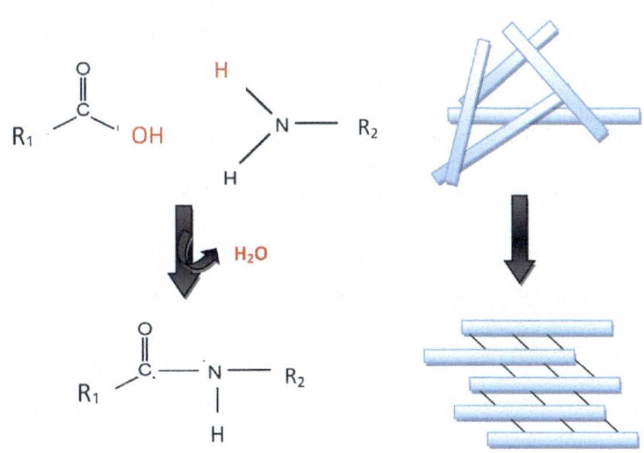


**FIG. 4.4**

Covalent binding: carboxylic and amino groups form a covalent bond are created; the remaining biorecognition molecules and the transducer surface or carrier surface are represented by R1 and R2 (A); cross-linking: single biomolecules are formed into three-dimensional aggregates.

producing metabolites (Waheed et al., 2019). It is one of the most extensively used methods for small-scale immobilization and has proven successful in several experimental and commercial trials. One of the most popular methods is encapsulation in beads or granules with the right size, mechanical strength, and porosity (Burgain et al., 2011). The membrane walls, which usually resemble a capsule, restrict the biocatalysts during this process, but they are still allowed to freely circulate within the cell's core region. Certain chemicals can flow through the semipermeable membrane of a biocatalyst cell. This preserves the integrity of the biocatalyst's position within the cell while permitting the passage of nutrients and substrates (Chen et al., 2015).

However, this phenomenon (permeability) is proportionate to the size of the core material and is dependent on the membrane pore size. The microencapsulation (Fig. 4.5), which shields the biocatalyst from adverse environmental conditions by preventing access to the capsule's interior, is the method's most important advantage (Suzana et al., 2013). The method's advantages include immobilizing a large number of cells in a small volume, protecting cells from harsh abiotic factors, and providing controlled release of cells or enzymes, all of which help to increase bioremediation efficiency and prolong operational lifespan (Picos-Corrales et al., 2023).

Encapsulation technology is a viable way to address the need for more environmentally friendly wastewater treatment methods and to comply with new or proposed stricter laws on nutrient discharges. Emerging biocatalysis and biomedical application approaches include whole-tissue cell encapsulation (Kang et al., 2014) and enzyme encapsulation (Sheldon and van Pelt, 2013). By encasing microorganisms in a permeable matrix, it is possible to retain specific microbial populations, prevent cell washout, and shield cells from unfavorable growth conditions (Manonmani and Joseph, 2018). Encapsulants can also create microenvironments by allowing chemical species to diffuse differently through various matrices. For example, in trials using encapsulants as a barrier, denitrification and electron donor usage were enhanced due to the differential diffusion of nitrogen species and organic substrates.

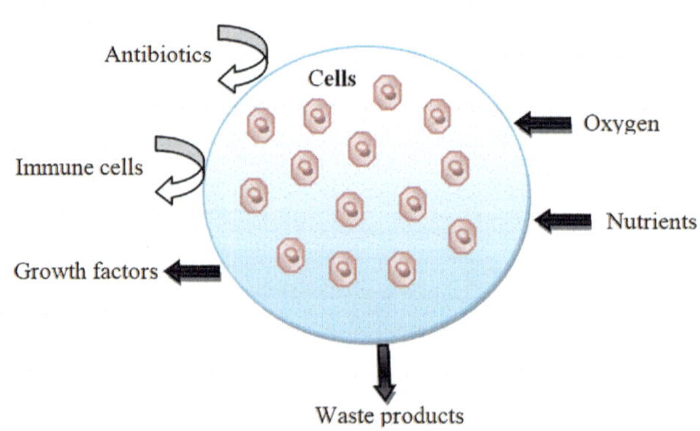


**FIG. 4.5**

Schematic illustrating cell microencapsulation.

In contrast to biofilm-based biotechnologies, the buildup of extracellular polymeric substances (EPS) is not necessary for encapsulation to form and immobilize cells. Thus particular bacteria of interest can be encapsulated to carry out specific biochemical activities, allowing for a quicker startup, particularly with autotrophic microorganisms that grow slowly (Manonmani and Joseph, 2018). While preserving anaerobic zones within, encapsulation can also prevent cells from separating due to shear stress during operation (Ahmad et al., 2020). Furthermore, customized encapsulants, rather than EPS, can control the environment surrounding the chosen and encapsulated microorganisms. For example, encapsulants' physical and chemical characteristics can be used to chelate heavy metals to reduce their inhibition (Zhu et al., 2020) or concentrate substrates by adsorption (Putra and Lee, 2020). Encapsulated cells' size, surface area, and starting concentration are additional engineering characteristics that encapsulated systems can regulate.

The following sections address stability, permeability, and sustainability, three crucial characteristics of encapsulating materials for wastewater treatment.

### 2.4.1 Encapsulation material stability

The most frequently utilized encapsulating materials are chemically manufactured polyvinyl alcohol (PVA) and naturally occurring polysaccharides. Alginate, carrageenan, and chitosan are examples of polysaccharide-based compounds derived from marine organisms and are extensively utilized in the food industry since they are nontoxic. The simplicity of the encapsulating method makes these materials appealing for use in wastewater treatment; however, they are unstable (Wang et al., 2021).

a. **Alginate:** Because it is relatively easy to cross-link utilizing the diffusion method of divalent cations in solution, alginate has been thoroughly researched as an encapsulant (Pawar and Edgar, 2012). The durability of alginate is problematic, though, since it can degrade microbially over time or chemically when the cross-linked polysaccharide networks dissolve in wastewater containing high levels of phosphate, ethylenediaminetetraacetic acid, or bicarbonate (Cruz et al., 2013).
b. **PVA:** It is another popular encapsulant, is made chemically from ethylene and acetic acid, and has lesser elasticity than alginate but better tearing toughness (Li et al., 2018). PVA can also be made via a straightforward gelation technique; however, it usually uses boric acid or a freeze-thaw cycle, which can compromise the survival of the microorganisms that are encapsulated. Other techniques employing various cross-linking agents, including phosphate, nitrate, or sulfate, were developed to overcome this drawback (Takei et al., 2011).
c. **Waterborne polyurethane (WPU)**: when used to encapsulate microorganisms in synthetic waterproofing for wastewater treatment, WPU demonstrated superior mechanical strength and reduced swelling compared to PVA and alginate (Chen et al., 2015), guaranteeing the encapsulated system's long-term structural integrity.

d. **PEG**: the expected lifespan of PEG is over ten years, demonstrating its exceptional stability.

Although these synthetic materials are more stable than those produced from nature, their synthesis procedures are more intricate. Furthermore, these synthetic materials are more expensive and less sustainable, which has limited their widespread use as encapsulants.

### 2.4.2 Sustainability of encapsulating materials

A cradle-to-grave approach must be used to analyze the sustainability of encapsulant materials. This involves assessing how the encapsulant affects human and environmental health throughout the sourcing and synthesis phases, how long it lasts, and what disposal alternatives are available when the product is finished being used. Encapsulant materials derived from nonpetrochemical feedstocks are preferred under green chemistry principles, while polymeric encapsulants that contain petroleum-based additives or components are viewed as less sustainable (Schneiderman and Hillmyer, 2017). Naturally produced polysaccharide-based materials are sustainable encapsulant possibilities. According to a recent study, EPS isolated from granular sludge had physical characteristics comparable to those of alginate, suggesting that EPS could be used as a substitute for inexpensive encapsulating materials. It was demonstrated that the theoretical capacity of wastewater treatment facilities globally to produce EPS is 10 times greater than the present annual output of 30,000 metric tons of alginate from brown algae (Kehrein et al., 2020). Even though cross-linked PVA and PEG polymers are believed to have low toxicity and are ecologically benign, PVA, PEG, and WPU are all produced using fossil fuels (Chen et al., 2005). It is feasible to source precursors for these polymers from renewable feedstocks. For example, using the "vinyl acetate cycle" to produce PVA from bioethanol has been suggested, which can increase the synthesis process's sustainability (Kawabata and Kurooka, 1995).

Also, it is essential to conduct a thorough assessment before implementing encapsulated microorganisms on a wide scale because several of these sustainability indicators are directly at odds with one another (e.g., durability versus degradability of these metrics). Loss of encapsulated microorganisms and a decline in effluent quality might result from unintentional encapsulant degradation during treatment or use. The stability and sustainability benefits of both natural and synthetic materials can be combined in composite or hybrid materials (Wang et al., 2021).

### 2.4.3 Mass transfer of nutrients within encapsulating materials

While diffusion controls mass transport within the encapsulant due to the obstruction of the polymer network, cell clusters, and surrounding EPS, convection provides microorganisms at the encapsulant's surface with easy access to substrates and electron acceptors. Thus effective substrate penetration into the encapsulating material is required **for encapsulated microorganisms to proliferate.** In addition to obstructing substrate and oxygen diffusion channels, encapsulated microorganisms also alter the internal composition and characteristics of the encapsulation material (Zhao et al., 2013).

This immobilized method's shortcomings include a poor cell recovery rate, a complex immobilization process (multi-step emulsification, gelation), and the creation of a mass transfer barrier (encapsulating material may prevent diffusion of oxygen or essential substrates), which raises costs and decreases treatment effectiveness and industry-scale use (Najim et al., 2024). According to a study, adding silica to alginate increased the laccase encapsulation efficiency to 70% from 59% when alginate was used alone. Once more, the immobilized enzyme's catalytic activity and efficiency were improved by laccase extract encapsulation in silica sol-gel as opposed to its free form (kcat of 89.9 for alginate, 63.7 for alginate–silica, and 56.9 $min^{-1}$ for silica sol–gel). However, the enzyme's catalytic activity encapsulated in alginate-silica was diminished by 50% after three reaction cycles. Therefore the ability to reuse the enzyme in many reaction cycles is limited by this technique (Gill et al., 2018).

There have been reports on the benefits of applying various cell immobilization techniques to bioreactor systems. Successful cell encapsulation in a polymer gel matrix has been demonstrated in both commercial and experimental settings. Across bioreactors, it was shown that encapsulated cells performed better than free cells across a wide range of environmental conditions. For instance, when compared to lost cells, encapsulated cells exhibit higher levels of plasmid stability, metabolic activity, and metabolite synthesis. This is because the encapsulation provides plasmids with greater protection, increasing the level of security that plasmids can enjoy (Cassidy et al., 1996).

## 2.5 Advantages and disadvantages of microbial immobilization techniques

As given in Table 4.1, every method of microbial immobilization—including encapsulation, covalent binding, entrapment, and adsorption—used to produce biocatalysts has advantages and disadvantages.

In general, microbial immobilization methods confront several obstacles that must be overcome to increase their efficacy and suitability for environmental remediation, notwithstanding the potential of each technology. Maintaining microbial viability, choosing suitable matrix materials, scaling up from laboratory to field applications, controlling expenses, addressing safety and environmental issues, filling knowledge and technology gaps, navigating regulatory challenges, and implementing efficient performance monitoring are some of these difficulties. To advance these technologies and increase their application in environmental protection, addressing these issues through research, innovation, and development remains imperative.

## 2.6 Major applications of immobilization

Physical, chemical, and biological treatments are the categories into which conventional wastewater treatment technologies like filtration, sedimentation, flocculation, chlorination, and activated sludges fall. These techniques, however, concentrate only on separating or eliminating contaminants from wastewater. Therefore the pollutants

**Table 4.1** Advantages and disadvantages of microbial immobilization techniques.

| Methods | Advantages | Disadvantages | References |
|---|---|---|---|
| Entrapment | Formation of a barrier to protect biomass, extending its viability and increasing its mechanical strength. | Diffusion restrictions, cell leakage, low loading capacity, cost, and deactivation due to immobilization. | (Berillo et al., 2021; Thomas et al., 2018) |
| Adsorption | It is inexpensive, quick, easy, gentle, and effective. In addition to the possibility of carrier reloading, it does not require chemical additives. | Because the adsorbed cell leaks off the carrier at a high rate, the binding force between bacteria and carriers is weak. Strong adsorption between cells and carriers must be established based on the type of suitable adsorbents to prevent microorganisms from desorbing from the matrix. | (Blanco et al., 2023; Saini et al., 2023) |
| Cross-linking or covalent bonding | By keeping the active conformation stable, it can increase the catalytic activity. | Cell damage during immobilization is difficult to avoid. By changing the structure of proteins, the complex immobilization conditions cause a notable decrease in enzyme activity. | (Mahmoudi et al., 2024; Alkayyali et al., 2019) |
| Encapsulation | The correct activity of the inactivated microorganisms is achieved without the need for chemical modification of the raw material. | Reduced mass transfer, fragile immobilization capsule formation, and instability at specific acidities. | (Picos-Corrales et al., 2023; Najim et al., 2024) |

that were isolated in the previous phases must be eliminated by subsequent processes (Alitaleshi et al., 2024). Treatment for oily, heavy metal, agricultural, and industrial wastewater pollutants is conducted using a more advanced immobilized microbial technology (Karimi, 2016; Li et al., 2023).

When wastewater contains refractory organics, such as cyanide, aniline, and phenol, among other contaminants, the effectiveness of traditional biological treatment techniques is reduced. This is because microorganisms struggle to survive in conventional biological treatment systems and require more time to complete their life cycles. Instead, an immobilized technique can solve this situation. In addition, the approach that employs immobilized microbes exhibits comparatively high treatment effectiveness when treating wastewater with high organic concentrations (Ogundolie et al., 2024).

In a previous investigation, wastewater containing acetonitrile was treated by immobilizing three distinct bacterial consortium strains on a new biofilm and allowing them to develop. Laboratory tests of these bacteria have shown that the technique successfully enhances the oxidation of toxicants in wastewater (Ismail and Khudhair, 2015). Polyurethane foam was created as a carrier to immobilize *Fusarium* species and enable the breakdown of phenols in a full mixer. Agar, polyacrylamide, carrageenan, and alginate were studied as possible carriers for immobilized microbes in phenol degradation (Zhang et al., 2015).

For the treatment of oil field wastewater in a bioreactor, the potential application of a consortium of actinobacteria from *Rhodococcus opacus* IEGM 263 and *Rhodococcus ruber* IEGM 231 immobilized on a polymeric carrier has been studied (Serebrennikova et al., 2017). According to a recent study by Choi et al. (2020), immobilizing anammox bacteria in flat-type cryogel films utilizing polyvinyl alcohol (FT-CPVAF) instantly revived the nitrogen removal that had originally been suppressed by an oxygen shock, with only a brief lag period. For the treatment of ammonium-rich synthetic wastewater, a new two-stage partial nitrification (PN)-anammox process was run in continuous flow employing FT-CPVAF. The nitrogen removal rate was $89.6\% \pm 0.76\%$ at a short hydraulic retention period (7.7 h).

The impact of immobilized *Bacillus* sp. P1's degradation of phenanthrene in the presence of varying Cd(II) concentrations was examined by Liu et al. (2019a). PVA and sodium alginate were used to immobilize Bacillus sp. P1, and PVA-SA-cell cryogel beads were made. According to the findings, using gel beads increased the number of adsorption sites and subsequently accelerated the breakdown of phenanthrene. According to another study by the same authors, biotrickling filters (BTFs) were packed with environmentally friendly cell-immobilized bamboo biochar or cell biochar beads. Cell-biochar beads increase the effectiveness of BTFs' pollution removal by acting as both an adsorber of pollutants and a carrier for developing attached bacteria. Over 99% of the toluene was eliminated (Liu et al., 2019b).

Cryogels are employed in many biotechnology applications, such as chromatographic materials and carriers for cell and molecular immobilization. *Trichoderma* spp. embedded supermacroporous cryogels based on poly(2-hydroxyethyl methacrylate) were created in a study to destroy cyanide. To determine their appropriateness for this use, the produced polymeric materials were evaluated for cyanide degradation at various pH levels, temperatures, and starting concentrations (Denizli and Aslıyüce, 2017).

Selecting affordable and environmentally friendly support matrices is essential because those used for immobilization may be costly. *Aspergillus terreus* has been successfully immobilized using inexpensive matrices such as bagasse, coconut coir, loofah sponge, and gunny bag (jute) (Ismail and Khudhair, 2015). *A. terreus* dramatically reduced chemical oxygen demand and biological oxygen demand, two important metrics that gauge how well microbes in wastewater use oxygen, after being immobilized on the matrices mentioned. Gunny bags were the most successful immobilizing medium for *A. terreus* in terms of lowering pollutant levels and enhancing effluent quality when compared to other matrices. Accordingly, the *A. terreus* strain effectively removed colors and broke down organic waste in effluents (Jogdand et al., 2012).

## 2.7 Minor applications of immobilization

There are several reasons to immobilize various cell cultures with an immobilization matrix. For instance, *Chlorella pyrenoidosa* and *Gluconobacter oxydans*, which are algae and bacteria, respectively, were coimmobilized in calcium alginate beads to produce dihydroxyacetone continuously (Ali et al., 2019; Alitaleshi et al., 2024; Alromi et al., 2021; Bekhit et al., 2022; Al Meani et al., 2020). Another study explored the possibility of coimmobilizing the bacterium *Bacillus subtilis* and the microalga *Scenedesmus obliquus* in carrageenan beads and placing them within air lift reactors to enhance the production of the alpha-amylase enzyme. Alpha-amylase activity increased by 20% due to coimmobilization, which addresses the oxygen diffusion issues (Chevalier and de la Noüe, 1988).

Further, encapsulating the green algae *Dunaliella tertiolecta* in alginate and *Dunaliella salina* in agar-agar increased glycerol production (Ismail and Khudhair, 2015). The use of continuous immobilized yeast cell technology to speed up conventional fermentation processes has been studied and assessed. Yeast metabolism can be altered by immobilized cell systems, which are heterogeneous systems where significant mass transfer constraints may arise. The design of the accessory reactor should be coordinated with the appropriate choice and layout of the carrier material. A discussion of the several carrier materials that have been tried and tested for beer manufacturing took place (Willaert, 2000).

Immobilization of cells and enzymes is frequently employed in the industrial manufacturing of valuable goods. Value-added compounds like 1,3-propanediol have been produced recently using immobilization techniques (Gungormusler-Yilmaz et al., 2016). *Komagataeibacter xylinum* B-12429 cells fixed in poly(vinyl alcohol) cryogel were used as a novel biocatalyst to produce bacterial cellulose. By immobilizing the cells in the poly(vinyl alcohol) cryogel, it was possible to "deceive" them. The cellulose-producing bacteria pushed it out, then went through the cryogel matrix's pores and accumulated in the medium without covering the cells. As a result, the cells were prevented from potentially going into inactivity and continued to work on synthesizing bacterial cellulose. For at least ten working cycles (60 days), immobilized cells were repeatedly used while maintaining 100% of their metabolic activity for cellulose production (Stepanov and Efremenko, 2018). A study used immobilization to investigate and assess sodium alginate at varying concentrations to synthesize protease enzymes from *P. aeruginosa*. For repeated batch fermentation and immobilized cells, sodium alginate is an appropriate matrix (Al Meani et al., 2020). Further examples of microbial immobilized cell applications are summarized in Table 4.2.

**Table 4.2** Examples of the application of immobilized microbial cells.

| Microorganism | Degraded/produced material | Immobilizing matrix | References |
|---|---|---|---|
| *Bacillus* cells | Phenol | Polyvinyl alcohol–sodium alginate (PVA–SA) | (Ismail and Khudhair, 2015) |
| *Bacillus* species | Phenol | Tea waste biomass | (Gupta and Balomajumder, 2015) |
| *Pseudomonas putida* | Phenol *o*-, *p*-, and *m*-cresols | Polyvinyl alcohol (PVA) | (El-Naas et al., 2017) |
| Bacterial strain (*Corynebacterium glutamicum*) | As(III), As(V) | Neem leaves/$MnFe_2O_4$ composite | (Podder and Majumder, 2019) |
| *Bacillus* species | Cr(VI) | Tea waste biomass | (Gupta and Balomajumder, 2015) |
| Modified sludge biomass | (MSB) Cu(II) Cr(VI) | Calcium alginate polyvinyl alcohol (PVA) | (Ramteke and Gogate, 2016) |
| Microalgae | Nutrients | Agar–alginate algal blocks (AAAB) | (Kumar et al., 2016) |
| Nitrifier cells | NH4-N | Polyvinyl alcohol and sodium alginate | (Wang and Chen, 2016) |
| *Pseudomonas putida* and *Bacillus licheniformis* | Reactive dyes (RR195, RO 72, RY 17, RB 36) | Sodium alginate and polyacrylamidegel beads | (Suganya and Revathi, 2016) |
| *Bacillus subtilis* | Methylene blue | Calcium alginate bead | (Upendar et al., 2016) |
| *Bacillus* spp., *Rhodococcus* spp. | Removal of ammonia | Polyvinyl alcohol (PVA) cryogel | (Serebrennikova et al., 2017; Choi et al., 2020) |
| *Actinobacillus succinogenes*, *Rhizopus oryzae* spp. | Succinic, fumaric, as well as lactic acid removal. | PVA cryogel | (Bouabidi et al., 2019) |
| *Bacillus cereus* | Degradation of crude oil | Polyacrylamide (PAAam)–N,N'-methylenebisacrylamide (BisAAm) cryogel | (Tekere, 2019) |
| *Bacillus* spp. | Removal of PAH | PVA–$H_3BO_3$–Ca-Alg beads | (Liu et al., 2019a) |
| *Chlorella* spp. | Removes ions Zn, Pb, Ni, Cu, Cd, Hg, nitrite, $NH_4$, and uranium | Agar beads, agarose, carrageenan, polyurethane foam, and alginate | (Liu et al., 2019b; Baimenov et al., 2020) |

*Continued*

**Table 4.2** Examples of the application of immobilized microbial cells—cont'd

| Microorganism | Degraded/produced material | Immobilizing matrix | References |
|---|---|---|---|
| *Komagataei bacterxylinum* | Water treatment, antimicrobial activity *E. coli* | Polylysine-b-polyvaline GA cryogel.<br>Gum Arabic linked with divinyl sulfone | (Denizli and Aslıyüce, 2017) |
| *Pseudomonas citronellolis* | Toluene and hydrocarbons | PVA bamboo–biochar beads | (Suner and Sahiner, 2018) |
| *Trichoderma* spp. | Cyanide removal | HEMA cryogel | (Stepanov and Efremenko, 2018) |
| *Pseudomonas aeruginosa* | Protease | Sodium alginate | (Al Meani et al., 2020) |
| *Escherichia coli* | Cylclodectrin (CD) production | Not recorded | (Zakaria et al., 2023) |
| *Saccharomyces cerevisiae* | 2G ethanol production | Ca-alginate and Ca-alginate–chitosan hybrid beads | (Soares et al., 2022) |

## 3 Microbial cells supported or coated by nanomaterials

Like nanomaterials technology, nanotechnology is a rapidly developing science that is propelling an industrial revolution, with NPs finding ever-increasing industrial applications. Nanotechnologies provide high-quality materials with unique properties and assist scientific research in developing specific procedures and machinery for various applications, including energy, health, environmental treatment, and agriculture (He et al., 2019). NPs have recently been used to support the immobilization of microbes (Zaki et al., 2022). NPs' distinct physicochemical characteristics, such as their large surface area, resistance to mass transfer, high mechanical strength, and effective diffusion, make them an intriguing and novel matrix for immobilizing microbial cells.

A study used cross-linking agents, adsorption, and precipitation (GA and ammonium sulfate) to immobilize laccase on a cheap nanosized bagasse magnetic biochar. The charcoal carrier increased catalytic activity, storage stability, and pH and temperature flux tolerance by improving enzyme immobilization. However, at pH values of 5.5 and 6, the immobilized laccase catalytic activity diminished by over 50%, demonstrating the method's vulnerability to certain environmental factors (Zhang et al., 2020). Alpha-amylase immobilized on the surface of metal oxide NPs was successfully used as a biosorbent for pollutant removal (Długosz et al., 2021). A biosorbent created by immobilizing *Sphingomonas* sp. on $Fe_3O_4$ NPs and gellan gum allowed (Wang et al., 2007) to bioremediate wastewater using a modified entrapment technique. Apart from the density of immobilized enzymes, Ding et al. (2015)

showed that the NP size and surface chemistry preserved or even enhanced immobilized enzyme performance in immobilization schemes.

## 3.1 Nanoencapsulation microbial cells

Agglomerated microorganisms can be preserved and their stability increased by coating MNPs with organic compounds like alginate. In addition, the coating may supply functional groups for efficient biosorption (Alromi et al., 2021). When Yao et al. (2018) encapsulated the probiotics (*Pediococcus pentosaceus* $Li0_5$) in MgO NPs and alginate–gelatin microbeads, they demonstrated excellent vitality. However, Suvarli et al. (2022) were able to encapsulate the enzyme-conjugated nanoparticles after effectively conjugating β-galactosidase onto the surface of polymer NPs.

After being separated through continuous enrichment culturing, bacterial strains that are effective at breaking down target pollutants are often coated (encapsulated) to create NP-bacteria assemblies (Wang et al., 2020). In general, the following traits should be present in bacteria that are utilized to immobilize using nanoforms:

1. A cell surface that readily accepts NP attachment or adhesion.
2. A powerful ability to break down specific organic contaminants.
3. The capacity to stimulate the production of valuable enzymes following NP decorating.

The secret to a coated or encapsulated composite lies in selecting a stable, cost-effective support with excellent biocompatibility. Graphene–magnetite nanocomposites (Ramalingam et al., 2018), α-$Fe_2O_3$ NPs (Khatoon and Rai, 2018), γ-$Al_2O_3$ NPs (Zhang et al., 2011), and $Fe_3O_4$ NPs (Wang et al., 2020) are frequently utilized. Iron oxide NPs have demonstrated significant potential for use in the encapsulation process due to their distinct magnetic characteristics, high surface-to-volume ratio, surface functional groups, shape, and enhanced biocompatibility (Khatoon and Rai, 2018). However, due to their high reactivity, naked metallic NPs are easily oxidized, resulting in the loss of their dispersibility and magnetism (Lu et al., 2007). To achieve adequate bioremediation effectiveness, NP surfaces have been modified using various substances, including glycine and polydopamine (Bai et al., 2018; Bestawy et al., 2020). The following are the primary benefits of modified nanomaterials for use in encapsulation techniques for practical bioremediation:

1. They can shield the magnetic core from external disturbances.
2. They can stop MNPs from clumping together and promote the creation of smaller NPs (Bardania et al., 2013).
3. They can improve NPs' surface functionalization (Xiao and Wiesner, 2013).

However, metal NPs, such as those of Ag, Au, Cu, and Cd, and metal oxides ($TiO_2$, $CeO_2$, ZnO, CuO, $FeO_2$, $Al_2O_3$, and $SiO_2$), generate metal complexes (Co–Zn–Fe oxide) that have been shown to have harmful effects on microorganisms. According to data, the toxicity of NPs to microorganisms is caused in three primary ways (Fig. 4.6):

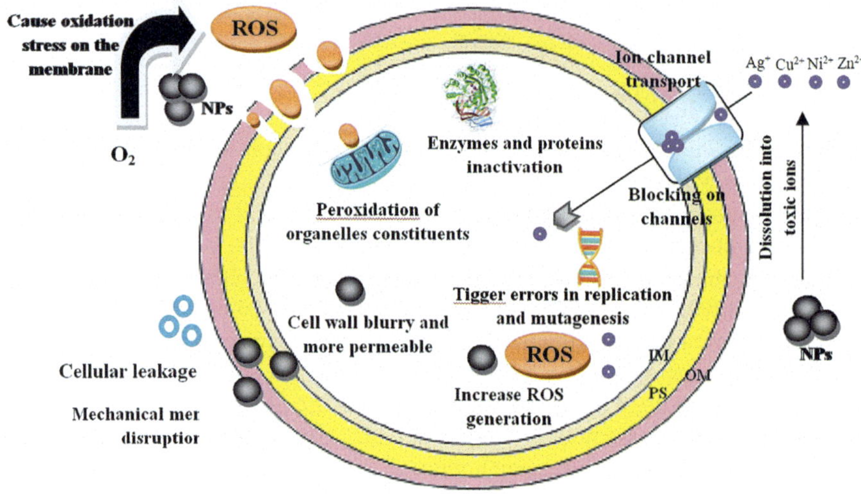


FIG. 4.6

Nanoparticles' toxicity to microorganisms' cells caused by several primary ways.

1. Metal NPs dissolve into hazardous ions. Silver, copper, nickel, and zinc are examples of metal ions that can (a) cause the inactivation of enzymes and proteins, (b) cause errors in replication and mutagenesis, (c) affect enzymatic reactions during the production of reactive oxygen species (ROS), and (d) cause ion imbalance by blocking ion channels (Pal et al., 2022).
2. Producing ROS, such as superoxide and hydroxyl radicals, occurs through a reaction with molecular oxygen (Pathakoti et al., 2020; Pathakoti et al., 2020). Overproduction of ROS can lead to homeostatic imbalance, oxidative stress on the membrane, peroxidation of several organelle contents, cellular metabolic disruption, and even death (Wang and Chen, 2016).
3. Disrupting mechanical membranes: depolarization, blurriness, increased permeability, and even cellular leakage are caused by the adsorption of certain NPs (Pagnout et al., 2021; Slavin et al., 2017).

Therefore before encapsulation, testing the bacteria's tolerance to the target NPs is required. Environmental factors, exposure duration, and the NPs' characteristics (size, shape, and type) all affect how harmful they are. Gold, palladium, silver sulfide, platinum, and other weakly soluble NPs are generally less toxic than metal or metal oxide NPs (Suresh et al., 2011).

Encapsulating microbial NPs is as easy as adding nanomaterials to bacterial solutions and thoroughly mixing them for a while. It is essential to observe the ratios of nanomaterials to microbial biomass. When the NP dose exceeded 100 μg/L, high NP encapsulation efficiencies were observed (Ebrahiminezhad et al., 2014). However, bacterial growth will be inhibited when the concentration of NPs exceeds the bacterium-specific acceptable limit (Khatoon and Rai, 2018). For example, when the ratio of magnetic particles to cells was 1:100 (g/g), the desulfurization activity of coated *Rhodococcus erythropolis* LSSE8-1-vgb was marginally suppressed, and at 1:25 (g/g), it was severely inhibited. This may be explained by the possibility that overly dense covering could function as a barrier to the mass transfer process (Wang et al., 2017).

The idea behind using NP-encapsulated microorganisms to biodegrade pollutants in situ is that the bacteria–NP adhesion strength should be high enough to withstand environmental interference. Washing the cell surfaces with ethanol, saline water, phosphate buffer, or deionized water did not significantly reduce the biomass loading of NP-coated cells (Li et al., 2009). Furthermore, the relationship between the coated bacteria and the carrier is severely harmed by aeration operations, a frequent biodegradation technique in wastewater treatment (Wang et al., 2020). Nearly every cell was still attached to the carrier, suggesting that the bond was strong.

### 3.1.1 Performance of NP-encapsulated and free bacteria

A comparison was made between the pollutant degradation performance of free and NP-encapsulated cells. The NP-encapsulated cells showed comparable activity to free cells for the degradation of 2-hydroxy biphenyl (Bardania et al., 2013), carbazole (Zakaria et al., 2015), and di-n-butyl phthalate (Wang et al., 2017), indicating that mass transfer hindrance did not occur in this system. Furthermore, compared to free microorganisms, NP-encapsulated microorganisms exhibit more excellent resistance to changing environments. Atrazine concentration (50–300 mg/L), temperature (20°C–45°C), agitation speed (50–300 rpm), pH (4–9), and other reaction conditions were used to compare the degradation efficiency of free and $\alpha$-$Fe_2O_3$-NP-encapsulated bacterial cells. The $\alpha$-$Fe_2O_3$–NP-encapsulated cells' atrazine degradation efficiency, which is maintained at a high level under all reaction conditions, varies only slightly. However, when external environmental parameters were measured outside the microbial optimum conditions, the breakdown capacity of free cells dramatically declined. In addition, after being encapsulated with MNPs, the inhibitory effect of heavy metals such as $Cr^{6+}$, $Pb^{2+}$, $Cu^{2+}$, and $Ni^{2+}$ was lessened and nearly returned to control levels (Wang et al., 2017). Once again, NP-coated cells are more stable in storage than uncoated cells. The efficiency of the free cells dropped to 55% after 30 days of storage at 4°C, but the $Fe_3O_4$ NP-coated cells continued to maintain a steady carbazole denitrogenation rate (Zakaria et al., 2015). Few studies have been published on the mechanism underlying NP-encapsulated cells' enhanced resistance to acidity and salinity. However, bacteria and enzymes coated with NPs or encapsulated in alginate beads showed resistance to unfavorable environments, and the mechanism of resistance involving them was suggested (Tarasi et al., 2018; Ghorbannezhad et al., 2018). Examining the resistance mechanism is required to improve the NP-bacteria assemblies' capacity to degrade pollutants in challenging environments. During DBP breakdown, the same metabolites were found in cultures of free and NP-encapsulated cells, indicating that the DBP degradation pathway remained unchanged upon strain encapsulation (Wang et al., 2017). However, this was not always the case, and further research is required to determine how NP-bacteria assemblages degrade pollutants. NP-encapsulated microorganisms have significant advantages in practical pollutant treatment due to their persistent degrading activity for numerous uses and ease of recovery when compared with free cells. The ability of microorganisms in nanomaterials to degrade pollutants remained high after multiple use cycles. Furthermore, during the sixth cycle, $Fe_3O_4$-coated *Sphingomonas* eliminated 80% of nonylphenol polyethoxylates in two days, surpassing the clearance rate of free cells (67%); it also exhibited higher activity overall (89% for NP-encapsulated cells compared to 77% for free cells) (Bai et al., 2018). There are, nevertheless, restrictions on the number of reaction cycles. Overall, after four cycles, the NP-encapsulated *Acinetobacter* sp. strain LMB-5 showed high activity (>90%); however, as the number of reaction cycles increased, the DBP reduction rate became disappointing (Wang et al., 2017). Another study examined the DBP degradation efficiency of *Pseudomonas* sp. W1 that had been immobilized using $Fe_3O_4$ iron nanoparticles (IONPs) coated with polydopamine (PDA). Immobilization enhanced W1's DBP degradation capability; on the sixth day after three cycles, the DBP degradation efficiency was 99.69%, 25.68% greater than that of unimmobilized W1 (Wang et al., 2020).

### 3.1.2 Factors that influence bioremediation using encapsulated bacteria

The initial pollutant concentration, temperature, pH, agitation, and other environmental conditions significantly impact how well NP-coated microbes remove pollutants. Altering some characteristics of NPs or preventing the growth of microbes that break down pollutants can affect the degradation process.

When the pH changed to an alkaline or acidic state, the $\alpha$-$Fe_2O_3$-coated bacterial cells' capacity to degrade atrazine dramatically dropped (Khatoon and Rai, 2018). Too high or too low pH will affect the physicochemical properties of microorganisms and the speciation of pollutants. Moreover, the pH constantly modifies the bacterial cell surfaces' or NPs' interfacial electrostatics (Ansari et al., 2009; Mehrotra et al., 2021). SMFS and DFS show that altering the pH of the solution controls fimbrial extension and fimbrial binding force to the mannosylated surface (Jacquot et al., 2014). In other words, the strength of the electrostatic force between the bacterial cells and the availability of attachment on their surfaces are both impacted by the pH of the solution. For example, measuring the contact force between dendrimer NPs and *Escherichia coli* surfaces at pH 3.5 revealed adhesion forces (Ismail and Khudhair, 2015) ranging from 100 to 500 pN, whereas at pH 6.0, nearly no adhesion forces were detected (Beaussart et al., 2018).

Temperature still affects how well the coated cells degrade, even though NP-coated cells can withstand a wider range of temperatures than free cells (Zhuang et al., 2015). NPs have an excellent adsorption capability at higher temperatures. Higher temperatures, however, have a detrimental impact on bacterial activity and can reduce oxygen solubility, further impairing the bacteria's capacity for biodegradation. The effect of varying agitation rates (50–300 rpm) on atrazine breakdown by NP-coated bacteria was examined (50–300 rpm) (Khatoon and Rai, 2018).

Bacterial cells coated with α-$Fe_2O_3$ showed significant atrazine breakdown at 150 rpm. The degradation rate dramatically dropped as the agitation speed moved toward higher and lower ranges. This suggested that it's critical to determine the proper agitation speeds for applying NP-coated bacteria. In addition, other factors influence the degrading ability of NP-coated bacteria and the effectiveness of its pollutant removal (Zdarta et al., 2018): (1) the surface properties of the coated NP and its surface charge have an impact on adherence to carrier surfaces, (2) the surface structure of microbes, and (3) characteristics of coated NP: organic types are chemically stable, whereas inorganic types are soluble in alkaline solutions and can be readily regenerated.

### 3.1.3 Pollutant removal mechanism by NP-encapsulated bacteria

Generally speaking, NP adsorption and microbial degradation are the primary mechanisms by which NP–bacteria assemblies remove contaminants. In addition, the constraints on the mass transfer rate to the cell body through the ambient phase and the catalytic activity of associated enzymes in cells are counteracted by the combined effects of bacteria and NPs (Fig. 4.7). The adsorption of NPs speeds up the transfer of pollutants from the environment to the cell surfaces because of their large specific surface area and high surface activity. This increases the rate of biodegradation.

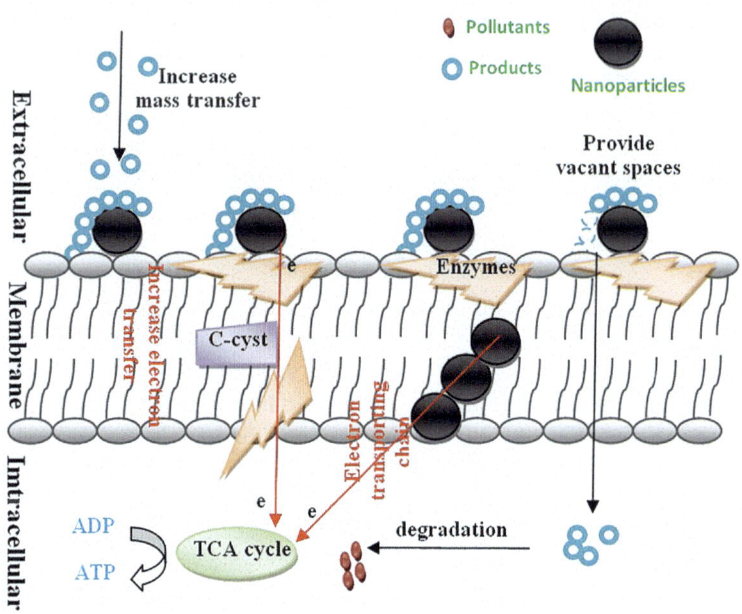


**FIG. 4.7**

Mechanical removal of contaminants, including the cooperation of C-type cytochromes, or C-cyts, between bacteria and nanoparticles.

Furthermore, metal NPs (such as Pd(0) NPs and FeS NPs) have been widely used to catalyze a variety of chemical reactions due to their diversified catalytic surfaces and greater chemical stereoselectivity (Huo et al., 2016; Mertens et al., 2007).

Microbial redox reactions, in which bacteria extract electrons from contaminants, are frequently the driving force behind biological treatments of pollutants. Electron transfer efficiency and cellular cytochrome activity were enhanced by the addition of NPs to bacterial surfaces (Wu et al., 2011). For instance, when *Pseudomonas putida* was decorated with Pd(0) NPs, the phenol biodegradation rate constant (8%–189%) and electron transfer rates (63.97% or 33.79%) were greater than those of native *P. putida*. Furthermore, the electrical conductivity and physical permeability of certain NPs significantly accelerated the electron-transporting chain and proton motive force, facilitating metabolic activity (including ATP synthesis and enzyme catalytic activities) (Niu et al., 2018). The pollutants that have been adsorbed on or desorbed from the NPs can be broken down by bacteria at the same time, causing naturally occurring holes in the NPs that enable toxins to continuously adsorb.

## 4 Biotechnology application of nanocell encapsulation

NP-encapsulated bacteria have demonstrated excellent environmental adaptability and application potential, particularly in challenging conditions such as high salinity, heavy metals, and extreme temperatures (both low and high). Research on NP-encapsulated bacteria is still in its early stages, and numerous challenges remain when employing them to remediate contaminated areas. Selecting an effective NP surface modification can enhance biological activities and induce unique physicochemical properties (Tarasi et al., 2018). The two main categories of coating materials for $Fe_3O_4$ NPs are coatings with organic shells, such as polymers (Yang et al., 2014) and surfactants (Liang et al., 2016), or coatings with inorganic components, such as carbon (Xiong et al., 2021), silica (Vishnu et al., 2021), and precious metals (e.g., Ag, Au).

NP-bacteria assemblies' protective shells must be biocompatible, nontoxic, and suitable for further functionalization with specific binding sites. To date, NPs have been modified using various materials for bacterial encapsulation technology. Finding the best-adapted materials for certain bacteria is still crucial, though. Laccase immobilization was effectively generated and evaluated on $Fe_3O_4$ NPs coated with poly(amidoisophthalic acid) (Fe@PA), poly(amidoisophthalic acid) and cyclodextrin (Fe@PA-CD), or poly(amidoisophthalic acid), cyclodextrin, and chitosan (Fe@PACD-Cs). Compared to the other two carriers, laccase immobilization on Fe@PACD-Cs shows more acceptable removal efficiencies for phenol and bisphenol, along with improved heat and pH stabilities (Feng et al., 2023).

One key tactic for making many processes economically feasible may be the use of cell nanocarriers in biotechnology applications to create nanobiocatalysts

(Shahbaz et al., 2022). Green energy can be produced by biotransforming lignocellulosic materials into bioethanol (Azeez and Al-Zuhairi, 2020). Immobilized enzymes are preferable over MNPs for hydrolyzing lignocellulosic materials because they are quickly recovered and reusable, extending the biocatalyst's life and lowering the cost of biocatalytic production (Zdarta et al., 2018). Singhvi and Kim (2020) examined how immobilized enzymes on MNPs biotransform lignocellulose into monosaccharides. A viable method for achieving effective biomass degradation and increasing the efficiency of biofuel generation is the incorporation of NPs.

Ivanova et al. (2011) discussed the use of trapped yeast cells with alginate/MNPs, which improved the generation of bioethanol from maize starch to 264.0 g/L/h while lowering the sugar concentration by 200 g/L. Sarwar et al. (2022) found that immobilized *Saccharomyces cerevisiae* SS-4 in calcium alginate/MNPs improved bioethanol production from wheat straw, achieving an optimal yield of 49.71 g/L. Meanwhile, immobilized yeast cells in calcium alginate produced 45.66 g/L of bioethanol, and free cells produced 36.52 g/L under optimal conditions of pH 4.5, 72 h at 28°C. The biofuel market has emerged as the most energy-efficient and environmentally friendly fuel option due to the production of biodiesel with significant qualities. Wang et al. (2011) used a four-packed-bed reactor with immobilized lipase on $Fe_3O_4$ NPs to convert over 88% of feedstock to biodiesel in 192 h. However, using immobilized *Thermomyces lanuginosus* lipase on MNPs, Raita et al. (2015) achieved 97.2% biodiesel synthesis from palm oil under ideal conditions of 23.2 wt% enzyme loading, a 4.7:1 methanol-to-oil ratio, 50°C, and 24 h. Furthermore, Karimi (2016) found that immobilized lipase on superparamagnetic iron oxide NPs produced 91% biodiesel from cooking waste in just 35 h.

NP-bacteria assemblies exhibit strong environmental adaptability and potential for use. In situ applications, however, have been the subject of relatively few reports. There is an urgent need to investigate the following aspects: (1) Cocontaminant inhibitory effects: The sole goal of research on NP-encapsulated microbial technology in the laboratory is to eliminate particular contaminants. The contaminants are complex for any polluted sites (farmland or coastal restoration, urban or industrial wastewater treatment, etc.). Regretfully, bioremediation is typically significantly inhibited by compound contaminants. (2) Concentration in the environment: Even though most NP-encapsulated bacteria can withstand concentrations of 150–1000 mg/L or higher, the actual environmental concentrations of organic pollutants are usually <1 mg/L, which may impede pollutant biodegradation (e.g., bioavailability of pollutants and concentrations too low to sustain cell growth or to induce the expression of functional enzymes) (Ren et al., 2018). In addition to being able to withstand high pollutant concentrations, the NPs–bacteria assemblies should also trigger the degradation process at trace ambient concentrations. In the meantime, more research is needed on other aspects, such as biological competitiveness, biological state, interactions between different contaminants, and the bioavailability of pollutants. Table 4.3 summarizes the use of different nanomicrobial encapsulated techniques regarding previous studies.

**Table 4.3** Previous studies related to pollutant removal using the nanomicrobial encapsulation approach.

| Pollutants | Microorganisms | Nanocoated | References |
|---|---|---|---|
| Sr(II) | *Kluyveromyces fragilis yeast* | Magnetic $Fe_3O_4$ NPs | (Ji et al., 2010) |
| Pb(II) | *Penicillium funiculosum (enzymes)* | Silicon dioxide NPs | (Mahmoud et al., 2012) |
| Pb(II) | *Phanerochaete chrysosporium (fungus)* | Ca-alginate–$Fe_3O_4$ NPs | (Xu et al., 2013) |
| Cu(II) | *Saccharomyces cerevisiae (yeast)* | Water-based magnetic fluid stabilized with perchloric acid | (Uzun et al., 2011) |
| Ni(II) | *Saccharomyces cerevisiae (yeast)* | nZVI | (Guler and Ersan, 2016) |
| Sr(II) | *Saccharomyces cerevisiae (yeast)* | Chitosan–$Fe_3O_4$ NP | (Yin et al., 2017) |
| Co(II)and Hg(II) | *Pleurotus eryngii (fungus)* | Maghemite ($\gamma$-$Fe_2O_3$) | (Ozdemir et al., 2018) |
| $Cd^{2+}$and $Pb^{2+}$ | *Lactic acid bacteria cells* | Magnetic NPs | (Li et al., 2020) |
| Five types of organic dyes | *Sargassum muticum (macroalgae)* | Iron oxide (NPs) | (Angelova et al., 2016) |
| Methylene blue (MB) | *Agrobacterium fabrum (bacteria)* | Iron oxide (NPs) | (Sharma et al., 2018) |
| Azo dyes (DO 34, CR, Orange II, and Red MX-5B) | *Bacillus subtilis (bacteria)* | Superparamagnetic $Fe_3O_4$ NPs | (Nadi et al., 2019) |
| Basic Red 46 azo dye (BR 46) | *Enterobacter cloacae* sp. *(bacteria)* | Coated cells with magnetic $Fe_3O_4$ NPs | (Bekhit et al., 2020) |
| Methyl orange (M.O.) | *Saccharomyces cerevisiae (yeast)* | $Fe_3O_4$ NPs | (Azeez and Al-Zuhairi, 2020) |
| Basic Blue 41 azo dyes | *Raoultella ornithinolytica* sp. *(bacteria)* | Magnetic $Fe_3O_4$ NPs | (Bekhit et al., 2022) |
| RO16 and Methylene Blue | *Manganese peroxidase (MnP) (enzyme)* | Iron oxide–chitosan | (Siddeeg et al., 2020) |

## 5 Conclusions

One interesting approach that shows promise for treating waste sites free of contaminants is the immobilization of microbial cells. The future of environmental bioremediation and protection will be shaped by several immobilization technologies that are now under development, in addition to the traditional methods of immobilizing microbial cells (encapsulation, trapping, cross-linking, and physical adsorption). The most popular small-scale microbial immobilization method for trapping is encapsulation, which has shown promise in numerous commercial and experimental studies. Similarly, although there are advantages and disadvantages to microbial cell immobilization methods, many of the disadvantages are worth investigating scientifically. In biotechnology applications, microbial immobilization on NPs works incredibly well. NPs enhance microbial performance because of their unique physicochemical characteristics, including high surface area, efficient diffusion, and biocompatibility. Biotechnology applications include the use of nanoencapsulation in a variety of methods to remove pollutants from industrial effluents. Consequently, various biotechnology applications have used the microbial NP encapsulation technique, a developing bioremediation method with enormous potential. However, an extensive research gap needs to be filled to increase scientific understanding of the underlying mechanisms of action of these developing techniques.

## References


Ahmad, H.A., Ni, S.Q., Ahmad, S., Zhang, J., Ali, M., Ngo, H.H., Guo, W., Tan, Z., Wang, Q., 2020. Gel immobilization: a strategy to improve the performance of anaerobic ammonium oxidation (anammox) bacteria for nitrogen-rich wastewater treatment. Bioresour. Technol. 1873-2976. 313. Available from: www.elsevier.com/locate/biortech. doi:10.1016/j.biortech.2020.123642.

Al Meani, S.A.L., Al-Duliamy, E.W., Al-Shojiary, A.S., Turky, N.M., Ahmed, Y.K., 2020. Production of protease using immobilization cells technique. Biochem. Cell. Arch. 09761772. 20(1), 1791–1795. Available from: http://www.connectjournals.com/useradmin/toc2.php?fulltext=3142201H_1791-1795.pdf&&volume=20&&bookmark=CJ-033216&&article=Production%20of%20protease%20using%20immobilization%20cells%20technique. doi:10.35124/bca.2020.20.1.1791.

Al-Rubaiey, N., 2022. A trends in ozone treatment of wastewater: a review. Iraqi J. Oil Gas Res. 2790-2137. 2 (1), 55–64. doi:10.55699/ijogr.2022.0201.1016.

Ali, H., Khan, E., Ilahi, I., 2019. Environmental chemistry and ecotoxicology of hazardous heavy metals: environmental persistence, toxicity, and bioaccumulation. J. Chem. 2090-9063. 2019, 1–14. doi:10.1155/2019/6730305.

Alitaleshi, F., Daghbandan, A., Pendashteh, A., 2024. Performance of rice husk biocarrier on ammonia nitrogen removal in the MBBR treating aquaculture wastewater using biological attached growth process: performance and kinetic study. J. Environ. Chem. Eng. 22133437. 12 (1), 111446. doi:10.1016/j.jece.2023.111446.

Alkayyali, T., Cameron, T., Haltli, B., Kerr, R.G., Ahmadi, A., 2019. Microfluidic and cross-linking methods for encapsulation of living cells and bacteria—a review. Anal. Chim. Acta. 00032670. 1053, 1–21. doi:10.1016/j.aca.2018.12.056.

Alromi, D.A., Madani, S.Y., Seifalian, A., 2021. Emerging application of magnetic nanoparticles for diagnosis and treatment of cancer. Polymers. 20734360. 13 (23). Available from: https://www.mdpi.com/2073-4360/13/23/4146/pdf. doi:10.3390/polym13234146.

Angelova, R., Baldikova, E., Pospiskova, K., Maderova, Z., Safarikova, M., Safarik, I., 2016. Magnetically modified Sargassum horneri biomass as an adsorbent for organic dye removal. J. Clean. Prod. 09596526. 137, 189–194. doi:10.1016/j.jclepro.2016.07.068.

Ansari, F., Grigoriev, P., Libor, S., Tothill, I.E., Ramsden, J.J., 2009. DBT degradation enhancement by decorating Rhodococcus erythropolis IGST8 with magnetic $Fe_3O_4$ nanoparticles. Biotechnol. Bioeng. 0006-3592. 102 (5), 1505–1512. doi:10.1002/bit.22161.

Azeez, R.A., Al-Zuhairi, F.K., 2020. Biofuels (bioethanol, biodiesel,and biogas) from lignocellulosic biomass: a review. J. Univ. Babylon Eng. Sci. 28, 202–215.

Bai, N., Wang, S., Sun, P., Abuduaini, R., Zhu, X., Zhao, Y., 2018. Degradation of nonylphenol polyethoxylates by functionalized $Fe_3O_4$ nanoparticle-immobilized Sphingomonas sp. Y2. Sci. Total Environ. 00489697. 615, 462–468. doi:10.1016/j.scitotenv.2017.09.290.

Baimenov, A.Z., Berillo, D.A., Moustakas, K., Inglezakis, V.J., 2020. Efficient removal of mercury (II) from water by use of cryogels and comparison to commercial adsorbents under environmentally relevant conditions. J. Hazard. Mater. 03043894. 399, 123056. doi:10.1016/j.jhazmat.2020.123056.

Bardania, H., Raheb, J., Mohammad-Beigi, H., Rasekh, B., Arpanaei, A., 2013. Desulfurization activity and reusability of magnetite nanoparticle–coated Rhodococcus erythropolis FMF and R. erythropolis IGTS8 bacterial cells. Biotechnol. Appl. Biochem. 0885-4513. 60 (3), 323–329. doi:10.1002/bab.1090.

Beaussart, A., Beloin, C., Ghigo, J.M., Chapot-Chartier, M.P., Kulakauskas, S., Duval, J.F.L., 2018. Probing the influence of cell surface polysaccharides on nanodendrimer binding to Gram-negative and Gram-positive bacteria using single-nanoparticle force spectroscopy. Nanoscale. 20403372. 10 (26), 12743–12753. Available from: http://pubs.rsc.org/en/journals/journal/nr. doi:10.1039/c8nr01766b.

Bekhit, F., Farag, S., Attia, A.M., 2020. Decolorization and degradation of the Azo dye by bacterial cells coated with magnetic iron oxide nanoparticles. Environ. Nanotechnol. Monit. Manag. 22151532. 14, 100376. doi:10.1016/j.enmm.2020.100376.

Bekhit, F., Farag, S., Attia, A.M., 2022. Characterization of immobilized magnetic $Fe_3O_4$ nanoparticles on Raoultella Ornithinolytica sp. and Its application for Azo Dye removal. Appl. Biochem. Biotechnol. 15590291. 194 (12), 6068–6090. Available from: https://www.springer.com/journal/12010. doi:10.1007/s12010-022-04076-3.

Berillo, D., Al-Jwaid, A., Caplin, J., 2021. Polymeric materials used for immobilisation of bacteria for the bioremediation of contaminants in water. Polymers. 2073-4360. 13 (7). Available from: https://www.mdpi.com/2073-4360/13/7/1073/pdf. doi:10.3390/polym13071073.

Bestawy, E.E., El-Shatby, B.F., Eltaweil, A.S., 2020. Integration between bacterial consortium and magnetite ($Fe_3O_4$) nanoparticles for the treatment of oily industrial wastewater. World J. Microbiol. Biotechnol. 1573-0972. 36 (9). Available from: www.wkap.nl/journalhome.htm/0959-3993. doi:10.1007/s11274-020-02915-1.

Blanco, A., Sampedro, M.A., Sanz, B., Llama, M.J., Serra, J.L. , 2023. Immobilization of non-viable cyanobacteria and their use for heavy metal adsorption from water. In: Sanchez, G., Hernandez, E. (Eds.), Environmental Biotechnology and Cleaner Bioprocesses, 1st ed. CRC Press, Spain, pp. 135–154. Available from: http://www.tandfebooks.com/doi/book/9781000944747. doi:10.1201/9781003417163-14.

Bouabidi, Z.B., El-Naas, M.H., Zhang, Z., 2019. Immobilization of microbial cells for the biotreatment of wastewater: a review. Environ. Chem. Lett. 16103661. 17 (1), 241–257. Available from: https://www.springer.com/journal/10311. doi:10.1007/s10311-018-0795-7.

Burgain, J., Gaiani, C., Linder, M., Scher, J., 2011. Encapsulation of probiotic living cells: from laboratory scale to industrial applications. J. Food Eng. 02608774. 104 (4), 467–483. doi:10.1016/j.jfoodeng.2010.12.031.

Cassidy, M.B., Lee, H., Trevors, J.T., 1996. Environmental applications of immobilized microbial cells: a review. J. Ind. Microbiol. 01694146. 16 (2), 79–101. doi:10.1007/BF01570068.

Chen, G., Li, J., Tabassum, S., Zhang, Z., 2015. Anaerobic ammonium oxidation (ANAMMOX) sludge immobilized by waterborne polyurethane and its nitrogen removal performance-a lab scale study. RSC Adv. 20462069. 5 (32), 25372–25381. Available from: http://pubs.rsc.org/en/journals/journalissues. doi:10.1039/c4ra14451a.

Chen, J., Spear, S.K., Huddleston, J.G., Rogers, R.D., 2005. Polyethylene glycol and solutions of polyethylene glycol as green reaction media. Green Chem. 1463-9262. 7 (2), 64–82. Available from: http://pubs.rsc.org/en/journals/journal/gc. doi:10.1039/b413546f.

Chevalier, P., de la Noüe, J., 1988. Behavior of algae and bacteria co-immobilized in carrageenan, in a fluidized bed. Enzyme Microb. Technol. 01410229. 10 (1), 19–23. doi:10.1016/0141-0229(88)90093-2.

Choi, M., Chaudhary, R., Lee, M., Kim, J., Cho, K., Chung, Y.-C., Bae, H., Park, J., 2020. Enhanced selective enrichment of partial nitritation and anammox bacteria in a novel two-stage continuous flow system using flat-type poly (vinylalcohol) cryogel films. Bioresour. Technol. 09608524. 300, 122546. doi:10.1016/j.biortech.2019.122546.

Cottet, C., Ramirez-Tapias, Y.A., Delgado, J.F., la Osa, O.D., Salvay, A.G., Peltzer, M.A., 2020. Biobased materials from microbial biomass and its derivatives. Materials. 19961944. 13 (6). Available from: https://res.mdpi.com/d_attachment/materials/materials-13-01263/article_deploy/materials-13-01263.pdf. doi:10.3390/ma13061263.

Cruz, I., Bashan, Y., Hernàndez-Carmona, G., De-Bashan, L.E., 2013. Biological deterioration of alginate beads containing immobilized microalgae and bacteria during tertiary wastewater treatment. Appl. Microbiol. Biotechnol. 14320614. 97 (22), 9847–9858. doi:10.1007/s00253-013-4703-6.

Curtean-Bănăduc, A., Burcea, A., Mihuţ, C.M., Berg, V., Lyche, J.L., Bănăduc, D., 2020. Bioaccumulation of persistent organic pollutants in the gonads of Barbus barbus (Linnaeus, 1758). Ecotoxicol. Environ. Saf. 10902414. 201. Available from: http://www.elsevier.com/inca/publications/store/6/2/2/8/1/9/index.htt. doi:10.1016/j.ecoenv.2020.110852.

Denizli, A., Aslıyüce, S., 2017. Design of PHEMA cryogel as bioreactor matrices for biological cyanide degradation from wastewater. Hacettepe J. Biol. Chem. 1303-5002. 4 (45), 639–645. doi:10.15671/hjbc.2018.208.

Ding, S., Cargill, A.A., Medintz, I.L., Claussen, J.C., 2015. Increasing the activity of immobilized enzymes with nanoparticle conjugation. Curr. Opin. Biotechnol. 18790429. 34, 242–250. Available from: http://www.elsevier.com/locate/copbio. doi:10.1016/j.copbio.2015.04.005.

Długosz, O., Matysik, J., Matyjasik, W., Banach, M., 2021. Catalytic and antimicrobial properties of α-amylase immobilised on the surface of metal oxide nanoparticles. J. Clust. Sci. 15728862. 32 (6), 1609–1622. Available from: https://www.springer.com/journal/10876. doi:10.1007/s10876-020-01921-5.

Dzionek, A., Wojcieszyńska, D., Guzik, U., 2022. Use of xanthan gum for whole cell immobilization and its impact in bioremediation—a review. Bioresour. Technol. 09608524. 351, 126918. doi:10.1016/j.biortech.2022.126918.

Ebrahiminezhad, A., Rasoul-Amini, S., Davaran, S., Barar, J., Ghasemi, Y., 2014. Impacts of iron oxide nanoparticles on the invasion power of listeria monocytogenes. Curr. Nanosci. 15734137. 10 (3), 382–388. doi:10.2174/15734137113096660109.

El-Naas, M.H., Alhaija, M.A., Al-Zuhair, S., 2017. Evaluation of an activated carbon packed bed for the adsorption of phenols from petroleum refinery wastewater. Environ. Sci.

Pollut. Res. 16147499. 24 (8), 7511–7520. Available from: http://www.springerlink.com/content/0944-1344. doi:10.1007/s11356-017-8469-8.

Feng, J.R., Deng, Q.X., Han, S.K., Ni, H.G., 2023. Use of nanoparticle-coated bacteria for the bioremediation of organic pollution: a mini review. Chemosphere. 18791298. 313. Available from: www.elsevier.com/locate/chemosphere. doi:10.1016/j.chemosphere.2022.137391.

Fortman, D.J., Brutman, J.P., De Hoe, G.X., Snyder, R.L., Dichtel, W.R., Hillmyer, M.A., 2018. Approaches to sustainable and continually recyclable cross-linked polymers. ACS Sustain. Chem. Eng. 21680485. 6 (9), 11145–11159. Available from: http://pubs.acs.org/journal/ascecg. doi:10.1021/acssuschemeng.8b02355.

Ghorbannezhad, H., Moghimi, H., Taheri, R.A., 2018. Enhanced biodegradation of phenol by magnetically immobilized Trichosporon cutaneum. Ann. Microbiol. 18692044. 68 (8), 485–491. Available from: http://springerlink.com/content/121707/. doi:10.1007/s13213-018-1353-z.

Giese, E.C., Silva, D.D.V., Costa, A.F.M., Almeida, S.G.C., Dussán, K.J., 2020. Immobilized microbial nanoparticles for biosorption. Crit. Rev. Biotechnol. 15497801. 40 (5), 653–666. Available from: http://www.tandfonline.com/loi/ibty20. doi:10.1080/07388551.2020.1751583.

Gill, J., Orsat, V., Kermasha, S., 2018. Optimization of encapsulation of a microbial laccase enzymatic extract using selected matrices. Process Biochem. 13595113. 65, 55–61. doi:10.1016/j.procbio.2017.11.011.

Górecka, E., Jastrzębska, M., 2011. Immobilization techniques and biopolymer carriers. Biotechnol. Food Sci. 75, 65–86. doi:10.34658/bfs.2011.75.1.65-86.

Guisan, J.M., Fernandez-Lorente, G., Rocha-Martin, J., Moreno-Gamero, D., 2022. Enzyme immobilization strategies for the design of robust and efficient biocatalysts. Curr. Opin. Green Sustain. Chem. 24522236. 35, 100593. doi:10.1016/j.cogsc.2022.100593.

Gül, Ü.D., 2018. Bioremediation of dyes in textile wastewater. Türk Bilim. Derlem. Derg. 11, 24–28.

Guler, U.A., Ersan, M., 2016. S. cerevisiae cells modified with nZVI: a novel magnetic biosorbent for nickel removal from aqueous solutions. Desalination Water Treat. 19443986. 57 (16), 7196–7208. Available from: http://www.tandfonline.com/toc/tdwt20/current. doi:10.1080/19443994.2015.1013992.

Gungormusler-Yilmaz, M., Cicek, N., Levin, D.B., Azbar, N., 2016. Cell immobilization for microbial production of 1,3-propanediol. Crit. Rev. Biotechnol. 15497801. 36 (3), 482–494. doi:10.3109/07388551.2014.992386.

Gupta, A., Balomajumder, C., 2015. Simultaneous removal of Cr(VI) and phenol from binary solution using Bacillus sp. immobilized onto tea waste biomass. J. Water Process Eng. 22147144. 6, 1–10. doi:10.1016/j.jwpe.2015.02.004.

Han, M., Zhang, C., Ho, S.-H., 2023. Immobilized microalgal system: an achievable idea for upgrading current microalgal wastewater treatment. Environ. Sci. Ecotechnol. 26664984. 14, 100227. doi:10.1016/j.ese.2022.100227.

He, X., Deng, H., Hwang, H.M., 2019. The current application of nanotechnology in food and agriculture. J. Food Drug Anal. 10219498. 27 (1), 1–21. Available from: http://www.elsevier.com/journals/journal-of-food-and-drug-analysis/1021-9498#. doi:10.1016/j.jfda.2018.12.002.

Huo, Y.C., Li, W.W., Chen, C.B., Li, C.X., Zeng, R., Lau, T.C., Huang, T.Y., 2016. Biogenic FeS accelerates reductive dechlorination of carbon tetrachloride by Shewanella putrefaciens CN32. Enzyme Microb. Technol. 18790909. 95, 236–241. Available from: www.elsevier.com/locate/enzmictec. doi:10.1016/j.enzmictec.2016.09.013.

Ismail, Z.Z., Khudhair, H.A. 2015. Recycling of immobilized cells for aerobic biodegradation of phenol in a fluidized bed bioreactor. In: Proceedings of 19th World Multi-Conference on Systemics, Cybernetics and Informatics—WMSCI 2015, vol. 2. International Institute of Informatics and Systemics, IIIS, Iraq. pp. 75–80. 9781941763254.

Ivanova, V., Petrova, P., Hristov, J., 2011. Application in the ethanol fermentation of immobilized yeast cells in matrix of alginate/magnetic nanoparticles, on chitosan-magnetite microparticles and cellulose-coated magnetic nanoparticles. Int. Rev. Chem. Eng. 3, 289–299. doi:10.48550/arXiv.1105.0619.

Jacquot, A., Sakamoto, C., Razafitianamarahavo, A., Caillet, C., Merlin, J., Fahs, A., Ghigo, J.M., Duval, J.F.L., Beloin, C., Francius, G., 2014. The dynamics and pH-dependence of Ag43 adhesins' self-association probed by atomic force spectroscopy. Nanoscale. 20403372. 6 (21), 12665–12681. Available from: http://pubs.rsc.org/en/journals/journal/nr. doi:10.1039/c4nr03312d.

Ji, Y.-Q., Huu, Y.-T., Tian, Q., Shao, X.-Z., Li, J., Safarikova, M., Safarik, I., 2010. Biosorption of strontium ions by magnetically modified yeast cells. Sep. Sci. Technol. 45, 499–1504. doi:10.1016/j.biortech.2011.11.081.

Jiang, J.Q., 2015. The role of coagulation in water treatment. Curr. Opin. Chem. Eng. 22113398. 8, 36–44. Available from: http://www.elsevier.com/wps/find/journaldescription.cws_home/725837/description#description. doi:10.1016/j.coche.2015.01.008.

Jogdand, V.G., Chavan, P.A., Ghogare, P.D., Jadhav, A.G., 2012. Remediation of textile industry waste-water using immobilized Aspergillus terreus. Eur. J. Exp. Biol. 2, 1550–1555.

Kang, A.R., Park, J.S., Ju, J., Jeong, G.S., Lee, S.H., 2014. Cell encapsulation via microtechnologies. Biomaterials. 18785905. 35 (9), 2651–2663. doi:10.1016/j.biomaterials.2013.12.073.

Karimi, M., 2016. Immobilization of lipase onto mesoporous magnetic nanoparticles for enzymatic synthesis of biodiesel. Biocatal. Agric. Biotechnol. 18788181. 8, 182–188. doi:10.1016/j.bcab.2016.09.009.

Karm, Z., Subhi, A.D., Hamied, R.S., 2020. Synthesis, characterization and application of gamma-alumina as adsorbent material to enhance iron removal from produced water. UPB Sci. Bull. Ser. B: Chem. Mater. Sci. 22863680. 82 (1), 237–246. Available from: https://www.scientificbulletin.upb.ro/rev_docs_arhiva/rezc04_909803.pdf.

Kawabata, N., Kurooka, T., 1995. Biodegradability of poly(vinyl acetate) containing a pyridinium group. J. Appl. Polym. Sci. 10974628. 56 (4), 509–516. doi:10.1002/app.1995.070560413.

Kehrein, P., Van Loosdrecht, M., Osseweijer, P., Garfí, M., Dewulf, J., Posada, J., 2020. A critical review of resource recovery from municipal wastewater treatment plants-market supply potentials, technologies and bottlenecks. Environ. Sci. Water Res. Technol. 20531419. 6 (4), 877–910. Available from: http://pubs.rsc.org/en/journals/journal/ew. doi:10.1039/c9ew00905a.

Khader, E.H., Mohammed, T.J., Mirghaffari, N., Salman, A.D., Juzsakova, T., Abdullah, T.A., 2022. Removal of organic pollutants from produced water by batch adsorption treatment. Clean Technol. Environ. Policy 16189558. 24 (2), 713–720. Available from: https://link.springer.com/journal/10098. doi:10.1007/s10098-021-02159-z.

Khatoon, H., Rai, J.P.N., 2018. Augmentation of Atrazine biodegradation by two Bacilli immobilized on α-Fe2O3 magnetic nanoparticles. Sci. Rep. 2045-2322. 8 (1). doi:10.1038/s41598-018-36296-1.

Krishnamoorthi, S., Banerjee, A., Roychoudhury, A., 2015. Immobilized enzyme technology: potentiality and prospects. J. Enzymol. Metab. 1 (1), 104.

Kubota, M., Nakabayashi, T., Matsumoto, Y., Shiomi, T., Yamada, Y., Ino, K., Yamanokuchi, H., Matsui, M., Tsunoda, T., Mizukami, F., Sakaguchi, K., 2008. Selective adsorption of bacterial cells onto zeolites. Colloids Surf. B Biointerfaces. 09277765. 64 (1), 88–97. doi:10.1016/j.colsurfb.2008.01.012.

Kumar, S.D., Santhanam, P., Park, M.S., Kim, M.K., 2016. Development and application of a novel immobilized marine microalgae biofilter system for the treatment of shrimp culture effluent. J. Water Process Eng. 22147144. 13, 137–142. Available from: http://www.journals.elsevier.com/journal-of-water-process-engineering/. doi:10.1016/j.jwpe.2016.08.014.

Li, D., Shan, R., Gu, J., Zhang, Y., Zeng, X., Lin, L., Yuan, H., Chen, Y., 2023. Co-pyrolysis of textile dyeing sludge/litchi shell and CaO: immobilization of heavy metals and the analysis of the mechanism. Waste Manag. 0956053X. 171, 382–392. doi:10.1016/j.wasman.2023.09.024.

Li, X., Ming, Q., Cai, R., Yue, T., Yuan, Y., Gao, Z., Wang, Z., 2020. Biosorption of $Cd^{2+}$ and $Pb^{2+}$ from apple juice by the magnetic nanoparticles functionalized lactic acid bacteria cells. Food Control. 09567135. 109, 106916. doi:10.1016/j.foodcont.2019.106916.

Li, X., Shu, M., Li, H., Gao, X., Long, S., Hu, T., Wu, C., 2018. Strong, tough and mechanically self-recoverable poly(vinyl alcohol)/alginate dual-physical double-network hydrogels with large cross-link density contrast. RSC Adv. 20462069. 8 (30), 16674–16689. Available from: http://pubs.rsc.org/en/journals/journal/ra. doi:10.1039/c8ra01302k.

Li, Y.G., Gao, H.S., Li, W.L., Xing, J.M., Liu, H.Z., 2009. In situ magnetic separation and immobilization of dibenzothiophene-desulfurizing bacteria. Bioresour. Technol. 09608524. 100 (21), 5092–5096. doi:10.1016/j.biortech.2009.05.064.

Liang, H., Li, D., Zhang, Z., 2016. One pot" synthesis of oleic acid-coated $Fe_3O_4$ and the effect of oleic acid dosage. J. Funct. Mater. 47 (4), 4045–4049.

Liu, S.H., Lin, H.H., Lai, C.Y., Lin, C.W., Chang, S.H., Yau, J.T., 2019b. Microbial community in a pilot-scale biotrickling filter with cell-immobilized biochar beads and its performance in treating toluene-contaminated waste gases. Int. Biodeterior. Biodegrad. 09648305. 144. Available from: www.elsevier.com/inca/publications/store/4/0/5/8/9/9. doi:10.1016/j.ibiod.2019.104743.

Liu, S.H., Zeng, Z.T., Niu, Q.Y., Xiao, R., Zeng, G.M., Liu, Y., Cheng, M., Hu, K., Jiang, L.H., Tan, X.F., Tao, J.J., 2019a. Influence of immobilization on phenanthrene degradation by Bacillus sp. P1 in the presence of Cd(II). Sci. Total Environ. 18791026. 655, 1279–1287. Available from: www.elsevier.com/locate/scitotenv. doi:10.1016/j.scitotenv.2018.11.272.

Lu, A.H., Salabas, E.L., Schüth, F., 2007. Magnetic nanoparticles: synthesis, protection, functionalization, and application. Angew. Chem. Int. Ed. 14337851. 46 (8), 1222–1244. doi:10.1002/anie.200602866.

Lu, Y., Song, S., Wang, R., Liu, Z., Meng, J., Sweetman, A.J., Jenkins, A., Ferrier, R.C., Li, H., Luo, W., Wang, T., 2015. Impacts of soil and water pollution on food safety and health risks in China. Environ. Int. 18736750. 77, 5–15. Available from: www.elsevier.com/locate/envint. doi:10.1016/j.envint.2014.12.010.

Mahmoud, M.E., Yakout, A.A., Abdel-Aal, H., Osman, M.M., 2012. High performance SiO2-nanoparticles-immobilized-Penicillium funiculosum for bioaccumulation and solid phase extraction of lead. Bioresour. Technol. 18732976. 106, 125–132. doi:10.1016/j.biortech.2011.11.081.

Mahmoudi, C., Douma, N.T., Mahmoudi, H., Iurciuc, C.E., Popa, M., 2024. Hydrogels based on proteins cross-linked with carbonyl derivatives of polysaccharides, with biomedical applications. Int. J. Mol. Sci. 14220067. 25 (14). Available from: http://www.mdpi.com/journal/ijms. doi:10.3390/ijms25147839.

Manisalidis, I., Stavropoulou, E., Stavropoulos, A., Bezirtzoglou, E., 2020. Environmental and health impacts of air pollution: a review. Front. Public Health. 22962565. 8. Available from: http://journal.frontiersin.org/journal/public-health/section/public-health-education-and-promotion#archive. doi:10.3389/fpubh.2020.00014.

Manonmani, U., Joseph, K., 2018. Research advances and challenges in anammox immobilization for autotrophic nitrogen removal. J. Chem. Technol. Biotechnol. 10974660. 93 (9), 2486–2497. Available from: http://onlinelibrary.wiley.com/journal/10.1002/(ISSN)1097-4660. doi:10.1002/jctb.5632.

Martinkova, P., Kostelnik, A., Valek, T., Pohanka, M., 2017. Main streams in the construction of biosensors and their applications. Int. J. Electrochem. Sci. 14523981. 12 (8), 7386–7403. doi:10.20964/2017.08.02.

Mehrotra, T., Dev, S., Banerjee, A., Chatterjee, A., Singh, R., Aggarwal, S., 2021. Use of immobilized bacteria for environmental bioremediation: a review. J. Environ. Chem. Eng. 22133437. 9 (5), 105920. doi:10.1016/j.jece.2021.105920.

Mertens, B., Blothe, C., Windey, K., De Windt, W., Verstraete, W., 2007. Biocatalytic dechlorination of lindane by nano-scale particles of Pd(0) deposited on Shewanella oneidensis. Chemosphere. 00456535. 66 (1), 99–105. doi:10.1016/j.chemosphere.2006.05.018.

Mohidem, N.A., Mohamad, M., Rashid, M.U., Norizan, M.N., Hamzah, F., Mat, H.B., 2023. Recent advances in enzyme immobilisation strategies: an overview of techniques and composite carriers. J. Compos. Sci. 2504477X. 7 (12). Available from: www.mdpi.com/journal/jcs. doi:10.3390/jcs7120488.

Monsan, P., Durand, G., Navarro, J.M., 1987. Immobilization of microbial cells by adsorption to solid supports. Methods Enzymol. 15577988. 135 (C), 307–318. doi:10.1016/0076-6879(87)35088-8.

Nadi, A., Boyer, D., Charbonnel, N., Boukhriss, A., Forestier, C., Gmouh, S., 2019. Immobilisation of bacteria onto magnetic nanoparticles for the decolorisation and degradation of azo dyes. IET Nanobiotechnol. 1751-8741. 13 (2), 144–149. doi:10.1049/iet-nbt.2018.5026.

Najim, A.A., Radeef, A.Y., al-Doori, I., Jabbar, Z.H., 2024. Immobilization: the promising technique to protect and increase the efficiency of microorganisms to remove contaminants. J. Chem. Technol. Biotechnol. 10974660. 99 (8), 1707–1733. Available from: http://onlinelibrary.wiley.com/journal/10.1002/(ISSN)1097-4660. doi:10.1002/jctb.7638.

Navarro, J.M., Durand, G., 1977. Modification of yeast metabolism by immobilization onto porous glass. Eur. J. Appl. Microbiol. 14320614. 4 (4), 243–254. doi:10.1007/BF00931261.

Nguyen, H.H., Kim, M., 2017. An overview of techniques in enzyme immobilization. Appl. Sci. Converg. Technol. 2288-6559. 26 (6), 157–163. doi:10.5757/asct.2017.26.6.157.

Niu, Z., Jia, Y., Chen, Y., Hu, Y., Chen, J., Lv, Y., 2018. Positive effects of bio-nano Pd (0) toward direct electron transfer in Pseudomonas putida and phenol biodegradation. Ecotoxicol. Environ. Saf. 10902414. 161, 356–363. Available from: http://www.elsevier.com/inca/publications/store/6/2/2/8/1/9/index.htt. doi:10.1016/j.ecoenv.2018.06.011.

Ogundolie, F.A., Babalola, O.O., Adetunji, C.O., Aruwa, C.E., Manjia, J.N., Muftaudeen, T.K., 2024. A review on bioremediation by microbial immobilization-an effective alternative for wastewater treatment. AIMS Environ. Sci. 23720352. 11 (6), 918–939. Available from:

https://www.aimspress.com/aimspress-data/aimses/2024/6/PDF/Environ-11-06-046.pdf. doi:10.3934/environsci.2024046.

Ozdemir, S., Mohamedsaid, S.A., Kilinc, E., Yıldırım, A., Soylak, M. 2018. Application of magnetized fungal solid phase extractor with Fe2O3 nanoparticle for determination and preconcentration of Co(II) and Hg(II) from natural water samples. Microchem. J. 0026265X. (143), 198–204. doi:10.1016/j.microc.2018.07.032.

Pagnout, C., Razafitianamaharavo, A., Sohm, B., Caillet, C., Beaussart, A., Delatour, E., Bihannic, I., Offroy, M., Duval, J.F.L., 2021. Osmotic stress and vesiculation as key mechanisms controlling bacterial sensitivity and resistance to TiO2 nanoparticles. Commun. Biol. 23993642. 4 (1). Available from: https://www.springer.com/journal/11852. doi:10.1038/s42003-021-02213-y.

Pal, A., Bhattacharjee, S., Saha, J., Sarkar, M., Mandal, P., 2022. Bacterial survival strategies and responses under heavy metal stress: a comprehensive overview. Crit. Rev. Microbiol. 1040-841X. 48 (3), 327–355. doi:10.1080/1040841x.2021.1970512.

Pathakoti, K., Manubolu, M., Hwang, H.M., 2020. Mechanistic insights into TiO2 and ZnO nanoparticle-induced metabolic changes in *Escherichia coli* under solar simulated light irradiation. Water Air Soil Pollut. 15732932. 231 (1). Available from: http://www.kluweronline.com/issn/0049-6979/. doi:10.1007/s11270-019-4388-2.

Pawar, S.N., Edgar, K.J., 2012. Alginate derivatization: a review of chemistry, properties and applications. Biomaterials. 18785905. 33 (11), 3279–3305. doi:10.1016/j.biomaterials.2012.01.007.

Picos-Corrales, L.A., Morales-Burgos, A.M., Ruelas-Leyva, J.P., Crini, G., García-Armenta, E., Jimenez-Lam, S.A., Ayón-Reyna, L.E., Rocha-Alonzo, F., Calderón-Zamora, L., Osuna-Martínez, U., Calderón-Castro, A., De-Paz-Arroyo, G., Inzunza-Camacho, L.N., 2023. Chitosan as an outstanding polysaccharide improving health-commodities of humans and environmental protection. Polymers. 20734360. 15 (3). Available from: http://www.mdpi.com/journal/polymers. doi:10.3390/polym15030526.

Podder, M.S., Majumder, C.B., 2019. Bacteria immobilization on neem leaves/MnFe2O4 composite surface for removal of As(III) and As(V) from wastewater. Arab. J. Chem. 18785352. 12 (8), 3263–3288. doi:10.1016/j.arabjc.2015.08.025.

Putra, R.N., Lee, Y.H., 2020. Entrapment of micro-sized zeolites in porous hydrogels: strategy to overcome drawbacks of zeolite particles and beads for adsorption of ammonium ions. Sep. Purif. Technol. 18733794. 237. Available from: http://www.journals.elsevier.com/separation-and-purification-technology/. doi:10.1016/j.seppur.2019.116351.

Qiao, K., Tian, W., Bai, J., Wang, L., Zhao, J., Song, T., Chu, M., 2020. Removal of high-molecular-weight polycyclic aromatic hydrocarbons by a microbial consortium immobilized in magnetic floating biochar gel beads. Mar. Pollut. Bull. 18793363. 159. Available from: www.elsevier.com/locate/marpolbul. doi:10.1016/j.marpolbul.2020.111489.

Raita, M., Arnthong, J., Champreda, V., Laosiripojana, N., 2015. Modification of magnetic nanoparticle lipase designs for biodiesel production from palm oil. Fuel Process. Technol. 03783820. 134, 189–197. doi:10.1016/j.fuproc.2015.01.032.

Ramalingam, B., Parandhaman, T., Choudhary, P., Das, S.K., 2018. Biomaterial functionalized graphene-magnetite nanocomposite: a novel approach for simultaneous removal of anionic dyes and heavy-metal ions. ACS Sustain. Chem. Eng. 21680485. 6 (5), 6328–6341. Available from: http://pubs.acs.org/journal/ascecg. doi:10.1021/acssuschemeng.8b00139.

Ramteke, L.P., Gogate, P.R., 2016. Treatment of water containing heavy metals using a novel approach of immobilized modified sludge biomass based adsorbents. Sep. Purif. Technol.

18733794. 163, 215–227. Available from: http://www.journals.elsevier.com/separation-and-purification-technology/. doi:10.1016/j.seppur.2016.02.047.

Rasheed, A., Ghous, T., Mumtaz, S., Zafar, M.N., Akhter, K., Shabir, R., Abdin, Z.U., Shafqat, S.S., 2020. Immobilization of Pseudomonas aeruginosa static biomass on eggshell powder for on-line preconcentration and determination of Cr (VI). Open Chem. 23915420. 18 (1), 303–313. Available from: http://www.degruyter.com/view/j/chem. doi:10.1515/chem-2020-0031.

Ren, L., Lin, Z., Liu, H., Hu, H., 2018. Bacteria-mediated phthalic acid esters degradation and related molecular mechanisms. Appl. Microbiol. Biotechnol. 14320614. 102 (3), 1085–1096. Available from: https://www.springer.com/journal/253. doi:10.1007/s00253-017-8687-5.

Rusu, L., Grigoraş, C.G., Simion, A.I., Suceveanu, E.M., Şuteu, D., Harja, M., 2021. Application of saccharomyces cerevisiae/calcium alginate composite beads for cephalexin antibiotic biosorption from aqueous solutions. Materials. 19961944. 14 (16). Available from: https://www.mdpi.com/1996-1944/14/16/4728/pdf. doi:10.3390/ma14164728.

Saberi Riseh, R., Skorik, Y.A., Thakur, V.K., Moradi Pour, M., Tamanadar, E., Noghabi, S.S., 2021. Encapsulation of plant Biocontrol Bacteria with alginate as a main polymer material. Int. J. Mol. Sci. 1422-0067. 22 (20), 11165. doi:10.3390/ijms222011165.

Saha, P., Rao, K.V.B., 2021. Immobilization as a powerful bioremediation tool for abatement of dye pollution: a review. Environ. Rev. 11818700. 29 (2), 277–299. Available from: https://www.nrcresearchpress.com/loi/er. doi:10.1139/er-2020-0074.

Saini, S., Tewari, S., Dwivedi, J., Sharma, V., 2023. Biofilm-mediated wastewater treatment: a comprehensive review. Mater. Adv. 2633-5409. 4 (6), 1415–1443. doi:10.1039/d2ma00945e.

Sampaio, C.S., Angelotti, J.A.F., Fernandez-Lafuente, R., Hirata, D.B., 2022. Lipase immobilization via cross-linked enzyme aggregates: problems and prospects—a review. Int. J. Biol. Macromol. 18790003. 215, 434–449. Available from: www.elsevier.com/locate/ijbiomac. doi:10.1016/j.ijbiomac.2022.06.139.

Sarwar, S., Tahir, A., Haq, I.U., Anum, F., 2022. Alkaline pre-treatment of wheat straw for production of ethanol using saccharomyces cerevisiae SS-4 immobilized in nanoparticles. Pakistan J. Bot. 05563321. 54 (1), 309–315. Available from: https://www.pakbs.org/pjbot/papers/1641323183.pdf. doi:10.30848/PJB2022-1(45).

Schneiderman, D.K., Hillmyer, M.A., 2017. 50th anniversary perspective: there is a great future in sustainable polymers. Macromolecules. 15205835. 50 (10), 3733–3749. Available from: http://pubs.acs.org/journal/mamobx. doi:10.1021/acs.macromol.7b00293.

Serebrennikova, M.K., Golovina, E.E., Kuyukina, M.S., Ivshina, I.B., 2017. A consortium of immobilized rhodococci for oilfield wastewater treatment in a column bioreactor. Appl. Biochem. Microbiol. 00036838. 53 (4), 435–440. Available from: http://www.kluweronline.com/issn/0003-6838. doi:10.1134/S0003683817040123.

Shahbaz, A., Hussain, N., Saleem, M.Z., Saeed, M.U., Bilal, M., Iqbal, H.M.N., 2022. Nanoparticles as stimulants for efficient generation of biofuels and renewables. Fuel. 0016-2361. 319. Available from: http://www.journals.elsevier.com/fuel/. doi:10.1016/j.fuel.2022.123724.

Sharma, S., Hasan, A., Kumar, N., Pandey, L.M., 2018. Removal of methylene blue dye from aqueous solution using immobilized Agrobacterium fabrum biomass along with iron oxide nanoparticles as biosorbent. Environ. Sci. Pollut. Res. 16147499. 25 (22), 21605–21615. Available from: http://www.springerlink.com/content/0944-1344. doi:10.1007/s11356-018-2280-z.

Sheldon, R.A., van Pelt, S., 2013. Enzyme immobilisation in biocatalysis: why, what and how. Chem. Soc. Rev. 0306-0012. 42 (15), 6223–6235. doi:10.1039/C3CS60075K.

Shi, S., Qu, Y., Ma, F., Zhou, J., 2014. Bioremediation of coking wastewater containing carbazole, dibenzofuran and dibenzothiphene by immobilized naphthalene-cultivated Arthrobacter sp. W1 in magnetic gellan gum. Bioresour. Technol. 18732976. 166, 79–86. Available from: www.elsevier.com/locate/biortech. doi:10.1016/j.biortech.2014.05.036.

Siddeeg, S.M., Tahoon, M.A., Mnif, W., Rebah, F.B., 2020. Iron oxide/chitosan magnetic nanocomposite immobilized manganese peroxidase for decolorization of textile wastewater. Processes. 22279717. 8 (1). Available from: https://res.mdpi.com/d_attachment/processes/processes-08-00005/article_deploy/processes-08-00005-v2.pdf. doi:10.3390/pr8010005.

Singhvi, M., Kim, B.S., 2020. Current developments in lignocellulosic biomass conversion into biofuels using nanobiotechology approach. Energies. 19961073. 13 (20). Available from: https://www.mdpi.com/1996-1073/13/20/5300. doi:10.3390/en13205300.

Slavin, Y.N., Asnis, J., Häfeli, U.O., Bach, H., 2017. Metal nanoparticles: understanding the mechanisms behind antibacterial activity. J. Nanobiotechnol. 14773155. 15 (1). Available from: http://www.jnanobiotechnology.com/start.asp. doi:10.1186/s12951-017-0308-z.

Soares, R.C., Zangirolami, T.C., Giordano, R.L.C., Demeke, M.M., Thevelein, J.M., Milessi, T.S., 2022. Cell immobilization using alginate-based beads as a protective technique against stressful conditions of hydrolysates for 2G ethanol production. Polymers. 20734360. 14 (12). Available from: https://www.mdpi.com/2073-4360/14/12/2400/pdf?version=1655191087. doi:10.3390/polym14122400.

Stepanov, N., Efremenko, E., 2018. "Deceived" concentrated immobilized cells as biocatalyst for intensive bacterial cellulose production from various sources. Catalysts. 2073-4344. 8 (1), 33. doi:10.3390/catal8010033.

Suganya, K., Revathi, K., 2016. Decolorization of reactive dyes by immobilized bacterial cells from textile effluents. Int. J. Curr. Microbiol. Appl. Sci. 23197692. 5 (1), 528–532. doi:10.20546/ijcmas.2016.501.053.

Suner, S.S., Sahiner, N., 2018. Humic acid particle embedded super porous gum Arabic cryogel network for versatile use. Polym. Adv. Technol. 10991581. 29 (1), 151–159. Available from: http://onlinelibrary.wiley.com/journal/10.1002/(ISSN)1099-1581. doi:10.1002/pat.4097.

Suresh, A.K., Doktycz, M.J., Wang, W., Moon, J.W., Gu, B., Meyer, H.M., Hensley, D.K., Allison, D.P., Phelps, T.J., Pelletier, D.A., 2011. Monodispersed biocompatible silver sulfide nanoparticles: facile extracellular biosynthesis using the γ-proteobacterium, Shewanella oneidensis. Acta Biomater. 18787568. 7 (12), 4253–4258. Available from: http://www.journals.elsevier.com/acta-biomaterialia. doi:10.1016/j.actbio.2011.07.007.

Suvarli, N., Wenger, L., Serra, C., Perner-Nochta, I., Hubbuch, J., Wörner, M., 2022. Immobilization of β-galactosidase by encapsulation of enzyme-conjugated polymer nanoparticles inside hydrogel microparticles. Front. Bioeng. Biotechnol. 2296-4185. 9. doi:10.3389/fbioe.2021.818053.

Suzana, C.U.S.M., Claudia, M., a, M., Larissa Guedes Fiuacute Za, M.C.O., ra, T.D.S., 2013. Immobilization of microbial cells: a promising tool for treatment of toxic pollutants in industrial wastewater. Afr. J. Biotechnol. 1684-5315. 12 (28), 4412–4418. doi:10.5897/AJB12.2677.

Taghizadeh, S.M., Ebrahiminezhad, A., Ghoshoon, M.B., Dehshahri, A., Berenjian, A., Ghasemi, Y., 2020. Magnetic immobilization of pichia pastoris cells for the production of recombinant human serum albumin. Nanomaterials. 2079-4991. 10 (1). Available from: https://www.mdpi.com/2079-4991/10/1/111/pdf. doi:10.3390/nano10010111.

Takei, T., Ikeda, K., Ijima, H., Kawakami, K., 2011. Fabrication of poly(vinyl alcohol) hydrogel beads crosslinked using sodium sulfate for microorganism immobilization. Process Biochem. 13595113. 46 (2), 566–571. doi:10.1016/j.procbio.2010.10.011.

Tarasi, R., Alipour, M., Gorgannezhad, L., Imanparast, S., Yousefi-Ahmadipour, A., Ramezani, A., Ganjali, M.R., Shafiee, A., Faramarzi, M.A., Khoobi, M., 2018. Laccase immobilization onto magnetic β-cyclodextrin-modified chitosan: improved enzyme stability and efficient performance for phenolic compounds elimination. Macromol. Res. 20927673. 26 (8), 755–762. Available from: http://www.springerlink.com/content/121720/. doi:10.1007/s13233-018-6095-z.

Teh, C.Y., Budiman, P.M., Shak, K.P.Y., Wu, T.Y., 2016. Recent advancement of coagulation-flocculation and its application in wastewater treatment. Ind. Eng. Chem. Res. 15205045. 55 (16), 4363–4389. Available from: http://pubs.acs.org/journal/iecred. doi:10.1021/acs.iecr.5b04703.

Tekere, M., 2019. Microbial bioremediation and different bioreactors designs applied. In: Jacob-Lopes, E., Zepka, L.Q. (Eds.), Biotechnology and Bioengineering. IntechOpen, pp. 1–19. doi:10.5772/intechopen.83661.

Thomas, D., O'Brien, T., Pandit, A., 2018. Toward customized extracellular niche engineering: progress in cell-entrapment technologies. Adv. Mater. 0935-9648. 30 (1). doi:10.1002/adma.201703948.

Tong, C.Y., Derek, C.J.C., 2023. Bio-coatings as immobilized microalgae cultivation enhancement: a review. Sci. Total Environ. 00489697. 887, 163857. doi:10.1016/j.scitotenv.2023.163857.

Torres, E., 2020. Biosorption: a review of the latest advances. Processes. 2227-9717. 8 (12), 1584. doi:10.3390/pr8121584.

Trelles, J.A., Rivero, C.W., 2020. Whole cell entrapment techniques. Methods Mol. Biol. 2100, 385–394. Available from: http://www.springer.com/series/7651. doi:10.1007/978-1-0716-0215-7_25.

Tripathi, A., Melo, J.S., 2021. Immobilization Strategies: Biomedical, Bioengineering and Environmental Applications Gels Horizons: From Science to Smart Materials. Springer Singapore. doi:10.1007/978-981-15-7998-1.

Upendar, G., Dutta, S., Chakraborty, J., Bhattacharyya, P., 2016. Removal of methylene blue dye using immobilized bacillus subtilis in batch & column reactor. Mater. Today Proc. 22147853. 3 (10), 3467–3472. doi:10.1016/j.matpr.2016.10.029.

Uzun, L., Sağlam, N., Safarikova, M., Safarik, I., Denizli, A., 2011. Copper biosorption on magnetically modified yeast cells under magnetic field. Sep. Sci. Technol. 0149-6395. 46 (6), 1045–1051. doi:10.1080/01496395.2010.541400.

Vallejo, J.A., Serrat, M., Pérez-Portuondo, I., Sánchez-Pérez, A., Ageitos, J.M., Villa, T.G., 2012. A novel Kluyveromyces marxianus strain with an inducible flocculation phenotype. AMB Express. 21910855. 2 (1). Available from: http://www.amb-express.com/. doi:10.1186/2191-0855-2-38.

Vishnu, D., Dhandapani, B., Santhiya, K., 2021. The symbiotic effect of integrated Muraya koenigii extract and surface-modified magnetic microspheres—a green biosorbent for the removal of Cu(II) and Cr(VI) ions from aqueous solutions. Chem. Eng. Commun. 0098-6445. 208 (6), 851–862. doi:10.1080/00986445.2019.1691538.

Waheed, Mazumder, M.A.J., Al-Ahmed, A., Ullah, R.P., 2019. Functional Biopolymers, Polymers and Polymeric Composites: A Reference Series. Springer. doi:10.1007/978-3-319-95990-04.

Wang, D., Chen, Y., 2016. Critical review of the influences of nanoparticles on biological wastewater treatment and sludge digestion. Crit. Rev. Biotechnol. 15497801. 36 (5), 816–828. doi:10.3109/07388551.2015.1049509.

Wang, J., Jiang, L.H., Zhou, Y., Ye, B.C., 2017. Enhanced biodegradation of di-n-butyl phthalate by Acinetobacter species strain LMB-5 coated with magnetic nanoparticles. Int. Biodeterior. Biodegrad. 09648305. 116, 184–190. Available from: www.elsevier.com/inca/publications/store/4/0/5/8/9/9. doi:10.1016/j.ibiod.2016.10.024.

Wang, Q., Wu, X., Jiang, L., Fang, C., Wang, H., Chen, L., 2020. Effective degradation of Di-n-butyl phthalate by reusable, magnetic $Fe_3O_4$ nanoparticle-immobilized Pseudomonas sp. W1 and its application in simulation. Chemosphere. 00456535. 250, 126339. doi:10.1016/j.chemosphere.2020.126339.

Wang, X., Liu, X., Zhao, C., Ding, Y., Xu, P., 2011. Biodiesel production in packed-bed reactors using lipase–nanoparticle biocomposite. Bioresour. Technol. 09608524. 102 (10), 6352–6355. doi:10.1016/j.biortech.2011.03.003.

Wang, Y., Tian, Y., Han, B., Zhao, H.B., Bi, J.N., Cai, B.L., 2007. Biodegradation of phenol by free and immobilized Acinetobacter sp. strain PD12. J. Environ. Sci. 10010742. 19 (2), 222–225. doi:10.1016/S1001-0742(07)60036-9.

Wang, Z., Ishii, S., Novak, P.J., 2021. Encapsulating microorganisms to enhance biological nitrogen removal in wastewater: recent advancements and future opportunities. Environ. Sci. Water Res. Technol. 20531419. 7 (8), 1402–1416. Available from: http://pubs.rsc.org/en/journals/journal/ew. doi:10.1039/d1ew00255d.

Willaert, R., 2000. Beer production using immobilized cell technology. Minerva Biotecnol. 12 (4), 319–330.

Wu, X., Zhao, F., Rahunen, N., Varcoe, J.R., Avignone-Rossa, C., Thumser, A.E., Slade, R.C.T., 2011. A role for microbial palladium nanoparticles in extracellular electron transfer. Angew. Chem. Int. Ed. 15213773. 50 (2), 427–430. doi:10.1002/anie.201002951.

Xiao, Y., Wiesner, M.R., 2013. Transport and retention of selected engineered nanoparticles by porous media in the presence of a biofilm. Environ. Sci. Technol. 15205851. 47 (5), 2246–2253. doi:10.1021/es304501n.

Xiong, H., Fu, J., Li, J., Ali, R., Wang, H., Liu, Y., Su, H., Li, Y., Lau, W.M., Mahmood, N., Mu, C., Jian, X., 2021. Strain-regulated sensing properties of α-Fe2O3 nano-cylinders with atomic carbon layers for ethanol detection. J. Mater. Sci. Technol. 10050302. 68, 132–139. Available from: http://www.sciencedirect.com/science/journal/10050302. doi:10.1016/j.jmst.2020.08.009.

Xu, P., Zeng, G., Huang, D., Hu, S., Feng, C., Lai, C., Zhao, M., Huang, C., Li, N., Wei, Z., Xie, G., 2013. Synthesis of iron oxide nanoparticles and their application in Phanerochaete chrysosporium immobilization for Pb(II) removal. Colloids Surf. A Physicochem. Eng. Asp. 09277757. 419, 147–155. doi:10.1016/j.colsurfa.2012.10.061.

Yang, J., Zou, P., Yang, L., Cao, J., Sun, Y., Han, D., Yang, S., Wang, Z., Chen, G., Wang, B., Kong, X., 2014. A comprehensive study on the synthesis and paramagnetic properties of PEG-coated $Fe_3O_4$ nanoparticles. Appl. Surf. Sci. 01694332. 303, 425–432. Available from: http://www.journals.elsevier.com/applied-surface-science/. doi:10.1016/j.apsusc.2014.03.018.

Yao, M., Li, B., Ye, H., Huang, W., Luo, Q., Xiao, H., McClements, D.J., Li, L., 2018. Enhanced viability of probiotics (Pediococcus pentosaceus Li05) by encapsulation in microgels doped with inorganic nanoparticles. Food Hydrocoll. 0268005X. 83, 246–252. Available from: https://www.journals.elsevier.com/food-hydrocolloids. doi:10.1016/j.foodhyd.2018.05.024.

Yin, Y., Wang, J., Yang, X., Li, W., 2017. Removal of strontium ions by immobilized saccharomyces cerevisiae in magnetic chitosan microspheres. Nucl. Eng. Technol. 2234358X. 49 (1), 172–177. Available from: http://www.kns.org/. doi:10.1016/j.net.2016.09.002.

Zakaria, B.S., Nassar, H.N., Saed, D., El-Gendy, N.S., 2015. Enhancement of carbazole denitrogenation rate using magnetically decorated Bacillus clausii BS1. Pet. Sci. Technol. 15322459. 33 (7), 802–811. Available from: www.tandf.co.uk/journals/titles/10916466.asp. doi:10.1080/10916466.2015.1014966.

Zakaria, N.N.M., Man, R.C., Sulaiman, S.Z., Mudalip, S.K.A., Manas, N.H.A., Yamani, L.N., 2023. Optimization of process parameters of immobilized *Escherichia coli* for cyclodextrin production. J. Adv. Res. Appl. Mech. 22897895. 107 (1), 1–10. Available from: www.akademiabaru.com/submit/index.php/aram/index. doi:10.37934/aram.107.1.110.

Zaki, A.G., Hasanien, Y.A., Abdel-Razek, A.S., 2022. Biosorption optimization of lead(II) and cadmium(II) ions by two novel nanosilica-immobilized fungal mutants. J. Appl. Microbiol. 13652672. 133 (2), 987–1000. Available from: https://academic.oup.com/jambio. doi:10.1111/jam.15624.

Zdarta, J., Meyer, A.S., Jesionowski, T., Pinelo, M., 2018. A general overview of support materials for enzyme immobilization: characteristics, properties, practical utility. Catalysts. 20734344. 8 (2). Available from: http://www.mdpi.com/2073-4344/8/2/92/pdf. doi:10.3390/catal8020092.

Zhang, K., Luo, X., Yang, L., Chang, Z., Luo, S., 2021. Progress toward hydrogels in removing heavy metals from water: problems and solutions—a review. ACS EST Water. 2690-0637. 1 (5), 1098–1116. doi:10.1021/acsestwater.1c00001.

Zhang, T., Li, W.L., Chen, X.X., Tang, H., Li, Q., Xing, J.M., Liu, H.Z., 2011. Enhanced biodesulfurization by magnetic immobilized Rhodococcus erythropolis LSSE8-1-vgb assembled with nano-γ-Al2O3. World J. Microbiol. Biotechnol. 09593993. 27 (2), 299–305. Available from: www.wkap.nl/journalhome.htm/0959-3993. doi:10.1007/s11274-010-0459-7.

Zhang, X., You, S., Ma, L., Li, C.C., 2015. The application of immobilized microorganism technology in wastewater treatment. In: 2nd International Conference on Machinery, Materials Engineering, Chemical Engineering and Biotechnology. pp. 103–106. doi: 10.2991/mmeceb-15.2016.22.

Zhang, Y., Piao, M., He, L., Yao, L., Piao, T., Liu, Z., Piao, Y., 2020. Immobilization of laccase on magnetically separable biochar for highly efficient removal of bisphenol A in water. RSC Adv. 2046-2069. 10 (8), 4795–4804. doi:10.1039/c9ra08800h.

Zhao, W., Zhang, Y., Liu, Y., Tan, M., Yu, W., Xie, H., Ma, Y., Sun, G., Lv, G., Zhao, S., Ma, X., 2013. Oxygen diffusivity in alginate/chitosan microcapsules. J. Chem. Technol. Biotechnol. 10974660. 88 (3), 449–455. doi:10.1002/jctb.3845.

Zhu, K., Davis, C.W., Novak, P.J., Arnold, W.A., 2020. Effects of encapsulation on the chemical inhibition of anaerobic hydrogen- and methane-producing microbial cells. Bioresour. Technol. Rep. 2589014X. 11. Available from: https://www.journals.elsevier.com/bioresource-technology-reports. doi:10.1016/j.biteb.2020.100451.

Zhuang, H., Han, H., Xu, P., Hou, B., Jia, S., Wang, D., Li, K. 2015. Biodegradation of quinoline by Streptomyces sp. N01 immobilized on bamboo carbon supported $Fe_3O_4$ nanoparticles. Biochem. Eng. J. 1369703X. (99), 44–47. doi:10.1016/j.bej.2015.03.004.

CHAPTER

# 5 Bioprospecting microbes for polyvinyl chloride plastic biodegradation: Potential and challenges

**Neha Hatwar and Asifa Qureshi**

*CSIR-National Environmental Engineering Research Institute (NEERI), Bioprospecting Microbes for PVC Plastic Biodegradation: Potential and Challenges, Nagpur, India*

## 1 Introduction

The growing accumulation of plastic waste, especially polyvinyl chloride (PVC), poses a severe threat to global ecosystems due to its nonbiodegradable nature and toxic by-products (Peng et al., 2025). PVC, widely used in construction materials, packaging, medical devices, and toys, accounts for a significant portion of plastic waste that persists in the environment for decades (Evode et al., 2021). During its degradation, PVC releases hazardous chemicals such as dioxins, phthalates, and other endocrine-disrupting compounds that severely affect human health and biodiversity (Kudzin et al., 2024). Furthermore, the fragmentation of PVC through physical and chemical processes results in the formation of microplastics (MPs), which infiltrate terrestrial and marine ecosystems, ultimately entering the food chain and causing widespread ecological damage (Khandare et al., 2021; Prakash and Palkar, 2021). Due to the inherent resistance of PVC to natural decomposition, developing sustainable methods for its degradation has become a priority in environmental management. PVC's durability stems from its high chlorine content and complex polymeric structure, which hinder conventional degradation methods. Chemical and thermal approaches, although effective, release harmful by-products and generate secondary pollution, making them unsustainable for long-term waste management (Lewandowski and Skórczewska, 2022). Mechanical recycling, while commonly used, is ineffective for PVC MPs due to their small size and heterogeneous nature. Consequently, biological degradation using microorganisms has gained traction as an eco-friendly, cost-effective alternative for addressing PVC pollution (Mnyango and Hlangothi, 2024). Microbial biodegradation harnesses the metabolic and enzymatic potential of microorganisms to break down PVC into harmless by-products, providing a sustainable and nontoxic approach to PVC waste management (Yang et al., 2024). Microorganisms play a crucial role in

Microbial Technologies for Sustainable Waste Management. https://doi.org/10.1016/B978-0-443-44093-9.00014-1

plastic biodegradation by utilizing enzymes that break down complex polymers into simpler compounds (Mohanan et al., 2020). In the case of PVC, bacteria, fungi, and actinomycetes possess the metabolic capability to degrade its structure through enzymatic processes, including dechlorination, oxidation, and polymer chain scission. These microbes produce specialized enzymes such as hydrolases, esterases, and dehydrochlorinases, which target the chemical bonds within PVC and facilitate its degradation (Akovali, 2012). The microbial degradation of PVC typically involves a series of stages, including the fragmentation of polymer chains, bioassimilation of smaller molecules, and mineralization into $CO_2$, water, and biomass (Gilani et al., 2023).

This book chapter highlights microbial species identified for PVC degradation and evaluates their metabolic capabilities. In addition, it explores the bioprospecting of microbes from diverse environments, including landfill sites, industrial wastewater, marine sediments, compost, and insect guts, to identify novel strains with enhanced PVC degradation potential. Furthermore, it addresses the challenges associated with microbial degradation, including the persistence of PVC in the environment, the slow rate of degradation, and the potential release of toxic byproducts. Finally, it outlines future research directions, such as genetic engineering, microbial consortia, and bioreactor technologies to enhance PVC biodegradation efficiency (Fig. 5.1).

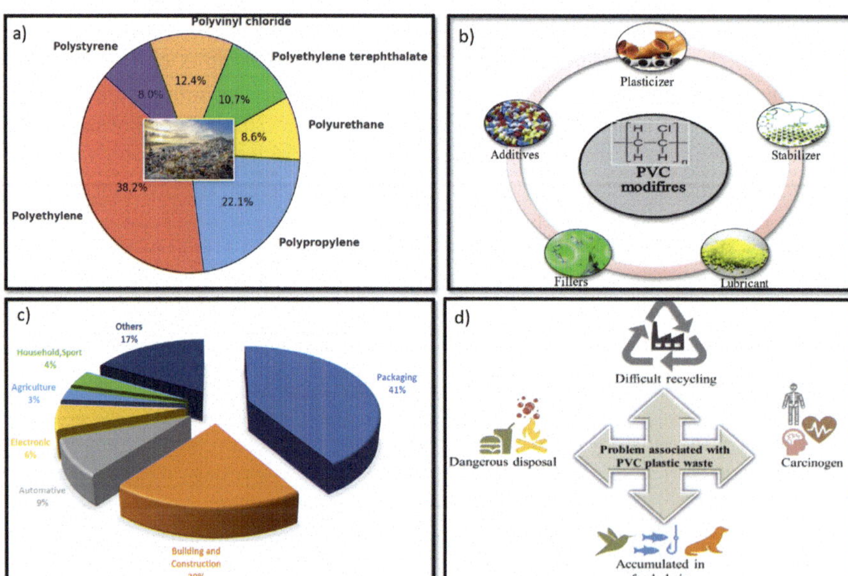


**FIG. 5.1**

Overview of PVC plastic waste occurrence in the environment: (a) distribution, (b) modifiers, (c) major sectors, and (d) environmental and health issues.

# 2 Biodegradation approaches involved in PVC degradation

Biotic degradation of PVC polymers involves the action of bacteria, fungi, and actinomycetes capable of metabolizing complex polymer structures. Recent studies have demonstrated the role of microbial consortia in degrading PVC by enzymatic cleavage of polymer chains, leading to a reduction in molecular weight and fragmentation of the polymer (Ali et al., 2009, 2014; Novotný et al., 2022). The role of biofilms formed on PVC surfaces has been highlighted in enhancing microbial attachment, altering the polymer's hydrophobic properties, and facilitating enzymatic degradation (Ghosh et al., 2019; Hatwar and Qureshi, 2025). In addition, gut microbiota from insect larvae, such as waxworms and mealworms, have shown potential in depolymerizing PVC, although the mineralization process remains incomplete (Zhang et al., 2022; Peng et al., 2023). Microcosm experiments have been used to assess the impact of PVC on the chemical and microbiological parameters of different environments. Studies employing 16S rRNA gene sequencing and metagenomic analysis have identified microbial taxa capable of degrading PVC under various environmental conditions, including soil, marine environments, and landfill sites (Lu et al., 2023; Barili et al., 2023). Multiomics approaches have further elucidated the metabolic pathways involved in PVC-MP biodegradation, providing insights into key enzymes and gene clusters responsible for polymer breakdown (Tiwari et al., 2022). PVC biodegradation occurs through several approaches, each leveraging the unique capabilities of microbial strains and environmental conditions.

## 2.1 Pure culture isolation and identification

Pure bacterial and fungal cultures, such as *Pseudomonas*, *Bacillus*, and *Aspergillus* species, have been shown to degrade PVC efficiently. These microbes produce hydrolases and esterases that target the polymer backbone, breaking it down into low-molecular-weight intermediates, which are subsequently mineralized (Novotný et al., 2022; Giacomucci et al., 2019). However, this approach often faces limitations due to the slow degradation rate and incomplete mineralization of PVC.

## 2.2 Microbial consortia designing

Mixed microbial communities from diverse microenvironments, such as soil, marine ecosystems, and compost, exhibit enhanced degradation due to their diverse enzymatic capabilities. Consortia of bacteria and fungi synergistically degrade PVC by secreting a broad spectrum of catabolic enzymes, resulting in more effective fragmentation of polymer chains (Anwar et al., 2016). The presence of diverse microbial communities contributes to the increased bioavailability of PVC, accelerating the degradation process.

### 2.3 Insect gut microbiota

Waxworms and mealworms have demonstrated the ability to depolymerize PVC through their gut microbiota, which harbors PVC-degrading microorganisms. These insect larvae metabolize PVC to sustain their energy requirements, although the process primarily leads to depolymerization rather than complete mineralization (Peng et al., 2025; Zhang et al., 2022).

### 2.4 Protoplast fusion and genetic engineering

Genetic engineering approaches have enhanced the PVC degradation potential of microorganisms by introducing recombinant strains with improved enzymatic activity. Techniques such as CRISPR–Cas9 and electroporation have enabled the modification of microbial genomes to express high-efficiency PVC-degrading enzymes, thereby enhancing the biodegradation capacity of selected strains (Patil and Bagde, 2012).

### 2.5 Multienvironment pretreatment approach for PVC degradation

Utilizing soil, compost, marine water, and sludge environments to pretreat and degrade PVC has proven to be effective in accelerating the biodegradation process. Combining these environments with diverse microbial communities and engineered strains offers a holistic approach to PVC-MP degradation (Ali et al., 2009, 2014; Sudhakar et al., 2023).

### 2.6 Microbial enzyme application for PVC biodegradation

Microbial enzymes such as esterases, hydrolases, and dechlorinases play a vital role in degrading PVC by breaking down polymer chains and releasing less toxic byproducts (Othman et al., 2021). Some key microbial strains identified for PVC degradation include *Pseudomonas aeruginosa, Bacillus subtilis, Aspergillus niger,* and *Phanerochaete chrysosporium*. These microorganisms utilize enzymatic pathways that involve oxidative and reductive dechlorination, ultimately converting PVC MPs into simpler molecules that can be assimilated and mineralized (Kumari et al., 2019) (Fig. 5.2).

## 3 Bioprospecting microbes from diverse environmental niches for PVC degradation

Bioprospecting involves exploring diverse ecosystems to identify microbial species with unique metabolic capabilities, making it a powerful tool for discovering potential PVC-degrading microbes. Various environmental niches, including soil ecosystems, marine environments, landfill sites, and extreme habitats, harbor diverse

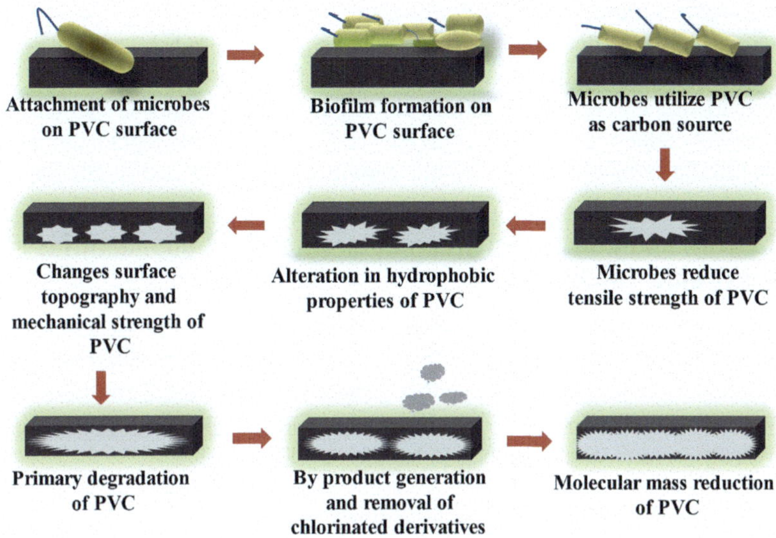


**FIG. 5.2**

Schematic representation of different steps in the biodegradation process of PVC.

microbial communities capable of adapting to challenging conditions and metabolizing recalcitrant compounds like PVC.

## 3.1 Soil ecosystems: Reservoirs of PVC-degrading microbes

Soil ecosystems, particularly those in landfill sites and industrial waste zones, serve as hotspots for plastic-degrading microbes. Soil harbors a diverse range of microbial populations, including bacteria, fungi, and actinomycetes, that can adapt to harsh conditions and metabolize polymeric waste such as PVC (Agarwal et al., 2024). The prolonged exposure of soil to PVC contamination provides an opportunity for the enrichment of microbial species capable of degrading complex polymers (Kudzin et al., 2024).

### 3.1.1 Diversity of bacterial community in soil ecosystems

Several bacterial species from soil, including *Pseudomonas*, *Bacillus*, and *Rhodococcus* species, have been identified as potent PVC degraders due to their ability to produce extracellular enzymes such as dehalogenases and oxidoreductases (Hatwar and Qureshi, 2025). These enzymes break down PVC by cleaving carbon–chlorine bonds and facilitating oxidative degradation (Auta and Emenike, 2017). The degradation potential of *Pseudomonas* and *Bacillus* species was demonstrated in landfill soils contaminated with plastic waste, highlighting a 25%–30% reduction in PVC polymer mass over 120 days (Giacomucci et al., 2019). Similarly, Gao et al. (2024) identified *Rhodococcus* strains from contaminated soil samples with strong enzymatic capabilities for PVC degradation (Gao et al., 2024).

#### 3.1.2 Fungal species for PVC degradation

Soil fungi, such as *A. niger* and *Penicillium chrysogenum,* produce lignin-degrading enzymes like laccases and peroxidases that aid in PVC polymer degradation by inducing oxidative cleavage (Srikanth et al., 2022). A study reported that soil fungi can effectively adhere to PVC film surfaces and initiate the biodegradation process through biofilm formation and enzymatic activity (Kaczmarek and Bajer, 2007).

### 3.2 Marine and coastal environments: Microbial communities under extreme conditions

Marine environments, including deep-sea sediments, coastal waters, and estuarine ecosystems, harbor diverse microbial communities capable of metabolizing persistent pollutants, including plastics like PVC. These environments are nutrient-limited but enriched with microbes that can break down complex polymers through oxidation, hydrolysis, and dehalogenation (Khandare et al., 2021).

#### 3.2.1 Marine bacteria with PVC-degrading capabilities

Marine bacteria such as *Alcanivorax*, *Vibrio*, and *Alteromonas* species exhibit hydrocarbon-degrading abilities and have shown potential to degrade polymeric materials like PVC (Khandare et al., 2021, 2021). The degradation potential of *Alcanivorax* species, isolated from deep-sea sediments, has been observed in degrading PVC films by 1.23%–1.73% over 90 days. In addition, *Vibrio* and *Alteromonas* species, isolated from coastal environments, demonstrate biofilm formation and enzymatic oxidation on PVC surfaces, enhancing polymer degradation. These microbes produce oxygenases and hydrolases that assist in PVC breakdown through oxidative cleavage. Arctic microbes exhibit an efficient ability to degrade MPs due to the polar atmosphere (Entezari et al., 2022).

#### 3.2.2 Role of biofilms in marine PVC degradation

Biofilms formed by marine bacteria on PVC surfaces enhance the degradation process by producing extracellular polymeric substances, which trap the polymer and facilitate enzymatic attack. Hatwar and Qureshi (2025) highlighted the role of biofilms in accelerating PVC degradation by marine microbial consortia, reducing the polymer's hydrophobicity and facilitating microbial colonization.

### 3.3 Landfill sites and industrial waste zones: Microbial adaptation to polymeric waste

Landfill sites and industrial waste zones, where synthetic polymer waste accumulates over time, provide a unique environment for the proliferation of PVC-degrading microbes. These environments facilitate microbial adaptation to plastic waste through prolonged exposure to diverse synthetic polymers (Malakhova et al., 2023).

1. **Enrichment of microbial consortia:** Microbial consortia from landfill sites exhibit enhanced PVC degradation potential due to their ability to form

biofilms and produce extracellular enzymes capable of breaking down PVC polymers. A microbial consortium from landfill soil exhibited a profound reduction in PVC after 120 days of incubation under aerobic conditions. The consortium included *P. aeruginosa*, *B. subtilis*, and *Achromobacter* species (Anwar et al., 2016).

2. **Enzymatic and metabolic mechanisms:** Microbes from landfills secrete hydrolytic enzymes such as cutinases, dehalogenases, and esterases, which cleave PVC polymer chains and facilitate oxidative depolymerization. These enzymes enhance microbial colonization on PVC surfaces, resulting in increased biofilm formation and accelerated degradation.
3. **Extreme environments (extremophiles): Adaptation and PVC biodegradation**: Extremophiles, including thermophiles, halophiles, and psychrophiles, thrive in extreme environmental conditions such as high temperatures, high salinity, and low oxygen levels. These organisms possess highly adaptable enzymatic systems that can degrade recalcitrant compounds, including PVC.

### *3.3.1 Thermophiles and halophiles in PVC degradation*

Thermophilic bacteria, such as *Thermus thermophilus* and *Geobacillus* species, isolated from geothermal springs and hydrothermal vents, exhibit enhanced enzymatic activity at high temperatures (Margaryan et al., 2018). These bacteria produce halotolerant enzymes, including esterases and oxidoreductases, which catalyze the hydrolysis of polymer chains (Fig. 5.3).

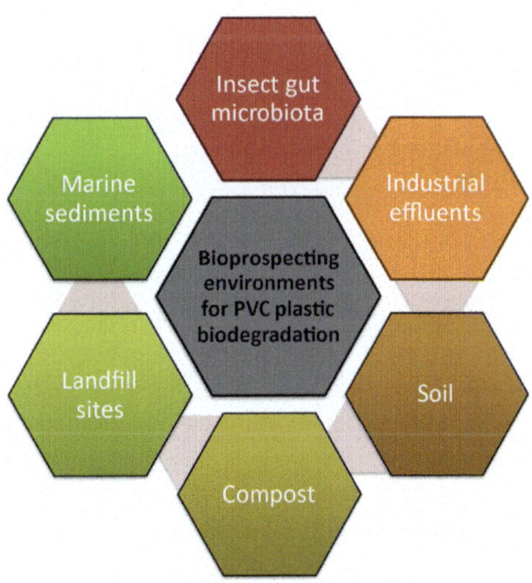


**FIG. 5.3**

Potential bioprospecting environments for PVC degradation, highlighting diverse ecological niches.

#### *3.3.2 Advances in metagenomics and bioinformatics in microbial screening*

Recent advancements in metagenomics and bioinformatics have revolutionized the discovery of PVC-degrading microorganisms by allowing researchers to analyze microbial communities from diverse environmental samples (Peng et al., 2024). Metagenomic sequencing enables the identification of genes encoding enzymes responsible for PVC degradation, while bioinformatics tools facilitate the annotation and functional characterization of these genes. High-throughput screening and genome mining techniques have also led to the discovery of novel microbial strains capable of degrading PVC polymers, further enhancing the understanding of microbial plastic degradation mechanisms (Huang et al., 2023).

## 4 Analytical methods used for determining PVC biodegradation

### 4.1 Molecular weight determination

The density and high molecular weight of PVC are among the most critical obstacles to its biodegradation (Novotný et al., 2022; Saeed et al., 2022). The most widespread method for analyzing the distribution of molecular weight is the combination of rheological measurements at high temperatures and gas chromatography techniques (Yadav et al., 2022; Luisa Pardo-rodríguez and Zorro-Mateus, 2021) (Table 5.1).

**Table 5.1** Microbial species reported for PVC biodegradation.

| Microbial species | Substrate for study | Degradation potential (%) | Incubation time (days) | References |
|---|---|---|---|---|
| *Chaetomium globosum ATCC 16021* | PVC film | 9 | 28 | (Akash et al., 2024) |
| *Phanerochaete chrysosporium PV1* | Starch-blended PVC sheet ~1–2 | ~1–2 | 49 | (Ali et al., 2014) |
| *Trichoderma hamatum* | Compost-treated PVC film | 19.9 ± 0.4 | 60 | (Novotný et al., 2022) |
| *Byssochlamys nivea* | Compost-treated PVC film | 15.5 ± 0.9 | 60 | (Novotný et al., 2022) |
| *Trametes suaveolens* | Compost-treated PVC film | 18.3 ± 0.7 | 60 | (Novotný et al., 2022) |
| *Pseudallescheria ellipsoidea* | Compost-treated PVC film | 18.3 ± 0.7 | 60 | (Novotný et al., 2022) |

| Microbial species | Substrate for study | Degradation potential (%) | Incubation time (days) | References |
|---|---|---|---|---|
| *Aspergillus terreus* | Compost-treated PVC film | 10 ± 1.7 | 60 | (Novotný et al., 2022) |
| *Graphium sp.* | Compost-treated PVC film | 10.5 ± 3.9 | 60 | (Novotný et al., 2022) |
| *Phanerochaete chrysosporium* | PVC powder film | 31 | 28 | (Khatoon et al., 2019) |
| *Phanerochaete chrysosporium PV1* | Soil burial PVC film | ~2 | 121 | (Ali et al., 2009) |
| *Aspergillus niger and A. glaucus* | Commercial polymer strips | 40 | 28 | (Jianquan and Dong, 2024) |
| *Penicillium sp. and Mucor sp.* | PVC film from PVC resin | ~15 and ~30 | 112 | (Jin et al., 2023) |
| *Vibrio, Alteromonas, Cobetia spp.* | Unplasticized PVC film | 1.23–1.73 | 60 | (Khandare et al., 2021) |
| *Bacillus amyloliquefaciens* | Compost-treated PVC film | 17.5 ± 0.2 | 60 | (Novotný et al., 2022) |
| *Pseudomonas, E. coli, Bacillus, Klebsiella* | Cellulose acetate butyrate blends PVC | 0.6 – 4.42 | 60 | (Sudhakar et al., 2023) |
| *Pseudomonas citronellolis* | Petroleum-derived PVC films | 19 | 30 | (Giacomucci et al., 2019) |
| *Bacillus and Micrococcus* | PVC film | 12.7 and 8.4 | 180 | (Yadav et al., 2022) |
| *Micrococcus species* | Soil burial PVC film | 8.87 | 70 | (Patil and Bagde, 2012) |
| *P. aeruginosa and Achromobacter sp.* | Epoxidized seed oil blends PVC | 53 and 43 | 180 | (Khatua et al., 2024) |
| *B. licheniformis and A. xylosoxidans* | Commercial polymer strips | 32.2 | 28 | (Akash et al., 2024) |
| *Bacillus amyloliquefaciens* | Bacillus amyloliquefaciens Gruppo Fabbri Spa PVC | 15–20 | 60 | (Jianquan and Dong, 2024) |

### 4.2 Functional group analysis

Fluorescence infrared (FTIR) spectroscopy is typically applied to examine the type and presence of functional groups on the surface of PVC substrates (Anwar et al., 2016). FTIR methodology shows the degree of degradation, which is calculated based on functional group alterations on the surface of the polymer. Several reports suggest that when microbes interact with PVC, they cause modifications in the vinyl, carbonyl, ester, and its double bond (Malakhova et al., 2023). Combinations of spectroscopic methods and chromatography analysis, such as thermal extraction–desorption (TED-GC/MS), are often used in MP identification approaches. Both methods distinguish between different polymers and additives. While the TED-GC/MS method is used to determine the total mass fraction of the polymer concerned, RS and FTIR, coupled with an optical microscope, provide information on particle size and number (Yuan et al., 2022). Attenuated total reflection (ATR-FTIR) provides results for samples that are >500 μm in particulate size (thin or thick fibers, films, or powder). The provided FTIR results represent the concentrations of functional groups on PVC substrates that alter during the interaction with microbes (Wu et al., 2014).

### 4.3 Surface topography

The growth of different microorganisms on the polymer surface can be detected in biodegradation (Das and Karak, 2012). Invasion of hyphal structures and biofilm persistence can be observed by microscopic methods. For instance, scanning electron microscopy has been used to observe biofilms that have formed on PVC plastic surfaces. It has been reported that PVC surfaces, once exposed to biodegradation, will undergo surface alteration (Zhi Xiang et al., 2023; Xiang et al., 2023).

### 4.4 Mineralization

The biodegradability of PVC due to bacterial respiration can be evaluated through mineralization. The mineralization of carbon through bacterial respiration has a significant impact on the production of $CO_2$ (Patil and Bagde, 2012). PVC biodegradability examines the carbon tracing of polymeric materials into $CO_2$. The emission of $CO_2$ during biodegradation can be measured by the Sturm test (Hatwar and Qureshi, 2025; Khandare et al., 2021).

## 5 Challenges for development of biobased PVC bioremediation process

**Limited biodegradation efficiency**: Unlike other plastic polymers, PVC is highly resistant to microbial degradation due to its complex molecular structure, which includes C–C, C–Cl, and C–H bonds that are difficult to break down. The microbial degradation of PVC is significantly slower than that of other plastics, such as polyethylene and polystyrene.

**Formation of MPs**: The natural degradation of PVC results in the formation of MPs (PVC-MPs), which further contribute to environmental pollution and complicate waste management strategies.

**Toxic by-products**: Nonbiological degradation methods, such as chemical and physical treatments, often lead to the release of harmful compounds, including dioxins, furans, and hydrochloric acid, thereby posing serious environmental and health risks.

**Knowledge gaps in microbial degradation**: While certain bacteria and fungi have shown potential to degrade PVC, the specific enzymes and metabolic pathways involved are not yet well understood.

**Lack of sustainable waste management solutions**: Current PVC waste management strategies rely heavily on landfill disposal and incineration, both of which have serious environmental consequences.

**Inadequate research on biodegradation mechanisms**: The role of enzymes like lignin peroxidase and catalase peroxidase in PVC depolymerization is not fully understood, requiring further investigation into microbial and enzymatic pathways.

**Limited understanding of PVC-MP impact on ecosystems**: The effects of PVC MPs on soil and aquatic microbes, plants, and micro/macroanimals remain largely unknown.

**Cost-intensive nature of biotechnological solutions**: While microbial and enzymatic degradation methods have potential, the cost of scaling up these processes remains a significant barrier to implementation.

**Sustainable development challenges**: Addressing PVC pollution aligns with multiple Sustainable Developmental Goals (SDGs) (e.g., SDG 12—Responsible Consumption and Production, SDG 15—Life on Land). However, integrating sustainable solutions into existing waste management frameworks faces financial, regulatory, and technical hurdles.

**Ecological surveillance gaps**: Effective monitoring of PVC degradation and its environmental impact is hindered by insufficient surveillance techniques, limited standardized methodologies, and gaps in long-term ecological data collection.

## 6 Future scope of research on PVC bioremediation in the environment

To attain effective, scalable, and sustainable PVC degradation, a bioremediation approach in the environment becomes imperative, which needs to focus on integrating pretreatment innovations, the usage of biodegradable blends, optimal enzymatic reactions, and genetic engineering insights. Microbial enzymes like dehalogenases and peroxidases could be applied for pretreatment and biodegradation, which could be improved by advances in synthetic biology and CRISPR-based gene editing, increasing their ability to cleave the resistant C–Cl and C–C bonds in PVC. A crucial next step would be to create engineered microbial consortia with optimized metabolic pathways and test their ability in sizable bioreactors. Other pretreatment options, such as UV irradiation, oxidative treatments, and thermal hydrolysis, could also be used to weaken the polymeric structure of PVC and increase its accessibility to microorganisms. More research on systems of combined chemical and biological degradation could be proposed for the efficient mitigation of PVC. Life-cycle assessment studies and mechanical durability evaluations have supported the idea that biodegradable polymers like polyhydroxy alkanoates (PHA), poly lactic acid (PLA), and starch-based additives could be used to create environmentally friendly PVC blends, which would preserve usability while encouraging biodegradation. The catalytic performance could be improved by protein engineering, immobilizing enzyme systems, and stabilization aided by nanotechnology. Overall, converting laboratory results into practical environmental solutions would be required for sustainable waste management strategies, such as PVC-specific bioreactors, plastic-degrading insect larvae, public-private partnerships, artificial intelligence (AI)-assisted sorting technologies, and supportive regulatory frameworks.

## 7 Conclusions

PVC waste is a persistent environmental pollutant with complex degradation challenges. Microbial degradation is a promising, eco-friendly solution, but its slow rate and knowledge gaps in enzymatic mechanisms hinder its widespread adoption. Biopolymer blending, genetic engineering, and OMICS-based approaches offer potential improvements, though scalability and toxicity concerns remain. To mitigate PVC pollution, an integrated approach combining biological, chemical, and physical degradation methods becomes imperative. Collaborative efforts between researchers, industries, and policymakers may provide sustainable solutions for PVC waste management. Further research into microbial interactions, enzyme pathways, and environmental impacts would be promising for designing potential bioremediation strategies.

## References

Agarwal, T., Atray, N., Sharma, J.G., 2024. A critical examination of advanced approaches in green chemistry: microbial bioremediation strategies for sustainable mitigation of plastic pollution. Future J. Pharm. Sci. 2314-7253. 10 (1). doi:10.1186/s43094-024-00645-x.

Akash, K., Parthasarathi, R., Elango, R., Bragadeeswaran, S., 2024. Characterization of Priestia megaterium S1, a polymer degrading gut microbe isolated from the gut of Tenebrio molitor larvae fed on Styrofoam. Arch. Microbiol. 1432072X. 206 (1). Available from: https://www.springer.com/journal/203. doi:10.1007/s00203-023-03785-5.

Akovali, G., 2012. Plastic Materials: Polyvinyl Chloride (PVC) Toxicity of Building Materials. Elsevier, Turkey, pp. 23–53. Available from https://www.sciencedirect.com/book/9780857091222. doi:10.1016/B978-0-85709-122-2.50002-4.

Ali, M.I., Ahmed, S., Javed, I., Ali, N., Atiq, N., Hameed, A., Robson, G., 2014. Biodegradation of starch blended polyvinyl chloride films by isolated Phanerochaete chrysosporium PV1. Int. J. Environ. Sci. Technol. 1735-1472. 11 (2), 339–348. doi:10.1007/s13762-013-0220-5.

Ali, M.I., Perveen, Q., Ahmad, B., Javed, I., Razi-Ul-Hussnain, R., Andleeb, S., Atique, N., Ghumro, P.B., Ahmed, S., Hameed, A., 2009. Studies on biodegradation of cellulose blended polyvinyl chloride films. Int. J. Agric. Biol. 18149596. 11 (5), 577–580. Available from http://www.fspublishers.org/ijab/past-issues/IJABVOL_11_NO_5/13.pdf.

Anwar, M.S., Kapri, A., Chaudhry, V., Mishra, A., Ansari, M.W., Souche, Y., Nautiyal, C.S., Zaidi, M.G.H., Goel, R., 2016. Response of indigenously developed bacterial consortia in progressive degradation of polyvinyl chloride. Protoplasma. 0033183X. 253 (4), 1023–1032. Available from: www.springer.co.at/springer.py?Page=40&Key=456&cat=9&id_journal=4. doi:10.1007/s00709-015-0855-9.

Auta, H.S., Emenike, C.U., 2017. S.H Fauziah, Distribution and importance of microplastics in the marine environment: a review of the sources, fate, effects, and potential solutions. Environ. Int. 01604120. 102, 165–176. doi:10.1016/j.envint.2017.02.013.

Barili, S., Bernetti, A., Sannino, C., Montegiove, N., Calzoni, E., Cesaretti, A., Pinchuk, I., Pezzolla, D., Turchetti, B., Buzzini, P., Emiliani, C., Gigliotti, G., 2023. Impact of PVC microplastics on soil chemical and microbiological parameters. Environ. Res. 00139351. 229, 115891. doi:10.1016/j.envres.2023.115891.

Das, G., Karak, N., 2012. Epoxidized Mesua ferrea L. seed oil-plasticized thermostable PVC and PVC-clay nanocomposites. J. Vinyl Addit. Technol. 15480585. 18 (3), 168–177. doi:10.1002/vnl.20297.

Entezari, S., Al, M.A., Mostashari, A., Ganjidoust, H., Ayati, B., Yang, J., 2022. Microplastics in urban waters and its effects on microbial communities: a critical review. Environ. Sci. Pollut. Res. 16147499. 29 (59), 88410–88431. Available from: https://www.springer.com/journal/11356. doi:10.1007/s11356-022-23810-2.

Evode, N., Qamar, S.A., Bilal, M., Barceló, D., Iqbal, H.M.N., 2021. Plastic waste and its management strategies for environmental sustainability. Case Stud. Chem. Environ. Eng. 26660164. 4. Available from: www.journals.elsevier.com/case-studies-in-chemical-and-environmental-engineering/. doi:10.1016/j.cscee.2021.100142.

Gao, W., Xu, M., Zhao, W., Yang, X., Xin, F., Dong, W., Jia, H., Wu, X., 2024. Microbial Degradation of (Micro)plastics: Mechanisms, Enhancements, and Future Directions. Fermentation 10 (9), 441. doi:10.3390/fermentation10090441.

Ghosh, S., Qureshi, A., Purohit, H.J., 2019. D-Tryptophan governs biofilm formation rates and bacterial interaction in P. mendocina and S. aureus. J. Biosci. 09737138. 44 (1). Available from: http://www.ias.ac.in/jbiosci/. doi:10.1007/s12038-018-9841-7.

Giacomucci, L., Raddadi, N., Soccio, M., Lotti, N., Fava, F., 2019. Polyvinyl chloride biodegradation by Pseudomonas citronellolis and Bacillus flexus. New Biotechnol. 18764347. 52, 35–41. Available from: http://www.elsevier.com. doi:10.1016/j.nbt.2019.04.005.

Gilani, I.E., Sayadi, S., Zouari, N., Al-Ghouti, M.A., 2023. Plastic waste impact and biotechnology: exploring polymer degradation, microbial role, and sustainable development implications. Bioresour. Technol. Rep. 2589014X. 24. Available from: https://www.journals.elsevier.com/bioresource-technology-reports. doi:10.1016/j.biteb.2023.101606.

Hatwar, N., Qureshi, A., 2025. Biodegradation of PVC by novel bacterial consortia isolated from municipal solid waste dumpsite. Journal of Hazardous Materials 500, 140589. doi:10.1016/j.jhazmat.2025.140589.

Hatwar, N., Qureshi, A., 2025. Comprehensive Review on Bio-Based Treatments for Polyvinyl Chloride Plastic. Appl. Biochem. Biotechnol. 15590291. 197 (5), 2769–2798. Available from: https://www.springer.com/journal/12010. doi:10.1007/s12010-024-05174-0.

Huang, S., Zhang, B., Zhao, Z., Yang, C., Zhang, B., Cui, F., Lens, P.N.L., Shi, W., 2023. Metagenomic analysis reveals the responses of microbial communities and nitrogen metabolic pathways to polystyrene micro(nano)plastics in activated sludge systems. Water Res. 00431354. 241, 120161. doi:10.1016/j.watres.2023.120161.

Jianquan, X., Dong, Y., 2024. Analysis of the gut microbiome associated to PVC biodegradation in yellow mealworms. Ecotoxicol. Environ. Saf. 01476513. 272, 116046. doi:10.1016/j.ecoenv.2024.116046.

Jin, L., Feng, P., Cheng, Z., Wang, D., 2023. Effect of biodegrading polyethylene, polystyrene, and polyvinyl chloride on the growth and development of yellow mealworm (Tenebrio molitor) larvae. Environ. Sci. Pollut. Res. 16147499. 30 (13), 37118–37126. Available from: https://www.springer.com/journal/11356. doi:10.1007/s11356-022-24957-8.

Kaczmarek, H., Bajer, K., 2007. Biodegradation of plasticized poly(vinyl chloride) containing cellulose. Journal of Polymer Science, Part B: Polymer Physics. 08876266. 45 (8), 903–919. doi:10.1002/polb.21100.

Khandare, S.D., Chaudhary, D.R., Jha, B., 2021. Bioremediation of polyvinyl chloride (PVC) films by marine bacteria. Mar. Pollut. Bull. 18793363. 169. Available from: www.elsevier.com/locate/marpolbul. doi:10.1016/j.marpolbul.2021.112566.

Khatoon, N., Jamal, A., Ali, M.I., 2019. Lignin peroxidase isoenzyme: a novel approach to biodegrade the toxic synthetic polymer waste. Environ. Technol. 0959-3330. 40 (11), 1366–1375. doi:10.1080/09593330.2017.1422550.

Khatua, S., Simal-Gandara, J., Acharya, K., 2024. Myco-remediation of plastic pollution: current knowledge and future prospects. Biodegradation. 15729729. 35 (3), 249–279. Available from: https://www.springer.com/journal/10532. doi:10.1007/s10532-023-10053-2.

Kudzin, M.H., Piwowarska, D., Festinger, N., Chruściel, J.J., 2024. Risks Associated with the Presence of Polyvinyl Chloride in the Environment and Methods for Its Disposal and Utilization. Materials. 19961944. 17 (1). Available from: http://www.mdpi.com/journal/materials. doi:10.3390/ma17010173.

Kumari, A., Chaudhary, D.R., Jha, B., 2019. Destabilization of polyethylene and polyvinylchloride structure by marine bacterial strain. Environ. Sci. Pollut. Res. 16147499. 26 (2), 1507–1516. Available from: http://www.springerlink.com/content/0944-1344. doi:10.1007/s11356-018-3465-1.

Lewandowski, K., Skórczewska, K., 2022. A Brief Review of Poly(Vinyl Chloride) (PVC) Recycling. Polymers. 2073-4360. 14 (15), 3035. doi:10.3390/polym14153035.

Lu, L., Li, W., Cheng, Y., Liu, M., 2023. Chemical recycling technologies for PVC waste and PVC-containing plastic waste: a review. Waste Manag. 18792456. 166, 245–258. Available from: www.elsevier.com/locate/wasman. doi:10.1016/j.wasman.2023.05.012.

Luisa Pardo-rodríguez, M., Zorro-Mateus, P.J.P., 2021. Biodegradation of polyvinyl chloride by Mucor s.p. and Penicillium s.p. isolated from soil. Revista de Investigación, Desarrollo E Innovación. 2389-9417. 11 (2), 387–400. doi:10.19053/20278306.v11.n2.2021.12763.

Malakhova, D.V., Egorova, M.A., Leontieva, M.R., Elcheninov, A.G., Panova, T.V., Yu, D., Aleksandrov, E.A.T., 2023. Anaerobic Microbial Degradation of Polypropylene and Polyvinyl Chloride Samples. Microbiology. 0026-2617. 92 (1), 83–93. doi:10.1134/s0026261722602706.

Margaryan, A., Shahinyan, G., Hovhannisyan, P., Panosyan, H., Birkeland, N.K., Trchounian, A., 2018. Geobacillus and Anoxybacillus spp. from Terrestrial Geothermal Springs Worldwide: diversity and Biotechnological Applications. Microorganisms Sustain. 8, 119–166. Available from https://www.springer.com/series/14379. doi:10.1007/978-981-13-0329-6_5.

Mnyango, J.I., Hlangothi, S.P., 2024. Polyvinyl chloride applications along with methods for managing its end-of-life items: a review. Progress in Rubber, Plastics and Recycling Technology. 14782413. Available from: http://journals.sagepub.com/home/PRP. doi:10.1177/14777606241308652.

Mohanan, N., Montazer, Z., Sharma, P.K., Levin, D.B., 2020. Microbial and Enzymatic Degradation of Synthetic Plastics. Front. Microbiol. 1664302X. 11. Available from: https://www.frontiersin.org/journals/microbiology#. doi:10.3389/fmicb.2020.580709.

Novotný, Č., Fojtík, J., Mucha, M., Malachová, K., 2022. Biodeterioration of Compost-Pretreated Polyvinyl Chloride Films by Microorganisms Isolated From Weathered Plastics. Front. Bioeng. Biotechnol. 22964185. 10. Available from: http://journal.frontiersin.org/journal/bioengineering-and-biotechnology#archive. doi:10.3389/fbioe.2022.832413.

Othman, A.R., Hasan, H.A., Muhamad, M.H., Ismail, N., Abdullah, S.R.S., 2021. Microbial degradation of microplastics by enzymatic processes: a review. Environ. Chem. Lett. 16103661. 19 (4), 3057–3073. Available from: https://www.springer.com/journal/10311. doi:10.1007/s10311-021-01197-9.

Patil, R., Bagde, U.S., 2012. Isolation of polyvinyl chloride degrading bacterial strains from environmental samples using enrichment culture technique. Afr. J. Biotechnol. 11, 7947–7956.

Peng, B.-Y., Sun, Y., Zhang, X., Sun, J., Yazhou, X., Xiao, S., Chen, J., Zhou, X., Zhang, Y., 2023. Unveiling the residual plastics and produced toxicity during biodegradation of polyethylene (PE), polystyrene (PS), and polyvinyl chloride (PVC) microplastics by mealworms (Larvae of Tenebrio molitor). J. Hazard. Mater. 03043894. 452, 131326. doi:10.1016/j.jhazmat.2023.131326.

Peng, H., Zhang, Z., Kang, X., Zhang, Y., Zhang, H., Wang, Y., Yang, D., Zhang, J., Wang, Y., Cui, L., Zhu, Y.G., Ju, F., 2025. Unveiling gut microbiota and metabolic functions contributed to polyvinyl chloride degradation in Spodoptera frugiperda larvae. J. Hazard. Mater. 18733336. 492. Available from: https://www.sciencedirect.com/science/journal/03043894. doi:10.1016/j.jhazmat.2025.138209.

Peng, H., Zhang, Z., Kang, X., Zhang, Y., Zhang, H., Wang, Y., Yang, D., Zhang, J., Wang, Y., Zhu, Y.G., Ju, F., 2024. Comparative Metagenomics Provide Mechanistic Insights into the Biodegradation of the Non-hydrolysable Plastic Polyvinyl Chloride in Gut Microbiota of Insect Larvae. bioRxiv. 26928205. Available from: https://www.biorxiv.org. doi:10.1101/2024.02.06.579071.

Prakash, A., Palkar, R.R., 2021. Co-processing of plastic waste in a cement kiln: a better option. Environ. Sci. Pollut. Res., 1–11.

Saeed, S., Iqbal, A., Deeba, F., 2022. Biodegradation study of Polyethylene and PVC using naturally occurring plastic degrading microbes. Arch. Microbiol. 1432072X. 204 (8). Available from: http://link.springer.de/link/service/journals/00203/index.htm. doi:10.1007/s00203-022-03081-8.

Srikanth, M., Sandeep, T.S.R.S., Sucharitha, K., Godi, S., 2022. Biodegradation of plastic polymers by fungi: a brief review. Bioresour. Bioprocess. 21974365. 9 (1). Available from: https://bioresourcesbioprocessing.springeropen.com. doi:10.1186/s40643-022-00532-4.

Sudhakar, Y.N., Selvakumar, M., Bhat, D.K., 2023. Enhancement and investigation of biodegradability of poly (methyl methacrylate) and poly (vinyl chloride) by blending with biodegradable polymer. Polym. Bull. 14362449. 80 (5), 5623–5639. Available from: https://www.springer.com/journal/289. doi:10.1007/s00289-022-04319-7.

Tiwari, N., Bansal, M., Santhiya, D., Sharma, J.G., 2022. Insights into microbial diversity on plastisphere by multi-omics. Arch. Microbiol. 1432072X. 204 (4). Available from: http://link.springer.de/link/service/journals/00203/index.htm. doi:10.1007/s00203-022-02806-z.

Wu, Y., Wang, G., Wang, Z., Liu, Y., Gu, P., Sun, D., 2014. Comparative study on the efficiency and environmental impact of two methods of utilizing polyvinyl chloride waste based on life cycle assessments. Front. Environ. Sci. Eng. 2095221X. 8 (3), 451–462. Available from: http://www.springerlink.com/content/2095-2201/. doi:10.1007/s11783-013-0614-0.

Xiang, P., Zhang, Y., Zhang, T., Qian, W., Zhao, C., 2023. Qiang Li, A novel bacterial combination for efficient degradation of polystyrene microplastics. J. Hazard. Mater. 03043894. 458, 131856. doi:10.1016/j.jhazmat.2023.131856.

Yadav, M., Manderia, S., Singh, S., Deva, M.A., 2022. Isolation and Characterization of Polyvinyl Chloride (PVC) Degrading Bacteria from Polluted Sites of Gwalior City. Nat. Environ. Pollut. Technol. 23953454. 21 (1), 201–207. doi:10.46488/nept.2022.v21i01.022.

Yang, J., Duan, A., Zhiling, L., Chen, Z., Qiongying, X., Huang, T., Liu, W., Wang, A., 2024. Critical advances and assessment on chemo-biological conversions of waste polyvinyl chloride. Sci. Total Environ. 00489697. 956, 177170. doi:10.1016/j.scitotenv.2024.177170.

Yuan, Z., Nag, R., Cummins, E., 2022. Human health concerns regarding microplastics in the aquatic environment - From marine to food systems. Sci. Total Environ. 00489697. 823, 153730. doi:10.1016/j.scitotenv.2022.153730.

Zhang, Z., Peng, H., Yang, D., Zhang, G., Zhang, J., Ju, F., 2022. Polyvinyl chloride degradation by a bacterium isolated from the gut of insect larvae. Nat. Commun. 20411723. 13 (1). Available from: http://www.nature.com/ncomms/index.html. doi:10.1038/s41467-022-32903-y.

Zhi Xiang, J.K., Bairoliya, S., Cho, Z.T., Cao, B., 2023. Plastic-microbe interaction in the marine environment: research methods and opportunities. Environ. Int. 01604120. 171, 107716. doi:10.1016/j.envint.2022.107716.

CHAPTER

# 6

# Microbial approaches for electronic waste decomposition and metal recovery

**Ifrah Rashid, Tasaduq Hussain Shah, Sobiya Gul, and Ahali Jahan**

*Faculty of Fisheries, Sher-e-Kashmir University of Agricultural Sciences and Technology of Kashmir, Srinagar, India*

## 1 Introduction

### 1.1 Background on electronic waste

Rapid advancements in science and technology are transforming human life, making it more digital, convenient, and fast-paced. Electronic devices integrated with advanced programming have enhanced comfort but have also introduced serious environmental issues (Pinto, 2008; Babu et al., 2007). One of the critical issues is electronic waste (e-waste), which is growing rapidly and posing major risks to ecosystems and human welfare due to the release of toxic entities like heavy metals and hazardous chemicals (Rosariastuti et al., 2019; Mudila et al., 2019; Needhidasan et al., 2014). E-waste arising from outdated or obsolete electronic devices becomes superfluous quickly due to rapid technological changes. Sadly, only a small percentage is recycled, while the remainder gets dumped into oceans, landfills, or open areas (Monika, 2010). Around 2 million metric tons (MMt) of e-waste are generated annually, with global totals expected to exceed 50 MMt by 2019 (41.8 MMt in 2014, 44.7 MMt in 2016) (Mmereki et al., 2016). Personal digital devices account for around 20%, though figures may vary due to inconsistent definitions across studies. This problem is aggravated by industrial goals like Moore's law, which promotes constant technological changes and increased output (Poppenheimer, 2013), further escalating environmental damage alongside innovation. Small electronic devices accounted for the greatest portion of the global e-waste in 2019, while large equipment and household appliances came in second and third, respectively, reaching a record of 53.6 MMt. By 2030, this amount is expected to increase to 74 MMt due to rising obsolescence rates, quick product replacement in developed countries, and growing electronic use in developing countries (Forti et al., 2020a).

Microbial Technologies for Sustainable Waste Management. https://doi.org/10.1016/B978-0-443-44093-9.00017-7

## 1.2 Composition and classification of e-waste

E-waste includes a broad category of discarded electrical and electronic items. Annexure III of the EU Waste Electrical and Electronic Equipment (WEEE) Directive divides e-waste into six categories: cooling and refrigeration devices, display equipment (screens and monitors), lighting products, large household appliances (washing machines and dryers), small devices (electric kettles and calculators), and small IT and telecommunications equipment (Baldé et al., 2024; Erkmen et al., 2025). The COVID-19 epidemic, which caused many people to adjust to working remotely and learning online, further increased the integration of electronic gadgets in daily life (Botelho Junior et al., 2024). The global e-waste issue has escalated due to increased reliance on electronic technologies. The global e-waste generated in 2020 reached approximately 62 million tons, of which only 13.8 million tons were appropriately collected and recycled (Baldé et al., 2024). By 2030, global e-waste production is predicted to increase to 82 million tons, highlighting the urgent need for sustainable waste treatment strategies.

## 1.3 Environmental and economic concerns

According to Tesfaye et al. (2017), e-waste is composed of a variety of material components, such as glass, ceramics, plastics, and ferrous and nonferrous metals, with their proportions varying depending on the type of device and per unit mass. In addition to base metals like iron (Fe), copper (Cu), and zinc (Zn), e-waste is typically a rich source of valuable materials, including precious metals like silver (Ag), gold (Au), platinum (Pt), and palladium (Pd), as well as technology-critical elements (TCEs) like gallium (Ga), indium (In), and rare earth elements (REEs) (Fornalczyk et al., 2022; Sahu et al., 2022). Regulatory frameworks in the EU and United States classify elements like Ga, In, tantalum (Ta), niobium (Nb), and REEs as essential components in electronic circuits, with their concentrations varying widely, often ranging from 0.01 mg/kg up to 100 mg/kg, with neodymium (Nd) occasionally exceeding 100 mg/kg (Işıldar et al., 2018; Ballinger et al., 2020). Due to the restricted natural availability of these critical elements and their reliance on uncertain geopolitical and environmental conditions, their global supply is becoming increasingly vulnerable (Abdou et al., 2023; Walton et al., 2021). The recovery of TCEs from e-waste has become a strategic priority in sustainable waste management due to the intentionally short lifespans of devices and the growing demand for these metals (Işıldar et al., 2018; Bakas et al., 2016; Helena Xavier and Ottoni, 2023). Managing e-waste is extremely challenging because of its enormous volume, complex composition, and hazardous components. In a circular economy, e-waste is a reliable secondary source of raw materials, as electronics contain up to 69 elements from the periodic table, including valuable metals like Au, Ag, Pt, and Pd (Kaya, 2016; Tsydenova and Bengtsson, 2011). Despite its potential, only 17.4% of the e-waste produced globally was properly recycled in 2019, resulting in the loss of these valuable resources, estimated at $57 billion (Forti et al., 2020ab). As the demand for electronic devices increases and natural resources become harder to find, urban mining—recovering valuable metals from e-waste—is becoming an important way to secure these materials. It is often cheaper and more efficient than mining metals from the earth because e-waste can contain higher concentrations of metals than natural ores (Zeng et al., 2018).

## 2 Conventional e-waste management techniques

E-waste management relies on a combination of physical, chemical, and biological methods to recover valuable metals and reduce environmental hazards. Each of these approaches contributes uniquely to the overall efficiency and sustainability of e-waste treatment.

### 2.1 Physical and chemical methods

Physical methods, primarily mechanical in nature, include crushing, shredding, and various separation processes (magnetic, electrostatic, and density-based separation). These procedures are crucial for separating metal-rich fractions from their nonmetallic and plastic counterparts. Reducing particle size and simplifying composition increases the efficiency of later recovery operations. Metals are then extracted from e-waste matrices using chemical processes, such as leaching agents. Metals such as copper, nickel, lead, and gold are dissolved using eco-friendly organic and inorganic acids (e.g., sulfuric and nitric acid). Techniques like acid leaching, solvent extraction, and electrochemical methods are mainly employed to attain high metal recovery rates. However, factors like pH, temperature, contact time, and reagent concentrations control the effectiveness of these processes (Siwi et al., 2018; Pant et al., 2018; Gayatri et al., 2017).

Pyrometallurgical methods have long been used to recover metals from e-waste through high-temperature treatments, including smelting, roasting, and incineration (Thakur and Kumar, 2020). One widely used method is the ISASMELT process, implemented in industrial-scale facilities such as Aurubis AG in Germany and Umicore in Belgium (Alvear and Nikolic, 2013). This approach uses selective oxidation of metal complexes to separate particular metals from impurities. While it performs efficiently at large scales, the process typically focuses on base metals, often overlooking technology-critical elements. These valuable metals, due to their specific thermodynamic behaviors, frequently end up in the slag and are not recovered (Chu et al., 2022; Hasan et al., 2022). Hydrometallurgy utilizes chemical leaching to dissolve metals from solid materials and provides an alternative to this energy-intensive method (Thakur and Kumar, 2020). Particle size, temperature, agitation, and the solvent type are some of the known factors that affect the effectiveness of this approach (Ray, 2023). Usually, techniques like ion exchange, solvent extraction, or electrowinning are used to recover the dissolved metals after leaching (Murali et al., 2022). Sulfuric acid, nitric acid, hydrochloric acid, and aqua regia are examples of frequently used solvents (Xie et al., 2023; Illés and Kékesi, 2023). Although these reagents are able to break down complex materials, they lack selectivity, making the process inefficient for recovering trace metals and necessitating multiple stages of purification. Moreover, the handling and disposal of the resulting toxic chemical waste pose serious environmental and safety challenges (Rice, 2016; Cheng et al., 2024).

### 2.2 Limitations of conventional approaches

Conventional methods for recycling e-waste are proving increasingly inadequate, limited by a mix of technological, environmental, economic, and regulatory challenges. With the rapid advancement of electronics, modern devices are now built with miniaturized components and intricate material blends, such as embedded batteries and complex alloys, which make both manual and automated disassembly costly, time-consuming, and often inefficient (Debnath et al., 2023). In many cases, especially within informal recycling sectors, outdated and dangerous practices such as open burning, acid baths, and primitive dismantling techniques are still in use. These methods release a range of harmful toxins, including lead, mercury, cadmium, dioxins, and persistent organic pollutants, into the surrounding environment (Gupta et al., 2014; Little, 2021). As a result, the land, water, and air in the vicinity remain contaminated, posing major health hazards, especially for the workers and the populations surrounding them. Cases of developmental delays and neurological issues have been well-documented, particularly in youngsters exposed to these chemicals (Mingyue et al., 2022).

Despite the availability of modern technologies capable of recovering valuable metals (cobalt, gold, and REEs) with better efficiency, actual recovery rates often remain quite low (30%). The main constraint is the high cost of these advanced technologies and the lack of appropriate infrastructure to support them (The Growing Environmental Risks of E-Waste, 2024). Globally, only around 20% of e-waste is collected and recycled through formal systems; the vast majority ends up in landfills or passes through unregulated channels that lack environmental oversight (Debnath et al., 2023). Regulatory inconsistencies add to the problem. Weak enforcement of extended producer responsibility (EPR) laws and fragmented international policies enable the illegal export of millions of tons of e-waste to countries ill-equipped to handle it safely (Gupta et al., 2014). All of these issues highlight a pressing need for new solutions, ones that are not only environmentally friendly but also economically feasible.

## 3 Microbial approaches in e-waste management

### 3.1 Overview of biological techniques

In response to the growing challenges, researchers and environmental conservationists are shifting their attention to biologically driven solutions that are both scalable and eco-efficient. Microbial methods, in particular, have proven to be a sustainable alternative by operating under mild conditions without using aggressive chemicals or energy-intensive methods. This shift is driven by the need not only to improve metal recovery rates, but also to reduce the ecological footprint of various recycling practices. One such promising area of innovation is the use of microbial processes for managing the growing e-waste problem in a sustainable manner. Metal-rich wastes like electronic scrap, tailings, slags, and industrial dusts are often stored in stockpiles or sent to landfills, mainly because there are still no widely adopted, efficient, and sustainable methods for recovering the valuable metals they contain (Binnemans et al., 2020).

Conventional techniques like pyrometallurgy and hydrometallurgy consume large amounts of energy and often rely on toxic chemicals, leading to dangerous by-products and secondary pollution (Jaiswal and Srivastava, 2024). Microbial technologies, such as bioleaching and biosorption, present a greener and more efficient alternative. These methods harness the natural metabolic functions of certain bacteria and fungi to recover valuable metals from e-waste without the need for harsh reagents or extreme conditions (Han et al., 2022). Microorganisms like *Acidithiobacillus* and *Pseudomonas* are particularly effective; they can break down complex e-waste materials and release metals like gold, copper, and REEs through processes like acidolysis and redoxolysis (Adetunji et al., 2023).

## 3.2 Bioleaching: Microbial metal extraction

Bioleaching has been considered a sustainable and effective method for extracting metals from ores and e-wastes by harnessing the natural metabolic activities of certain microbes. The process involves several stages of biochemical reactions in which bacteria and fungi convert metals from their solid mineral forms to soluble compounds (Fig. 6.1). In bacterial bioleaching, acidophilic organisms such as *Acidithiobacillus*

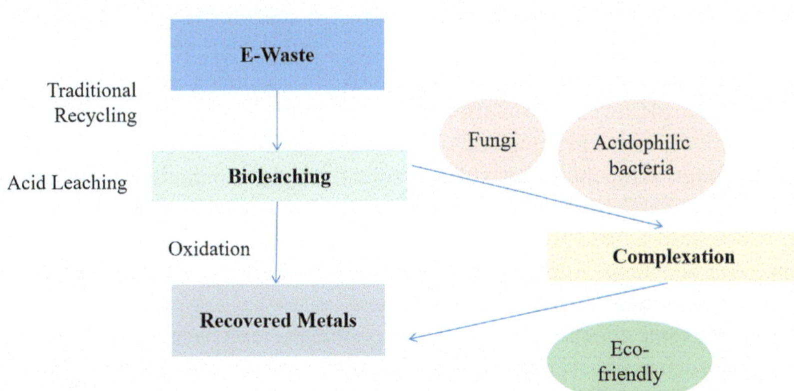


**FIG. 6.1**

Bioleaching of e-waste using microbes for eco-friendly metal recovery.

*ferrooxidans* and *Leptospirillum ferrooxidans* cause metal release through oxidation processes (Valix, 2017; Needhidasan et al., 2014). These bacteria oxidize ferrous iron ($Fe^{2+}$) into ferric iron ($Fe^{3+}$) while also oxidizing sulfur compounds to form sulfuric acid. These chemical alterations decompose metal sulfides, thereby solubilizing the metals for extraction. There are direct and indirect kinds of bioleaching. In direct bioleaching, bacteria attach to the surface of minerals and, via enzymatic action, induce the release of metals; indirectly, ferric ions produced by the metabolic activity of microorganisms continuously oxidize the mineral matrix (Hussien, 2025).

Many fungi are known to perform eco-friendly and sustainable reclamation of metals from ores and e-waste. This is mainly attributed to their ability to accumulate organic acids, which help dissolve metals. Examples include citric, oxalic, and gluconic acids. Two major routes are employed by these organic acids: acidolysis, in which protons are released by the organic acids and attack metal oxides, and complexolysis, where the organic acid molecules bond with metal ions to form stable, water-soluble complexes. Fungi belonging to the *Aspergillus* and *Penicillium* genera appear well-suited for metal leaching under these mechanisms for metals such as uranium, copper, zinc, and REEs (Bhatti and Tuovinen, 2025; Hussien, 2025). The leaching ability of these fungi is most effective under acidic conditions, particularly within a pH range of 1.5–3.5, which supports metal dissolution and prevents the re-formation of insoluble compounds (Schmeide and Bernhard, 2008). Optimal bioleaching performance depends on carefully controlled environmental conditions, such as using glucose or agricultural waste (e.g., wheat bran) as carbon sources, maintaining temperatures between 30°C and 45°C, and ensuring good aeration. Under such conditions, over 80% uranium recovery has been achieved in laboratory-scale tests with *Aspergillus niger* and *Aspergillus lentulus* by adopting both direct (one-step) and indirect (two-step) leaching methods. Beyond leaching, certain fungi also contribute to bioremediation by transforming toxic forms of metals into less harmful ones, such as converting hexavalent chromium into the relatively safe trivalent state (Bhatti and Tuovinen, 2025). Their dual potential for metal recovery and detoxification makes filamentous fungi highly valuable in sustainable waste management strategies. Environmental factors greatly influence microbial metal recovery. While acidophilic bacteria such as *A. ferrooxidans* thrive by oxidizing iron and sulfur compounds at lower pH levels (1.5–3.0), fungi tend to perform better at slightly milder acidic conditions, with a pH range of 4–6, as these conditions favor organic acid production by fungi. Temperature also plays a role: mesophilic organisms grow best between 25°C and 35°C, while thermophiles such as *L. ferrooxidans* can tolerate much higher temperatures, up to 80°C. Interestingly, mixed microbial consortia comprising both bacteria and fungi often achieve better metal recovery, as their metabolisms complement each other (Table 6.1).

Despite the toxic effects exerted by metals, some bacteria have evolved resistance mechanisms, such as metal efflux pumps, to cope, survive, and function in heavy-metal environments. The efficiency of bioleaching is also dependent on the particle size (since smaller particles provide more surface area for microbial action) and the availability of oxygen, which promotes aerobic microbial oxidation (Castro et al., 2024).

**Table 6.1** Microbial species involved in the bioleaching of e-waste and their associated metal recovery efficiencies.

| Microbial species | Type of e-waste | Conditions | Leaching agents/ metabolites | Recovered metals (% or mg/kg) | References |
|---|---|---|---|---|---|
| *A. niger* | Fluorescent lamp powder (<60 μm) | 30°C, pH 5.3, 1% pulp, 150 rpm | Oxalic, citric, gluconic acids | Y (13.4), Eu (12), La (0.9), Tb (0.7) | (Castro et al., 2023) |
| *A. niger* + *A. tubingensis* | PCBs (<1 mm) | 30°C, pH 5.7, 3% pulp | Citric, oxalic, malic acids | Cu (86), Ni (51), Zn (100) | (Trivedi et al., 2022) |
| *A. niger* | LCDs (≤74 μm) | 70°C, pH 7.5 | High oxalic, gluconic, citric content | Al (81), As (69), In (60) | (Parsa et al., 2024) |
| *A. ferrooxidans* + *A. thiooxidans* | LED monitor waste (<75 μm) | 30°C, pH 2 | Biogenic sulfate | In (99) | (Khezerloo et al., 2023) |
| *Alcaligenes aquatilis* | PCBs (d50 = 0.175 mm) | 28°C | Not specified | Cu (48) | (Madhavan et al., 2024) |
| *A. ferrooxidans* | PCBs (75–100 μm) | 30°C, pH 1.8 | Not specified | Cu (84), Pb (98), Al (100) | (Silva et al., 2023) |
| *Pseudomonas aeruginosa* | PCBs (<2 mm) | 30°C, pH 8 | Cyanide (biogenic from glycine) | Ag (90), Au (20) | (Merli et al., 2022) |
| *A. thiooxidans* | WEEE | 30°C, pH 3.5 | Sulfuric acid (biogenic) | Ce, Eu, Nd (99), La, Y (~80) | (Marra et al., 2018) |
| *Penicillium citrinum* | LIB powder | 40°C, pH ~ 5 | Citric acid | Mn (79), Li (90) | (Naseri and Mousavi, 2022) |
| *A. niger* | Spent LIB | 30°C, pH 6 | Mixed organic acids | Cu (100), Li (95), Mn (70), Co (45) | (Horeh et al., 2016) |
| *Bacillus megaterium* | LED waste | 30°C, pH 7 | Biogenic cyanide | Ag (93), Au (91), Cu (87), Ga (84) | (Pourhossein et al., 2021) |
| *Penicillium expansum* | PCBs | 25°C, pH 7.5 | Mixed acids (gluconic, citric, malic, etc.) | REEs: Pr, Nd, Gd (~70), Tb (50), La (40) | (Gonzalez Baez et al., 2024) |
| *Cellulosimicrobium funkei* | GaAs solar waste | 30°C, pH 7 | Supernatant with amino acids | Ga (70) | (Maneesuwannarat et al., 2016) |

A further step involves exploring the benefits that bacterial and fungal systems could bring together for maximized metal recovery. Hybrid approaches that utilize *A. ferrooxidans* with *A. niger* have demonstrated efficiencies in copper recovery of over 90% under controlled conditions, consuming far less energy than conventional extraction processes (Castro et al., 2024; Harshan and Rajan, 2025). Although great potential exists, it is hindered by barriers to industrial-scale application, including microbial viability maintenance in toxic conditions, variability in mixed waste materials, and fine-tuning environmental conditions to achieve maximum performance. Aligned with the principles of the "circular economy," this biological process offers an environmentally friendly and sustainable way of recovering critical and strategic metals from low-grade ore and e-waste, while also decreasing dependence on traditional mining and minimizing environmental degradation (Rendón-Castrillón et al., 2023; Hussien, 2025).

## 3.3 Biosorption and bioaccumulation

Another biological technique, biosorption, exploits the ability of microbial cell walls to bind metal ions. Organisms like *Saccharomyces cerevisiae* and *Chlorella* species have shown notable capacity for copper uptake, especially under optimal temperature and pH conditions (Siwi et al., 2018). On the contrary, bioaccumulation refers to metals being absorbed and internalized within the microbial cells. Different bacteria, including *Bacillus licheniformis* and *Pseudomonas aeruginosa,* have been able to efficiently remove heavy metals like copper, lead, and zinc from polluted sites (Banerjee et al., 2015; Aderonke et al., 2017). Genetic engineering has been utilized in this process to its advantage, such as the engineering of *Chromobacterium violaceum* to produce cyanide lixiviants that greatly increased the rate of gold leaching (Tay et al., 2013).

In addition, biological strategies such as biostimulation and biotransformation are being explored. Biostimulation focuses on improving environmental conditions and nutrient availability to enhance microbial activity and degradation potential. On the contrary, biotransformation uses microbial enzymes to chemically alter toxic substances into less harmful forms, facilitating safer and more effective waste remediation (Diep et al., 2018).

## 3.4 Enzymatic degradation

Enzymatic action plays a pivotal role in the microbial breakdown of synthetic materials embedded in e-waste. Certain bacteria isolated from contaminated soils produce enzymes that can efficiently degrade multiple types of e-waste plastics. Both individual strains, such as *P. aeruginosa* PE10, and microbial consortia have been shown to degrade polymers under natural conditions. With the help of an analytical approach, methods like Fourier transform infrared spectroscopy and scanning electron microscopy have evidenced the major alteration of the plastic surfaces after treatment, thus indicating their degradation (Debbarma et al., 2024). This demonstrates how important microbial enzymes—specifically hydrolases, oxidases, and depolymerases—are in breaking down complicated polymer structures. Another dimension is added by the use of synthetic biology tools, which empower microbial engineering toward the development of engineered strains that are fine-tuned for enhanced degradation (Valix, 2017; Islam et al., 2020).

## 4 Microbial mechanisms for e-waste decomposition

### 4.1 Microbial strains and e-waste degradation

Microorganisms, especially bacteria and fungi, have great potential in addressing the complex nature of e-waste, breaking down its organic and inorganic components. The usual course of action may demand more energy and often involves the use of chemicals hazardous to health; fungi and bacteria, on the contrary, take advantage of their life cycle or metabolic liquefaction to convert toxic e-waste into harmless or even beneficial materials as by-products. Studies have demonstrated that these microbes can act on a variety of substances commonly found in e-waste, including synthetic polymers, resins, and heavy metals (Valix, 2017; Han et al., 2022). Research has identified multiple bacterial strains like *Alcaligenes*, *Enterobacter*, *Citrobacter sedlakii*, *Brevundimonas diminuta*, *Pseudomonas*, and *Bacillus* that are capable of degrading high-impact polystyrene and related plastic compounds used in electronics (Sekhar et al., 2016). Gut microbes of wax moths (*Galleria mellonella*), such as *Enterococcus* and *Enterobacter,* further help in the degradation of plastics that contain challenging polymers such as polyurethane, polystyrene, and acrylonitrile butadiene styrene. Generally, these microbes degrade plastics by colonizing plastic surfaces and releasing enzymes that fragment large polymers into smaller ones that are further metabolized and mineralized. With this, pollution is lessened, as well as establishing a great premise for green recycling techniques (Zhu et al., 2021).

### 4.2 Metal recovery through microbes

Recovery of metals through microbial methods adds an important adjunct to e-waste management. Very often, microbes best suited for biorecovery are found in severely contaminated environments and have evolved under extreme conditions of metal toxicity. The bacteria *Acidithiobacillus, Sulfobacillus,* and *Pseudomonas,* along with fungi such as *Aspergillus* and *Purpureocillium lilacinum*, have shown much promise in the recovery of base metals like Cu, Fe, Ni, Pb, Zn, and precious metals like Ag, Au, Co, Pd, and Pt from waste electronics (Valix, 2017; Islam et al., 2020). Acidophilic bacteria are especially efficient owing to their endurance against high concentrations of several toxic metals (Zhao and Wang, 2019; Mudila et al., 2021).

At the core of microbial metal extraction lies a series of biologically driven mechanisms like bioleaching, biosorption, and bioaccumulation. Bioleaching, also referred to as biohydrometallurgy, uses the metabolic capabilities of certain microbes

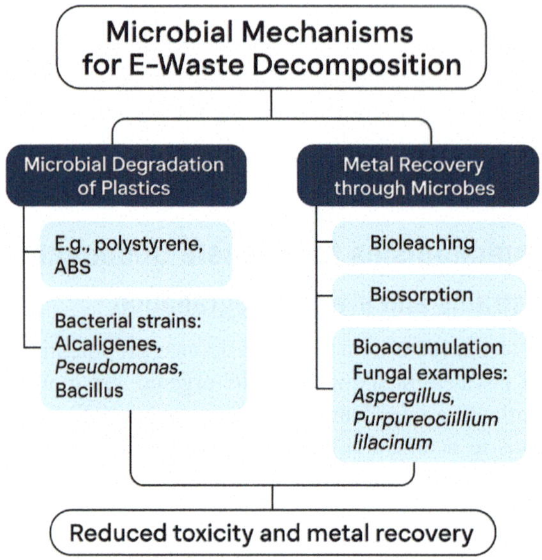


**FIG. 6.2**

Flowchart illustrating microbial mechanisms for e-waste decomposition.

to transform insoluble metal compounds into extractable ionic forms (Valix, 2017; Islam et al., 2020). The transformation occurs through microorganisms directly acting on mineral surfaces or indirectly regenerating ferric ions that act as oxidants. Typical processes include acidolysis, redoxolysis, and complexolysis. Biosorption refers to the passive binding of metal ions by microbial cell walls or extracellular compounds, while bioaccumulation involves the active uptake and storage of metals by microbial cells (Fig. 6.2). Jointly, these biological processes assist not only in the recovery of metals but also in detoxifying the leachates of e-waste, thereby leading to safer and more viable options for downstream waste treatment (Rana et al., 2021).

# 5 Advances in microbial technologies for e-waste recycling

## 5.1 Genetic engineering

Advancements in synthetic biology have revolutionized metal extraction from e-waste by making it possible to genetically tailor microbes. One striking example is the engineering of *Cupriavidus metallidurans* to capture gold ions from e-waste. This engineered strain was able to recover almost 150 g of gold per metric ton of circuit boards in industrial-scale tests (Han et al., 2022). Along similar lines, *Pseudomonas putida 06909* was genetically engineered to express the metal-binding EC20 peptide, enhancing cadmium uptake and decreasing toxicity inside the cell (Zhu et al., 2020). This modification endows the microbes not only with an increased ability for metal uptake but also with improved resistance in hostile waste environments. Other researchers have gone further by utilizing cutting-edge genetic modeling tools like genome-scale metabolic modeling (GEM) to engineer microbial pathways for metal extraction. Using GEM, it has been possible to pinpoint genetic modifications to enhance the production of metal-leaching compounds such as cyanide, achieving a 30%–40% increase in the recovery of precious metals, including copper and gold (Han et al., 2022). Such advances in bioengineering can thus lead to less resource-intensive, sustainable, and efficient e-waste recycling technologies, aligning well with circular economy goals.

## 5.2 Nanotechnology integration

Nanotechnology is progressively being combined with microbial technology to provide the much-needed enhancement for extracting metals from e-waste. One of the latest promising approaches involves the use of magnetic nanoparticles along with genetically engineered bacteria expressing metal-binding peptides such as SynHMB. These hybrids have been reported to remove 85%–90% of toxic metals like cadmium and lead from e-waste leachate (Zhu et al., 2020). The nanoparticles provide a large surface area for microorganism attachment, offering an easy route for downstream processing where magnet-based separation is used to recover the metal-laden biomass. On the contrary, nanoscale zero-valent iron assists bioleaching by providing a continuous supply of ferrous ions to keep key acidophiles such as *A. ferrooxidans* active in degrading metal-containing compounds in e-waste (Lefevre et al., 2015). Therefore these nanomaterials appear to be upgrading and operationalizing biologically assisted systems of metal recovery.

## 5.3 Microbial consortia

Synergistic microbial consortia are emerging as tools to address the compositional complexity of e-waste. These mixed communities pool their general strengths from different microbes, thereby allowing for a better metal recovery rate. A mixture of *Acidithiobacillus ferrivorans* and *Acidithiobacillus thiooxidans* can recover almost 98% of copper from printed circuit boards (PCBs), while cocultures of bacteria producing biosurfactants and sulfur oxidation have been able to enhance chromium extraction by up to 50% (Han et al., 2022). Of course, fungal–bacterial consortia are also of noteworthy importance. The pairing of fungi, such as *A. niger*, with bacteria like *Pseudomonas fluorescens* has led to better extraction of REEs through complementary mechanisms that involve acidolysis and redoxolysis (Adetunji et al., 2023). These consortia work around the limits imposed by single species. Some archaea, such as certain *Ferroplasma* spp., act in support by decomposing organic compounds that could inhibit bacterial oxidation, thereby maintaining active metal solubilization (Lefevre et al., 2015).

### 5.4 Bioelectrochemical systems

Bioelectrochemical system (BES) has emerged as a promising approach, combining microbial processes with electrochemical methods for the recovery of valuable metals from e-waste. In an experiment, *A. ferrooxidans* was utilized to facilitate bioleaching of copper from shredded PCBs, after which electrowinning was performed on this bioleachate to yield copper foil with > 99% purity (Han et al., 2022). Another application of BES is microbial fuel cells, wherein electroactive bacteria such as *Shewanella oneidensis* generate electrical currents to induce the cathodic reduction of metal ions. This method has been successfully applied for the recovery of precious metals from e-waste leachates (Rendón-Castrillón et al., 2023). BES has attracted attention due to its lower energy consumption (typically 40%–60% less than traditional extractive methods) and reduced environmental impact, with very few secondary wastes generated.

## 6 Challenges and limitations

Although microbial technologies offer a viable route to sustainable e-waste recycling, several hindrances still limit their widespread use and large-scale implementation. One of the main obstacles is that microbial processes are relatively slow and less effective in comparison with conventional chemical or thermal methods. Extracting metals from PCBs using acid-loving bacteria takes days or even weeks, whereas chemical leaching typically yields comparable results within a few hours. In industrial conditions, the actual recovery rates for critical metals often remain under 30%, a sharp contrast to the over 90% efficiencies achieved in controlled laboratory environments. Variable environmental conditions, inconsistent microbial performance in complex, diverse e-waste, and the challenge of maintaining productive microbial communities at industrial scales are the main causes of this gap (Han et al., 2022).

Toxicity is another issue of concern. The leachate of e-waste usually contains heavy metals (lead, cadmium, and mercury) and organic contaminants (brominated flame retardants and PCBs). These chemicals can cause significant harm to microbial health, hindering growth, influencing metabolic activity, and inhibiting bioleaching. Copper occurs in concentrations >10 g/L, which may seriously interfere with the activities of *A. ferrooxidans*, an essential bacterium in metal extraction. Likewise, fungi such as *A. niger* are vulnerable under mixed-metal stresses; this vulnerability inhibits redox balance and affects the bioleaching potential even of genetically modified strains that are supposed to thrive in toxic environments (Lin et al., 2022).

Challenges are further magnified when discussing the upscaling of processes. For industrial bioleaching, low e-waste loading rates between 1% and 10% w/v are typically maintained to avoid overwhelming the microbes, thereby constraining overall throughput capacity and economic viability. E-waste itself is a complex mixture of plastics, metals, and other materials that require highly adaptable microbial consortia. The cultivation and maintenance of such mixed microbial populations in bioreactor systems, however, prove to be troublesome and, at times, resource-intensive. Suitable reactor systems must ensure proper oxygen flow and nutrient supply while also requiring efficient separation of biomass from leachates. This approach inevitably adds layers of complexity and costs. Even the so-called two-step bioleaching method, wherein microbial growth is separated from metal extraction, does not necessarily reduce costs and introduces its own infrastructure and monitoring requirements (Debbarma et al., 2024).

Finally, regulatory and financial challenges also hamper progress. Few established standards or regulatory guidelines for microbial recycling of e-waste actually exist; hence, industries find it hard to even consider investing in it. Here, the associated costs include everything involved in the development of engineered microbial strains, from specialized reactor design to preprocessing of e-waste, more so than even classic or informal recycling procedures. Chemical and thermal methods, particularly from an economic perspective for valued metals such as copper and gold, still remain the traditional options and are proposed alternatives only from an environmental perspective (Lin et al., 2022; Han et al., 2022). In addition, poor enforcement of EPR laws and ongoing global issues concerning illegal e-waste dumping further discourage efforts toward the commercialization of greener microbial alternatives (Lin et al., 2022; Han et al., 2022).

## 7 Integration of microbial processes into circular economy models

Microbial recycling is gaining attention as an important component of sustainable waste management practices, particularly in the context of the circular economy. These biological processes, which often rely on the inherent abilities of bacteria and fungi to break down plastics and extract metals, offer an eco-friendly and cost-effective alternative to traditional e-waste recycling methods. Microbial approaches not only allow the recovery of valuable metals like copper, gold, and palladium, but also promote the degradation and reuse of synthetic polymers. This aligns with the circular economy goals of resource recovery, waste reduction, and material reuse (Valix, 2017; Han et al., 2022).

*Microbial consortium* is another interesting strategy, where carefully chosen microbial combinations are allowed to work together to manage the complex mixture of elements found in e-waste. Multiple substrates can be degraded simultaneously, and different metals can be recovered in a single, integrated process since various microorganisms target different components in e-waste. This strategy has already been adopted in industrial sectors. Mint Innovation has devised a hybrid system combining chemical and microbial steps to extract metals from e-waste at demonstration-scale facilities. Similarly, **Network 2 Supplies Ltd.** (N2S), a UK-based IT lifecycle management and sustainable e-waste recycling firm founded in 2002, has also created a bioleaching-electrowinning process to recover high-purity copper from PCBs (Valix, 2017; Islam et al., 2020; Han et al., 2022).

The industrial scaling of novel developments is being facilitated by advances in microbial engineering, synthetic biology, and reactor design. The Horizon 2020 MIX-UP project is one of the EU's flagship initiatives focusing on the microbial-based upcycling of engineered plastic waste, and pilot plants capable of processing several metric tons of e-waste per day are already operational. Yet, some problems remain. The balance of continuous cultures cannot be easily automated; more precise control of pH, temperature, and waste loading is necessary for consistent performance at scale. In addition, the microbial populations have to be resistant to changes in the composition of the waste stream (Valix, 2017; Islam et al., 2020).

The impact of microbial recycling systems on the e-waste management sector is noteworthy, especially regarding the environment and economy. These systems mitigate reliance on primary mining, lower the costs of traditional recycling methods, and reduce the emissions of fossil fuels and toxic waste. Microbial recycling also cuts down on toxic waste generation and greenhouse gas emissions. This process, aside from its environmental perks, can spur job creation within the green tech sector and, importantly, reduce the volume of dangerous waste that winds up in landfills, a clear win for both the environment and overall public health. As regulations shift and incentives for sustainable actions grow, microbial recycling is well-placed to become a critical element in circular e-waste handling (Han et al., 2022; Lee et al., 2024).

## 8 Mint innovation: A microbial solution to urban e-waste mining

Mint Innovation is a New Zealand-based clean technology company that has become a global contender in using microbes to recover valuable metals from e-waste. Many traditional recycling techniques are ineffective in recovering these valuable metals and often rely on ecologically damaging procedures, as e-waste continues to increase, surpassing 53 million tons worldwide in 2019 alone. Mint Innovation offers an alternative approach that is both sustainable and scalable (World-first Biotechnology, 2024). Their process begins with the mechanical shredding of e-waste, such as discarded PCBs. After this, common metals like copper and tin are removed through chemical leaching. The leftover residue, containing valuable metals, is subsequently treated using naturally occurring microorganisms. *C. metallidurans*, one of the microbes, is well-known for its ability to directly absorb gold ions from solution. This biological stage enables the selective and efficient recovery of these valuable metals without the need for harsh chemicals or energy-intensive smelting processes (World-first Biotechnology, 2024).

The approach developed by Mint Innovation is notable for its emphasis on small, urban biorefineries. Unlike conventional recycling methods that involve shipping waste to large, often overseas, smelting facilities, Mint Innovation aims at setting up localized plants capable of processing thousands of tons of e-waste each year. Up to 3500 tons of waste is treated annually in its Sydney facility (The Venture Using Microbes to Turn E-Waste into Precious Metals, 2021). This regional approach reduces transport emissions, accelerates the recycling process, and enables recovered metals to be recycled directly into local manufacturing industries ranging from electronics to jewelry production. This not only promotes sustainability but also strengthens the circular economy by reusing resources within local communities (World-first Biotechnology, 2024). Mint Innovation's approach has considerable environmental and economic benefits. The company uses low-cost, nontoxic chemicals and naturally existing microbes, resulting in lower energy consumption and reduced emissions compared to traditional methods. Instead of exporting waste or dumping it in landfills, this approach focuses on extracting valuable metals on-site, preserving natural resources, and creating skilled jobs in green technology sectors. Their innovative model has received international recognition, including a spot in the Circulars Accelerator program of the World Economic Forum, and has attracted significant investments to support their global expansion (C&EN, 2024).

# 9 Future directions and research perspectives

Enhancing the efficiency and scalability of microbial recycling technologies for e-waste is an emerging research focus area. Advances in synthetic biology and metabolic engineering are making it possible to design microbial strains that are better suited to the growing demands of e-waste processing. These modified strains not only have higher enzymatic activity and offer a wider range of substrate utilization but also exhibit improved resistance to the toxic compounds found in e-waste (Valix, 2017). Such improvements enable microbes to break down complex polymers more effectively, extract metals efficiently, and convert waste into valuable by-products like biobased chemicals and industrial feedstocks.

One of the renowned efforts in this field is the *EU Horizon 2020 MIX-UP project*, with a goal to develop engineered microbial consortia and customized enzyme amalgamations to degrade and upcycle mixed plastic waste into commercially valuable compounds (Han et al., 2022). This underlines the growing potential of integrating microbial engineering with industrial biotechnology workflows to recover resources in accordance with circular economy concepts. Microbial recycling, to flourish at an industrial scale, requires careful optimization of processing parameters. pH, temperature, aeration, nutrient balance, and pulp density must be carefully adjusted to sustain microbial activity and maximize recovery rates. The two-step bioleaching strategy, which separates microbial growth from the actual metal recovery process, has shown enhanced yields and operational control in large-scale settings (Gumulya et al., 2018). However, the toxicity and chemical diversity of e-waste can disturb microbial populations and diminish their efficacy, making real-world applications challenging (Valix, 2017; Han et al., 2022; Rana et al., 2021).

Another promising avenue lies in the discovery and use of novel microbial species with specialized metabolic abilities. While acidophilic bacteria and fungi continue to play a major role, researchers are now looking into extremophiles and less-studied environmental isolates, such as *Pseudarthrobacter oxydans*, which have demonstrated potential for recovering REEs under harsh conditions. Mixed microbial consortia, combinations of microbes that carry out complementary functions, are increasingly being used to handle the compositional complexity of e-waste. These consortia benefit from metabolic cross-feeding and coordinated enzyme production, leading to better degradation rates and metal recovery compared to single-strain approaches (Han et al., 2022).

The role of computational tools in strengthening this field is expanding rapidly. Several techniques, such as GEMs, artificial intelligence, and bioinformatics, are being employed to identify gene targets for strain enhancement and pathway optimization (Liow et al., 2020; Buekens and Yang, 2014; Slijkhuis, 2018). Together with high-throughput screening methods and synthetic biology platforms, these strategies considerably accelerate the development of customized microbial systems for specific waste profiles.

Progress from the policy and regulatory perspective is equally important. Financial support, policy frameworks, and standardized protocols are required to adopt microbial technologies for e-waste recycling. Further enhancements in process dependability and environmental effectiveness can be achieved via improved logistics of e-waste management, such as sorting, pretreatment, and incorporation into existing recycling frameworks (Dixit et al., 2023). A balanced policy toward maintaining the ecology and economy will also need to be complemented by cooperation from researchers, industry, and policymakers. Multidisciplinary cooperation will be needed to translate microbial recycling from promising laboratory techniques into effective, large-scale industrial applications. Microbial technologies integrated with traditional recycling techniques might provide the optimal balance on the most forward pathway for maximizing resource recovery and minimizing harm to the environment. The advancement of policies and public investment, along with innovation, will enable the management of e-waste and the circular economy to enhance the central role of microbial technologies (Hahladakis et al., 2018; Jiakang et al., 2020).

## 10 Conclusions

Microbial processes for managing e-waste represent a practical and environmentally conscious alternative to traditional recycling methods. Through the natural activity of bacteria, fungi, and mixed microbial communities, it is possible to break down synthetic materials and extract valuable metals such as copper, gold, and REEs. Techniques like bioleaching, biosorption, and bioaccumulation have demonstrated the ability to handle complex e-waste while reducing the energy use and harmful emissions typically associated with chemical or thermal treatments. Beyond their lower environmental impact, these biological methods contribute to the recovery of important raw materials that are increasingly scarce and in high demand. Microbial systems offer a more comprehensive approach to handling the diverse composition of e-waste by supporting the breakdown of polymers and the extraction of metals within the same process. Pilot-scale projects have demonstrated that microbial recycling can be used both outside the laboratory and scaled up for real-world conditions, even in the face of fluctuating and contaminated waste streams. Recent developments, including the use of specially selected microbial strains, genetically improved organisms, and mixed cultures, have helped address key challenges such as slow processing, sensitivity to toxic substances, and difficulties in maintaining system stability. Microbes have also been used in more advanced recovery systems, including hybrid methods that combine biological and electrochemical steps for more precise metal recovery. Looking ahead, broader adoption of these technologies will depend not only on research progress but also on infrastructure development, economic support, and cooperation between public and private sectors. With rising global concern over the environmental impact of e-waste and the need to secure access to critical materials, microbial recycling stands out as a promising way forward. It offers a cleaner, more adaptable, and more sustainable solution to one of the most pressing waste problems of our time.

## References

Abdou, M., Cobelo-Garcia, A., Santos-Echeandía, J., Schäfer, J., 2023. Technology-critical Elements Contaminants of Emerging Concern in the Marine Environment: Current Challenges in Marine Pollution. Elsevier, Portugal, pp. 401–438. Available from: https://www.sciencedirect.com/book/9780323902977. doi:10.1016/B978-0-323-90297-7.00007-X.

Aderonke, A.K., Oladimeji, O.O., Shittu, O.B., Okeyode, I.C., Taiwo, M.O., 2017. Bioaccumulation of heavy metals using selected organisms isolated from electronic waste dumpsite of two south-western states in nigeria. Appl. Environ. Res. 2287-075X. 29–40. doi:10.35762/aer.2017.39.2.3.

Adetunji, A.I., Oberholster, P.J., Erasmus, M., 2023. Bioleaching of metals from E-waste using microorganisms: a review. Minerals 13 (6). doi:10.3390/min13060828.

Alvear, F., Nikolic, S., 2013. ISASMELT™ for recycling of valuable elements contributing to a more sustainable society. TMS Annu. Meet. 100–109. doi:10.1007/978-3-319-48763-2_11.

Babu, B.R., Parande, A.K., Basha, C.A., 2007. Electrical and electronic waste: a global environmental problem. Waste Manag. Res. 25 (4), 307–318. doi:10.1177/0734242x07076941.

Baez, A.G., Muñoz, L.P., Timmermans, M.J., Garelick, H., Purchase, D., 2024. Molding the future: optimization of bioleaching of rare earth elements from electronic waste by Penicillium expansum and insights into its mechanism. Bioresour. Technol. 402.

Bakas, I.C., Herczeg, M.C., Vea, E.B.C., Youhanan, L.I., Baxter, J.Ø., e, A., 2016. Critical metals in Discarded Electronics: Mapping Recycling Potentials from Selected Waste Electronics in the Nordic Region, Nordic Council of Ministers.

Baldé, C.P., Kuehr,, Yamamoto, T., McDonald, R., D'Angelo, E., Althaf, S., Bel, G., Deubzer, O., Fernandez-Cubillo, E., Forti, V., Gray, V., 2024. Global E-waste Monitor, UNITAR.

Ballinger, B., Schmeda-Lopez, D., Kefford, B., Parkinson, B., Stringer, M., Greig, C., Smart, S., 2020. The vulnerability of electric-vehicle and wind-turbine supply chains to the supply of rare-earth elements in a 2-degree scenario. Sustain. Prod. Consum. 22, 68–76. doi:10.1016/j.spc.2020.02.005.

Banerjee, S., Gothalwal, R., Sahu, P.K., Sao, S., 2015. Microbial observation in bioaccumulation of heavy metals from the Ash Dyke of thermal power plants of Chhattisgarh, India. Adv. Biosci. Biotechnol. 2156-8456. 06 (02), 131–138. doi:10.4236/abb.2015.62013.

Bhatti, T.M., Tuovinen, O.H., 2025. Bioleaching of uranium from ores and rocks using filamentous fungi. Front. Microbiol. 1664302X. 16. Available from: https://www.frontiersin.org/journals/microbiology#. doi:10.3389/fmicb.2025.1523962.

Binnemans, K., Jones, P.T., Manjón Fernández, Á., Torres, V.M., 2020. Hydrometallurgical processes for the recovery of metals from steel industry by-products: a critical review. J. Sustain. Metall. 21993831. 6 (4), 505–540. Available from: https://link.springer.com/journal/40831. doi:10.1007/s40831-020-00306-2.

Botelho Junior, A.B., da Silva, M.D.C.R., Camargo, P.S.S., Munchen, D.D., Cenci, M.P., Bertuol, D.A., Veit, H.M., Tenório, J.A.S., Espinosa, D.C.R., 2024. Electronic waste in emerging countries: current scenario of generation, policies, and recycling technologies regarding the coronavirus pandemic. Int. J. Environ. Sci. Technol. 1735-1472. 21 (1), 1121–1140. doi:10.1007/s13762-023-05128-9.

Buekens, A., Yang, J., 2014. Recycling of WEEE plastics: a review. J. Mater. Cycles Waste Manag. 1438-4957. 16 (3), 415–434. doi:10.1007/s10163-014-0241-2.

C&EN, Electronic waste is a gold mine waiting to be tapped. (2024).

Castro, L., Gómez-Álvarez, H., González, F., Muñoz, J.A., 2023. Biorecovery of rare earth elements from fluorescent lamp powder using the fungus Aspergillus Niger in batch and semicontinuous systems. Miner. Eng. 08926875. 201, 108215. doi:10.1016/j.mineng.2023.108215.

Castro, L., Zhang, R., Muñoz, J.A., Sand, W., 2024. Editorial: bioleaching and biorecovery of critical raw materials from secondary sources. Front. Microbiol. 1664302X. 15. Available from: https://www.frontiersin.org/journals/microbiology#. doi:10.3389/fmicb.2024.1395820.

Cheng, Q., Wang, Z., Wang, Y., Jun-Tao, L., Haiyan, F., 2024. Recent advances in preferentially selective Li recovery from spent lithium-ion batteries: a review. J. Environ. Chem. Eng. 22133437. 12 (3), 112903. doi:10.1016/j.jece.2024.112903.

Chu, H., Qian, C., Tian, B., Shiyue, Q., Wang, J., Xin, B., 2022. Pyrometallurgy coupling bioleaching for recycling of waste printed circuit boards. Resources, Conservation and Recycling. 09213449. 178 (106018). doi:10.1016/j.resconrec.2021.106018.

Debbarma, P., Suyal, D.C., Kumar, S., Zaidi, M.G.H., Goel, R., 2024. Comparative e-waste plastics biodegradation efficacy of monoculture Pseudomonas aeruginosa strain PE10 and bacterial consortium under in situ condition. Front. Microbiol. 1664-302X. 14. doi:10.3389/fmicb.2023.1277186.

Debnath, B., Das, A., Chowdary, P.A., Bhattacharyya, S., 2023. Technological Advancement in E-waste Management: Towards Smart, Sustainable, and Intelligent Systems Technological Advancement in E-waste Management: Towards Smart, Sustainable, and Intelligent Systems. CRC Press, India, pp. 1–122. Available from: http://www.tandfebooks.com/doi/book/9781003804635. doi:10.1201/9781003317050.

Diep, P., Mahadevan, R., Yakunin, A.F., 2018. Heavy Metal Removal by Bioaccumulation Using Genetically Engineered Microorganisms. Front. Bioeng. Biotechnol. 6. 10.3389/fbioe.2018.00157

Dixit, R., Kumar, S., Pandey, G., 2023. Biological approaches for E-waste management: a green-go to boost circular economy. Chemosphere. 00456535. 336, 139177. doi:10.1016/j.chemosphere.2023.139177.

Erkmen, A.N., Ulber, R., Jüstel, T., Altendorfner, M., 2025. Towards sustainable recycling of critical metals from e-waste: bioleaching and phytomining. Resources, Conservation and Recycling. 09213449. 215, 108057. doi:10.1016/j.resconrec.2024.108057.

Fornalczyk, A., Willner, J., Jablonska-Czapla, M., Grygoyc, K., Rachwal, M. 2022 Analysis of the content of technology critical metals in the e-waste 2022/01 31st International Conference on Metallurgy and Materials, METAL 2022 10.37904/metal.2022.4516 9788088365068 703–707 TANGER Ltd. Poland.

Forti, V., Balde, C.P., Kuehr, R., Bel, G., The Global E-waste Monitor 2020: quantities, flows and the circular economy potential. (2020a),

Gayatri, Y., Shailaja Raj, M., Vijayalakshmi, B., 2017. Biosorption of lead by Bacillus licheniformis isolated from E-waste landfill, Hyderabad, Telangana, India. Int. J. Bioassays. 6 (2), 5240–5244. doi:10.21746/IJBIO.2017.02.003.

Gumulya, Y., Boxall, N.J., Khaleque, H.N., Santala, V., Carlson, R.P., Kaksonen, A.H., 2018. In a quest for engineering acidophiles for biomining applications: challenges and opportunities. Genes. 20734425. 9 (2). Available from: http://www.mdpi.com/2073-4425/9/2/116/pdf. doi:10.3390/genes9020116.

Gupta, S., Modi, G., Saini, R., Agarwala, V., 2014. A review on various electronic waste recycling techniques and hazards due to its improper handling. Int. Ref. J. Eng. Sci. 3 (5), 5–17.

Hahladakis, J.N., Velis, C.A., Weber, R., Iacovidou, E., Purnell, P., 2018. An overview of chemical additives present in plastics: migration, release, fate and environmental impact during their use, disposal and recycling. J. Hazard. Mater. 18733336. 344, 179–199. Available from: www.elsevier.com/locate/jhazmat. doi:10.1016/j.jhazmat.2017.10.014.

Han, P., Teo, W.Z., Yew, W.S., 2022. Biologically engineered microbes for bioremediation of electronic waste: wayposts, challenges and future directions. Eng. Biol. 6 (1), 23–34. doi:10.1049/enb2.12020.

Harshan, K., Rajan, A.P., 2025. Unveiling the potential of microbial biominers in bioleaching for heavy metal recovery from E-waste – a comprehensive review. J. Hazard. Mater. Adv. 27724166. 18, 100691. doi:10.1016/j.hazadv.2025.100691.

Hasan, M.M., Rhamdhani, M.A., Brooks, G.A., 2022. Thermodynamics of Gallium (Ga) at Black Copper Smelting Conditions Relevant to E-Waste Processing. Metall. Mater. Trans. B. 53 (5), 3136–3146. doi:10.1007/s11663-022-02593-4.

Helena Xavier, L., Ottoni, M., 2023. Leonardo Picanço Peixoto Abreu, A comprehensive review of urban mining and the value recovery from e-waste materials. Resources, Conservation and Recycling 190. 10.1016/j.resconrec.2022.106840

Horeh, N.B., Mousavi, S.M., Shojaosadati, S.A., 2016. Bioleaching of valuable metals from spent lithium-ion mobile phone batteries using Aspergillus Niger. J. Power Sources. 03787753. 320, 257–266. doi:10.1016/j.jpowsour.2016.04.104.

Hussien, S.S., 2025. Advances in Microbial Leaching as a Non-conventional Technique for Metal Recovery from Ores: a Review. Geomicrobiol. J. 15210529. 42 (4), 255–267. Available from: http://www.tandf.co.uk/journals/titles/01490451.asp. doi:10.1080/01490451.2025.2460506.

Illés, I.B., Kékesi, T., 2023. The production of high-purity gallium from waste LEDs by combining sulfuric acid digestion, cation-exchange and electrowinning. J. Environ. Chem. Eng. 22133437. 11 (5), 110391. doi:10.1016/j.jece.2023.110391.

Işıldar, A., Rene, E.R., van Hullebusch, E.D., Lens, P.N.L., 2018. Electronic waste as a secondary source of critical metals: management and recovery technologies. Resources, Conservation and Recycling 135, 296–312. doi:10.1016/j.resconrec.2017.07.031.

Islam, A., Ahmed, T., Awual, M.R., Rahman, A., Sultana, M., Aziz, A.A., Monir, M.U., Teo, S.H., Hasan, M., 2020. Advances in sustainable approaches to recover metals from e-waste-A review. J. Clean. Prod. 09596526. 244. Available from: https://www.journals.elsevier.com/journal-of-cleaner-production. doi:10.1016/j.jclepro.2019.118815.

Jaiswal, M., Srivastava, S., 2024. A review on sustainable approach of bioleaching of precious metals from electronic wastes. J. Hazard. Mater. Adv. 27724166. 14, 100435. doi:10.1016/j.hazadv.2024.100435.

Jiakang, R., Huo, Y., Yang, Y., 2020. Microbial degradation and valorization of plastic wastes. Front. Microbiol. 1664-302X. 11. doi:10.3389/fmicb.2020.00442.

Kaya, M., 2016. Recovery of metals and nonmetals from electronic waste by physical and chemical recycling processes. Waste Manag. 0956053X. 57, 64–90. doi:10.1016/j.wasman.2016.08.004.

Khezerloo, S., Nasirpour, N., Pourhossein, F., Mousavi, S.M., 2023. Bioleaching of indium from spent light-emitting diode monitors and selective recovery followed by solvent extraction. J. Environ. Manag. 03014797. 335, 117520. doi:10.1016/j.jenvman.2023.117520.

Lee, J., Choi, H., Kim, J., 2024. Environmental and economic impacts of e-waste recycling: a systematic review. Chem. Eng. J. 13858947. 494, 152917. doi:10.1016/j.cej.2024.152917.

Lefevre, E., Bossa, N., Wiesner, M.R., Gunsch, C.K., 2015. A review of the environmental implications of in situ remediation by nanoscale zero valent iron (nZVI): behavior, transport and impacts on microbial communities. Sci. Total Environ. 18791026. 565, 889–901. Available from: www.elsevier.com/locate/scitotenv. doi:10.1016/j.scitotenv.2016.02.003.

Lin, S., Ali, M.U., Zheng, C., Cai, Z., Wong, M.H., 2022. Toxic chemicals from uncontrolled e-waste recycling: exposure, body burden, health impact. J. Hazard. Mater. 18733336. 426. Available from: www.elsevier.com/locate/jhazmat. doi:10.1016/j.jhazmat.2021.127792.

Liow, L.T., Go, M.K., Chang, M.W., Yew, W.S., 2020. Toolkit development for cyanogenic and gold biorecovery chassis Chromobacterium violaceum. ACS Synth. Biol. 21615063. 9 (4), 953–961. Available from: http://pubs.acs.org/journal/asbcd6. doi:10.1021/acssynbio.0c00064.

Little, P.C., 2021. Burning Matters: Life, Labor, and E-Waste Pyropolitics in Ghana. Oxford University Press, New York. doi:10.1093/oso/9780190934545.001.0001.

Madhavan, M., Kodialbail, V.S., Saidutta,, 2024. Performance of fluidized-bed bioreactor in copper bioleaching from printed circuit boards using alcaligenes aquatilis. Waste Biomass Valorization. 15 (2), 1213–1224. doi:10.1007/s12649-023-02202-8.

Maneesuwannarat, S., Vangnai, A.S., Yamashita, M., Thiravetyan, P., 2016. Bioleaching of gallium from gallium arsenide by Cellulosimicrobium funkei and its application to semiconductor/electronic wastes. Process Saf. Environ. Prot. 09575820. 99, 80–87. Available from: http://www.elsevier.com/wps/find/journaldescription.cws_home/713889/description#description. doi:10.1016/j.psep.2015.10.008.

Marra, A., Cesaro, A., Rene, E.R., Belgiorno, V., Lens, P.N.L., 2018. Bioleaching of metals from WEEE shredding dust. J. Environ. Manag. 10958630. 210, 180–190. Available from: https://www.sciencedirect.com/journal/journal-of-environmental-management. doi:10.1016/j.jenvman.2017.12.066.

Merli, G., Becci, A., Amato, A., 2022. Recovery of precious metals from printed circuit boards by cyanogenic bacteria: optimization of cyanide production by statistical analysis. J. Environ. Chem. Eng. 10 (3). doi:10.1016/j.jece.2022.107495.

Mingyue, X., Yang, M., Sun, H., Gao, M., Wang, Q., Chuanfu, W., 2022. Bioconversion of biowaste into renewable energy and resources: a sustainable strategy. Environ. Res. 00139351. 214 (Pt 2), 113929. doi:10.1016/j.envres.2022.113929.

Mmereki, D., Baizhan, L., Baldwin, A., Hong, L., 2016. The Generation, Composition, Collection, Treatment and Disposal System, and Impact of E-Waste. InTech. 10.5772/61332.

Monika, J.K., 2010. E-waste management: as a challenge to public health in India. Indian J. Community Med. 35 (3), 382–385. doi:10.4103/0970-0218.69251.

Mudila, H., Prasher, P., Kumar, A., Sharma, M., Verma, A., Verma, S., Khati, B., 2021. E-waste and its hazard management by specific microbial bioremediation processes. Microorganisms Sustain. 26, 139–166. Available from: https://www.springer.com/series/14379. doi:10.1007/978-981-15-7455-9_6.

Mudila, H., Prasher, P., Kumar, M., Kapoor, H., Kumar, A., Zaidi, M.G.H., Verma, A., 2019. An insight into Cadmium poisoning and its removal from aqueous sources by Graphene Adsorbents. Int. J. Environ. Health Res. 29 (1), 1–21. doi:10.1080/09603123.2018.1506568.

Murali, A., Zhang, Z., Shine, A.E., Free, M.L., Sarswat, P.K., 2022. E-wastes derived sustainable Cu recovery using solvent extraction and electrowinning followed by thiosulfate-based gold and silver extraction. J. Hazard. Mater. Adv. 27724166. 8. Available from: https://www.journals.elsevier.com/journal-of-hazardous-materials-advances. doi:10.1016/j.hazadv.2022.100196.

Naseri, T., Mousavi, S.M., 2022. Insights into the polysaccharides and proteins production from Penicillium citrinum during bioleaching of spent coin cells. Int. J. Biol. Macromol. 209 (Pt A), 1133–1143.

Needhidasan, S., Samuel, M., Chidambaram, R., 2014. Electronic waste - An emerging threat to the environment of urban India. J. Environ. Health Sci. Eng. 12 (1). doi:10.1186/2052-336X-12-36.

Pant, D., Giri, A., Dhiman, V., 2018. Bioremediation techniques for E-waste management. Energy, Environment, and Sustainability 105–125. doi:10.1007/978-981-10-7413-4_5.

Parsa, A., Horeh, N.B., Mousavi, S.M., 2024. A hybrid thermal-biological recycling route for efficient extraction of metals and metalloids from end-of-life liquid crystal displays (LCDs). Chemosphere 352.

Pinto, V.N., 2008. E-waste hazard: the impending challenge. Indian J. Occup. Environ. Med. 09732284. 12 (2), 65–70. doi:10.4103/0019-5278.43263.

Poppenheimer, L., (2013),

Pourhossein, F., Rezaei, O., Mousavi, S.M., Beolchini, F., 2021. Bioleaching of critical metals from waste OLED touch screens using adapted acidophilic bacteria. J. Environ. Health Sci. Eng. 2052336X. 19 (1), 893–906. Available from: https://rd.springer.com/journal/40201. doi:10.1007/s40201-021-00657-2.

Rana, A., Sindhu, M., Kumar, A., Dhaka, R.K., Chahar, M., Singh, S., Nain, L., 2021. Restoration of heavy metal-contaminated soil and water through biosorbents: a review of current understanding and future challenges. Physiol. Plant. 173 (1), 394–417. doi:10.1111/ppl.13397.

Ray, A.K., 2023. Leaching, Vol. 2B. Elsevier BV, pp. 727–756. doi:10.1016/b978-0-08-101097-6.00012-2.

Rendón-Castrillón, L., Ramírez-Carmona, M., Ocampo-López, C., Gómez-Arroyave, L., 2023. Bioleaching techniques for sustainable recovery of metals from solid matrices. Sustainability. 2071-1050. 15 (13), 10222. doi:10.3390/su151310222.

Rice, N.M., 2016. A hydrochloric acid process for nickeliferous laterites. Miner. Eng. 08926875. 88, 28–52. Available from: http://www.journals.elsevier.com/minerals-engineering/. doi:10.1016/j.mineng.2015.09.017.

Rosariastuti, R., Mahardika, D.P., Supriyadi, P., Hartati, S., 2019. Soil Bioremediation of lead (Pb) polluted paddy field using Mendong (Fimbristylis globulosa), Rhizobium SpI 3, compost, and inorganic fertilizer. IOP Conf. Ser. Earth Environ. Sci. 1755-1315. 230 (1), 012014. doi:10.1088/1755-1315/230/1/012014.

Sahu, S., Mohapatra, M., Devi, N., 2022. Effective hydrometallurgical route for recovery of energy critical elements from E-wastes and future aspects. Mater. Today Proc. 67, 1016–1023. doi:10.1016/j.matpr.2022.05.491.

Schmeide, K., Bernhard, G., 2008. Spectroscopic study of the uranium(IV) complexation by organic model ligands in aqueous solution. Uranium, Mining and Hydrogeology 591–598. doi:10.1007/978-3-540-87746-2_73.

Sekhar, V.C., Nampoothiri, K.M., Mohan, A.J., Nair, N.R., Bhaskar, T., Pandey, A., 2016. Microbial degradation of high impact polystyrene (HIPS), an e-plastic with decabromodiphenyl oxide and antimony trioxide. J. Hazard. Mater. 18733336. 318, 347–354. Available from: www.elsevier.com/locate/jhazmat. doi:10.1016/j.jhazmat.2016.07.008.

Silva, J.G., da Silva, M.T., Dias, R.M., Cardoso, V.L., de Resende, M.M., 2023. Biolixiviation of metals from computer printed circuit boards by acidithiobacillus ferrooxidans and bioremoval of metals by mixed culture subjected to a magnetic field. Curr. Microbiol. 14320991. 80 (6). Available from: https://www.springer.com/journal/284. doi:10.1007/s00284-023-03307-y.

Siwi, W.P., Rinanti, A., Silalahi, M.D.S., Hadisoebroto, R., Fachrul, M.F., 2018. Effect of immobilized biosorbents on the heavy metals (Cu 2+) biosorption with variations of temperature and initial concentration of waste. IOP Conf. Ser. Earth Environ. Sci. 1755-1307. 106 (1), 012113. doi:10.1088/1755-1315/106/1/012113.

Slijkhuis, C., 2018. Recycling plastics from WEEE requiring a sensible and practical approach on POPs. Going Green Care Innov.

Tay, S.B., Natarajan, G., Rahim, M.N.B.A., Tan, H.T., Chung, M.C.M., Ting, Y.P., Yew, W.S., 2013. Enhancing gold recovery from electronic waste via lixiviant metabolic engineering in Chromobacterium violaceum. Sci. Rep. 20452322. 3. doi:10.1038/srep02236.

Tesfaye, F., Lindberg, D., Hamuyuni, J., 2017. Valuable Metals and Energy Recovery from Electronic Waste Streams. In: Zhang, L (Ed.), San Diego, California, USA, TMS 2017 Annual Meeting & Exhibition February 26 – March 2, 2017, Springer, Cham, Switzerland, pp. 103–116. doi:10.1007/978-3-319-521923_11

Thakur, P., Kumar, S., 2020. Metallurgical processes unveil the unexplored "sleeping mines" e- waste: a review. Environ. Sci. Pollut. Res. 0944-1344. 27 (26), 32359–32370. doi:10.1007/s11356-020-09405-9.

The Growing Environmental Risks of E-Waste. (2024), Available from: https://www.genevaenvironmentnetwork.org/resources/updates/the-growing-environmental-risks-of-e-waste/.

The Venture Using Microbes to Turn E-Waste into Precious Metals. (2021), Available from: https://www.cie.auckland.ac.nz/newsroom/the-venture-using-microbes-to-turn-e-waste-into-precious-metals/.

Trivedi, A., Vishwakarma, A., Saawarn, B., Mahanty, B., Hait, S., 2022. Fungal biotechnology for urban mining of metals from waste printed circuit boards: a review. J. Environ. Manag. 323.

Tsydenova, O., Bengtsson, M., 2011. Chemical hazards associated with treatment of waste electrical and electronic equipment. Waste Manag. 0956053X. 31 (1), 45–58. doi:10.1016/j.wasman.2010.08.014.

Valix, M., 2017. Bioleaching of Electronic Waste: Milestones and Challenges Current Developments in Biotechnology and Bioengineering: Solid Waste Management. Elsevier Inc., Australia, pp. 407–442. Available from: http://www.sciencedirect.com/science/book/9780444636645. doi:10.1016/B978-0-444-63664-5.00018-6.

Walton, A., Anderson, P., Harper, G., Mann, V., Beddington, J., Abbott, A., Bloodworth, A., OudeNijeweme, D., Schofield, E., Wall, F., Glover, N., Chaddock, R., Gross, R., Lee, R., Grimes, R., Lewis, R., Gibson, V., McGuiness, P., 2021. Securing Technology-critical Metals for Britain. University of Birmingham. 10.25500/epapers.bham.00003450.

World-first Biotechnology. (2024), Available from: https://www.mint.bio/.

Xie, Y., Chenquan, N., Han, Z., Zhong, H., Zhiguo, H., Sun, W., 2023. High recovery of lithium from coal residue by roasting and sulfuric acid leaching. Miner. Eng. 08926875. 202, 108284. doi:10.1016/j.mineng.2023.108284.

Zeng, X., Mathews, J.A., Jinhui, L., 2018. Urban mining of E-waste is becoming more cost-effective than virgin mining. Environ. Sci. Technol. 52 (8), 4835–4841. doi:10.1021/acs.est.7b04909.

Zhao, F., Wang, S., 2019. Bioleaching of Electronic Waste Using Extreme Acidophiles Electronic Waste Management and Treatment Technology. Elsevier, China, pp. 154–174. Available from: http://www.sciencedirect.com/science/book/9780128161906. doi:10.1016/B978-0-12-816190-6.00007-8.

Zhu, N., Zhang, B., Yu, Q., 2020. Genetic engineering-facilitated coassembly of synthetic bacterial cells and magnetic nanoparticles for efficient heavy metal removal. ACS Appl. Mater. Interfaces. 19448252. 12 (20), 22948–22957. Available from: http://pubs.acs.org/journal/aamick. doi:10.1021/acsami.0c04512.

Zhu, P., Pan, X., Xiaowei, L., Liu, X., Liu, Q., Zhou, J., Dai, X., Qian, G., 2021. Biodegradation of plastics from waste electrical and electronic equipment by greater wax moth larvae (Galleria mellonella). J. Clean. Prod. 09596526. 310, 127346. doi:10.1016/j.jclepro.2021.127346.

CHAPTER

# Nanotechnology-enabled microbial approaches in agricultural waste management

7

**Sevinç Yeşilyurt[a] and Muazzez Gürgan[b]**

[a]*Department of Soil Science and Plant Nutrition, Tekirdağ Namık Kemal Üniversitesi, Tekirdağ, Turkey*

[b]*Department of Biology, Tekirdağ Namık Kemal Üniversitesi, Tekirdağ, Turkey*

## 1 Nanotechnology and its importance in agriculture

Agriculture can be defined as the art of obtaining the highest yield from a unit area in the most economical, sustainable, and healthy way. It plays a crucial role in providing food security and sustaining development. However, many factors threaten agricultural production, including the depletion of natural and water resources, a decrease in soil fertility, climate change, an increasing world population, and environmental pollution (FAO, 2017; Tilman et al., 2011). Innovative technologies should be developed to address the challenges faced in agriculture today. Nanotechnology has been considered an important solution in agriculture in recent years. By using nanotechnological approaches, effective fertilization and pesticide use can be achieved, along with the rapid detection of pests and diseases. Moreover, soil health can be improved in this way, and waste management can be enhanced, leading to overall environmental protection. Materials at the nanoscale can interact with plants and soil, enabling higher yields with fewer agricultural inputs, thereby contributing to agricultural sustainability. These technologies also support the biological structure of the soil by reducing the loss of plant nutrients and offering environmentally friendly production models (Sharma and Kumar, 2024).

The goal of these initiatives is increasing productivity and also creating environmentally friendly, resource-conserving, and resilient agricultural systems for the future. Multidisciplinary research is, therefore, critical for making nanotechnological applications safe, economically, and environmentally sustainable (Duhan et al., 2017). In this respect, nanotechnology is a particularly promising field for solving many of the problems encountered in agriculture. The nanoscale materials and systems offered by nanotechnology have the potential to increase yields, reduce costs, and minimize environmental impact at every stage of agricultural production (Chen

**Microbial Technologies for Sustainable Waste Management. https://doi.org/10.1016/B978-0-443-44093-9.00006-2**

and Yada, 2011; Nair et al., 2010). Specifically, nanofertilizers, nanopesticides, and nanocarrier systems provide essential substances to plants in a controlled manner, thereby contributing to both yield increases and helping to minimize damage to the environment (Kah et al., 2013).

In addition, nanosensors developed through nanotechnology allow parameters such as soil moisture, pH levels, and nutrients to be monitored continuously, thus enabling effective precision agricultural practices (Sekhon, 2014). Such innovative practices support the digitalization and sustainability of agriculture. By integrating nanosensors into cloud systems via wireless networks, farmers can monitor soil conditions and critical parameters such as plant health and pest populations (Garg et al., 2024). Hence, the use of inputs can be optimized while simultaneously decreasing costs and environmental impacts. Moreover, fertilizers and pesticides developed through nanoencapsulation can release an adequate amount of resources at the desired time. These smart systems also contribute to sustainable agriculture by reducing energy consumption and keeping maintenance and communication costs under control. Thus both productivity in agricultural production increases and natural resources are preserved (Finger, 2023). Therefore nanotechnology-supported digital agriculture practices increase productivity and contribute to the creation of a climate change-resilient agricultural infrastructure.

Nanobiosensors can detect plant pathogens, nutrient deficiencies, or environmental stressors at an early stage so that necessary interventions can be made appropriately and quickly. Hence, these systems can prevent yield loss (Vandana et al., 2025). Nanoparticles (NPs) can be used to remediate contaminated soils. For example, they can bind to heavy metals or pesticide residues; therefore they contribute to restoring soil health (Del Prado-audelo et al., 2021). Nanocoating technologies can increase the germination rate of seeds and improve the resistance of young plants against various stress conditions (Shah et al., 2024). NPs can be used as genetic material carriers or RNA-based pest control agents for crops, which serves as a faster and safer option compared to genetic engineering (Demirer et al., 2021).

## 2 Classification of NPs

NPs are nanomaterials with sizes of approximately 1–100 nm. They possess unique physical, chemical, and biological properties. For instance, they exhibit a higher surface area-to-volume ratio, quantum effects, and surface reactivity compared to micro and larger-scale materials. These properties give them significant potential for various applications, such as in biomedical engineering, electronics, energy production, environmental technologies, and materials science (Adiloğlu et al., 2019; Albanese et al., 2012; Salata, 2004). The classification of NPs depends on characteristics such as morphological structure, chemical composition, and application areas. For example, they are classified as organic or inorganic based on their chemical structure. Organic NPs are typically composed of biodegradable polymers or lipids, while inorganic NPs may contain metal, metal oxide, or carbon-based structures (Abbasi et al., 2023). Moreover, combinations of different structures, such as core–shell structures, have been actively researched in recent years (Bhattacharjee, 2016). They are also categorized based on their size, shape (e.g., spherical, rod, star-shaped), surface properties (e.g., charge, functional groups), crystal structure, and synthesis methods. These properties can directly influence the behavior, targeting capacity, and toxicity profile of particles in biological systems (Prabhu and Poulose, 2012).

## 3 Classification according to chemical composition

### 3.1 Organic NPs

Organic NPs are highly desirable for drug delivery systems and gene therapy since they are biodegradable and biocompatible. Polymeric NPs, such as poly(lactic-co-glycolic acid), polylactic acid, and chitosan, are used to deliver drugs to the target site by providing controlled release (Geszke-Moritz and Moritz, 2024). The cell membrane-like structure of liposomes offers advantages in drug transport and gene delivery. Moreover, dendrimers, which are highly branched, customizable structures with functional groups, represent another group of polymeric NPs important for drug delivery (Chaudhari et al., 2016).

### 3.2 Inorganic NPs

Metal and metal oxide NPs are invaluable for antimicrobial, magnetic, and optical applications. For instance, gold (Au) and silver (Ag) are frequently used in photothermal therapy and biosensors (Altammar, 2023). Among the metal oxides, silica ($SiO_2$) is suitable for high surface area, and iron oxide ($Fe_3O_4$) is suitable for magnetic targeting. Carbon-based structures, such as graphene and some nanotubes, are commonly used in electronics, energy, and environmental treatment applications (Chen et al., 2020).

### 3.3 Classification by structural properties

#### *3.3.1 Core–shell NPs*

These are structures in which one material is used as the core and a different coating is applied on top. This structure provides advantages in terms of surface modification, controlled drug release, and stability (Yang et al., 2022).

#### *3.3.2 Porous NPs*

In particular, mesoporous silica NPs have found wide application in the biomedical field due to their drug loading capacity and surface functionalizability (Wang et al., 2015).

### 3.4 Classification by physical characteristics

- Size: NP size directly affects cellular uptake, distribution, and bioavailability (Geszke-Moritz and Moritz, 2024).
- Shape: Forms such as spherical, rod, and star make a difference in cell interaction and targeting.
- Surface properties: Intracellular targeting can be achieved through surface charge and functional groups. Positively charged NPs generally cross the cell membrane more easily (Umut, 2013).

### 3.5 Classification by production method

Nanomaterials are classified by their production methods: top-down (from macro to nanoscale, e.g., milling, lithography) and bottom-up (NP production at the atomic/molecular level, e.g., sol–gel, chemical vapor deposition). More controlled and homogeneous structures are obtained with bottom-up production (Tripathy, 2023).

### 3.6 Classification according to usage areas

NPs can also be classified according to usage areas. They can be used for biomedical [targeted drug delivery, gene therapy, tumor imaging (Islam et al., 2025; Wang et al., 2024)], environmental [water treatment, pollutant adsorption, heavy metal removal (Mgadi et al., 2024)], energy [solar cells, supercapacitors, hydrogen production (Mohammed et al., 2025)], and electronics applications [sensors, transistors, memory systems (Payal, 2022)].

## 4 Agricultural use of nanomaterials

In recent years, nanotechnology-based biosensors have become an important tool for the rapid, sensitive, and economical detection of soil moisture, water content, nutrients, pesticide residues, and pathogen presence in agriculture. Based on the properties of nanomaterials, the sensors can offer different sensing methods. Gold NPs are highly biocompatible and can be easily functionalized; hence, they are widely used in biosensors. Their color-changing and electrochemical signal amplification properties make them effective for pesticide detection (Feng et al., 2020; Hegde et al., 2022).

Carbon-based nanomaterials, such as graphene and carbon nanotubes, have high surface areas and high conductivity, which is why they have been used to detect nutrients and metal ions in soil. Nitrate, ammonium, and heavy metal ions can be detected effectively by using these nanomaterials (Miguel-Rojas and Pérez-de-Luque, 2023). Moreover, semiconductor nanomaterials, such as quantum dots and carbon dots, are applicable in biological analysis, like pathogen detection, since they have multianalysis capabilities and low toxicity. Plant pathogens can be easily and simultaneously detected by quantum-based sensors (Mondal et al., 2022).

Nanobiosensors are advantageous because of their high sensitivity, fast response time, and low cost. However, more research is needed to improve the reliability of these technologies under field conditions and to disseminate their use (Khatri and Rathore, 2018). Nanotechnology is attracting increasing attention in line with sustainability goals in agriculture, and in this context, the use of nanomaterials is being evaluated in many fields. In particular, nanofertilizers enable plants to uptake nutrients more efficiently and increase agricultural productivity by reducing environmental losses (Upadhayay et al., 2023). Metal-based nanomaterials, such as zinc oxide (ZnO) and iron (Fe) NPs, significantly increase tolerance to abiotic stresses by inducing antioxidant enzyme activities in wheat and tomato plants (Tripathi et al., 2023). Nanopesticides and nanoherbicides can be produced to reach target organisms more effectively with lower doses (Shao et al., 2022). However, biosafety issues must be addressed, and toxicological assessments should be carried out together with regulatory frameworks to facilitate the spread of these applications (Kaningini et al., 2022).

Soil health protection is another area where nanotechnology can be applied. Nanomaterials, such as carbon NPs, increase soil fertility by improving nutrient transport in the soil and are also used to remove pollutants (El-Ramady et al., 2023). Both the use of nanomaterials as fertilizers and their use in environmental remediation processes provide many benefits. However, comprehensive toxicological studies should be conducted, and regulatory standards should be developed for the long-term and safe use of nanotechnology in agriculture (Tripathi et al., 2023).

## 5 Effects of nanotechnology in plant cultivation

Nanotechnology plays an important role as a transformational tool in agricultural production and has led to remarkable advances in agriculture, such as increasing productivity, stress tolerance, and nutrient quality. The high surface area of nanomaterials and their ability to interact with biological systems enhance nutrient uptake, photosynthetic efficiency, and resistance to abiotic stress in plants (Cao et al., 2024). Conventional fertilizers are inefficient in terms of sustainable agriculture since they have low nutrient use efficiency, cause nutrient loss through leaching or evaporation in arid regions, and need to be repeatedly applied, which increases their negative environmental impacts (Kah et al., 2019; Pooja, 2022). Nanotechnology-based fertilizers (also called nanofertilizers) stand out as an innovative and effective alternative in plant nutrition strategies to conventional fertilizers. Their small particle size and high specific surface area make nanofertilizers highly soluble in soil, more efficiently absorbed by plants, and capable of providing long-term and balanced use of nutrients through targeted, controlled release mechanisms (Pooja, 2022). Thus the bioavailability of nutrients to plants is increased, environmental losses and toxicity risks are reduced, and the efficiency of fertilization is significantly improved. More controlled and efficient nutrient release compared to conventional fertilizers allows nanofertilizers to support plant growth and reduce environmental losses. For example, zinc oxide (ZnO) NPs improved tolerance to drought stress by increasing zinc uptake in wheat and tomato plants (Dimkpa et al., 2019). Similarly, iron content in wheat was shown to increase by using nanoiron (Fe) fertilizers (Upadhyay et al., 2024).

Nanotechnological approaches can lead to increased photosynthesis in plants. An example is the use of titanium dioxide ($TiO_2$). This NP has been shown to enhance photosynthesis and plant growth by activating the Rubisco enzyme (Farahi et al., 2023). Similarly, silicon dioxide ($SiO_2$) NPs have been shown to have positive effects on plant health. Studies show that they can improve the photosynthesis process, mainly by increasing the activity of the carbonic anhydrase enzyme. In a study carried out with cotton plants, $SiO_2$ NPs were found to regulate stomatal opening, increase chlorophyll levels, enhance carbonic anhydrase activity, and hence improve photosynthesis (Liang et al., 2023).

Nanofertilizers also show their effects on seed germination by enhancing cell division. Interestingly, faster cell division means plants need less water to germinate, a big advantage when water is limited (Kumar et al., 2018; Kumari et al., 2023). Moreover, nanofertilizers seem to help regulate plant hormones, especially those that control growth, like auxin and cytokinin, making the whole development process more efficient (Fiol et al., 2021). Hence, nanofertilizers are not just about nutrients; they also act as plant growth regulators. However, caution is required in agricultural applications of nanotechnology. If applied in high doses, some NPs can actually harm plants, causing stress or stunting growth (Madanayake and Adassooriya, 2021). That is why it is so important to get the dosage, method of application, and environmental impact just right for nanomaterials to be both safe and effective in agriculture.

Nanopesticides are a recent development in agricultural nanotechnology, designed to improve how active compounds—such as antimicrobials, insecticides, and herbicides—are delivered to target organisms. These substances are incorporated into nanocarrier systems that allow for a slow and controlled release, improving the availability and efficiency of the compounds while reducing the amount needed and minimizing residual buildup in the environment (Pan et al., 2023; Singh et al., 2018). However, the antibacterial activity of NPs might disrupt soil microbiota and can lead to a decrease in the number and health of beneficial soil microbes, including nitrogen-fixing bacteria and symbiotic fungi like mycorrhizae (Eghbalinejad et al., 2024). The disruption of soil microbiota could damage nutrient cycling in the soil, and soil health may worsen. This can result in decreased plant growth and production. Therefore the effects of nanopesticides should be thoroughly assessed in terms of their impact on soil ecology, persistence in the environment, and effects on nontarget organisms (Shang et al., 2019). Sustainable agriculture can only be achieved with a careful balance between effectiveness and ecological responsibility.

Nanofertilizers can be produced using nanotechnology approaches. They have many advantages compared to conventional fertilizers, which are already known to decrease soil fertility. Nanofertilizers are highly soluble, and their release can be controlled. These characteristics enable them to assist plants in a more efficient way in terms of nutrient absorption. Hence, both nutrient loss in soil and an increase in plant production can be achieved. NPs also enhance seed germination, support cell division, modulate hormone production, and aid photosynthetic processes. Altogether, NPs can be considered active plant growth regulators. Although they offer many advantages, they need to be applied cautiously. Similar to conventional fertilizers, excessive use of nanofertilizers can cause toxicity to plants and the environment. Precise application strategies should be implemented to minimize risks and maximize benefits. Environmental conditions that affect plant growth, crop species, and growth stages of the plants should be taken into consideration for more efficient and environmentally conscious agricultural practices.

## 6 Nanotechnology in managing agricultural waste

Management of wastes originating from agricultural applications gains importance in terms of environmental sustainability and resource efficiency. Nanotechnology offers promising and innovative solutions for the reevaluation of waste, treating it, and reducing environmental impacts (Lou et al., 2023). Due to their high surface area and strong adsorption capacities, nanomaterials such as nanoactivated carbon, metal oxide NPs, and nanobiocomposites have been increasingly applied for the removal of organic pollutants from agricultural wastes (Mandal et al., 2024). Furthermore, the development of nanomaterials using agricultural biomass, such as silica NPs synthesized from rice husks, aligns with the principles of a circular economic approach (Ouyang et al., 2023). In addition, rapid and precise detection of agricultural contaminants, such as pesticide residues and heavy metals, has become possible with the help of nanobiosensors. Consequently, the efficiency of monitoring systems has been enhanced (Kulkarni et al., 2022).

Application of nanotechnology in agricultural waste management should be carried out after a detailed risk assessment. Nanomaterials might be persistent in the environment and accumulate in soil microorganisms, which are required for waste treatment (Rıza Demirkıran and Sohrabi, 2024). Therefore interdisciplinary research should be carried out to sustainably apply these technologies. Moreover, comprehensive regulatory guidelines should be formed to ensure environmental safety.

## 7 Nanotechnology and soil health

Recent advances in nanobiotechnology have led to the development of new and creative approaches to make the phytoremediation process, which is the use of plants to clean up contaminants in soil, more effective (Song et al., 2019). For this purpose, NPs are designed as effective agents to remediate contaminated soil and enhance soil quality (Karn et al., 2009; Zulfiqar et al., 2019). Nowadays, NPs play an important role in products like nanofertilizers, nanopesticides, and nanoherbicides, as well as in delivering agrochemicals directly to where they are needed. They can improve soil health by enhancing enzyme activity (such as urease, which is important for nitrogen use), balancing pH, and improving the structure and fertility of the soil. As a result, they help plants absorb nutrients more easily and support healthier soil overall. Hence, soil quality can be enhanced together with obtaining higher product yields (Kumari et al., 2023).

It is highly likely that these engineered NPs will pass into the environment in considerable quantities through direct release during their lifecycle (Keller, 2010). In the specific case of the soil environment, the key processes of transformation and potential risks associated with engineered NPs are dissolution, adsorption/agglomeration, bioaccumulation in plants, bioaccumulation in invertebrates and toxicity, microbial toxicity, direct particle uptake/toxicity, and particle migration (Klaine et al., 2008).

Nanotechnology offers significant opportunities to improve soil health and support sustainable agricultural practices. NPs help reduce chemical inputs by improving the efficacy of fertilizers and pesticides, while positively affecting soil structure, microbial activity, and enzymatic functions. However, alongside the benefits of this technology, the potential long-term impacts on the soil ecosystem need to be carefully assessed. We still do not fully understand how NPs build up in soil, how toxic they might be, or how they affect the balance of microbes. That is why it is crucial to continue interdisciplinary research in this area. To safely and effectively use nanotechnology in managing agricultural soils, it is essential to develop new technologies while also carefully assessing the environmental risks at the same time.

## 8 The future of nanotechnology in agriculture

Nanotechnology is promising to change the game in terms of agricultural practices for sustainability, efficiency, and environmental impact reduction. Recent studies show that the integration of nanotechnology into agricultural production processes is becoming more widespread. These developments are presented in the following sections.

### 8.1 Nanobiosensors and smart agriculture

Nanobiosensors are used in agriculture to monitor critical parameters such as soil moisture, pH, nutrients, and plant health in real time and with high precision. These sensors can provide real-time accurate data through their electrochemical and optical-based structures (Shaw and Honeychurch, 2022). Nanomaterials such as carbon nanotubes, graphene derivatives, and quantum dots increase the sensitivity of these sensors so that even very low levels of analytes can be detected (Malik et al., 2023).

Sensors can be combined with artificial intelligence to improve how pests, diseases, water use, and fertilizer needs are detected and managed (Ramachandran et al., 2022). Farmers can use portable biosensors connected to their smartphones to quickly collect data in the field, helping them make smarter decisions on the spot (Mansoor et al., 2025). Because of this, nanobiosensors are becoming a key part of modern digital farming. By using biosensors, soil health can be checked regularly, and the use of water and fertilizers can be controlled so that environmental standards are met, and even soil can be cleaned up efficiently. Harmful substances can be detected more quickly, and necessary precautions can be taken accordingly. This makes them especially useful when dealing with soil contamination problems (Thatoi et al., 2024).

## 8.2 Nanofertilizers and nutrient use efficiency

Nanofertilizers provide several benefits over traditional fertilizers, including greater solubility, controlled nutrient release, and improved plant absorption. These characteristics enhance crop yields by promoting better plant development. In precision agriculture, nanobiosensors play a key role in monitoring soil nutrient levels to maintain fertility. They are specially engineered to detect essential macronutrients like nitrogen, phosphorus, and potassium in soil. Depending on the nutrient, various sensor technologies can be employed, such as optical sensors, ion-selective electrodes, electrochemical detectors, and spectroscopy (Mansoor et al., 2025). These sensors facilitate the monitoring of nutrient concentrations in soil so that effective fertilization can be applied.

These approaches increase yields, optimize fertilizer use, reduce costs, and minimize negative environmental impacts. Hence, nutrient loss prevention and water preservation can be achieved via these approaches. Since resources can be managed effectively, sustainable agriculture is supported (Horváth et al., 2024). The widespread use of nanofertilizers in agriculture will both sustainably increase agricultural productivity by improving the efficiency of nutrient use and minimize environmental impacts. Eventually, climate-friendly agricultural systems can be maintained.

## 8.3 Soil health and nanoremediation

NPs are used to improve soil health and remove contaminants. Carbon NPs increase nutrient use efficiency in corn, while titanium dioxide NPs and plant growth-promoting rhizobacteria support plant growth in contaminated soils (Usman et al., 2020; Zaman et al., 2025).

Nanomaterials enhance plants' ability to cope with environmental stresses such as drought, salinity, heavy metal pollution, and extreme temperatures by regulating reactive oxygen species (ROS) levels, activating signaling pathways, and promoting osmolyte accumulation in cells and tissues. Heavy metals cause oxidative stress, disrupt enzymatic activity, and damage cellular structures, while NPs promote metal detoxification and enhance antioxidant defense systems (Ghori et al., 2019; Liu et al., 2021). For example, ZnO and cerium oxide ($CeO_2$) NPs have been shown to reduce heavy metal toxicity-induced ROS accumulation, thereby minimizing oxidative damage to cellular components. However, certain NPs can stimulate the synthesis of chelating agents such as phytochelatins, which bind to heavy metal ions and sequester them in vacuoles, resulting in effective detoxification of the cytosol (Xia et al., 2008).

Soil health is the foundation of sustainable agriculture and environmental management. In this context, nanotechnology offers effective and innovative solutions for the remediation of contaminated soils, especially through nanoremediation applications. Structures such as metal oxide NPs, carbon-based nanomaterials, and nanobiohybrid systems have attracted attention due to their high surface area and reactivity in the removal of heavy metals, pesticide residues, and organic pollutants from soils. Nanomaterials also have the potential to improve soil structure and nutrient cycling by supporting soil microbial activities. However, before the widespread use of these technologies, their environmental and ecotoxicological impacts need to be carefully evaluated. Nanoremediation strategies supported by proper material selection, controlled application methods, and long-term monitoring studies will make significant contributions to both protecting soil health and strengthening sustainable agriculture.

### 8.4 Resilience to climate stress

Nanotechnology enhances plant resilience to abiotic stresses such as drought, salinity, and heat stress. The individual and combined effects of drought and heat stress on vital processes such as stomatal conductance, photosynthetic capacity, cellular oxidative balance, metabolomic structure, and signal transduction mechanisms are comprehensively addressed. It also elaborates on key regulatory molecules, such as transcription factors and signal transduction proteins, that play important roles in plant adaptation to these stress conditions. Zinc oxide NPs alleviate drought stress and promote plant growth in wheat and maize (Sato et al., 2024).

As the negative impacts of climate change on agricultural production increase, enhancing the resilience of plants against drought, high temperatures, and other environmental stress factors becomes critically important for ensuring food security and establishing sustainable agricultural systems. At this point, nanotechnology stands out with its potential to increase plant resilience under stress conditions by providing solutions for plant stress physiology. Thanks to the carrier systems developed with nanoformulations, the controlled release of stress-related hormones, stimulation of antioxidant defense systems, and protection of photosynthetic efficiency become possible. Furthermore, nanobiosensor technologies enable the early detection of environmental stresses, allowing timely intervention and optimization of resource use. In the future, by integrating nanotechnology with advanced genetic and biotechnological approaches, it is expected that plants resistant to environmental stress conditions will be developed, and smart agricultural systems with climate adaptation will become widespread. Thus nanotechnology-supported solutions will lay the foundations for productive and environmentally friendly agricultural systems that are resistant to climate change.

## 9 Conclusions

Nanotechnology is becoming increasingly important as an interdisciplinary field of science with the potential to revolutionize the agricultural sector. Classifying NPs according to their size, structure, and chemical properties is critical to understanding the effects of these materials on plant growth, soil health, and agricultural production. Nanomaterials used in agriculture include a wide variety of products such as nanofertilizers, nanopesticides, nanobiosensors, and nanocomposite carriers. Thanks to these materials, nutrient utilization efficiency is increased, early diagnosis of diseases is provided, and plants resistant to environmental stress factors are developed. The application of nanotechnology in plant breeding positively affects yields through mechanisms such as increasing photosynthetic capacity, regulating hormone transmission pathways, and optimizing cellular stress responses. In addition, nanotechnological solutions in recycling agricultural waste contribute to environmental sustainability and make waste management more effective. Nanotechnology also plays an important role in monitoring and improving soil health, offering innovative applications that support soil enzymatic activity, microbial diversity, and nutrient cycling. Looking to the future, with the widespread integration of nanotechnology in agriculture, precision agriculture practices will become more effective, resource use will be optimized, and an agricultural infrastructure that is resilient to global threats such as climate change will be developed. However, it is of utmost importance that the environmental and biosafety aspects of these technologies are carefully evaluated and implemented in line with sustainability principles. The future of nanotechnology in agriculture depends on responsible innovation and holistic policy approaches, as well as scientific developments. It is necessary to draw attention to the development of state policies in this context by devoting more space to the subject in world literature.

## References


Abbasi, R., Shineh, G., Mobaraki, M., Doughty, S., Tayebi, L., 2023. Structural parameters of nanoparticles affecting their toxicity for biomedical applications: a review. J. Nanopart. Res. 1388-0764. 25 (3). doi:10.1007/s11051-023-05690-w.

Adiloğlu, S., Solmaz, Y., Çifçi, D.İ., Meriç, S., 2019. Comparative Impact Assessment of TiO2 and ZnO Nanoparticles to Rocket (Eruca Sativa L) Plant Nanotechnology in Water and Wastewater Treatment: Theory and Applications. Elsevier, Turkey, pp. 115–123. Available from: https://www.sciencedirect.com/book/9780128139028. doi:10.1016/B978-0-12-813902-8.00006-X.

Albanese, A., Tang, P.S., Chan, W.C.W., 2012. The effect of nanoparticle size, shape, and surface chemistry on biological systems. Annu. Rev. Biomed. Eng. 15454274. 14, 1–16. doi:10.1146/annurev-bioeng-071811-150124.

Altammar, K.A., 2023. A review on nanoparticles: characteristics, synthesis, applications, and challenges. Front. Microbiol. 1664302X. 14. Available from: https://www.frontiersin.org/journals/microbiology#. doi:10.3389/fmicb.2023.1155622.

Bhattacharjee, S., 2016. DLS and zeta potential – what they are and what they are not? J. Control. Release. 01683659. 235, 337–351. doi:10.1016/j.jconrel.2016.06.017.

Cao, Y., Turk, K., Bibi, N., Ghafoor, A., Ahmed, N., Azmat, M., Ahmed, R., Ghani, M.I., Ahanger, M.A., 2024. Nanoparticles as catalysts of agricultural revolution: enhancing crop tolerance to abiotic stress: a review. Front. Plant Sci. 1664462X. 15. Available from: https://www.frontiersin.org/journals/plant-science. doi:10.3389/fpls.2024.1510482.

Chaudhari, H.S., Popat, R.R., Adhao, V.S., Shrikhande, V.N., 2016. Dendrimers: Novel carriers for drug delivery. In: Journal of Applied Pharmaceutical Research 4 (1), 01–19.

Chen, H., Yada, R., 2011. Nanotechnologies in agriculture: new tools for sustainable development. Trends Food Sci. Technol. 09242244. 22 (11), 585–594. doi:10.1016/j.tifs.2011.09.004.

Chen, H., Zhang, L., Li, M., Xie, G., 2020. Synthesis of core–shell micro/nanoparticles and their tribological application: a review. Materials. 19961944. 13 (20), 1–29. Available from: https://www.mdpi.com/1996-1944/13/20/4590/pdf. doi:10.3390/ma13204590.

Del Prado-audelo, M.L.K., Escutia-Guadarrama, L.R.-G., Magaña, J.M., Leyva-Gómez, J.J., 2021. Nanoremediation: nanomaterials and nanotechnologies for environmental cleanup. Front. Environ. Sci. 9, 793765. doi:10.3389/FENVS.2021.793765/BIBTEX.

Demirer, G.S., Silva, T.N., Jackson, C.T., Thomas, J.B., Ehrhardt, D.W., Rhee, S.Y., Mortimer, J.C., Landry, M.P., 2021. Nanotechnology to advance CRISPR–Cas genetic engineering of plants. Nat. Nanotechnol. 17483395. 16 (3), 243–250. Available from: http://www.nature.com/nnano/index.html. doi:10.1038/s41565-021-00854-y.

Dimkpa, C.O., Singh, U., Bindraban, P.S., Elmer, W.H., Gardea-Torresdey, J.L., White, J.C., 2019. Zinc oxide nanoparticles alleviate drought-induced alterations in sorghum performance, nutrient acquisition, and grain fortification. Sci. Total Environ. 18791026. 688, 926–934. Available from: www.elsevier.com/locate/scitotenv. doi:10.1016/j.scitotenv.2019.06.392.

Duhan, J.S., Kumar, R., Kumar, N., Kaur, P., Nehra, K., Duhan, S., 2017. Nanotechnology: the new perspective in precision agriculture. Biotechnol. Rep. 2215017X. 15, 11–23. Available from: http://www.journals.elsevier.com/biotechnology-reports/. doi:10.1016/j.btre.2017.03.002.

Eghbalinejad, M., Hofman, J., Kotouček, J., Grillo, R., Hochmanová Bílková, Z., Reiff, N., Höss, S., 2024. Nano-enabled pesticides: a comprehensive toxicity assessment of tebuconazole nanoformulations with nematodes at single species and community level. Environ. Sci. Eur. 2190-4715. 36 (1). doi:10.1186/s12302-024-00879-9.

El-Ramady, H., Abdalla, N., Sári, D., Ferroudj, A., Muthu, A., Prokisch, J., Fawzy, Z.F., Brevik, E.C., Solberg, S., 2023. Nanofarming: promising Solutions for the future of the global agricultural industry. Agronomy. 20734395. 13 (6), 1600. Available from: www.mdpi.com/journal/agronomy/. doi:10.3390/agronomy13061600.

FAO, The future of food and agriculture -Trends and challenges. (2017),

Farahi, S.M.M., Yazdi, M.E.T., Einafshar, E., Akhondi, M., Ebadi, M., Azimipour, S., Mahmoodzadeh, H., Iranbakhsh, A., 2023. The effects of titanium dioxide (TiO2) nanoparticles on physiological, biochemical, and antioxidant properties of Vitex plant (*Vitex agnus - Castus* L). Heliyon. 24058440. 9 (11), e22144. Available from: http://www.journals.elsevier.com/heliyon/. doi:10.1016/j.heliyon.2023.e22144.

Feng, Z., Gao, N., Liu, J., Hongdong, L., 2020. Boron-doped diamond electrochemical aptasensors for trace aflatoxin B1 detection. Anal. Chim. Acta. 00032670. 1122, 70–75. doi:10.1016/j.aca.2020.04.062.

Finger, R., 2023. Digital innovations for sustainable and resilient agricultural systems. Eur. Rev. Agric. Econ. 0165-1587. 50 (4), 1277–1309. doi:10.1093/erae/jbad021.

Fiol, D.F., Terrile, M.C., Frik, J., Mesas, F.A., Álvarez, V.A., Casalongué, C.A., 2021. Nanotechnology in plants: recent advances and challenges. J. Chem. Technol. Biotechnol. 10974660. 96 (8), 2095–2108. Available from: http://onlinelibrary.wiley.com/journal/10.1002/(ISSN)1097-4660. doi:10.1002/jctb.6741.

Garg, S., Rumjit, N.P., Roy, S., 2024. Smart agriculture and nanotechnology: technology, challenges, and new perspective. Adv. Agrochem. 27732371. 3 (2), 115–125. Available from: https://www.sciencedirect.com/science/journal/27732371. doi:10.1016/j.aac.2023.11.001.

Geszke-Moritz, M., Moritz, M., 2024. Biodegradable polymeric nanoparticle-based drug delivery systems: comprehensive overview, perspectives and challenges. Polymers. 2073-4360. 16 (17), 2536. doi:10.3390/polym16172536.

Ghori, N.-H., Ghori, T., Hayat, M.Q., Imadi, S.R., Gul, A., Altay, V., Ozturk, M., 2019. Heavy metal stress and responses in plants. Int. J. Environ. Sci. Technol. 1735-1472. 16 (3), 1807–1828. doi:10.1007/s13762-019-02215-8.

Hegde, M., Pai, P., Shetty, M.G., Babitha, K.S., 2022. Gold nanoparticle based biosensors for rapid pathogen detection: a review. Environmental Nanotechnology, Monitoring and Management. 18, 100756. Available from: http://www.journals.elsevier.com/environmental-nanotechnology-monitoring-and-management/. doi:10.1016/j.enmm.2022.100756.

Horváth, J., Kátai, L., Szabó, I., Korzenszky, P., 2024. An electrical conductivity sensor for the selective determination of soil salinity. Sensors. 1424-8220. 24 (11), 3296. doi:10.3390/s24113296.

Islam, S., Ahmed, M.M.S., Islam, M.A., Hossain, N., Chowdhury, M.A., 2025. Advances in nanoparticles in targeted drug delivery–A review. Results Surf. Interfaces. 19 100529. Available from: https://www.sciencedirect.com/science/journal/26668459. doi:10.1016/j.rsurfi.2025.100529.

Kah, M., Beulke, S., Tiede, K., Hofmann, T., 2013. Nanopesticides: state of knowledge, environmental fate, and exposure modeling. Crit. Rev. Environ. Sci. Technol. 1064-3389. 43 (16), 1823–1867. doi:10.1080/10643389.2012.671750.

Kah, M., Tufenkji, N., White, J.C., 2019. Nano-enabled strategies to enhance crop nutrition and protection. Nat. Nanotechnol. 17483395. 14 (6), 532–540. Available from: http://www.nature.com/nnano/index.html. doi:10.1038/s41565-019-0439-5.

Kaningini, A.G., Nelwamondo, A.M., Azizi, S., Maaza, M., Mohale, K.C., 2022. Metal nanoparticles in agriculture: a review of possible use. Coatings. 20796412. 12 (10). Available from: www.mdpi.com/journal/coatings. doi:10.3390/coatings12101586.

Karn, B., Kuiken, T. and Otto, M., 2009. Nanotechnology and in Situ Remediation: A Review of the Benefits and Potential Risks. Environ Health Perspect 117 (12), 1813–1831. doi:10.1289/ehp.090079310.1289/ehp.0900793.s1.

Keller, M., 2010. Managing grapevines to optimise fruit development in a challenging environment: a climate change primer for viticulturists. Aust. J. Grape Wine Res. 13227130. 16 (1), 56–69. doi:10.1111/j.1755-0238.2009.00077.x.

Khatri, K., Rathore, M.S., 2018. Plant Nanobionics and Its Applications for Developing Plants with Improved Photosynthetic Capacity. IntechOpen. 10.5772/intechopen.76815.

Klaine, S.J., Alvarez, P.J.J., Batley, G.E., Fernandes, T.F., Handy, R.D., Lyon, D.Y., Mahendra, S., McLaughlin, M.J., Lead, J.R., 2008. Nanomaterials in the environment: behavior, fate, bioavailability, and effects. Environ. Toxicol. Chem. 07307268. 27 (9), 1825–1851. doi:10.1897/08-090.1.

Kulkarni, M.B., Ayachit, N.H., Aminabhavi, T.M., 2022. Recent advancements in nanobiosensors: current trends, challenges, applications, and future scope. Biosensors. 20796374. 12 (10), 892. Available from: http://www.mdpi.com/journal/biosensors/. doi:10.3390/bios12100892.

Kumar, V., Sachdev, D., Pasricha, R., Maheshwari, P.H., Taneja, N.K., 2018. Zinc-supported multiwalled carbon nanotube nanocomposite: a synergism to micronutrient release and a smart distributor to promote the growth of onion seeds in arid conditions. ACS Appl. Mater. Interfaces. 19448252. 10 (43), 36733–36745. Available from: http://pubs.acs.org/journal/aamick. doi:10.1021/acsami.8b13464.

Kumari, A., Rana, V., Yadav, S.K., Kumar, V., 2023. Nanotechnology as a powerful tool in plant sciences: recent developments, challenges and perspectives. Plant Nano Biol. 27731111. 5, 100046. doi:10.1016/j.plana.2023.100046.

Liang, Y., Liu, H., Yuanyuan, F., Penghui, L., Shuang, L., Gao, Y., 2023. Regulatory effects of silicon nanoparticles on the growth and photosynthesis of cotton seedlings under salt and low-temperature dual stress. BMC Plant Biol. 1471-2229. 23 (1). doi:10.1186/s12870-023-04509-z.

Liu, Y., Xiao, Z., Chen, F., Yue, L., Zou, H., Lyu, J., Wang, Z., 2021. Metallic oxide nanomaterials act as antioxidant nanozymes in higher plants: trends, meta-analysis, and prospect. Sci. Total Environ. 00489697. 780, 146578. doi:10.1016/j.scitotenv.2021.146578.

Lou, J., Babadi, M.R., Otadi, M., Tarahomi, M., Quyet Van, L., Khonakdar, H.A., Cheng, L., 2023. Agricultural waste valorization towards (nano)catalysts for the production of chemicals and materials. Fuel. 00162361. 351, 128935. doi:10.1016/j.fuel.2023.128935.

Madanayake, N.H., Adassooriya, N.M., 2021. Phytotoxicity of nanomaterials in agriculture. Open Biotechnol. J. 18740707. 15 (1), 109–118. Available from: https://openbiotechnologyjournal.com/contents/volumes/V15/TOBIOTJ-15-109/TOBIOTJ-15-109.pdf. doi:10.2174/1874070702115010109.

Malik, S., Singh, J., Goyat, R., Saharan, Y., Chaudhry, V., Umar, A., Ibrahim, A.A., Akbar, S., Ameen, S., Baskoutas, S., 2023. Nanomaterials-based biosensor and their applications: a review. Heliyon. 24058440. 9 (9), e19929. doi:10.1016/j.heliyon.2023.e19929.

Mandal, P., Yadav, M.K., Majumder, A., Ghosal, P.S., 2024. Application of Nanomaterials Synthesized Using Agriculture Waste for Wastewater Treatment Green and Sustainable Approaches Using Wastes for the Production of Multifunctional Nanomaterials. Elsevier, India, pp. 305–318. Available from: https://www.sciencedirect.com/book/9780443191831. doi:10.1016/B978-0-443-19183-1.00019-2.

Mansoor, S., Iqbal, S., Popescu, S.M., Kim, S.L., Chung, Y.S., Baek, J.-H., 2025. Integration of smart sensors and IOT in precision agriculture: trends, challenges and future prospectives. Front. Plant Sci. 1664-462X. 16. doi:10.3389/fpls.2025.1587869.

Mgadi, K., Ndaba, B., Roopnarain, A., Rama, H., Adeleke, R., 2024. Nanoparticle applications in agriculture: overview and response of plant-associated microorganisms. Front. Microbiol. 1664-302X. 15. doi:10.3389/fmicb.2024.1354440.

Miguel-Rojas, C., Pérez-de-Luque, A., 2023. Nanobiosensors and nanoformulations in agriculture: new advances and challenges for sustainable agriculture. Emerg. Top. Life Sci. 2397-8554. 7 (2), 229–238. doi:10.1042/etls20230070.

Mohammed, H., Mia, M.F., Wiggins, J., Desai, S., 2025. Nanomaterials for energy storage systems—a review. Molecules. 14203049. 30 (4), 883. Available from: http://www.mdpi.com/journal/molecules. doi:10.3390/molecules30040883.

Mondal, R., Dam, P., Chakraborty, J., Paret, M.L., Katı, A., Altuntas, S., Sarkar, R., Ghorai, S., Gangopadhyay, D., Mandal, A.K., Husen, A., 2022. Potential of nanobiosensor in sustainable agriculture: the state-of-art. Heliyon. 24058440. 8 (12), e12207. Available from: http://www.journals.elsevier.com/heliyon/. doi:10.1016/j.heliyon.2022.e12207.

Nair, R., Varghese, S.H., Nair, B.G., Maekawa, T., Yoshida, Y., Kumar, D.S., 2010. Nanoparticulate material delivery to plants. Plant Sci. 01689452. 179 (3), 154–163. doi:10.1016/j.plantsci.2010.04.012.

Ouyang, H., Safaeipour, N., Othman, R.S., Otadi, M., Sheibani, R., Kargaran, F., Van Le, Q., Khonakdar, H.A., Li, C., 2023. Agricultural waste-derived (nano)materials for water and wastewater treatment: current challenges and future perspectives. J. Clean. Prod. 09596526. 421, 138524. Available from: https://www.journals.elsevier.com/journal-of-cleaner-production. doi:10.1016/j.jclepro.2023.138524.

Pan, X., Guo, X., Zhai, T., Zhang, D., Rao, W., Cao, F., Guan, X., 2023. Nanobiopesticides in sustainable agriculture: developments, challenges, and perspectives. Environ. Sci. Nano. 2051-8153. 10 (1), 41–61. doi:10.1039/d2en00605g.

Payal, P.P., 2022. Role of nanotechnology in electronics: a review of recent developments and patents. Recent Pat. Nanotechnol. 18722105. 16 (1), 45–66. Available from: https://www.eurekaselect.com/article/113469. doi:10.2174/1872210515666210120114504.

Pooja, R.M., 2022. Agricultural Nanobiotechnology: Current Possibilities and Constraints Nanotechnology in Agriculture and Environmental Science. CRC Press, India, pp. 1–14. Available from: http://www.tandfebooks.com/doi/book/9781000793000. doi:10.1201/9781003323945-1.

Prabhu, S., Poulose, E.K., 2012. Silver nanoparticles: mechanism of antimicrobial action, synthesis, medical applications, and toxicity effects. Int. Nano Lett. 2008-9295. 2 (32). doi:10.1186/2228-5326-2-32.

Ramachandran, R.P., Erkinbaev, C., Vellaichamy, C., 2022. Smart Nano-biosensors in Sustainable Agriculture and Environmental Applications Food, Medical, and Environmental Applications of Nanomaterials. Elsevier, Canada, pp. 527–542. Available from: https://www.sciencedirect.com/book/9780128228586. doi:10.1016/B978-0-12-822858-6.00019-4.

Rıza Demirkıran, A., Sohrabi, M., 2024. The application of nanotechnology on plant nutrition and agriculture: a review. J. Agric. 2636-8757. 7 (1), 100–112. doi:10.46876/ja.1384560.

Salata, O.V., 2004. Applications of nanoparticles in biology and medicine. J. Nanobiotechnol. 14773155. 2 (3). Available from: http://www.jnanobiotechnology.com/content/2/1/3. doi: 10.1186/1477-3155-2-3.

Sato, H., Mizoi, J., Shinozaki, K., Yamaguchi-Shinozaki, K., 2024. Complex plant responses to drought and heat stress under climate change. Plant J. 0960-7412. 117 (6), 1873–1892. doi:10.1111/tpj.16612.

Sekhon, B.S., 2014. Nanotechnology in agri-food production: an overview. Nanotechnology, Science and Applications. 11778903. 7 (2), 31–53. Available from: http://www.dovepress.com/getfile.php?fileID=20118. doi:10.2147/NSA.S39406.

Shah, M.A., Shahnaz, T., Zehab-ud-Din, J.H.M., Nazir, S., Qurashi, A., Ahmed, G.H., 2024. Application of nanotechnology in the agricultural and food processing industries: a review. Sustain. Mater. Technol. 22149937. 39, e00809. Available from: https://www.sciencedirect.com/science/journal/22149937. doi:10.1016/j.susmat.2023.e00809.

Shang, Y., Hasan, M.K., Ahammed, G.J., Li, M., Yin, H., Zhou, J., 2019. Applications of nanotechnology in plant growth and crop protection: a review. Molecules. 14203049. 24 (14), 2558. Available from: https://www.mdpi.com/1420-3049/24/14/2558/pdf. doi:10.3390/molecules24142558.

Shao, C., Zhao, H., Wang, P., 2022. Recent development in functional nanomaterials for sustainable and smart agricultural chemical technologies. Nano Converg. 2196-5404. 9 (11). doi:10.1186/s40580-022-00302-0.

Sharma, R., Kumar, V., 2024. Nano enabled agriculture for sustainable soil. Waste Manag. Bull. 29497507. 2 (1), 152–161. doi:10.1016/j.wmb.2024.01.002.

Shaw, D.S., Honeychurch, K.C., 2022. Nanosensor Applications in Plant Science. Biosensors. 20796374. 12 (9), 675. Available from: http://www.mdpi.com/journal/biosensors/. doi:10.3390/bios12090675.

Singh, J., Dutta, T., Kim, K.H., Rawat, M., Samddar, P., Kumar, P., 2018. Green' synthesis of metals and their oxide nanoparticles: applications for environmental remediation. J. Nanobiotechnol. 16 (1), 1–24. doi:10.1186/S12951-018-0408-4.

Song, B., Piao, X., Chen, M., Tang, W., Zeng, G., Gong, J., Zhang, P., Shujing, Y., 2019. Using nanomaterials to facilitate the phytoremediation of contaminated soil. Crit. Rev. Environ. Sci. Technol. 1064-3389. 49 (9), 791–824. doi:10.1080/10643389.2018.1558891.

Thatoi, H., Mohapatra, S., Das, S.K., Pradhan, S.K., 2024. Biosensor Applied Biotechnology and Bioinformatics: Agriculture, Pharmaceutical Research and Environment. Wiley, pp. 395–406. doi:10.1002/9781119896869.ch17.

Tilman, D., Balzer, C., Hill, J., Befort, B.L., 2011. Global food demand and the sustainable intensification of agriculture. Proc. Natl. Acad. Sci. U. S. A. 10916490. 108(50), 20260–20264. Available from: http://www.pnas.org/content/108/50/20260.full.pdf+html. doi: 10.1073/pnas.1116437108.

Tripathi, S., Mahra, S., J, V., Tiwari, K., Rana, S., Tripathi, D.K., Sharma, S., Sahi, S., 2023. Recent advances and perspectives of nanomaterials in agricultural management and associated environmental risk: a review. Nanomaterials. 20794991. 13 (10). 1604. Available from: http://www.mdpi.com/journal/nanomaterials. doi:10.3390/nano13101604.

Tripathy, S., 2023. Top-down and Bottom-up Approaches for Synthesis of Nanoparticles. Materials Research Forum, LLC, pp. 92–130. doi:10.21741/9781644902370-4.

Umut, E., 2013. Surface Modification of Nanoparticles Used in Biomedical Applications. InTech. 10.5772/55746.

Upadhayay, V.K., Chitara, M.K., Mishra, D., Jha, M.N., Jaiswal, A., Kumari, G., Ghosh, S., Patel, V.K., Naitam, M.G., Singh, A.K., Pareek, N., Taj, G., Maithani, D., Kumar, A., Dasila, H., Sharma, A., 2023. Synergistic impact of nanomaterials and plant probiotics in agriculture: a tale of two-way strategy for long-term sustainability. Front. Microbiol. 1664302X. 14, 1133968. Available from: https://www.frontiersin.org/journals/microbiology#. doi:10.3389/fmicb.2023.1133968.

Upadhyay, R., Meena, B.R., Riya, Debapratim, B. Kumar, P., 2024. Nano Fertilizers for Agronomic Biofortification. Emerging Trends in Agriculture Sciences-Collaborative Insights, Priyanka Dubey, vol. 2. Adam's Book Company, Maharashtra, pp. 30–47.

Usman, M., Farooq, M., Wakeel, A., Nawaz, A., Cheema, S.A., Rehman, H.U., Ashraf, I., Sanaullah, M., 2020. Nanotechnology in agriculture: current status, challenges and future opportunities. Sci. Total Environ. 18791026. 721, 137778. Available from: www.elsevier.com/locate/scitotenv. doi:10.1016/j.scitotenv.2020.137778.

Vandana, J.N., Sharavanan, P.T., Johnson, I., Anandham, R., Raja, K., Karthikeyan, G., Angappan, K., Vaithiyalingan, M., 2025. Nanobiosensors for early detection of plant pathogens. Plant Sci. Today. 23481900. 12 (1), 1–12. Available from: https://horizonepublishing.com/journals/index.php/PST/article/view/4285. doi:10.14719/pst.4285.

Wang, B., Hu, S., Teng, Y., Chen, J., Wang, H., Xu, Y., Wang, K., Xu, J., Cheng, Y., Gao, X., 2024. Current advance of nanotechnology in diagnosis and treatment for malignant tumors. Signal Transduct. Target. Ther. 20593635. 9 (1), 200. Available from: https://www.nature.com/sigtrans. doi:10.1038/s41392-024-01889-y.

Wang, Y., Zhao, Q., Han, N., Bai, L., Jia, L., Liu, J., Che, E., Liang, H., Zhang, Q., Jiang, T., Wang, S., 2015. Mesoporous silica nanoparticles in drug delivery and biomedical applications. Nanomedicine: Nanotechnology, Biology and Medicine. 15499634. 11 (2), 313–327. doi:10.1016/j.nano.2014.09.014.

Xia, T., Kovochich, M., Liong, M., Mädler, L., Gilbert, B., Shi, H., Yeh, J.I., Zink, J.I., Nel, A.E., 2008. Comparison of the mechanism of toxicity of zinc oxide and cerium oxide nanoparticles based on dissolution and oxidative stress properties. ACS Nano. 1936086X. 2 (10), 2121–2134. doi:10.1021/nn800511k.

Yang, Y., Zeng, Z., Almatrafi, E., Huang, D., Zhang, C., Xiong, W., Cheng, M., Zhou, C., Wang, W., Song, B., Tang, X., Zeng, G., Xiao, R., Li, Z., 2022. Core-shell structured nanoparticles for photodynamic therapy-based cancer treatment and related imaging. Coord. Chem. Rev. 00108545. 458, 214427. Available from: http://www.journals.elsevier.com/coordination-chemistry-reviews/. doi:10.1016/j.ccr.2022.214427.

Zaman, W., Ayaz, A., Park, S., 2025. Nanomaterials in agriculture: a pathway to enhanced plant growth and abiotic stress resistance. Plants. 2223-7747. 14 (5), 716. doi:10.3390/plants14050716.

Zulfiqar, F., Navarro, M., Ashraf, M., Akram, N.A. and Munné-Bosch, S., 2019. Nanofertilizer use for sustainable agriculture: Advantages and limitations. Plant Science 289, 110270. doi:10.1016/j.plantsci.2019.110270.

CHAPTER

# 8 Enzyme-assisted biomass conversion for sustainable waste management

**Yehia A.G. Mahmoud and Aya Ahmed Beltagy**

*Faculty of Science, Botany and Microbiology Department, Tanta University, Tanta, Egypt*

## 1 Introduction

Biomass, in its broadest sense, is any material that was or is a part of a living species. However, biomass is usually limited to plant-derived materials when discussing renewable energy applications. These materials include agricultural residues (like corn stalks and wheat straw), by-products of industrial processes (like sawdust, sugar cane bagasse, pulp residues, and distiller's grains), or particular energy crops (like switchgrass, sorghum, Miscanthus, and short-rotation woody crops) (Hames, 2009). Aquatic plants and algae, animal waste, municipal solid waste, and timber and wood waste are also considered biomass resources (Demirbaş, 2001). Approximately 25% of the world's energy requirements can be met by biomass, a renewable resource obtained from the forestry or agricultural sectors. Furthermore, biomass can be used as the starting material for a variety of important chemicals, medications, and food additives (Demirbaş, 2001). The four major categories of biomass conversion technologies are direct combustion, thermochemical, biochemical, and agrochemical processes. Thermochemical conversion processes include direct liquefaction, pyrolysis, gasification, and supercritical fluid extraction. Biomass is used to meet a variety of energy needs, including producing electricity, heating homes, fueling vehicles, and providing process heat for industrial facilities. The most plentiful, pervasive, and renewable raw resource that can be used for human benefit is lignocellulosic biomass. Although there are alternative ways to use it, the most widely accepted method is to convert biomass using enzymes because it is environmentally sustainable. Because the structure of biomass is hard and complicated, no single enzyme component can fully deconstruct it; instead, a set of enzymes is required to degrade the biomass into simple monomers. Biomass must be pretreated before enzymes can increase their accessibility and hydrolyze it effectively. Enzyme performance has improved significantly in recent years due to genetic modification, including activities, output, characteristics, and effectiveness (Patel et al., 2020). Many efforts have been made in the last 6 years to reduce costs by focusing on increasing the efficiency of known enzymes, discovering new, more active enzymes, developing enzyme mixtures that are best suited for specific pretreatment substrates, and lowering the cost of enzyme production.

Microbial Technologies for Sustainable Waste Management. https://doi.org/10.1016/B978-0-443-44093-9.00013-X

Here, we discuss the most recent advances in enzyme technology for biofuel production, as well as the challenges that the industry continues to face (Merino and Cherry, 2007). This chapter describes biomass-processing enzymes, which can convert the world's most abundant raw material into valuable chemicals and products to meet human demand (Patel et al., 2020).

## 2 Biomass classification

Because biomass varies greatly in terms of variety and amount, as well as in terms of its composition, there is no clear-cut method to classify it; biomass is divided into two categories based on its source, purpose, and end products:

1. Sorting by the kinds of biomass found in nature (based on vegetation type or ecology).
2. Sorting by how biomass is used and applied as feedstock.

The most common categorization, which classifies biomass into different groups, is listed in Table 8.1.

### 2.1 Categories of biomass

#### *2.1.1 Wood and woody biomass*

The USDA Forest Service defines woody biomass as material derived from trees and woody plants, including branches, tops, needles, leaves, and other woody

**Table 8.1** Categories of different biomass nature.

| Organic wastes | Wastes nature |
|---|---|
| Agricultural residues | Flowers and grasses (alfalfa, arundo, bamboo, banana, brassica, cane, artichoke, Miscanthus, switchgrass, timothy, etc.); straw (barley, bean, flax, corn, mint, oat, rape, rice, rye, sesame, sunflower, wheat, etc.); and additional remnants (fruits, shells, husks, hulls, pits, pips, grains, seeds, coir, stalks, cobs, kernels, bagasse, food scraps, fodder, pulps, cakes, etc.). |
| Woody biomass | Leaves, branches, bark, and by-products like chips, lumps, pellets, sawdust, and even materials from sawmills. These can come from both angiosperms (hardwoods) and gymnosperms (softwoods). |
| Human waste biomass and animals | Fecal wastes from human and livestock animal populations, bones, various manures, meat-bone meal, etc. |
| Hydrophytes biomass | It can also refer to specific types, such as phytoplankton, benthic plants, floating plants, aquatic plants, hydrophytes, aquatic weeds, seaweed, macroalgae, microalgae, and aquatic microorganisms. |
| Biomass mixtures | A blend of different types of crop residues, wood, or waste materials. |

components, which are typically by-products of forest management in forests, woodlands, or rangelands. Examples of woody biomass used for bioenergy include pellets, chips, fuelwood logs, charcoal, briquettes, wood shavings, and waste wood from industrial processes. Bioenergy has replaced fossil fuels in three key areas: ethanol substituting gasoline in transportation, densified biomass replacing coal in electricity generation, and pelletized biomass used for various heating purposes (Pandey and Erbaugh, 2024). Typically, biomass used for the generation of both energy and fuel comes from four major sources: manufacturing residues, nonmerchant timber residues, postconsumer wood waste, and urban and agricultural debris.

### 2.1.2 Herbaceous biomass

According to the European standard EN 14961-1, which classifies solid biofuels, herbaceous biomass originates from plants with nonwoody stems that wither at the end of the growing season. This type of biomass includes cereal straw, by-products from food processing, and various grains or seeds. Herbaceous biomass resources typically fall into one of two primary categories: energy crops and agricultural residues.

1. Agricultural residues are by-products from the food, fiber, or agriculture industries. Some of these residues are collected for use in applications like animal feed. However, there is a lack of comprehensive tracking of these residues, including their locations and potential as sources of bioenergy.
2. In contrast, energy crops are specifically grown for use in the bioenergy industry.

### 2.1.3 Aquatic biomass

#### Aquatic plants

Aquatic biomass refers to macroalgae, microalgae, and emerging aquatic plants. Aquatic plants are one sustainable feedstock that is frequently disregarded in these biorefineries. With the exception of their lower lignin content, aquatic plants are comparable to terrestrial plants in composition and therefore require milder pretreatment conditions. Due to their high protein content, they can serve as potential feedstocks in biorefineries for carbohydrates, sugars, and lipids (Kaur et al., 2019). They also exhibit good biomass productivity, rapid growth rates, and increased photosynthetic efficiency. The surplus aquatic biomass produced can be used in biorefineries to create a variety of biocommodities, including fish feed, protein sources, soil compost, biofertilizer, biofuels, and value-added goods, provided that appropriate management procedures are followed. These aquatic plants also have the important advantage of not requiring arable land for cultivation, which allows their large root systems to be utilized in tertiary wastewater treatment systems to remove pollutants. In both tropical and temperate climates, aquatic plants are common. They can thrive in various environments, such as brackish and contaminated waters, and can exist as free-floating, submerged, or rooted (emerged) plants (Verma and Suthar, 2015). Because of their tremendous capacity for expansion, ease of adaptation, and low care needs, aquatic plants generate enormous amounts of biomass when nutrient availability is high. By making effective use of these nutrients, aquatic plants flourish when waste discharges into bodies of water increase. Compounds like proteins, lipids, and carbohydrates make up the biomass of plants (Alam et al., 2021). The carbohydrate component can be fermented to produce liquid fuels such as bioethanol, biomethanol, biobutanol, biogas, and biohydrogen (bio-$H_2$), along with other valuable materials and compounds. The protein fraction makes them an appropriate source of protein-rich animal feed, while the lipid fractions may be utilized as a possible source of biodiesel. As evidenced by the rise in demand for aquatic plants' biomass over the past few years, research on using these plants to produce biofuels and other value-added goods has received significant interest (Arefin et al., 2021).

#### Microalgae

They provide a range of commercially valuable by-products, such as sugars, fats, oils, and beneficial bioactive compounds, and may be cultivated in difficult agroclimatic conditions. Before the Taiwan Chlorella Manufacturing Company started growing the green alga *Chlorella* in 1964, microalgae were not widely grown economically, in contrast to macroalgae (seaweed). By 1980, 46 sizable businesses, mostly in Asia, were producing roughly 12 tons of dry matter annually, primarily from the cyanobacterium *Arthrospira* and *Chlorella*. Since then, the field of microalgal biotechnology has undergone significant growth and transformation, generating roughly 5000 tons of dry matter annually (Spolaore et al., 2006). In other words, algae are now economically produced in small amounts at high values.

#### Emerging plants grow partially submerged in marshes and swamps

There are approximately 55,000 species and over 100,000 strains of freshwater, terrestrial, and brackish water algae found in nature. One of the main advantages of using algae is their ability to convert sunlight, water, and $CO_2$ into a wide range of metabolites and compounds, which accumulate in algal biomass (Vassilev and Vassileva, 2016). Aquatic biomass is considered an ideal raw material for third-generation biodiesel production because it does not compete with food crops and produces significantly more biomass per hectare compared to land-based crops. However, before these biofuels can be successfully commercialized, there are still considerable technological and economic challenges to overcome. In addition, factors such as temperature, pH, salinity, $CO_2$ and $O_2$ concentrations, light levels, and nutrient availability all influence the performance of aquatic biomass (Tursi, 2019).

### 2.1.4 Animal and human waste biomass

The most common sources include bones, meat meal, various types of animal manures, and human feces. These wastes were previously recovered and sold as fertilizers or simply utilized on agricultural land, but appropriate waste management has been made possible by stricter environmental rules addressing issues of pollution, health, and odor. The most practical way to turn this waste into useful products is through anaerobic digestion. For instance, biogas, a by-product of the process, serves

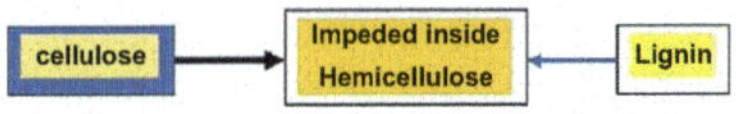


**FIG. 8.1**

Structure of lignocellulosic biomass.

as a direct fuel source for culinary applications and for heating indoor spaces and water systems, or it can be used to power internal combustion engines and turbines (Tursi, 2019).

#### *2.1.5 Biomass mixtures*

When many substrates from the above-mentioned classifications are present in mixed form, they are sometimes categorized under this heading (Tursi, 2019).

### 2.2 Chemical composition of biomass

Biomass exhibits a highly diverse composition. For example, plant-derived residues are predominantly composed of cellulose, hemicellulose, and lignin (Fig. 8.1). Cereals are mostly made of starch, although cattle manure is high in proteins in varying proportions. It goes without saying that different chemical structures provide diverse chemical characteristics (Tursi, 2019).

## 3 Enzymes involved in biomass conversion

Lignocellulosic biomass (LCB), a widely abundant resource globally, is recognized as an affordable and promising bioresource for producing valuable chemicals and metabolites through enzymatic processes (Hernández Fariñas et al., 2017; Saldarriaga-Hernández et al., 2020; Nargotra et al., 2022). LCB primarily consists of three main components and includes sources such as energy crops, weed biomass, softwood, hardwood, and agricultural residues. The key enzymes responsible for breaking down lignocellulose are cellulases, hemicellulases, and ligninolytic enzymes. Specifically, cellulases and hemicellulases break down cellulose and hemicellulose into simpler sugars, which can then be used to manufacture a variety of products, including hydrogels, biofilms, drug delivery coatings, organic acids, and biofuels like bioethanol and biodiesel (Sharma et al., 2022). Once thought to be a barrier in LCB biorefineries, lignin has recently been utilized to create a variety of goods, including carbon-based nanomaterials, bioplastics, and vanillin (Ashokkumar et al., 2022). The hydrolysis of lignin and its transformation into beneficial chemicals are accomplished by a range of ligninolytic enzymes (Singhania et al., 2022).

### 3.1 Cellulolytic enzymes

Cellulases are multienzyme complexes that hydrolyze cellulose. The three primary enzymes that make up the enzyme complex are β-glucosidases, cellobiohydrolases

(CBHs), and endoglucanases. To produce glucose monomers, these three enzymes work in concert with the cellulose polymers (Legodi et al., 2019).

### 3.1.1 Endoglucanases or endo-1,4-β-D-glucanase

Endoglucanase (EC 3.2.1.4) plays a crucial role in the biodegradation of cellulose and belongs to the glycoside hydrolase (GH) superfamily, which is classified based on amino acid sequences and three-dimensional structures (Akram et al., 2018). Also known as endo-1,4-β-D-glucan glucanohydrolases, endoglucanases initiate the enzymatic breakdown of cellulose by reducing its degree of polymerization. They act randomly on the β (1,4) glycosidic bonds within the amorphous regions of cellulose chains, producing oligosaccharides (Raghuwanshi et al., 2014; Houfani et al., 2020). Endoglucanases are distributed across 13 GH families, with most thermostable variants classified under the GH12 family. Their activity and specificity are influenced by the spatial arrangement of two key carboxylic acid residues within their active site. They can both invert and retain their catalytic mode of action (Akram et al., 2018). Numerous species, including bacteria (both aerobic and anaerobic), fungi, archaea, and protozoa, produce endoglucanases. However, fungi like *Trichoderma* and *Aspergillus* sp. are utilized to produce these enzymes for commercial purposes (Saldarriaga-Hernández et al., 2020).

### 3.1.2 Exoglucanase

Exoglucanases (EC 3.2.1.91), also known as CBHs such as CBHI and CBHII, target the reducing and nonreducing ends of cellulose chains, respectively, to release cellobiose and smaller amounts of cellotriose (Annamalai et al., 2016; Hiras et al., 2016; Houfani et al., 2020). Among the cellulases produced by cellulolytic fungi—accounting for approximately 70% of total cellulase production—CBHs are the most extensively studied exoglucanases. These enzymes belong primarily to GH families 6, 7, and 48 (Annamalai et al., 2016). Typically, GH6 exoglucanases cleave cellobiose units from the nonreducing ends of cellulose chains, while those from GH7 and GH48 mainly act at the reducing ends to release cellobiose (Hiras et al., 2016). Many different types of bacteria and fungi manufacture CBHs, although GH family 7 CBHs are mostly produced by fungi, while GH family 48 CBHs are predominantly produced by bacteria. CBHs are believed to effectively break down microcrystalline cellulose by removing cellulose chains from the microcrystalline cellulose structure. Cellobiose inhibits CBHs competitively (Annamalai et al., 2016).

### 3.1.3 β-D glucosidase

β-D glucosidase (BGL), also known as cellobiase (EC 3.2.1.21), is the key enzyme responsible for the final step in cellulose hydrolysis and acts as the rate-limiting factor in this process. It catalyzes the breakdown of soluble cellobiose and cellodextrins into D-glucose units (Yao et al., 2016). BGLs play a crucial role in the cellulase degradation system by producing gentiobiose and sophorose, compounds that serve as inducers for the cellulase enzyme complex (Naraian and Gautam, 2017). Based on amino acid sequence similarities, BGLs have been classified into several GH families, including GH1, GH3, GH5, GH9, and GH30. GH families 3 and 1 include fungal

BGL (Yao et al., 2016; Singhania et al., 2017). Furthermore, BGL action is reversibly limited by stored glucose, which makes BGL regulation challenging. However, glucose also stimulates the creation of BGL, which is classified into the GH1 and GH3 groups (Singhania et al., 2017; Srivastava et al., 2019). The synthesis of biofuel and the prevention of catabolite repression on endoglucanase and CBH depend heavily on BGLs, which are the rate-limiting enzymes. However, microorganisms or even commercial cellulase only produce extremely small amounts of them. Consequently, it is inevitable to add BGL to the enzyme cocktail to create an efficient cellulolytic cocktail, which could lead to a higher sugar yield (Singhania et al., 2017).

### 3.1.4 Cellobiose dehydrogenase

White-rot fungi produce extracellular enzymes known as cellobiose dehydrogenases (EC 1.1.99.18), which contain two distinct cofactors—heme and flavin—located in separate domains (Martínez et al., 2017). These enzymes consist of a heme-containing cytochrome domain and a flavodehydrogenase domain with a noncovalently bound flavin adenine dinucleotide (FAD) cofactor, connected by a flexible, proteolysis-sensitive linker (Sulej et al., 2019). Cellobiose dehydrogenases collaborate closely with lytic polysaccharide monooxygenases (LPMOs). Electrons are transferred from the flavin domain to the heme domain, where catalytic histidine and other residues facilitate the oxidation of cellobiose to cellobionolactone. This intramolecular electron transfer has been extensively studied from both structural and mechanistic perspectives (Martínez et al., 2017).

### 3.1.5 Lytic polysaccharide monooxygenase

Copper-containing reductases, known as LPMOs (EC 1.14.99.54), are a novel class of enzymes that work in concert with cellulases to break down cellulose. They degrade cellulose using an oxidative process that involves molecular oxygen and electron donors (Martínez et al., 2017; Houfani et al., 2020). LPMOs require a reducing agent to activate molecular oxygen at their copper-containing active site (Saldarriaga-Hernández et al., 2020). In some fungal LPMOs, activation involves the reduction of a catalytic cupric ion coordinated by ligands, including a tyrosine residue and a methylated terminal histidine. By reducing $Cu^{2+}$ to $Cu^{+}$ and then interacting with $O_2$ to generate a reactive copper–superoxide complex, an artificial reductant like ascorbic acid can be used to accomplish this activation. Through monooxygenase activity, this process promotes the oxidative breakdown of crystalline carbohydrate chains (Martínez et al., 2017). Oxidation occurring at the C1 carbon—or at the C4 carbon in some fungi—of glucose units within cellulose leads to the formation of aldonic acid and subsequent cellulose degradation (Arfi et al., 2014). Research suggests that LPMOs preferentially use hydrogen peroxide ($H_2O_2$) rather than molecular oxygen as a cosubstrate for cellulose oxidation (Bissaro et al., 2017). By carefully regulating the supply of $H_2O_2$, LPMOs can operate in the absence of oxygen, enabling these monocopper enzymes to carry out their biological functions (Kont et al., 2020). Comparative studies between reference enzymes and LPMOs using $H_2O_2$ as a cosubstrate have characterized the mechanism of LPMO-driven cellulose oxidation, revealing that both bacterial and fungal LPMOs exhibit high peroxygenase efficiency relative to the reference enzymes.

## 3.2 Hemicellulases

Hemicellulose, the second most abundant polysaccharide in plant cell walls, is a heterogeneous, branched, and amorphous heteropolymer (Houfani et al., 2020). It also contains uronic acids such as D-galacturonic acid, D-glucuronic acid, and methyl galacturonic acid (Houfani et al., 2020; Sharma et al., 2022). The polymers within hemicellulose are interconnected through covalent and hydrogen bonds, forming hydrogen bonds with cellulose and aromatic ester linkages with lignin, creating a strong connection between the two. In addition, hemicellulose exhibits hygroscopic and hydrophilic properties (Sethi and Scharf, 2013). Depending on the kind of plant, the chemical composition of hemicellulose in LCB varies; softwood or gymnosperms contain mostly glucomannans, whereas hardwood or angiosperms have mostly xylans (Albornoz-Palma et al., 2020). Many enzymes, most of which are hydrolytic and work similarly to cellulolytic enzymes, mediate the breakdown of hemicellulose (Saldarriaga-Hernández et al., 2020). Xylanase endoenzymes hydrolyze the β-D-xylopyranosyl bonds by targeting the main chain of xylan, which reduces the substrate's polymerization and releases xylobiose, xylooligosaccharides, and xylose (Sharma et al., 2018). Following this, exoenzymes called β-xylosidases initiate a stepwise breakdown process, starting at the nonreducing ends of the xylooligosaccharides and ultimately producing xylose (Najjarzadeh et al., 2020).

### *3.2.1 Endoxylanase*

Endoxylanases break the β-1,4 glycosidic bonds in the xylan backbone of hemicellulose (Saldarriaga-Hernández et al., 2020). Specifically, endo-1,4-beta-xylanase (EC 3.2.1.8) hydrolyzes xylan to produce xylooligosaccharides, while endo-1,3-beta-xylanase (EC 3.2.1.32) targets glycosidic bonds between xylans at carbons 1 and 3, generating xylobiose, xylotriose, and xylotetrose. In addition, α-D-xyloside xylohydrolase (EC 3.2.1.177) cleaves terminal residues of nonreducing α-D-xylose, releasing free α-D-xylose (Carli et al., 2016; Luo et al., 2016; Alokika and Singh, 2019; Saldarriaga-Hernández et al., 2020).

Endoxylanases primarily belong to GH families 10 and 11. GH11 enzymes preferentially act on longer xylan chains and typically do not produce xylose, whereas GH10 endoxylanases can process smaller substrates and often generate xylose (Sharma et al., 2018; Santibáñez et al., 2021). While xylanases are produced by a wide range of organisms, including bacteria, fungi, archaea, marine algae, and protozoans, filamentous fungi are the most commonly reported commercial source of these enzymes (Santibáñez et al., 2021).

### *3.2.2 β-D-xylosidase*

EC 3.2.1.37 β-D-xylosidases interact with xylose residues and other nonreducing ends of xylooligosaccharides. After xylanases gradually hydrolyze xylan, β-xylosidases are essential because they convert short oligomers into monomers. Xylobiose is an appropriate substrate for these enzymes, and the degree of polymerization is inversely connected with the oligosaccharide's affinity (Alokika and Singh, 2019; Binod et al., 2019). Most β-xylosidases are classified within the GH3 family. Despite variations in their three-dimensional structures, all β-xylosidases share a characteristic pocket-shaped active site, which is negatively charged due to acidic residues and features hydrophobic regions formed by aromatic amino acids (Rohman et al., 2019). When using p-nitrophenyl-β-D-xyloside as a substrate, the enzymatic activity of many β-xylosidases is competitively inhibited by their product, xylose (Rohman et al., 2019).

### 3.2.3 Arabinofuranosidase

Arabinofuranosidases, also known as arabinases, catalyze the release of arabinose from polysaccharides, arabinooligosaccharides, or synthetic substrates through both exo- and endo-acting mechanisms (Alokika and Singh, 2019). Specifically, α-L-arabinofuranosidases (EC 3.2.1.55) hydrolyze α-1,2-, α-1,3-, and α-1,5-L-arabinofuranosyl residues from the nonreducing ends of arabinose-containing oligo- and polysaccharides (Contesini et al., 2017). These enzymes can also remove 4-O-methyl glucuronic acid groups from xylan (Binod et al., 2019). Arabinofuranosidases are distributed among GH families GH30, GH43, GH51, GH54, and GH62, with GH62 being unique in comprising exclusively arabinofuranosidases (Contesini et al., 2017; Alokika and Singh, 2019). There are three main types of arabinofuranosidases: Type A enzymes preferentially hydrolyze p-nitrophenyl-α-L-arabinofuranoside and arabinooligosaccharides; Type B enzymes break down arabinoxylan into xylose and L-arabinose; and Type C, known as 1→4-β-D-arabinoxylan arabinofuranohydrolase, specifically targets arabinosidic linkages in various arabinoxylans to release L-arabinose. Notably, Type C enzymes are inactive on a range of synthetic substrates (Contesini et al., 2017).

### 3.2.4 Esterase

Acetyl xylanesterase (EC 3.1.1.72) catalyzes the hydrolysis of acetyl-ester bonds between xylose units in xylan, releasing O-acetyl groups and acetic acid (Saldarriaga-Hernández et al., 2020). This enzyme acts on acetyl groups in polymeric acetylated xylose, xylan, acetylated glucose, p-nitrophenyl acetate, and α-naphthyl acetate but does not affect acetylated mannan or pectin (Saldarriaga-Hernández et al., 2020). Another important esterase, ferulic acid esterase (EC 3.1.1.73), cleaves feruloyl ester bonds in xylan, liberating ferulic acid from the O-5 position of arabinose in glucuronoarabinoxylan found in monocots (Binod et al., 2019; Karnaouri et al., 2019). Acetoxylan esterases display a strong affinity for O-acetyl-4-O-methyl-D-glucurono-D-xylan and are classified into several carbohydrate esterase (CE) families, including CE1–7 and CE16 (Karnaouri et al., 2019). Most acetoxylan esterases possess acetyl esterase activity, except for those in the CE4 family (Sista Kameshwar and Qin, 2018). A subset of feruloyl esterases falls under the CE1 family, alongside other esterases. More recently identified, glucuronoyl esterases belong to the CE15 family and catalyze the hydrolysis of ester bonds linking lignin to 4-O-methyl-D-glucuronic acid residues within lignin–carbohydrate complexes (Karnaouri et al., 2019).

### 3.2.5 α-Glucuronidase

α-Glucuronidases (EC 3.2.1.131) hydrolyze the α-1,2-glycosidic bond between D-glucuronic acid or its 4-O-methyl ether and the nonreducing end xylopyranosyl unit of short xylooligosaccharides (Alokika and Singh, 2019). Typically, glucuronidases from the GH67 family are intracellular or membrane-associated enzymes that preferentially target short glucuronic acid-substituted xylooligosaccharides, while those belonging to the GH115 family are extracellular and more effective on polymeric glucuronic acid-containing xylans (Malgas et al., 2019). Given that hemicellulose is rich in xylan, xylanases are considered the most crucial group among hemicellulases. The xylanase market is projected to grow significantly, with an estimated demand of $500 million USD by 2023, up from the current range of $200–300 million USD (Irdawati et al., 2018; Chadha et al., 2019).

Xylanases have widespread applications beyond lignocellulosic biorefineries, including the feed and food industries—such as poultry, baking, coffee extraction, functional foods, and agricultural silage—as well as the commercial paper and pulp sectors. The global xylanase market is distributed across five main regions: Latin America, North America, Asia-Pacific, Europe, and Africa (Chadha et al., 2019). Numerous bacterial, yeast, fungal, and actinomycete species have been explored for their xylanase production capabilities (Alokika and Singh, 2019).

## 3.3 Ligninolytic enzymes

Coniferyl, p-coumaryl, and sinapyl alcohol are the three aromatic alcohols that make up lignin, the second most abundant heterogeneous polymer after cellulose. These alcohols are linked through carbon–carbon (C–C) and aryl ether bonds (Sharma et al., 2022; Chio et al., 2019). In hardwood lignin, coniferyl and sinapyl alcohols constitute approximately 25%–50% and 50%–75%, respectively, while coniferyl alcohol dominates softwood lignin, comprising about 90%–95%. Grass lignin typically contains all three monomeric alcohols (Houfani et al., 2020). These aromatic alcohols undergo free radical polymerization to form three related phenylpropanoid monomeric units: guaiacyl, p-hydroxyphenyl, and syringyl. The phenolic groups within these units serve as primary substrates for enzymatic activity (Houfani et al., 2020; Saldarriaga-Hernández et al., 2020). The monomeric lignin units are interconnected by condensed C–C bonds (β-5′, 5-5′, β-β, β-1) and noncondensed C–O–C (β-O-4) ether bonds. Among these, β-O-4 ether linkages are more susceptible to enzymatic hydrolysis compared to C–C bonds (Suseela, 2019). Ether bonds represent the predominant type of linkage in lignin, accounting for at least 56% of all connections. In addition, variations in the quantity and ratios of these monomeric units occur among different plant species, influencing lignin composition and structure. The percentage of connections also varies. The majority of connections in softwood and hardwood are β-O-4 couplings, which make up roughly 50% and 60% of all linkages, respectively. Because hemicellulose contains a significant amount of xylan, xylanases are the most significant of all the hemicellulases (Chio et al., 2019). Ester, hydrogen, and covalent bonds promote lignin's interaction with cellulose and hemicellulose (Houfani et al., 2020). The main cause of lignocellulosic biomass (LCB) resistance to degradation is lignin's complex, amorphous, and heterogeneous polymeric structure, which hinders its breakdown (Houfani et al., 2020). The diverse and intricate linkages within lignin demand the involvement of multiple enzymes for effective biological depolymerization, some of which are detailed in the following sections.

### 3.3.1 Laccase

Laccases (EC 1.10.3.2), a group of multicopper oxidases produced by fungi, bacteria, and plants, catalyze substrate oxidation by transferring electrons from a mononuclear copper site to a trinuclear copper center (Saldarriaga-Hernández et al., 2020; Weng et al., 2021). The laccase holoenzyme typically consists of glycoprotein dimers or tetramers, with each monomer containing four copper atoms associated with type 1, type 2, and type 3 redox sites (Binod et al., 2019). Exhibiting a redox potential ranging from 0.4 to 0.8 V relative to the standard hydrogen electrode, laccases exploit the redox properties of copper to oxidize aromatic compounds (Saldarriaga-Hernández et al., 2020; Mayr et al., 2021; Chio et al., 2019). Upon oxidizing phenolic substrates, laccase generates an unstable phenoxyl free radical intermediate, which promotes C$\alpha$ oxidation and cleavage of polymer linkages such as C$\alpha$–C$\beta$ and alkyl–aryl bonds (Chio et al., 2019; Weng et al., 2021). Besides phenolics, laccases can hydrolyze a broad spectrum of substrates, including aliphatic and aromatic amines, diphenols, and polyphenols (Binod et al., 2019). Although primarily acting on phenolic compounds, laccases can also oxidize nonphenolic substrates through the involvement of low-molecular-weight electron carriers called mediators—such as 2,2′-azinobis-(3-ethylbenzothiazoline-6-sulfonate), 1-hydroxybenzotriazole (HBT), and 3-hydroxyanthranilic acid (HAA)—which facilitate substrate–enzyme interaction and enhance the enzyme's oxidation potential (Saldarriaga-Hernández et al., 2020; Chio et al., 2019; Weng et al., 2021). With mediator assistance, laccases have been reported to degrade 80%–90% of lignin. White-rot fungi are among the most well-known producers of these enzymes (Chio et al., 2019).

### 3.3.2 Manganese peroxidase

Manganese peroxidases (EC 1.11.1.13) are glycosylated heme proteins that catalyze the oxidation of $Mn^{2+}$ ions to $Mn^{3+}$ using $Mn^{2+}$ as a reducing substrate. The generated $Mn^{3+}$ then forms complexes with $\alpha$-hydroxy acids and dicarboxylic acids (Saldarriaga-Hernández et al., 2020; Chio et al., 2019). These enzymes facilitate lignin depolymerization through coupled oxidation and reduction reactions. The catalytic cycle begins with manganese peroxidase mediating the binding of hydrogen peroxide ($H_2O_2$) to its native ferric form. This is followed by the oxidation of $Mn^{2+}$ to $Mn^{3+}$ in the presence of chelators. The highly reactive $Mn^{3+}$ species diffuses away from the enzyme and oxidizes phenolic lignin components into phenoxy radicals, which drive lignin breakdown. Organic acid chelators such as oxalate and malonate play a crucial role in stabilizing $Mn^{3+}$ and enhancing enzyme activity (Chio et al., 2019; Weng et al., 2021). Due to its higher redox potential, manganese peroxidase is particularly effective during the initial stages of lignin degradation, outperforming laccase in oxidizing phenolic compounds within lignin (Datta et al., 2017).

### 3.3.3 Lignin peroxidase

Lignin peroxidases (EC 1.11.1.14) are glycosylated enzymes featuring an iron protoporphyrin group at their active site, along with a heme cofactor (Datta et al., 2017). Owing to their high redox potential and capacity to generate cation radicals, these enzymes can directly oxidize both phenolic and nonphenolic substrates without requiring mediators (Datta et al., 2017; Binod et al., 2019). The catalytic activity of lignin peroxidases depends on hydrogen peroxide, which initiates the oxidation process for both substrate types. In addition, veratryl alcohol acts as a cofactor and electron donor, further enhancing enzyme function (Chio et al., 2019). The catalytic cycle of these enzymes typically unfolds in three stages: an initial oxidation step followed by two sequential one-electron reduction steps. In the first stage, lignin peroxidase catalyzes the reduction of $H_2O_2$ to water while oxidizing ferric iron [Fe(III)] to the radical cation intermediate [$Fe(IV) = O^{+}$]. Subsequently, this intermediate is converted into two [$Fe(IV) = O$] species through the two successive one-electron reductions (Weng et al., 2021).

### 3.3.4 Dye-decolorizing peroxidase

Dye-decolorizing peroxidases (DyPs, EC 1.11.1.19) are heme-containing enzymes that degrade lignin via radical-mediated oxidation. Although they share a hydrogen peroxide-dependent catalytic mechanism with class II heme peroxidases, DyPs differ in their structural and sequence characteristics (Rahmanpour et al., 2016; Datta et al., 2017). Unlike most peroxidases, which are primarily fungal, DyPs are widely distributed among various bacteria. One notable product of their activity on nonphenolic lignin-related compounds is p-anisaldehyde, formed through decarboxylation (de Gonzalo et al., 2016; Chio et al., 2019). DyPs are categorized into four types—A, B, C, and D—based on sequence differences. Types A, B, and C are mostly bacterial, whereas fungi predominantly produce type D. Types C and D share greater substrate oxidation activity, while types A and B are generally smaller enzymes with lower catalytic efficiency (Chio et al., 2019).

### 3.3.5 β-Etherase

β-Etherase is a glutathione-dependent enzyme that specifically targets the β position of an oxidized aryl unit containing an α-ketone group, facilitating the reductive cleavage of β-aryl ether linkages, which is vital for lignin degradation (Bugg et al., 2020). These enzymes play a key role in effective lignin depolymerization through the β-O-4 ether and biphenyl catabolic pathways (Weng et al., 2021). The breakdown of the most common β-O-4 ether bonds in lignin involves a group of enzymes, including glutathione-S-transferase, β-etherase, and Cα-dehydrogenase. First, the enzyme LigD (a Cα-dehydrogenase) oxidizes the hydroxyl group at the Cα position in lignin. Then, β-etherases LigE or LigF cleave the β-O-4 ether bond, producing intermediate compounds such as vanillin and α-glutathionyl-β-hydroxypropiovanillone (GS-HPV). In the final step, the glutathione-S-transferase LigG oxidizes GS-HPV, and after glutathione is removed, the resulting hydroxypropiovanillone can be further converted into vanillin (Picart et al., 2015; Chio et al., 2019).

#### 3.3.6 Biphenyl bond cleavage enzyme

Biphenyl linkages, which constitute about 10% of the bonds in softwood lignin, represent another important type of connection in lignin alongside the β-O-4 aryl-ether bonds (Weng et al., 2021). These biphenyl bonds commonly occur between two guaiacyl units. During the hydrolysis of 5,5′-dehydrodivanillate, the nonheme, iron-dependent demethylase enzyme LigX demethylates one methoxy group, resulting in the formation of a hydroxyl group. This hydroxylated product then serves as a substrate for oxidative cleavage by LigZ, an extradiol dioxygenase. The resulting ring-cleavage product is further broken down by LigY, a C–C hydrolase, producing 4-carboxy-2-hydroxypentadienoic acid and 5-carboxyvanillic acid. Finally, two decarboxylases, LigW and LigW2, convert 5-carboxyvanillic acid into vanillate or vanillic acid (Pollegioni et al., 2015). This sequence of enzymatic reactions highlights how the cleavage of biphenyl bonds facilitates lignin degradation.

## 4 Microbial production of lignocellulolytic enzymes

The extracellular biocatalysts known as lignocellulolytic enzymes are responsible for the complete fractionation of lignocellulose. Microbes, insects, and plants are just a few of the many organisms in nature that contain lignocellulolytic enzymes. Microbial enzyme production has garnered significant attention due to its economic advantages. Research has primarily focused on the diversity and quantity of enzymes generated by microorganisms under optimal growth conditions. These microorganisms, which can be free-living in the environment or reside in animal digestive systems, have the potential to degrade lignocellulosic biomass. Fungi and their enzymes offer several advantages over bacterial enzymes, including the production of extracellular enzymes, easier recovery, greater tolerance to harsh environmental conditions, the ability to grow on agroindustrial waste, biodiversity, higher enzyme yields, and the production of supplementary enzymes (Sharma et al., 2018).

### 4.1 Production of lignocellulolytic enzymes from fungi

Wood-degrading fungi are primarily responsible for lignocellulose decomposition. Their highly specialized lifestyles, reflected in significant phylogenetic and phenotypic diversity, enable them to break down and assimilate even the most resistant organic polymers. These fungi employ a broad spectrum of extracellular enzymes to disrupt the chemical and physical integrity of lignocellulosic biomass (Kracher and Ludwig, 2016). Fungal degradation occurs via two main pathways: extracellular processes essential for polysaccharide breakdown and intracellular mechanisms associated with the cell envelope. The extracellular enzymatic arsenal is composed of two categories: oxidative enzymes that target lignin and cleave phenyl rings, and hydrolytic enzymes responsible for polysaccharide degradation (Andlar et al., 2018). In addition, three types of fungi—brown-rot, white-rot, and soft-rot—have been identified, each exhibiting distinct modes and effects in lignocellulose degradation (Saldarriaga-Hernández et al., 2020).

### 4.1.1 Soft-rot fungi

Soft-rot fungi, called ascomycetes and deuteromycetes, first emerge before the brown and white-rot processes have begun. These slow decomposers have the capacity to break down wood in severe environments by hydrolyzing the polysaccharides found in the surface layers of plant material (Andlar et al., 2018; Saldarriaga-Hernández et al., 2020). Fungi produce enzymes such as cellulases, which primarily degrade cellulose in the cell wall by creating small openings in the secondary cell wall. These fungi preferentially target cellulose and require fixed nitrogen for enzyme production. In addition, fungal enzymes like laccases and peroxidases break down lignin in lignocellulosic biomass, leading to wood softening and darkening (Andlar et al., 2018; Saldarriaga-Hernández et al., 2020; Saini and Sharma, 2021). Soft-rot fungi, while strong producers of extracellular peroxidases and oxidases, are unable to oxidize guaiacyl lignin but can readily oxidize the syringyl lignin units (Saini and Sharma, 2021). Compared to brown-rot and white-rot fungi, their enzymes are less selective and have a more limited range of applications. These fungi typically belong to the genera Trichoderma, Aspergillus, and Neurospora (Andlar et al., 2018; Saini and Sharma, 2021).

### 4.1.2 Brown-rot fungi

Brown-rot fungi, which belong mainly to the Basidiomycetes and some Ascomycetes species, preferentially degrade cellulose and hemicellulose—two major polysaccharides in lignocellulosic biomass—more rapidly than lignin (Abbas et al., 2017; Saini and Sharma, 2021). Unlike other fungi, brown-rot fungi do not rely on enzymes for breaking down complex carbohydrates; instead, they use low-molecular-weight compounds like oxalic acid in their reduction mechanism and notably lack exoglucanase enzymes (Abbas et al., 2017; Altgen et al., 2020). These fungi employ the chelator-mediated Fenton system, an iron-dependent Fenton chemistry, to fractionate lignocellulosic biomass without enzymatic involvement. In addition, they contribute to lignin demethylation (Abbas et al., 2017; Andlar et al., 2018). Brown-rot fungi are primarily applied in the pretreatment of lignocellulosic biomass and subsequent saccharification processes during biofuel production to prevent the chemical conversion of complex sugars into monomers (Saini and Sharma, 2021). Various species of brown-rot fungi have been used for hydrolyzing lignocellulosic biomass to produce biofuels and other valuable products. *Serpula lacrymans* produced phenol and monosaccharides from five different agricultural wastes: rice straw, corn leaves, sugarcane bagasse, cocoa pods, and corn cobs. According to the findings, corn leaves produced the most total reducing sugars (207.37 mg $g^{-1}$) and rice straw produced the most total soluble phenols (0.140 mg $g^{-1}$) (Nurika et al., 2020). The enzymatic hydrolysis of *Eucalyptus globulus* and *Pinus radiata* was found to be enhanced by the basidiomycetous fungus *Laetiporus sulfurous* by 7% and 11%, respectively (Monrroy et al., 2011). Hemicellulose and lignin were disrupted, respectively, by a successive pretreatment of wheat straw with brown-rot fungus (*Gloeophyllum trabeum*) and white-rot fungi (*Ganoderma lobatum*). Compared to untreated wheat straw, the sequential pretreatment successfully produced a 2.8-fold higher glucose yield (Hermosilla et al., 2018).

### 4.1.3 White-rot fungi

White-rot fungi, belonging to the Basidiomycota phylum—the largest in the fungal kingdom—are capable of degrading the three primary polymers of lignocellulosic biomass: cellulose, hemicellulose, and lignin (Andlar et al., 2018). Their unique capacity to fully mineralize lignin into $CO_2$ allows them to break down lignin more efficiently and rapidly than other fungi, such as soft-rot and brown-rot species, using lignin as their sole carbon and energy source (Andlar et al., 2018; Saini and Sharma, 2021). To effectively degrade lignin, white-rot fungi produce various enzymes, including laccase, manganese peroxidase, lignin peroxidase, and multifunctional peroxidases (Martínez et al., 2017). Delignifying enzymes are categorized as selective or nonselective based on their substrate specificity and mode of action. Nonselective delignifiers degrade all lignocellulosic polymers simultaneously, whereas selective delignifiers mainly hydrolyze lignin while sparing cellulose and hemicellulose (Marinović et al., 2018; Zabed et al., 2019). Selective delignifiers have attracted more research interest due to their ability to separate lignin from intact cellulose (Saini and Sharma, 2021). Studies have explored the potential of white-rot fungi to convert lignocellulosic biomass into valuable products. For instance, pretreating radiata pine wood chips with *Stereum hirsutum* and *Trametes versicolor* enriched the biomass in cellulose and reduced crystallinity to 37% and 44%, respectively, compared to 46% in untreated samples (Shirkavand et al., 2017). Manganese-treated wheat straw from *Pleurotus ostreatus* increased the glucose yield by over 90% (Fu et al., 2021). Six distinct white-rot fungi—*Pleurotus sajor-caju, Tremella versicolor, Coriolopsis gallica, Pleurotu ostreatus, Penicillium chrysosporium,* and *Lentinula edodes*—were used to pretreat maize stover. Following *P. sajor-caju*-aided pretreatment, the greatest lignin degradation, sugar production, and ethanol yield were 38.29%, 71.24%, and 0.124 g/g maize stover biomass, respectively (Ding et al., 2019).

## 4.2 Bacterial lignocellulolytic enzymes

Several studies have reported the biological pretreatment of lignocellulosic biomass using bacterial enzymes to degrade lignin. Compared to fungi, bacteria exhibit faster growth rates and higher metabolic activity, which shortens the duration required for pretreatment. However, fungi generally possess a stronger ability to degrade lignin than bacteria (Zabed et al., 2019). Bacteria produce a group of ligninolytic enzymes believed to break lignin down into smaller aromatic compounds that can be transported into the cell for further aromatic degradation (Chauhan, 2020). Various bacterial taxa, primarily α-proteobacteria, actinomycetes, and γ-proteobacteria, have been identified as contributors to lignin degradation (Saldarriaga-Hernández et al., 2020). Actinomycetes, filamentous bacteria, produce a wide range of enzymes essential for biodegradation. Their unique morphology allows them to penetrate and colonize organic matter effectively, facilitating disruption of the substrate. In addition, they secrete extracellular enzymes capable of hydrolyzing cellulose, hemicellulose, and lignin (Limaye et al., 2017). Anaerobic cellulolytic bacteria produce multienzyme complexes known as cellulosomes, which degrade the polysaccharides in lignocellulosic biomass (Artzi et al., 2017). A comparative study of cellulosomes from *Clostridium clariflavum* DSM 19732 and *Clostridium thermocellum* growing on pretreated rice straw showed that the former was more efficient in substrate breakdown. Further research is needed to understand the synergistic effects of mixed-species cellulosomal complexes. Bacteria often produce intracellular laccases that lack carbohydrate groups and exist in monomeric, multimeric, or homotrimeric forms. These bacterial laccases serve functions such as pigment production and spore protection in natural environments (Plácido and Capareda, 2015). Notable lignin-degrading bacterial species include *actinobacteria, α-proteobacteria Ochrobactrum* sp., *γ-proteobacterium Pseudomonas putida KT2440, Microbacterium phyllosphaerae,* and *Rhodococcus jostii RHA1* (Granja-Travez et al., 2020). In one study, *Bacillus amyloliquefaciens* SL degraded 28.55% of lignin in tobacco straw over 15 days and showed significant production of manganese peroxidase, lignin peroxidase, and laccase (Mei et al., 2020). Another example includes *Bacillus subtilis*, which under microaerobic conditions improved methane production from corn straw. Pretreatment with *B. subtilis* followed by anaerobic digestion of maize straw enhanced methane yields (Xu et al., 2018). In addition, a consortium of twenty bacterial lignin degraders was used to delignify pine lignocellulose-containing soil (1%), leading to increased gas emissions from *P. putida*-treated samples (Rashid et al., 2017).

## 4.3 Lignocellulolytic enzyme production from recombinant microorganisms

Many microbial species have been explored for producing a variety of cellulolytic enzymes that remain stable under harsh conditions. The main challenges now lie in identifying and optimizing the production of recombinant cellulolytic enzymes from these microbes. Advances in cellulase engineering within fermentative microorganisms, such as *Saccharomyces cerevisiae,* have recently increased the economic viability of lignocellulosic biomass, especially through integrated bioprocessing approaches (Davison et al., 2020). For instance, a consolidated bioprocessing method was employed to develop yeast consortia displaying cellulolytic enzymes on their surfaces. This strategy significantly improved the release of reducing sugars from corn stover biomass due to the synergistic action and close proximity of cellulase and xylanase enzymes (Chen et al., 2018). Similarly, genetically engineered *B. subtilis* secreted xylanase that effectively depolymerized xylan in situ, yielding up to 7.1 g/L of xylose, which accounted for 66.7% of the total xylose present in 13.3 g/L of xylan (Mhatre et al., 2022). Thus the effective application of recombinant cellulolytic enzymes from microbes is crucial for the success of lignocellulosic biorefineries.

## 4.4 Techniques used for the production of lignocellulolytic enzymes

Fermentation is the primary technique used for producing these enzymes. Since lignocellulolytic enzymes are secreted extracellularly, submerged fermentation (SmF) is the most commonly employed method. In SmF, microorganisms are cultivated in a nutrient-rich liquid medium that supports their growth. However, the processes of enzyme extraction and concentration in SmF can be costly. To address this, solid-state fermentation (SSF) was developed, where microbes grow on moist solid substrates that mimic their natural environment (Hölker et al., 2004). Despite this, SmF offers advantages over SSF due to the easier handling and monitoring of the liquid culture medium.

### *4.4.1 Solid-state fermentation*

SSF, which involves growing microorganisms on solid substrates to produce metabolites, is rapidly gaining popularity as an alternative method to convert various organic wastes into value-added products. Biologically, SSF entails cultivating microbes on solid materials such as lignocellulosic biomass (LCB) or other biomasses with minimal or no free water (Sánchez et al., 2024). Compared to SmF, SSF has a lower risk of contamination due to its reduced water activity. Its many advantages—such as utilizing agricultural and agroindustrial residues as substrates for microbial growth and lowering the energy consumption associated with processing cellulosic waste—have made SSF a preferred technique for enzyme production (Patel et al., 2023). The efficiency of SSF and product recovery are influenced by the choice of natural substrate (see Table 8.2), as well as factors like the substrate's chemical composition, porosity, and particle size. In addition, the availability of substrate, the balance of macro- and micronutrients, and related costs play crucial roles in SSF (Rocha et al., 2024). Solid substrates act as a support matrix in SSF and may be either natural or synthetic. Because microorganisms metabolize organic compounds such as cellulose, starch, and lignin for carbon, these materials are commonly used as natural substrates. For example, starch-based substrates are employed for producing amylase and other enzymes, whereas cellulase enzymes are produced industrially using cellulosic or lignocellulosic substrates (Sánchez et al., 2024). SSF is particularly advantageous for enzyme synthesis as it mimics the natural conditions that favor robust growth and reproduction of filamentous fungi (Huervana et al., 2024).

**Table 8.2** List of enzymes produced from the growth of microorganisms on agroresidue as a substrate, as mentioned by several researchers.

| Enzyme(s) | Microorganism(s) | Substrate |
|---|---|---|
| Laccase | *Hexagonia hirta* MSF2 | Coir pith |
| Xylanase | *Streptomyces thermocarboxydus* | Corncob and cornstover |
| Xylanase | *Rasamsonia composticola* (XylRc) | Wheat bran |
| Xylanase | *Fusarium oxysporum* | Brewers' spent grain |
| Pectinase | *Aspergillus fumigatus* BT – 4 | Grapefruit peel |
| Cellulase | *Aspergillus niger* | Defatted rice bran |
| Cellulase, xylanase, amylase, pectinase, protease | *Bacillus subtilis* BC–00 | Rice straw |
| Inulinase | *Aspergillus niger* A42 | Wheat bran, barley husk, and oat husk |
| Amylases | *Pleurotus pulmonarius* CCB19 | Cocoa and peach palm wastes |

### *4.4.2 Submerged fermentation*

SmF is a widely utilized method for producing secondary metabolites and industrial enzymes by cultivating microorganisms in liquid nutrient media. In this approach, microbes such as bacteria, yeasts, and fungi are grown in closed vessels filled with nutrient-rich broth, supported by a gas supply system that delivers the oxygen necessary for aerobic microbial activity (Nazir et al., 2024). During aerobic SmF, enzymes are secreted into the medium after nutrient consumption. Several studies have demonstrated the aerobic degradation of lignin by microorganisms, including *Aspergillus*, *T. versicolor*, *Bacillus* species, and *P. ostreatus*, all of which produce lignocellulolytic enzymes under oxygenated conditions (Dukare et al., 2025; Nargotra et al., 2025; Zhi et al., 2025). In contrast, anaerobic fermentation occurs in the absence of oxygen, with gases like carbon dioxide and nitrogen creating a reducing environment. While less extensively studied, anaerobic lignin degradation shows promise. For instance, *Klebsiella aerogenes* TL3, isolated from a termite gut, was shown to degrade alkali lignin anaerobically through mechanisms such as side-chain oxidation, demethylation, and aromatic ring cleavage. The degradation rate increased from 14.8% to 17.4% over 72 h with the addition of glucose, which served as an electron donor, enhancing bond cleavage and overall lignin breakdown (Rocha et al., 2024). In addition, three facultative anaerobic lignin-degrading bacteria—*Klebsiella* sp. LEA1, *Pseudomonas* sp. LEA2, and *Burkholderia* sp. LEA3—were identified in Thailand's deciduous forest soils. These strains grew on alkali lignin and produced monomers like succinic acid, benzoic acid, 4-hydroxybenzoic acid, 4-vinylguaiacol, 2,6-dimethoxyphenol, and catechol (Sumranwanich et al., 2024). Although fungi predominantly degrade lignin under aerobic conditions, their effectiveness in anaerobic environments remains uncertain. However, some fungi, like Neocallimastigomycetes, have demonstrated anaerobic lignin bond cleavage in plant biomass such as grass and hardwood (Lankiewicz et al., 2023).

Advancements in large-scale fermentation have significantly improved microbial enzyme production, an essential component of the biotechnology industry. Industrial fermentation is conducted in large fermenters—often up to 1000 $m^3$—using renewable substrates such as maize, sugar, or soy (Taiwo et al., 2022). Microorganisms secrete enzymes into the culture broth as they metabolize these carbon and nitrogen sources. Standard fermentation methods include fed-batch and continuous modes. In fed-batch processes, sterilized nutrients are added incrementally as microbial biomass accumulates. Continuous fermentation maintains a steady state by continuously supplying fresh nutrients and withdrawing an equal volume of culture (Chai et al., 2024; Cheute et al., 2024). To optimize production, parameters such as temperature, pH, oxygen levels, and $CO_2$ generation are carefully monitored and adjusted. After fermentation, enzymes are recovered by removing insoluble materials like cells through centrifugation (Gan et al., 2024). Large-scale SmF systems are critical in various industries, including chemicals, food and feed, paper and pulp, detergents, textiles, and biofuel production (Malik and Javed, 2024).

## 4.5 Biomass conversion optimization for enhanced enzyme production

The production of enzymes by bacteria is influenced by various nutritional and physical factors, such as substrate type, medium composition (including salts), pH, temperature, light exposure, and agitation or shaking during cultivation (Zhang et al., 2019). Due to the high cost of commercial enzymes and increasing industrial demand, the traditional One-Factor-at-a-Time (OFAT) approach has been commonly used to optimize these parameters. However, OFAT is time-consuming and only evaluates the individual effects of each factor without accounting for interactions among them (Kumar et al., 2018; Deswal et al., 2011). To overcome these limitations, response surface methodology (RSM), a statistical optimization technique, was introduced. RSM enables the simultaneous evaluation of multiple variables and their interactions, making it more efficient and cost-effective compared to OFAT (Abdullah et al., 2018; Neelkant et al., 2019). Consequently, RSM is now widely preferred for enzyme production optimization.

### 4.5.1 Strain improvement for enhanced enzyme production

Various strategies have been employed to improve microbial strains for enhanced enzyme production, including genome shuffling, gene cloning, mutagenesis, and selective screening (Peterson and Nevalainen, 2012; Singhania et al., 2017). Strain improvement has become a fundamental approach in boosting enzyme yields and has advanced considerably over time. Researchers have cloned genes responsible for enzyme synthesis into different host systems, with *Escherichia coli* being particularly preferred due to its efficiency compared to other hosts such as yeast or plants.

For example, a mutant of *Trichoderma viride* QM6a was developed from over 100 wild-type strains to increase cellulase production. In addition, a recombinant strain of *S. cerevisiae* (HZ3001) was engineered to metabolize xylose by integrating genes for the D-xylose metabolic pathway (Sun et al., 2012). In another instance, a laccase mutant fused with a dockerin domain was engineered from the aerobic bacterium *Thermobifida fusca*, enhancing its activity (Davidi et al., 2016). These advancements illustrate how strain improvement remains one of the most widely applied and evolving techniques for maximizing enzyme productivity.

## 5 Recent advances in enzyme technology

Microbial enzymes often encounter limitations in industrial settings, such as denaturation in fermenters, low yield, and diminished activity under normal environmental conditions. To overcome these hurdles, enzyme engineering has emerged as a valuable approach. This field involves modifying natural enzymes or designing novel ones with superior characteristics by applying genetic engineering techniques. A deep understanding of enzyme structure and function enables methods like gene amplification, transferring genes into faster-growing hosts, and heterologous expression to significantly increase enzyme output. For example, genes from slow-growing filamentous organisms can be successfully expressed in more efficient hosts like *E. coli* or *Bacillus licheniformis* to enhance productivity (Singh et al., 2022). Enzyme properties such as stability, catalytic efficiency, specificity, and performance under extreme conditions can be improved using advanced engineering methods like directed evolution, site-directed mutagenesis, terminal fusion, and truncation. These strategies allow for the modification of existing enzymes and the development of robust industrial biocatalysts, ultimately reducing the cost of producing value-added products (Bozkurt et al., 2025). Recent advancements in genomic, transcriptomic, and proteomic analyses of laccase-producing microbes have provided insights into optimizing this critical enzyme for lignocellulosic biomass (LCB) treatment (Singh et al., 2023). Enzyme engineering, particularly through rational design and evolutionary techniques, is essential to meet the demands of bioprocessing and to drive the bioeconomy forward (Singh et al., 2025). In a notable study involving *Trichoderma harzianum* EM0925, researchers identified key amino acids (Asp389, Glu589, Gln185, Cys390, Tyr354, and Tyr526) responsible for the multifunctionality of the BX1 enzyme using proteomics, molecular docking, and mutagenesis, achieving a 90.23% increase in fermentable sugar yield at a 2% (w/v) solid substrate concentration (Liu et al., 2025). Covalent immobilization, which involves linking enzymes to carriers using cross-linking agents, has proven effective for improving enzyme recovery and reuse (Holyavka et al., 2025). For example, laccase immobilized on magnetic nanoparticles coated with nanocellulose retained 68.9% of its activity after 60 days, demonstrating improved stability and efficiency. This approach holds promise for applications like biomass delignification and the breakdown of phenolic compounds (Behram et al., 2024). Due to the high cost of enzyme use in various sectors, Singh and colleagues have proposed several cost-reduction strategies, including on-site production, improved enzyme immobilization, enhanced stability, and recyclability. They also emphasize the potential of AI and machine learning in optimizing enzyme production (Khan et al., 2024).

## 6 Case studies and industrial applications of lignocellulolytic enzymes

Lignocellulolytic enzymes have a wide range of applications across diverse industries, including bioremediation, textiles, pulp and paper, food and beverage processing, and biofuel production (Fig. 8.2). In the food sector, enzymes like cellulases, xylanases, and laccases are employed to increase product yield, improve texture stability, enhance flavor profiles, and support juice extraction processes. In baking, these enzymes help soften dough and improve the overall quality of baked goods. When used in animal feed, they promote better fiber breakdown and enhance the absorption of essential minerals. Within the beverage industry, they facilitate malt extraction for beer production, increase fermentation efficiency, and contribute to extending the shelf life of products such as wine and beer.

### 6.1 Biofuels

Lignocellulosic biomass serves as a sustainable, carbon-neutral alternative to fossil fuels for use in the transportation and energy industries. Biofuels derived from this biomass can be generated in both liquid forms—such as bioethanol, biodiesel, biobutanol, biooil, and aviation fuels—and gaseous forms, including biogas, syngas, biomethane, and biohydrogen. These fuels are produced through various conversion processes, which are broadly categorized into biochemical and thermochemical methods. Biochemical approaches involve techniques like enzymatic hydrolysis, fermentation, and anaerobic digestion, whereas thermochemical processes include

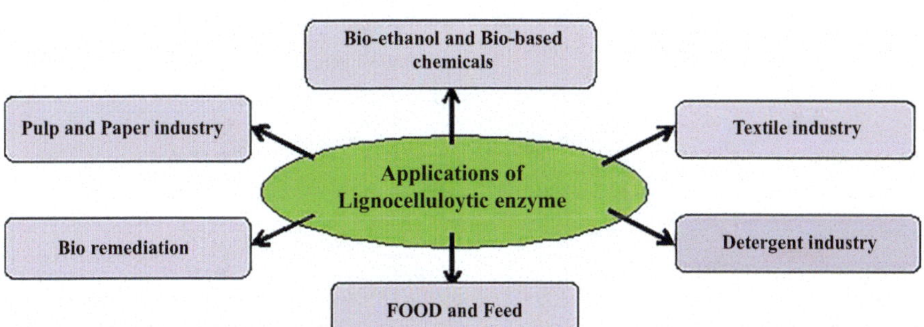


**FIG. 8.2**

Application of lignocellulolytic enzymes in various fields fitting into the concept of "reduce–reuse–recycle" and contributing toward "circular economy."

pyrolysis, gasification, liquefaction, torrefaction, Fischer–Tropsch synthesis, and transesterification (Nanda et al., 2014; Parakh et al., 2020). The thermochemical and biological conversion pathways, as well as a number of bioproducts with commercial significance, are depicted in Fig. 8.3.

Pyrolysis and liquefaction are the two primary thermochemical techniques used to convert biomass into liquid fuels, particularly for producing biooils. The efficiency and quality of the resulting biooil are heavily influenced by several operational parameters, including temperature, heating rate, pressure, residence time, catalyst presence, feedstock concentration, reactor design, and the specific characteristics of the biomass. Among these methods, fast pyrolysis is favored for its higher biooil yields due to rapid heating, shorter vapor residence times, and faster quenching of hydrocarbon vapors, unlike slow pyrolysis, which tends to produce more char. Compared to pyrolysis, liquefaction typically results in oils with lower oxygen content, thereby offering improved fuel properties and higher energy density. Biooils derived from lignocellulosic biomass contain a wide variety of oxygenated compounds depending on the degradation of cellulose, hemicellulose, and lignin. These include alcohols, phenols, carbonyls, vanillin, aldehydes, ketones, acids, ethers, catechols, and pyranones (Mohan et al., 2006). Specifically, the decomposition of cellulose and hemicellulose generates alcohols, aldehydes, esters, sugars, ketones, and furans (Joshi and Lawal, 2012).

## 6.2 Textile industry

Because of its selectivity, enzymes have become crucial biochemicals in the textile industry. Because they are very selective and harmless, enzymes have taken the role of labor-intensive traditional chemical processes in the textile industry. The cellulose

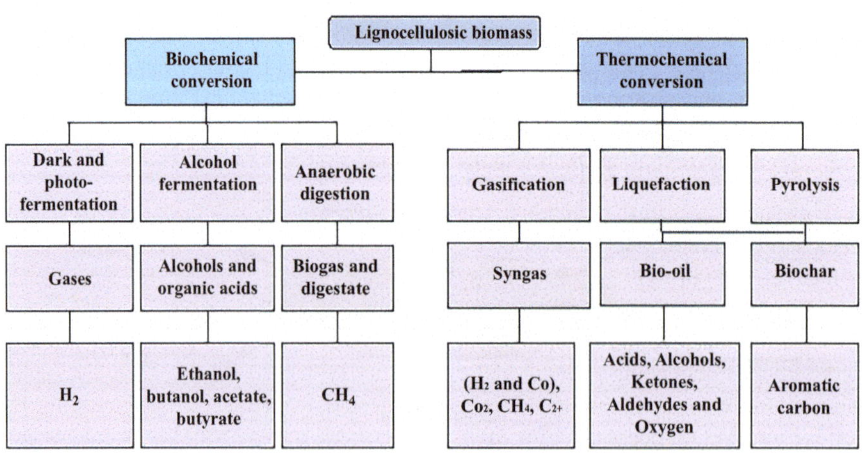


**FIG. 8.3**

Thermochemical and biological conversion of lignocellulosic biomass to different industrially relevant bioproducts.

thread's hairiness can be efficiently removed by cellulases and xylanases with minimal weight loss. They can also add brightness and give textiles a glossy, smooth texture. Acidic cellulases that were extracted from *Trichoderma reesei* (Kuhad et al., 2011), neutral cellulases from *Humicol insolens* (Anish et al., 2007), and xylanases from *Bacillus pumilus* (Battan et al., 2012; Singh et al., 2018) are widely used. In addition, laccase from *P. ostreatus* and *T. versicolor* (Mojsov, 2014; Anish et al., 2007) is extensively employed in the sectors that process textiles.

## 6.3 Pulp and paper industry

Due to the heavy reliance on chemicals in the pulp and paper industry, growing environmental concerns have prompted a shift toward more sustainable, biobased alternatives. Enzymes have emerged as a valuable and promising solution within this sector. Cellulases, either on their own or combined with xylanases, are widely used for deinking various types of recycled paper (Kumar et al., 2018). The action of cellulases and hemicellulases partially breaks down carbohydrate chains, facilitating the release of ink particles from fiber surfaces (Juturu and Wu, 2014; Kumar et al., 2018; Kuhad et al., 2011). In addition, laccase plays a key role in delignification and pulp brightening, while also removing pitch-causing lipophilic substances in both wood and nonwood pulps. This enzyme can improve the mechanical, chemical, and physical properties of pulp by modifying lignocellulosic fibers or generating reactive lignin radicals. Moreover, laccase contributes to the detoxification of harmful and colored compounds in industrial effluents by promoting their transformation through polymerization or depolymerization reactions (Virk et al., 2012).

## 6.4 Food and feed

Materials once considered waste are now being efficiently repurposed to produce valuable products such as chemicals, enzymes, fermentable sugars, biofuels, cost-effective energy sources for fermentation, enhanced animal feed, human nutritional supplements, and intermediate compounds for synthesizing high-value chemicals (Kuhad et al., 2011). In the food industry, enzymes serve diverse functions; for example, amylases and xylanases help maintain the softness and elasticity of bread. They are utilized in juice clarification by cellulase and laccase, which help remove polyphenols during filtration. These enzymes also enhance beer flavor. A deeper understanding of these enzymes could expand their applications in food. Enzyme combinations like pectinases, cellulases, and hemicellulases aid in more efficient olive oil extraction. Furthermore, enzymes such as xylanases and glucanases serve as feed additives in single-stomach animal diets (Kirk et al., 2002). Cellulases or xylanases have been used as a pretreatment to improve the nutritional value of grain feed and agricultural silage, remove antinutritional elements, and break down specific feed ingredients to increase the feed's total nutritional value for efficient utilization (Choct, 2006; Juturu and Wu, 2014; Vasconcellos et al., 2015).

### 6.5 Bioplastics and biocomposites

Bioplastics offer an eco-friendly alternative to conventional plastics in sectors such as packaging, electronics, healthcare, textiles, and everyday applications (Thakur et al., 2018). Derived from lignocellulosic biomass, bioplastics are biodegradable and environmentally safer than petroleum-based polymers, which can release toxic compounds like dioxins and polycyclic aromatic hydrocarbons (Mekonnen et al., 2013; Thakur et al., 2018). One kilogram of traditional plastic emits around 2.8 kg of $CO_2$ (Burgos et al., 2016), and of the 34 million tons of plastic waste produced annually, approximately 93% ends up in landfills or the ocean. Since many petrochemical plastics do not biodegrade, they persist in the environment for centuries, causing long-term ecological damage. This has sparked growing interest in bioplastics as sustainable alternatives (Mekonnen et al., 2013). Bioplastics are both biobased and biodegradable, breaking down naturally or under specific conditions. They can be sourced from microorganisms like polyhydroxyalkanoates, plant-based cellulose, pectin, proteins like wool and gelatin, or lipids from plant oils and animal fats (Burgos et al., 2016). In addition, biobased chemicals can be used to synthesize polymers such as polybutylene succinate and polylactic acid (PLA) (Song et al., 2009). Other biodegradable options include polycaprolactone, poly(butylene succinate-co-terephthalate), and polyglycolic acid, which vary in their degradability (Bátori et al., 2018). PLA, for example, is used in medical fields for sutures and implants (Pawar et al., 2014). "Drop-in" bioplastics like biopropylene, biopolyethylene terephthalate, and biopolyethylene replicate petrochemical plastics in structure and production methods but use renewable feedstocks. However, their higher production costs—due to technical limitations and scale—remain a challenge. Fig. 8.4 illustrates various approaches to producing bioplastics from lignocellulosic biomass.

Biocomposites made from lignocellulosic materials are increasingly attractive due to their sustainability, durability, and biodegradability (Trache et al., 2016). These composites integrate natural fibers with biopolymers or synthetic matrices, enhancing mechanical strength and environmental resistance. Materials such as hemp or wood fibers are used alongside bioplastics like PLA. Despite their advantages, recycling or separating these components at the end of their life remains technologically challenging (Ramamoorthy et al., 2015).

## 7 Challenges and limitations of enzymatic degradation

Enzymatic degradation offers great potential for the efficient management and valorization of biomass waste, but it faces several challenges and limitations that need to be overcome. To develop effective strategies and technologies, it is crucial to understand these obstacles. The primary challenges associated with enzymatic degradation include: Lignocellulosic biomass waste possesses a complex and resilient structure that hinders enzymatic degradation of cellulose and hemicellulose because of the presence of lignin. Lignin not only obstructs the breakdown process but also serves as a physical barrier. To overcome this issue, different pretreatment

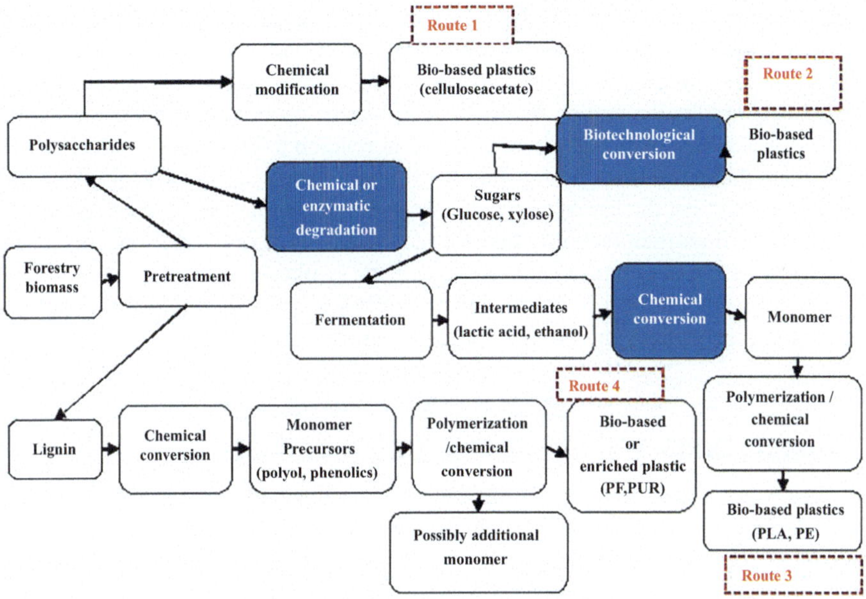


**FIG. 8.4**

Production routes of bioplastics from lignocellulosic biomass. PE, polyethylene; PF, phenol–formaldehyde resin; PLA, polylactic acid; PUR, polyurethane.

approaches—chemical, physical, or biological—are used to break down lignin and improve enzyme accessibility to the target polymers (Mishra et al., 2018). The composition and characteristics of biomass waste can vary widely depending on its source, which complicates enzymatic breakdown. This variability requires enzymes to be tailored to specific biomass types, potentially necessitating the use of enzyme mixtures or specialized enzymes. Developing effective enzyme systems that can process diverse biomass feedstocks remains a challenge (Sweeney and Xu, 2012). The high cost of producing enzymes, especially those with specific functions, poses challenges for large-scale enzymatic degradation processes. To enhance cost-effectiveness, it is essential to optimize the manufacturing and purification of enzymes. Techniques such as enzyme immobilization and recycling can reduce enzyme usage and overall production costs (Sakhuja et al., 2021). Extreme pH and high temperatures can lead to stability issues for enzymes used in biomass degradation, potentially causing denaturation or reduced activity over time. Inhibitors, including chemicals from lignin or biomass decomposition by-products, can further hinder enzyme activity. Ongoing research aims to enhance enzyme stability and develop enzymes that are more resistant to these inhibitors (Robinson, 2015). Integrating enzymatic degradation into existing biomass waste processing involves overcoming challenges related to compatibility with other process steps like pretreatment and fermentation. In addition, for large-scale enzymatic degradation, there is a need for affordable and scalable reactor systems, as well as optimized processes (Balan, 2014). To implement enzymatic degradation procedures effectively, it is crucial to evaluate their techno-economic viability. This involves considering costs associated with enzyme production, the availability and pricing of substrates, downstream processing requirements, and market demand for the end products. In addition, the economic competitiveness of biomass waste degradation must be assessed in relation to other waste management and value-adding methods (Robinson, 2015). Public acceptance and regulatory obstacles may prevent the use of enzymatic degrading technologies for biomass waste management. Important factors to take into account include resolving concerns about genetically modified organisms utilized in enzyme synthesis, guaranteeing the safety and environmental impact of the enzymatic processes, and adhering to laws controlling waste management and product quality (Wesseler et al., 2023).

## 8 Conclusions and future research prospects

Enzyme-based methods offer an efficient and sustainable approach for managing and valorizing biomass waste. This chapter has explored the diverse and promising field of enzyme-mediated strategies for the effective treatment and value addition of biomass residues. By examining various factors, we have gained valuable insights into how enzymes break down biomass waste and their potential role in promoting sustainable resource utilization. Biomass waste presents serious environmental challenges, such as waste accumulation, depletion of resources, and greenhouse gas emissions. However, through proper management and valorization techniques, biomass waste can be transformed into valuable products, helping to minimize environmental impact and support a circular bioeconomy. Enzymes play a crucial role in biomass processing due to their unique abilities to degrade, pretreat, hydrolyze, and convert biomass components. This chapter has detailed the types of biomass waste suitable for enzymatic degradation, key enzymes involved, and factors that influence enzyme activity. In addition, recent advances in enzyme engineering have been discussed, highlighting their potential to enhance degradation efficiency and enzyme utility. Hydrolysis and enzymatic pretreatment are vital steps in biomass valorization, as enzymes break down complex biomass structures into more soluble and usable forms. Various enzymatic pretreatment methods, types of hydrolytic enzymes, and factors affecting enzymatic hydrolysis efficiency have been thoroughly covered. The chapter also emphasizes enzyme-driven conversion of biomass waste into high-value products, including biofuels, specialty chemicals, bioplastics, and nutraceuticals, showcasing enzymes' versatility and specificity in targeting biomass components.

Future research should focus on discovering and utilizing new enzymes from diverse sources such as genetically engineered microbes, metagenomic libraries, and extremophilic organisms. By identifying enzymes with unique properties and high substrate specificity, it can broaden the range of enzymatic tools available for biomass waste valorization. In addition, enzyme engineering methods like directed evolution and protein design can be employed to enhance enzyme efficiency and tailor their functions to specific types of waste. Research should prioritize improving enzyme yield, stability, and adaptability to different substrates and operational conditions. This includes exploring alternative host organisms for enzyme production, developing innovative fermentation techniques, and optimizing enzyme immobilization strategies. Moreover, integrating on-site enzyme production within biorefinery systems could reduce expenses and enhance overall process efficiency. Biorefineries play a crucial role in converting biomass waste into value-added products. Therefore research should focus on exploring advanced biorefinery models, such as integrated systems, sequential utilization of biomass components, and hybrid processing approaches. Enzyme-based methods can be optimized and combined with other conversion technologies like electrochemical techniques, thermochemical processes, and microbial fermentation. These multidisciplinary strategies will enhance the production of fuels, chemicals, materials, and energy while enabling efficient utilization of diverse biomass feedstocks. Furthermore, research should incorporate comprehensive life-cycle assessments (LCA) and sustainability evaluations to ensure enzyme-based methods are truly sustainable. This involves examining the environmental, economic, and social impacts of enzyme-driven processes throughout their entire life cycle. LCA can guide the optimization of these processes by identifying critical impact areas and supporting informed decision-making to minimize environmental harm and promote sustainable waste management practices.

## References


Abbas, R., Shirvanimoghaddam, M., Li, Y., Vucetic, B. 2017. On the performance of massive grant-free NOMA. IEEE International Symposium on Personal, Indoor and Mobile Radio Communications, PIMRC. Institute of Electrical and Electronics Engineers Inc., Australia, 9781538635315, pp. 1–6. doi:10.1109/PIMRC.2017.8292475.

Abdullah, H., Christwardana, M., Hadiyanto, H., 2018. Optimization of cellulase production by *Aspergillus niger* ITBCC L74 with bagasse as substrate using response surface methodology. HAYATI J. Biosci. 20864094. 25 (3), 115–125. Available from: http://journal.ipb.ac.id/index.php/hayati/article/download/23824/15500. doi:10.4308/hjb.25.3.115.

Akram, F., Haq, I.U., Imran, W., Mukhtar, H., 2018. Insight perspectives of thermostable endoglucanases for bioethanol production: a review. Renew. Energy 18790682. 122, 225–238. Available from: http://www.journals.elsevier.com/renewable-and-sustainable-energy-reviews/. doi:10.1016/j.renene.2018.01.095.

Alam, S.N., Khalid, Z., Guldhe, A., Singh, B., Korstad, J., 2021. Harvesting and pretreatment techniques of aquatic macrophytes and macroalgae for production of biofuels. Environ. Sustain. 2523-8922. 4 (2), 299–316. doi:10.1007/s42398-021-00178-6.

Albornoz-Palma, G., Ching, D., Valerio, O., Mendonça, R.T., Pereira, M., 2020. Effect of lignin and hemicellulose on the properties of lignocellulose nanofibril suspensions. Cellulose 1572882X. 27 (18), 10631–10647. Available from: https://www.springer.com/journal/10570. doi:10.1007/s10570-020-03304-5.

Alokika, Singh, B., 2019. Production, characteristics, and biotechnological applications of microbial xylanases. Appl. Microbiol. Biotechnol. 0175-7598. 103 (21-22), 8763–8784. doi:10.1007/s00253-019-10108-6.

Altgen, M., Kyyrö, S., Paajanen, O., Rautkari, L., 2020. Resistance of thermally modified and pressurized hot water extracted Scots pine sapwood against decay by the brown-rot fungus *Rhodonia placenta*. Eur. J. Wood Wood Prod. 0018-3768. 78 (1), 161–171. doi:10.1007/s00107-019-01482-z.

Andlar, M., Rezić, T., Marđetko, N., Kracher, D., Ludwig, R., Šantek, B., 2018. Lignocellulose degradation: an overview of fungi and fungal enzymes involved in lignocellulose degradation. Eng. Life Sci. 1618-0240. 18 (11), 768–778. doi:10.1002/elsc.201800039.

Anish, R., Rahman, M.S., Rao, M., 2007. Application of cellulases from an alkalothermophilic *Thermomonospora* sp. in biopolishing of denims. Biotechnol. Bioeng. 10970290. 96 (1), 48–56. doi:10.1002/bit.21175.

Annamalai, N., Rajeswari, M.V., Balasubramanian, T., 2016. Thermostable and Alkaline Cellulases from Marine Sources New and Future Developments in Microbial Biotechnology and Bioengineering: Microbial Cellulase System Properties and Applications. Elsevier Inc., United States, pp. 91–98. Available from: http://www.sciencedirect.com/science/book/9780444635075. doi:10.1016/B978-0-444-63507-5.00009-5.

Arefin, M.A., Rashid, F., Islam, A., 2021. A review of biofuel production from floating aquatic plants: an emerging source of bio-renewable energy. Biofuels Bioprod. Biorefin. 19321031. 15 (2), 574–591. Available from: http://onlinelibrary.wiley.com/journal/10.1002/(ISSN)1932-1031. doi:10.1002/bbb.2180.

Arfi, Y., Shamshoum, M., Rogachev, I., Peleg, Y., Bayer, E.A., 2014. Integration of bacterial lytic polysaccharide monooxygenases into designer cellulosomes promotes enhanced cellulose degradation. Proc. Natl. Acad. Sci. U. S. A. 0027-8424. 111 (25), 9109–9114. doi:10.1073/pnas.1404148111.

Artzi, L., Bayer, E.A., Moraïs, S., 2017. Cellulosomes: bacterial nanomachines for dismantling plant polysaccharides. Nat. Rev. Microbiol. 17401534. 15 (2), 83–95. Available from: http://www.nature.com/nrmicro/index.html. doi:10.1038/nrmicro.2016.164.

Ashokkumar, V., Venkatkarthick, R., Jayashree, S., Huetor, S., Dharmaraj, S., Kumar, G., W., Chen, C., Ngamcharussrivichai., 2022. Recent advances in lignocellulosic biomass for biofuels and value-added bioproducts - A critical review, Bioresource Technology 344, 126195. doi:10.1016/j.biortech.2021.126195.

Balan, V., 2014. Current challenges in commercially producing biofuels from lignocellulosic biomass. ISRN Biotechnol. 2090-9403. 2014, 1–31. doi:10.1155/2014/463074.

Bátori, V., Åkesson, D., Zamani, A., Taherzadeh, M.J., Sárvári Horváth, I., 2018. Anaerobic degradation of bioplastics: a review. Waste Manag. 18792456. 80, 406–413. Available from: https://www.elsevier.com/locate/wasman. doi:10.1016/j.wasman.2018.09.040.

Battan, B., Dhiman, S.S., Ahlawat, S., Mahajan, R., Sharma, J., 2012. Application of thermostable xylanase of *Bacillus pumilus* in textile processing. Indian J. Microbiol. 00468991. 52 (2), 222–229. doi:10.1007/s12088-011-0118-1.

Behram, T., Pervez, S., Nawaz, M.A., Ullah, R., Khan, A.A., Ahmad, B., Alanzai, A.M., Ahmad, A., Jan, A.K., Rahman, H.U., Jamal, M., Tour Jan, Mohyuddin, A., Khan, N.M., Ahmad, S., 2024. Synthesis and analysis of silica nanocarriers for pectinase immobilization: enhancing enzymatic stability for continuous industrial applications. Heliyon 24058440. 10 (1). Available from: https://www.sciencedirect.com/science/journal/24058440. doi:10.1016/j.heliyon.2023.e23549.

Binod, P., Gnansounou, E., Sindhu, R., Pandey, A., 2019. Enzymes for second generation biofuels: recent developments and future perspectives. Bioresour. Technol. Rep. 2589014X. 5, 317–325. doi:10.1016/j.biteb.2018.06.005.

Bissaro, B., Røhr, Å.K., Müller, G., Chylenski, P., Skaugen, M., Forsberg, Z., Horn, S.J., Vaaje-Kolstad, G., Eijsink, V.G.H., 2017. Oxidative cleavage of polysaccharides by monocopper enzymes depends on $H_2O_2$. Nat. Chem. Biol. 15524469. 13 (10), 1123–1128. Available from: http://www.nature.com/nchembio. doi:10.1038/nchembio.2470.

Bozkurt, E.U., Ørsted, E.C., Volke, D.C., Nikel, P.I., 2025. Accelerating enzyme discovery and engineering with high-throughput screening. Nat. Prod. Rep. 0265–0568. doi:10.1039/d4np00031e.

Bugg, T.D.H., Williamson, J.J., Rashid, G.M.M., 2020. Bacterial enzymes for lignin depolymerisation: new biocatalysts for generation of renewable chemicals from biomass. Curr. Opin. Chem. Biol. 18790402. 55, 26–33. Available from: http://www.elsevier.com/locate/cbi. doi:10.1016/j.cbpa.2019.11.007.

Burgos, N., Valdés, A., Jiménez, A., 2016. Valorization of agricultural wastes for the production of protein-based biopolymers. J. Renew. Mater. 2164-6325. 4 (3), 165–177. doi:10.7569/jrm.2016.634108.

Carli, S., Meleiro, L.P., Rosa, J.C., Moraes, L.A.B., Jorge, J.A., Masui, D.C., Furriel, R.P.M., 2016. A novel thermostable and halotolerant xylanase from *Colletotrichum graminicola*. J. Mol. Catal. B Enzym. 18733158. 133, S508. Available from: https://www.elsevier.com/locate/molcatb. doi:10.1016/j.molcatb.2017.05.002.

Chadha, B.S., Kaur, B., Basotra, N., Tsang, A., Pandey, A., 2019. Thermostable xylanases from thermophilic fungi and bacteria: current perspective. Bioresour. Technol. 09608524. 277, 195–203. doi:10.1016/j.biortech.2019.01.044.

Chai, W.Y., Tan, M.K., Teo, K.T.K., Tham, H.J., 2024. Optimization of fed-batch Baker's yeast fermentation using deep reinforcement learning. Process Integr. Optim. Sustain. 25094246. 8 (2), 395–411. Available from: https://www.springer.com/journal/41660. doi:10.1007/s41660-024-00406-6.

Chauhan, P.S., 2020. Role of various bacterial enzymes in complete depolymerization of lignin: a review. Biocatal. Agric. Biotechnol. 18788181. 23, 101498. doi:10.1016/j.bcab.2020.101498.

Chen, L., Du, J.L., Zhan, Y.J., Li, J.A., Zuo, R.R., Tian, S., 2018. Consolidated bioprocessing for cellulosic ethanol conversion by cellulase–xylanase cell-surfaced yeast consortium. Prep. Biochem. Biotechnol. 15322297. 48 (7), 653–661. Available from: https://www.tandf.co.uk/journals/titles/10826068.asp. doi:10.1080/10826068.2018.1487846.

Cheute, V.M.S., Uber, T.M., Dos Santos, L.F.O., Backes, E., Dantas, M.P., Contato, A.G., Castoldi, R., de Souza, C.G.M., Corrêa, R.C.G., Bracht, A., Peralta, R.M., 2024. Biotransformation of pollutants by *Pycnoporus* spp. in submerged and solid-state fermentation: mechanisms, achievements, and perspectives. Biomass 26738783. 4 (2), 313–328. Available from: https://www.mdpi.com/journal/biomass. doi:10.3390/biomass4020015.

Chio, C., Sain, M., Qin, W., 2019. Lignin utilization: a review of lignin depolymerization from various aspects. Renew. Sustain. Energy Rev. 13640321. 107, 232–249. doi:10.1016/j.rser.2019.03.008.

Choct, M., 2006. Enzymes for the feed industry: past, present and future. World's Poult. Sci. J. 0043-9339. 62 (1), 5–16. doi:10.1079/wps200480.

Contesini, F.J., Liberato, M.V., Rubio, M.V., Calzado, F., Zubieta, M.P., Riaño-Pachón, D.M., Squina, F.M., Bracht, F., Skaf, M.S., Damasio, A.R., 2017. Structural and functional characterization of a highly secreted α-L-arabinofuranosidase (GH62) from *Aspergillus nidulans* grown on sugarcane bagasse. Biochim. Biophys. Acta Proteins Proteomics

18781454. 1865 (12), 1758–1769. Available from: http://www.elsevier.com/locate/bbapap. doi:10.1016/j.bbapap.2017.09.001.

Datta, R., Kelkar, A., Baraniya, D., Molaei, A., Moulick, A., Meena, R., Formanek, P., 2017. Enzymatic degradation of lignin in soil: a review. Sustainability. 2071-1050. 9 (7), 1163. doi:10.3390/su9071163.

Davidi, L., Moraïs, S., Artzi, L., Knop, D., Hadar, Y., Arfi, Y., Bayer, E.A., 2016. Toward combined delignification and saccharification of wheat straw by a laccase-containing designer cellulosome. Proc. Natl. Acad. Sci. U. S. A. 0027-8424. 113 (39), 10854–10859. doi:10.1073/pnas.1608012113.

Davison, S.A., den Haan, R., van Zyl, W.H., 2020. Exploiting strain diversity and rational engineering strategies to enhance recombinant cellulase secretion by *Saccharomyces cerevisiae*. Appl. Microbiol. Biotechnol. 0175-7598. 104 (12), 5163–5184. doi:10.1007/s00253-020-10602-2.

de Gonzalo, G., Colpa, D.I., Habib, M.H.M., Fraaije, M.W., 2016. Bacterial enzymes involved in lignin degradation. J. Biotechnol. 18734863. 236, 110–119. Available from: https://www.elsevier.com/locate/jbiotec. doi:10.1016/j.jbiotec.2016.08.011.

Demirbaş, A., 2001. Biomass resource facilities and biomass conversion processing for fuels and chemicals. Energy Convers. Manag. 01968904. 42 (11), 1357–1378. doi:10.1016/s0196-8904(00)00137-0.

Deswal, D., Khasa, Y.P., Kuhad, R.C., 2011. Optimization of cellulase production by a brown rot fungus *Fomitopsis* sp. RCK2010 under solid state fermentation. Bioresour. Technol. 09608524. 102 (10), 6065–6072. doi:10.1016/j.biortech.2011.03.032.

Ding, C., Wang, X., Li, M., 2019. Evaluation of six white-rot fungal pretreatments on corn stover for the production of cellulolytic and ligninolytic enzymes, reducing sugars, and ethanol. Applied Microbiology and Biotechnology 103 (14), 5641–5652.

Dukare, A., Sharma, K., Nehete, L., Dsouza, C., Nadanathangam, V., Saxena, S., 2025. Optimization of xylanase production by *Bacillus* sp. in submerged fermentation using cotton biomass as cost-effective feedstock and its potential application for bio-bleaching of cotton stalk pulp. Biomass Convers. Biorefin. 2190–6815. doi:10.1007/s13399-025-06522-8.

Fu, X., Zhang, J., Gu, X., Yu, H., Chen, S., 2021. A comprehensive study of the promoting effect of manganese on white rot fungal treatment for enzymatic hydrolysis of woody and grass lignocellulose. Biotechnology for Biofuels 14 (1), 176.

Gan, J.S., Iqbal, H.M.N., Show, P.L., Rahdar, A., Bilal, M., 2024. Upgrading recalcitrant lignocellulosic biomass hydrolysis by immobilized cellulolytic enzyme–based nanobiocatalytic systems: a review. Biomass Convers. Biorefin. 21906823. 14 (4), 4485–4509. Available from: https://www.springer.com/journal/13399. doi:10.1007/s13399-022-02642-7.

Granja-Travez, R.S., Persinoti, G.F., Squina, F.M., Bugg, T.D.H., 2020. Functional genomic analysis of bacterial lignin degraders: diversity in mechanisms of lignin oxidation and metabolism. Appl. Microbiol. Biotechnol. 14320614. 104 (8), 3305–3320. Available from: https://link.springer.de/link/service/journals/00253/index.htm. doi:10.1007/s00253-019-10318-y.

Hames, B.R., 2009. Biomass compositional analysis for energy applications. Methods Mol. Biol. (Clifton, N.J.). 19406029. 581, 145–167. doi:10.1007/978-1-60761-214-8_11.

Hermosilla, E., Rubilar, O., Schalchli, H., da Silva, A.S.A., Ferreira-Leitao, V., Diez, M.C., 2018. Sequential white-rot and brown-rot fungal pretreatment of wheat straw as a promising alternative for complementary mild treatments. Waste Manag. 18792456.

79, 240–250. Available from: https://www.elsevier.com/locate/wasman. doi:10.1016/j.wasman.2018.07.044.

Hernández Fariñas, T., Ribeiro, L., Soudant, D., Belin, C., Bacher, C., Lampert, L., Laurent Barillé, M.W., 2017. Contribution of benthic microalgae to the temporal variation in phytoplankton assemblages in a macrotidal system. J. Phycol. 0022-3646. 53 (5), 1020–1034. doi:10.1111/jpy.12564.

Hiras, J., Wu, Y.W., Deng, K., Nicora, C.D., Aldrich, J.T., Frey, D., Kolinko, S., Robinson, E.W., Jacobs, J.M., Adams, P.D., Northen, T.R., Simmons, B.A., Singer, S.W., 2016. Comparative community proteomics demonstrates the unexpected importance of actinobacterial glycoside hydrolase family 12 protein for crystalline cellulose hydrolysis. mBio 21507511. 7 (4). Available from: http://mbio.asm.org/content/7/4/e01106-16.full.pdf. doi:10.1128/mBio.01106-16.

Hölker, U., Höfer, M., Lenz, J., 2004. Biotechnological advantages of laboratory-scale solid-state fermentation with fungi. Appl. Microbiol. Biotechnol. 64 (2), 175–186. doi:10.1007/s00253-003-1504-3.

Holyavka, M.G., Goncharova, S.S., Artyukhov, V.G., 2025. Various options for covalent immobilization of cysteine proteases—ficin, papain, bromelain. Int. J. Mol. Sci. 14220067. 26 (2). Available from: http://www.mdpi.com/journal/ijms. doi:10.3390/ijms26020547.

Houfani, A.A., Anders, N., Spiess, A.C., Baldrian, P., Benallaoua, S., 2020. Insights from enzymatic degradation of cellulose and hemicellulose to fermentable sugars– a review. Biomass Bioenergy 18732909. 134. Available from: http://www.journals.elsevier.com/biomass-and-bioenergy/. doi:10.1016/j.biombioe.2020.105481.

Huervana, F.H., Traifalgar, R.F.M., Dionela, C.S., 2024. Solid-state fermentation converts rice bran into a high-protein feed ingredient for *Penaeus monodon*. Front. Mar. Sci. 22967745. 11. Available from: https://www.frontiersin.org/journals/marine-science#. doi:10.3389/fmars.2024.1384492.

Irdawati, I., Syamsuardi, S., Agustien, A., Rilda, Y., 2018. Screening of thermophilic bacteria produce xylanase from Sapan Sungai Aro hot spring, South Solok. IOP Conf. Ser. Mater. Sci. Eng. 1757-8981. 335 (1), 012021. doi:10.1088/1757-899x/335/1/012021.

Joshi, N., Lawal, A., 2012. Hydrodeoxygenation of pyrolysis oil in a microreactor. Chem. Eng. Sci. 00092509. 74, 1–8. doi:10.1016/j.ces.2012.01.052.

Juturu, V., Wu, J.C., 2014. Microbial cellulases: engineering, production and applications. Renew. Sustain. Energy Rev. 13640321. 33, 188–203. doi:10.1016/j.rser.2014.01.077.

Kameshwar, A.K.S., Qin, W., 2018. Understanding the structural and functional properties of carbohydrate esterases with a special focus on hemicellulose deacetylating acetyl xylan esterases. Mycology 21501211. 9 (4), 273–295. Available from: http://www.tandfonline.com/toc/tmyc20/current. doi:10.1080/21501203.2018.1492979.

Karnaouri, A., Antonopoulou, I., Zerva, A., Dimarogona, M., Topakas, E., Rova, U., Christakopoulos, P., 2019. Thermophilic enzyme systems for efficient conversion of lignocellulose to valuable products: structural insights and future perspectives for esterases and oxidative catalysts. Bioresour. Technol. 09608524. 279, 362–372. doi:10.1016/j.biortech.2019.01.062.

Kaur, M., Kumar, M., Singh, D., Sarita Sachdeva, S.K.P., 2019. A sustainable biorefinery approach for efficient conversion of aquatic weeds into bioethanol and biomethane. Energy Convers. Manag. 01968904. 187, 133–147. doi:10.1016/j.enconman.2019.03.018.

Khan, M., Singh, T., Pal, D.B., Khan, S., Ahmad, S., Jandrajupalli, S.B., Haque, S., Singh, R., Srivastava, N., 2024. Enhanced production of bacterial hydrolytic endoglucanase enzyme using waste leaves of water hyacinth and its thermal stability under the influence of $TiO_2$

nanoparticles. Biomass Convers. Biorefin. 21906823. 14 (2), 2185–2191. Available from: https://www.springer.com/journal/13399. doi:10.1007/s13399-022-02421-4.

Kirk, O., Borchert, T.V., Fuglsang, C.C., 2002. Industrial enzyme applications. Curr. Opin. Biotechnol. 09581669. 13 (4), 345–351. Available from: http://www.elsevier.com/locate/copbio. doi:10.1016/S0958-1669(02)00328-2.

Kont, R., Bissaro, B., Eijsink, V.G.H., Väljamäe, P., 2020. Kinetic insights into the peroxygenase activity of cellulose-active lytic polysaccharide monooxygenases (LPMOs). Nat. Commun. 2041-1723. 11 (1). doi:10.1038/s41467-020-19561-8.

Kracher, D., Ludwig, R., 2016. Cellobiose dehydrogenase: an essential enzyme for lignocellulose degradation in nature - A review. Bodenkultur 27195430. 67 (3), 145–163. Available from: https://www.degruyter.com/view/j/boku. doi:10.1515/boku-2016-0013Austria.

Kuhad, R.C., Gupta, R., Singh, A., 2011. Microbial cellulases and their industrial applications. Enzyme Res. 20900414. 2011 (1). doi:10.4061/2011/280696.

Kumar, B., Bhardwaj, N., Alam, A., Agrawal, K., Prasad, H., Verma, P., 2018. Production, purification and characterization of an acid/alkali and thermo tolerant cellulase from *Schizophyllum* commune NAIMCC-F-03379 and its application in hydrolysis of lignocellulosic wastes. AMB Express. 2191-0855. 8 (1). doi:10.1186/s13568-018-0696-y.

Lankiewicz, T.S., Choudhary, H., Gao, Y., Amer, B., Lillington, S.P., Leggieri, P.A., Brown, J.L., Swift, C.L., Lipzen, A., Na, H., Amirebrahimi, M., Theodorou, M.K., Baidoo, E.E.K., Barry, K., Grigoriev, I.V., Timokhin, V.I., Gladden, J., Singh, S., Mortimer, J.C., Ralph, J., Simmons, B.A., Singer, S.W., O'Malley, M.A., 2023. Lignin deconstruction by anaerobic fungi. Nat. Microbiol. 20585276. 8 (4), 596–610. Available from: https://www.nature.com/nmicrobiol/. doi:10.1038/s41564-023-01336-8.

Legodi, L.M., Grange, D.L., van Rensburg, E.L.J., Ncube, I., 2019. Isolation of cellulose degrading fungi from decaying banana pseudostem and *Strelitzia alba*. Enzyme Res. 2090-0406. 2019, 1–10. doi:10.1155/2019/1390890.

Limaye, L., Patil, R., Ranadive, P., Kamath, G., 2017. Application of potent actinomycete strains for bio-degradation of domestic agro-waste by composting and treatment of pulp-paper mill effluent. Adv. Microbiol. 2165-3402. 07 (01), 94–108. doi:10.4236/aim.2017.71008.

Liu, L., Zhang, Y., Huang, Y., Jiang, T., Qijun, Y., Yang, J., Yuan, H., 2025. Characterization of a multifunctional enzyme from *Trichoderma harzianum* and its application in enhanced enzymatic hydrolysis. Bioresour. Technol. 09608524. 415, 131701. doi:10.1016/j.biortech.2024.131701.

Luo, L., Cai, J., Wang, C., Lin, J., Xin, D., Zhou, A., Xiang, M., 2016. Purification and characterization of an alkaliphilic endo-xylanase from *Streptomyces althioticus* LMZM and utilization in the pulp paper industry. J. Chem. Technol. Biotechnol. 0268-2575. 91 (4), 1093–1098. doi:10.1002/jctb.4690.

Malgas, S., Mafa, M.S., Mkabayi, L., Pletschke, B.I., 2019. A mini review of xylanolytic enzymes with regards to their synergistic interactions during hetero-xylan degradation. World J. Microbiol. Biotechnol. 15730972. 35 (12). Available from: https://www.wkap.nl/journalhome.htm/0959-3993. doi:10.1007/s11274-019-2765-z.

Malik, W.A., Javed, S., 2024. Enhancement of cellulase production by cellulolytic bacteria SB125 in submerged fermentation medium and biochemical characterization of the enzyme. Int. J. Biol. Macromol. 18790003. 263. Available from: https://www.sciencedirect.com/science/journal/01418130. doi:10.1016/j.ijbiomac.2024.130415.

Marinović, M., Aguilar-Pontes, M.V., Zhou, M., Miettinen, O., de Vries, R.P., Mäkelä, M.R., Hildén, K., 2018. Temporal transcriptome analysis of the white-rot fungus *Obba rivulosa* shows expression of a constitutive set of plant cell wall degradation targeted genes during growth on solid spruce wood. Fungal Genet. Biol. 10960937. 112, 47–54. Available from: http://www.journals.elsevier.com/fungal-genetics-and-biology/. doi:10.1016/j.fgb.2017.07.004.

Martínez, A.T., Ruiz-Dueñas, F.J., Camarero, S., Serrano, A., Linde, D., Lund, H., Vind, J., Tovborg, M., Herold-Majumdar, O.M., Hofrichter, M., Liers, C., Ullrich, R., Scheibner, K., Sannia, G., Piscitelli, A., Pezzella, C., Sener, M.E., Kılıç, S., van Berkel, W.J.H., Guallar, V., Lucas, M.F., Zuhse, R., Ludwig, R., Hollmann, F., Fernández-Fueyo, E., Record, E., Faulds, C.B., Tortajada, M., Winckelmann, I., Rasmussen, J.A., Gelo-Pujic, M., Gutiérrez, A., Del Río, J.C., Rencoret, J., Alcalde, M., 2017. Oxidoreductases on their way to industrial biotransformations. Biotechnol. Adv. 07349750. 35 (6), 815–831. Available from: https://www.elsevier.com/inca/publications/store/5/2/5/4/5/5/index.htt. doi:10.1016/j.biotechadv.2017.06.003.

Mayr, S.A., Subagia, R., Weiss, R., Schwaiger, N., Weber, H.K., Leitner, J., Ribitsch, D., Nyanhongo, G.S., Guebitz, G.M., 2021. Oxidation of various kraft lignins with a bacterial laccase enzyme. Int. J. Mol. Sci. 14220067. 22 (23). Available from: https://www.mdpi.com/1422-0067/22/23/13161/pdf. doi:10.3390/ijms222313161.

Mei, J., Shen, X., Gang, L., Huajie, X., Fufang, W., Sheng, L., 2020. A novel lignin degradation bacteria-*Bacillus* amyloliquefaciens SL-7 used to degrade straw lignin efficiently. Bioresour. Technol. 09608524. 310, 123445. doi:10.1016/j.biortech.2020.123445.

Mekonnen, T., Mussone, P., Khalil, H., Bressler, D., 2013. Progress in bio-based plastics and plasticizing modifications. J. Mater. Chem. A 20507496. 1 (43), 13379–13398. Available from: http://pubs.rsc.org/en/journals/journalissues/ta. doi:10.1039/c3ta12555f.

Merino, S.T., Cherry, J., 2007. Progress and challenges in enzyme development for biomass utilization. Adv. Biochem. Eng. Biotechnol. 07246145. 108, 95–120. doi:10.1007/10_2007_066.

Mhatre, A., Kalscheur, B., Mckeown, H., Bhakta, K., Sarnaik, A.P., Flores, A., Nielsen, D.R., Wang, X., Soundappan, T., Varman, A.M., 2022. Consolidated bioprocessing of hemicellulose to fuels and chemicals through an engineered *Bacillus subtilis-Escherichia coli* consortium. Renew. Energy 18790682. 193, 288–298. Available from: http://www.journals.elsevier.com/renewable-and-sustainable-energy-reviews/. doi:10.1016/j.renene.2022.04.124.

Mishra, S., Singh, P.K., Dash, S., Pattnaik, R., 2018. Microbial pretreatment of lignocellulosic biomass for enhanced biomethanation and waste management. 3 Biotech 21905738. 8 (11). Available from: http://www.springerlink.com/content/2190-572x/. doi:10.1007/s13205-018-1480-z.

Mohan, D., Pittman, C.U., Steele, P.H., 2006. Pyrolysis of wood/biomass for bio-oil: a critical review. Energy Fuels. 08870624. 20 (3), 848–889. doi:10.1021/ef0502397.

Mojsov, K., 2014. Biotechnological applications of laccases in the textile industry. Savremene Tehnologije. 2217-9712. 3 (1), 76–79. doi:10.5937/savteh1401076M.

Monrroy, M., Ortega, I., Ramírez, M., Baeza, J., Freer, J., 2011. Structural change in wood by brown rot fungi and effect on enzymatic hydrolysis. Enzyme Microb. Technol. 01410229. 49 (5), 472–477. doi:10.1016/j.enzmictec.2011.08.004.

Najjarzadeh, N., Matsakas, L., Rova, U., Christakopoulos, P., 2020. Effect of oligosaccharide degree of polymerization on the induction of xylan-degrading enzymes by *Fusarium oxysporum* f. sp. *lycopersici*. Molecules. 1420-3049. 25 (24), 5849. doi:10.3390/molecules25245849.

Nanda, S., Mohammad, J., Reddy, S.N., Kozinski, J.A., Dalai, A.K., 2014. Pathways of lignocellulosic biomass conversion to renewable fuels. Biomass Convers. Biorefin. 21906823. 4 (2), 157–191. Available from: http://www.springer.com/engineering/energy+technology/journal/13399. doi:10.1007/s13399-013-0097-z.

Naraian, R., Gautam, R.L., 2017. Penicillium Enzymes for the Saccharification of Lignocellulosic Feedstocks New and Future Developments in Microbial Biotechnology and Bioengineering: Penicillium System Properties and Applications. Elsevier, India, pp. 121–136. Available from: http://www.sciencedirect.com/science/book/9780444635013. doi:10.1016/B978-0-444-63501-3.00006-5.

Nargotra, P., Sharma, V., Sharma, S., Kapoor, N., Bajaj, B.K., 2022. Development of consolidated bioprocess for biofuel-ethanol production from ultrasound-assisted deep eutectic solvent pretreated *Parthenium hysterophorus* biomass. Biomass Convers. Biorefin. 21906823. 12 (12), 5767–5782. Available from: https://www.springer.com/journal/13399. doi:10.1007/s13399-020-01017-0.

Nargotra, P., Sharma, V., Wang, H.M.D., Shieh, C.J., Liu, Y.C., Kuo, C.H., 2025. Biocatalysis for lignin conversion and valorization: driving sustainability in the circular economy. Catalysts 20734344. 15 (1). Available from: https://www.mdpi.com/journal/catalysts. doi:10.3390/catal15010091.

Nazir, M., Iram, A., Cekmecelioglu, D., Demirci, A., 2024. Approaches for producing fungal cellulases through submerged fermentation. Front. Biosci. Elite. 1945-0494. 16 (1). doi:10.31083/j.fbe1601005.

Neelkant, K.S., Shankar, K., Jayalakshmi, S.K., Sreeramulu, K., 2019. Optimization of conditions for the production of lignocellulolytic enzymes by *Sphingobacterium* sp. ksn-11 utilizing agro-wastes under submerged condition. Prep. Biochem. Biotechnol. 15322297. 49 (9), 927–934. Available from: https://www.tandf.co.uk/journals/titles/10826068.asp. doi:10.1080/10826068.2019.1643735.

Nurika, I., Suhartini, S., Barker, G.C., 2020. Biotransformation of tropical lignocellulosic feedstock using the brown rot fungus *Serpula lacrymans*. Waste Biomass Valor. 1877265X. 11 (6), 2689–2700. Available from: http://www.springer.com/engineering/journal/12649. doi:10.1007/s12649-019-00581-5.

Pandey, S., Erbaugh, J.T., 2024. Driving sustainable uptake: a systematic review of global literature on policies governing woody biomass for energy. Discov. Sustain. 26629984. 5 (1). Available from: https://www.springer.com/journal/43621. doi:10.1007/s43621-024-00205-6.

Parakh, P.D., Nanda, S., Kozinski, J.A., 2020. Eco-friendly transformation of waste biomass to biofuels. Curr. Biochem. Eng. 22127119. 6 (2), 120–134. doi:10.2174/2212711906999200425235946.

Patel, A.K., Dixit, P., Pandey, A., Singhania, R.R., 2020. Promising enzymes for biomass processing. In: Singh, S.P., Singhania, R.R., Zhi, L.A., Larroche, C. (Eds.), Biomass, Biofuels, Biochemicals: Advances in Enzyme Catalysis and Technologies. Elsevier, South Korea, pp. 245–271. Available from: https://www.sciencedirect.com/book/9780128198209. doi: 10.1016/B978-0-12-819820-9.00013-2.

Patel, A.K., Dong, C.D., Chen, C.W., Pandey, A., Singhania, R.R., 2023. Production, Purification, and Application of Microbial Enzymes Biotechnology of Microbial Enzymes: Production, Biocatalysis, and Industrial Applications, second ed., Elsevier, Taiwan, pp. 25–57. Available from: https://www.sciencedirect.com/book/9780443190599. doi:10.1016/B978-0-443-19059-9.00019-0.

Pawar, R.P., Tekale, S.U., Shisodia, S.U., Totre, J.T., Domb, A.J., 2014. Biomedical applications of poly(lactic acid). Recent Pat. Regen. Med. 22102973. 4 (1), 40–51. Available from: http://www.benthamscience.com/rpgm/index.htm. doi:10.2174/22102 96504666140402235024.

Peterson, R., Nevalainen, H., 2012. Trichoderma reesei RUT-C30 – thirty years of strain improvement. Microbiology. 1350-0872. 158 (1), 58–68. doi:10.1099/mic.0.054031-0.

Picart, P., De María, P.D., Schallmey, A., 2015. From gene to biorefinery: microbial β-etherases as promising biocatalysts for lignin valorization. Front. Microbiol. 1664302X. 6. Available from: http://journal.frontiersin.org/article/10.3389/fmicb.2015.00916/full. doi:10.3389/fmicb.2015.00916.

Plácido, J., Capareda, S., 2015. Ligninolytic enzymes: a biotechnological alternative for bioethanol production. Bioresour. Bioprocess. 2197-4365. 2 (1). doi:10.1186/s40643-015-0049-5.

Pollegioni, L., Tonin, F., Rosini, E., 2015. Lignin-degrading enzymes. FEBS J. 1742-464X. 282 (7), 1190–1213. doi:10.1111/febs.13224.

Raghuwanshi, S., Deswal, D., Karp, M., Kuhad, R.C., 2014. Bioprocessing of enhanced cellulase production from a mutant of *Trichoderma asperellum* RCK2011 and its application in hydrolysis of cellulose. Fuel. 00162361. 124, 183–189. doi:10.1016/j.fuel.2014.01.107.

Rahmanpour, R., Rea, D., Jamshidi, S., Fülöp, V., Bugg, T.D.H., 2016. Structure of *Thermobifida fusca* DyP-type peroxidase and activity towards Kraft lignin and lignin model compounds. Arch. Biochem. Biophys. 10960384. 594, 54–60. Available from: http://www.elsevier.com/inca/publications/store/6/2/2/7/8/7/index.htt. doi:10.1016/j.abb.2016.02.019.

Ramamoorthy, S.K., Skrifvars, M., Persson, A., 2015. A review of natural fibers used in biocomposites: plant, animal and regenerated cellulose fibers. Polym. Rev. 15583716. 55 (1), 107–162. Available from: http://www.tandf.co.uk/journals/titles/15583724.asp. doi:10.1080/15583724.2014.971124.

Rashid, G.M.M., Durán-Peña, M.J., Rahmanpour, R., Sapsford, D., Bugg, T.D.H., 2017. Delignification and enhanced gas release from soil containing lignocellulose by treatment with bacterial lignin degraders. J. Appl. Microbiol. 13645072. 123 (1), 159–171. doi:10.1111/jam.13470.

Robinson, P.K., 2015. Enzymes: principles and biotechnological applications. Essays Biochem. 00711365. 59, 1–41. Available from: http://essays.biochemistry.org/content/59/1.full-text.pdf. doi:10.1042/BSE0590001.

Rocha, T.M., Marcelino, P.R.F., Costa, R.A.M.D., Rubio-Ribeaux, D., Barbosa, F.G., da Silva, S.S., 2024. Agricultural bioinputs obtained by solid-state fermentation: from production in biorefineries to sustainable agriculture. Sustainability 20711050. 16 (3). Available from: http://www.mdpi.com/journal/sustainability/. doi:10.3390/su16031076.

Rohman, A., Dijkstra, B.W., Puspaningsih, N.N.T., 2019. β-xylosidases: structural diversity, catalytic mechanism, and inhibition by monosaccharides. Int. J. Mol. Sci. 14220067. 20 (22). Available from: https://www.mdpi.com/1422-0067/20/22/5524/pdf. doi:10.3390/ijms20225524.

Saini, S., Sharma, K.K., 2021. Fungal lignocellulolytic enzymes and lignocellulose: a critical review on their contribution to multiproduct biorefinery and global biofuel research. Int. J. Biol. Macromol. 18790003. 193, 2304–2319. Available from: https://www.elsevier.com/locate/ijbiomac. doi:10.1016/j.ijbiomac.2021.11.063.

Sakhuja, D., Ghai, H., Rathour, R.K., Kumar, P., Bhatt, A.K., Bhatia, R.K., 2021. Cost-effective production of biocatalysts using inexpensive plant biomass: a review. 3 Biotech 21905738. 11 (6). Available from: https://www.springer.com/journal/13205. doi:10.1007/s13205-021-02847-z.

Saldarriaga-Hernández, S., Velasco-Ayala, C., Flores, P.L.-I., de Jesús Rostro-alanis, M., Parra-Saldivar, R., Iqbal, H.M.N., Carrillo-Nieves, D., 2020. Biotransformation of lignocellulosic biomass into industrially relevant products with the aid of fungi-derived lignocellulolytic enzymes. Int. J. Biol. Macromol. 18790003. 161, 1099–1116. Available from: https://www.elsevier.com/locate/ijbiomac. doi:10.1016/j.ijbiomac.2020.06.047.

Sánchez, A., Oiza, N., Artola, A., Font, X., Barrena, R., Moral-Vico, J., Gea, T., 2024. Solid-state fermentation: a review of its opportunities and challenges in the framework of circular bioeconomy. Afinidad. 2339-9686. 81 (601), 51–57. doi:10.55815/424209.

Santibáñez, L., Henríquez, C., Corro-Tejeda, R., Bernal, S., Armijo, B., Salazar, O., 2021. Xylooligosaccharides from lignocellulosic biomass: a comprehensive review. Carbohydr. Polym. 01448617. 251, 117118. doi:10.1016/j.carbpol.2020.117118.

Sethi, A., Scharf, M.E., 2013. Biofuels: Fungal, Bacterial and Insect Degraders of Lignocellulose Encyclopedia of Life Sciences. Wiley. doi:10.1002/9780470015902.a0020374.

Sharma, S., Sharma, V., Nargotra, P., Bajaj, B.K., 2018. Process desired functional attributes of an endoxylanase of GH10 family from a new strain of *Aspergillus terreus* S9. Int. J. Biol. Macromol. 18790003. 115, 663–671. Available from: https://www.elsevier.com/locate/ijbiomac. doi:10.1016/j.ijbiomac.2018.04.096.

Sharma, V., Tsai, M.L., Chen, C.W., Sun, P.P., Patel, A.K., Singhania, R.R., Nargotra, P., Dong, C.D., 2022. Deep eutectic solvents as promising pretreatment agents for sustainable lignocellulosic biorefineries: a review. Bioresour. Technol. 18732976. 360. Available from: https://www.elsevier.com/locate/biortech. doi:10.1016/j.biortech.2022.127631.

Shirkavand, E., Baroutian, S., Gapes, D.J., Young, B.R., 2017. Pretreatment of radiata pine using two white rot fungal strains *Stereum hirsutum* and *Trametes versicolor*. New Zealand Energy Convers. Manag. 01968904. 142, 13–19. Available from: https://www.journals.elsevier.com/energy-conversion-and-management. doi:10.1016/j.enconman.2017.03.021.

Singh, A., Kaur, A., Patra, A.K., Mahajan, R., 2018. A sustainable and green process for scouring of cotton fabrics using xylano-pectinolytic synergism: switching from noxious chemicals to eco-friendly catalysts. 3 Biotech 21905738. 8 (4). Available from: http://www.springerlink.com/content/2190-572x/. doi:10.1007/s13205-018-1193-3.

Singh, G., Kumar, S., Afreen, S., Bhalla, A., Khurana, J., Chandel, S., Aggarwal, A., Arya, S.K., 2023. Laccase mediated delignification of wasted and non-food agricultural biomass: recent developments and challenges. Int. J. Biol. Macromol. 01418130. 235, 123840. doi:10.1016/j.ijbiomac.2023.123840.

Singh, T.A., Sharma, M., Sharma, M., Sharma, G.D., Passari, A.K., Bhasin, S., 2022. Valorization of agro-industrial residues for production of commercial biorefinery products. Fuel. 00162361. 322, 124284. doi:10.1016/j.fuel.2022.124284.

Singh, V., Shirbhate, E., Kore, R., Vishwakarma, S., Parveen, S., Veerasamy, R., Tiwari, A.K., Rajak, H., 2025. Microbial metabolites-induced epigenetic modifications for inhibition of colorectal cancer. Curr. Status Future Perspect. Mini Rev. Med. Chem. 18755607. 25 (1), 76–93. doi:10.2174/0113895575320344240625080555.

Singhania, R.R., Adsul, M., Pandey, A., Patel, A.K., 2017. Cellulases. Elsevier BV, pp. 73–101. doi:10.1016/b978-0-444-63662-1.00004-x.

Singhania, R.R., Patel, A.K., Raj, T., Chen, C.W., Ponnusamy, V.K., Tahir, N., Kim, S.H., Dong, C.D., 2022. Lignin valorisation via enzymes: a sustainable approach. Fuel 00162361. 311. Available from: http://www.journals.elsevier.com/fuel/. doi:10.1016/j.fuel.2021.122608.

Song, J.H., Murphy, R.J., Narayan, R., Davies, G.B.H., 2009. Biodegradable and compostable alternatives to conventional plastics. Philos. Trans. R. Soc. Lond. B Biol. Sci. 0962-8436. 364 (1526), 2127–2139. doi:10.1098/rstb.2008.0289.

Spolaore, P., Joannis-Cassan, C., Duran, E., Isambert, A., 2006. Commercial applications of microalgae. J. Biosci. Bioeng. 13891723. 101 (2), 87–96. doi:10.1263/jbb.101.87.

Srivastava, N., Rathour, R., Jha, S., Pandey, K., Srivastava, M., Thakur, V.K., Sengar, R.S., Gupta, V.K., Mazumder, P.B., Khan, A.F., Mishra, P.K., 2019. Microbial beta glucosidase enzymes: recent advances in biomass conversation for biofuels application. Biomolecules 2218273X. 9 (6). Available from: https://www.mdpi.com/2218-273X/9/6/220/pdf. doi:10.3390/biom9060220.

Sulej, J., Osińska-Jaroszuk, M., Jaszek, M., Grąz, M., Kutkowska, J., Pawlik, A., Chudzik, A., Bancerz, R., 2019. Antimicrobial and antioxidative potential of free and immobilised cellobiose dehydrogenase isolated from wood degrading fungi. Fungal Biol. 18786146. 123 (12), 875–886. doi:10.1016/j.funbio.2019.09.007.

Sumranwanich, T., Amosu, E., Chankhamhaengdecha, S., Phetruen, T., Loktumraks, W., Ounjai, P., Harnvoravongchai, P., 2024. Evaluating lignin degradation under limited oxygen conditions by bacterial isolates from forest soil. Sci. Rep. 2045-2322. 14 (1). doi:10.1038/s41598-024-64237-8.

Sun, J., Wen, F., Si, T., Xu, J.H., Zhao, H., 2012. Direct conversion of xylan to ethanol by recombinant *Saccharomyces cerevisiae* strains displaying an engineered minihemicellulosome. Appl. Environ. Microbiol. 10985336. 78 (11), 3837–3845. Available from: http://aem.asm.org/content/78/11/3837.full.pdf. doi:10.1128/AEM.07679-11.

Suseela, V., 2019. Potential roles of plant biochemistry in mediating ecosystem responses to warming and drought. In: Mohan, J.E., Shefferson, R.P., Melillo, J.M. (Eds.), Ecosystem Consequences of Soil Warming: Microbes, Vegetation, Fauna and Soil Biogeochemistry. Elsevier, United States, pp. 103–124. Available from: https://www.sciencedirect.com/book/9780128134931. doi: 10.1016/B978-0-12-813493-1.00005-3.

Sweeney, M.D., Xu, F., 2012. Biomass converting enzymes as industrial biocatalysts for fuels and chemicals: recent developments. Catalysts 20734344. 2 (2), 244–263. Available from: http://www.mdpi.com/2073-4344/2/2/244/pdf. doi:10.3390/catal2020244.

Taiwo, A.E., Tom-James, A., Falowo, O.A., Okoji, A., Adeyi, O., Olalere, A.O., Eloka-Eboka, A., 2022. Techno-economic analysis of cellulase production by *Trichoderma reesei* in submerged fermentation processes using a process simulator. S. Afr. J. Chem. Eng. 10269185. 42, 98–105. Available from: https://www.journals.elsevier.com/south-african-journal-of-chemical-engineering/. doi:10.1016/j.sajce.2022.08.001.

Thakur, S., Chaudhary, J., Sharma, B., Verma, A., Tamulevicius, S., Thakur, V.K., 2018. Sustainability of bioplastics: opportunities and challenges. Curr. Opin. Green Sustain. Chem. 24522236. 13, 68–75. Available from: http://www.journals.elsevier.com/current-opinion-in-green-and-sustainable-chemistry. doi:10.1016/j.cogsc.2018.04.013.

Trache, D., Hussin, M.H., Chuin, C.T.H., Sabar, S., Fazita, M.R.N., Taiwo, O.F.A., Hassan, T.M., Haafiz, M.K.M., 2016. Microcrystalline cellulose: isolation, characterization and bio-composites application—A review. Int. J. Biol. Macromol. 18790003. 93, 789–804. Available from: https://www.elsevier.com/locate/ijbiomac. doi:10.1016/j.ijbiomac.2016.09.056.

Tursi, A., 2019. A review on biomass: importance, chemistry, classification, and conversion. Biofuel Res. J. 22928782. 6 (2), 962–979. doi:10.18331/brj2019.6.2.3.

Vasconcellos, V.M., Tardioli, P.W., Giordano, R.L.C., Farinas, C.S., 2015. Production efficiency versus thermostability of (hemi)cellulolytic enzymatic cocktails from different cultivation systems. Process Biochem. 13595113. 50 (11), 1701–1709. doi:10.1016/j.procbio.2015.07.011.

Vassilev, S.V., Vassileva, C.G., 2016. Composition, properties and challenges of algae biomass for biofuel application: an overview. Fuel 00162361. 181, 1–33. Available from: http://www.journals.elsevier.com/fuel/. doi:10.1016/j.fuel.2016.04.106.

Verma, R., Suthar, S., 2015. Utility of duckweeds as source of biomass energy: a review. BioEnergy Res. 1939-1234. 8 (4), 1589–1597. doi:10.1007/s12155-015-9639-5.

Virk, A.P., Sharma, P., Capalash, N., 2012. Use of laccase in pulp and paper industry. Biotechnol. Prog. 15206033. 28 (1), 21–32. doi:10.1002/btpr.727.

Weng, C., Peng, X., Han, Y., 2021. Depolymerization and conversion of lignin to value-added bioproducts by microbial and enzymatic catalysis. Biotechnol. Biofuels. 1754-6834. 14 (1). doi:10.1186/s13068-021-01934-w.

Wesseler, J., Kleter, G., Meulenbroek, M., Purnhagen, K.P., 2023. EU regulation of genetically modified microorganisms in light of new policy developments: possible implications for EU bioeconomy investments. Appl. Econ. Perspect. Policy 20405804. 45 (2), 839–859. Available from: https://onlinelibrary.wiley.com/journal/20405804. doi:10.1002/aepp.13259.

Xu, W., Fu, S., Yang, Z., Lu, J., Guo, R., 2018. Improved methane production from corn straw by microaerobic pretreatment with a pure bacteria system. Bioresource Technology 259, 18–23.

Yao, G., Ruimei, W., Kan, Q., Gao, L., Liu, M., Yang, P., Jian, D., Zhonghai, L., Yinbo, Q., 2016. Production of a high-efficiency cellulase complex via β-glucosidase engineering in *Penicillium oxalicum*. Biotechnol. Biofuels. 1754-6834. 9 (1). doi:10.1186/s13068-016-0491-4.

Zabed, H.M., Akter, S., Yun, J., Zhang, G., Awad, F.N., Qi, X., Sahu, J.N., 2019. Recent advances in biological pretreatment of microalgae and lignocellulosic biomass for biofuel production. Renew. Sustain. Energy Rev. 18790690. 105, 105–128. Available from: https://www.journals.elsevier.com/renewable-and-sustainable-energy-reviews. doi:10.1016/j.rser.2019.01.048.

Zhang, F., Bunterngsook, B., Li, J.X., Zhao, X.Q., Champreda, V., Liu, C.G., Bai, F.W., 2019. Regulation and production of lignocellulolytic enzymes from *Trichoderma reesei* for biofuels production. Adv. Bioenergy 4, 79–119. Available from: https://www.sciencedirect.com/bookseries/advances-in-bioenergy. doi:10.1016/bs.aibe.2019.03.001.

Zhi, T.C., Chen, I.T.S., Peng, T.Y., Chang, L.K., Balakrishnan, K., Xian, O.Z., 2025. Process optimization studies on xylanase production and bioethanol fermentation. 603, 03005.

CHAPTER

# Anaerobic digestion: Microbial powerhouses for waste-to-energy conversion

9

**Rahul Mondal and Subodh Kumar Maiti**

*Department of Environmental Science and Engineering, Indian Institute of Technology (Indian School of Mines) Dhanbad, Dhanbad, India*

## 1 Introduction

Anaerobic digestion (AD), or anaerobic codigestion if multiple substrates are used together, refers to the conversion of biodegradable organic waste into biogas by microbes in the absence of free molecular oxygen. The biogas produced typically comprises (by volume) around 55%–75% methane ($CH_4$), 25%–45% carbon dioxide ($CO_2$), 2%–7% water vapor ($H_2O$), and trace amounts of other gases like hydrogen sulfide ($H_2S$), nitrogen ($N_2$), ammonia ($NH_3$), and hydrogen ($H_2$). Ranking among the oldest instances of applied microbiology, AD offers several advantages over aerobic treatment of organics, including a high level of waste valorization at a high loading rate, with substantially less microbial sludge production. The requirements for nutrients and aeration are also minimal. The produced biogas can be processed to fuel vehicles and used to generate heat and electricity at both domestic and industrial levels. Valorization of the digestate has also been recognized as a promising source for producing organic fertilizers and other horticultural products, supporting a circular economy. Fig. 9.1 illustrates a schematic flowchart for the AD process, highlighting some of the applications of the product and the by-product formed.

## 2 Feedstock suitable for AD

### 2.1 Biochemical methane potential

For organic waste to be considered for AD, the most important factor to be taken into account is its biochemical methane potential (BMP). BMP is defined as the maximum amount of $CH_4$ that can be produced from it under anaerobic conditions (Yasim and Buyong, 2023). Theoretically, the $CH_4$ yield of a substrate can be calculated using the following three methods (Tarvin and Buswell, 1934).

Microbial Technologies for Sustainable Waste Management. https://doi.org/10.1016/B978-0-443-44093-9.00011-6

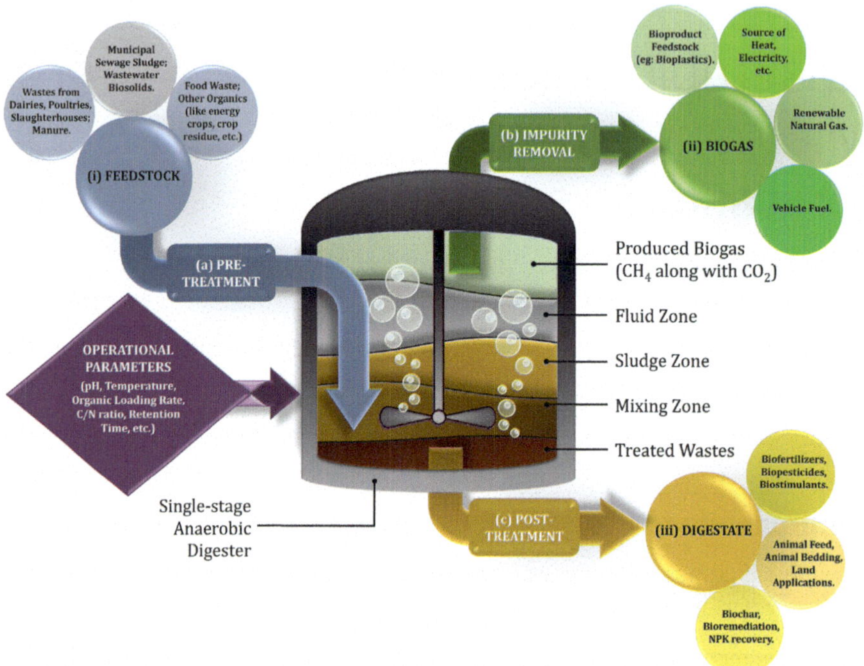


**FIG. 9.1 Procedural diagram enumerating the steps involved in the anaerobic digestion of organic feedstock.**

The figure depicts a typical single-stage anaerobic digestion process, showing: (i) the different types of feedstocks that can be used, (ii) biogas, the main product formed, and (iii) the digestate produced as a by-product, noting their potential uses. The operational parameters mentioned are crucial to be engineered for the AD process. Pretreatment of feedstock is imperative in enhancing the efficiency of AD, while impurity removal and posttreatment of the product and by-product formed are essential to render them useful.

a. **Elemental composition analysis:**

$$(BMP_{th_{AIC}})(mLCH_4/gVS) = \frac{22400\left(\frac{n}{2}+\frac{a}{8}-\frac{b}{4}-\frac{3c}{4}\right)}{12n+a+16b+14c}$$

where ***n***, ***a***, ***b***, and ***c*** are the number of moles of C, H, O, and N in one mole of the substrate, and the factor 22,400 represents the molar volume (in mL/mol) of a gas at standard temperature and pressure conditions. The stoichiometric equation used for this analysis is as follows:

$$C_nH_aO_bN_c+\left(n-\frac{a}{4}-\frac{b}{2}+\frac{3c}{4}\right)H_2O\rightarrow\left(\frac{n}{2}+\frac{a}{8}-\frac{b}{4}-\frac{3c}{4}\right)CH_4+\left(\frac{n}{2}-\frac{a}{8}+\frac{b}{4}+\frac{3c}{4}\right)CO_2+cNH_3$$

**b. Chemical oxygen demand analysis:**

$$(BMP_{th_{COD}})(mLCH_4/gVS) = \frac{1000(n_{CH_4}RT)}{pVS_{added}}$$

where R is the universal gas constant (R = 0.082 atm L/mol K), T is the temperature of the glass bottle (T = 308 K), P is the atmospheric pressure (P = 1 atm), $VS_{added}$ (in g) is the amount of volatile solids (VS) of the substrate, 1000 is the conversion factor for L to mL, and $n_{CH_4}$ (in mol) is the amount of molecular $CH_4$, determined as:

$$n_{CH_4} = \frac{COD}{64(g/mol)}$$

**c. Organic fraction composition analysis:**

$$(BMP_{th_{OFC}})(mLCH_4/gVS) = 415 \times \%\ carbohydrates + 496 \times \%\ proteins + 1014 \times \%\ lipids$$

where the factors 415, 496, and 1014 are, respectively, the $CH_4$ yields (based on the average molecular formulae) from 1 g each of carbohydrates ($C_6H_{10}O_5$), proteins ($C_5H_7O_2N$), and lipids ($C_{57}H_{104}O_6$), and the percentages are defined in terms of their VS.

Theoretical calculation of BMP is cost-effective and convenient. A high stoichiometric composition of hydrogen and nitrogen is evidently not favored. This is also in accordance with the fact that the formation of gases like $H_2S$ and $NH_3$ inhibits digester efficiency. The COD analysis also shows a positive correlation between COD and BMP (Yasim and Buyong, 2023). However, the theoretical yield can substantially overestimate the biogas yield in AD. This is because these empirical methods do not take into consideration that not the whole organic fraction of the feedstock is biodegradable, and some can even act as inhibitors (Raposo et al., 2011). Thus experimental values of BMP, as determined from BMP assays, are more reliable. However, BMP tests can be extremely time-consuming.

The first step of a BMP assay involves treating the sample with a seed inoculum of anaerobic bacteria culture, typically sourced from an existing digester. The assay bottle is then flushed with a 70:30 (v/v) gas stream of $N_2$ and $CO_2$ (or, in some cases, pure $N_2$) to purge out any existing $O_2$. Several such replicates are made for a given substrate, and blanks with the AD seed in distilled water are also prepared. These assay bottles are then incubated at 35°C or 55°C for around 30–60 days, and the amounts of $CO_2$ and $CH_4$ formed in the substrate samples are periodically monitored against their corresponding concentrations in the blanks (Filer et al., 2019; Donatelli and Chang, 2024). The BMP can then be expressed in terms of the cumulative volume of $CH_4$ per unit mass of VS or chemical oxygen demand (COD), as per the following equations (Donatelli and Chang, 2024).

$$Methane\ yield\,(BMP_{exp})(mLCH_4/gVS) = \frac{V_{sCH_4}(t) - V_{bCH_4}(t)}{W_{VS}}$$

or

$$Methane\,yield\left(BMP_{exp}\right)\left(mLCH_4/gCOD\right)=\frac{V_{sCH_4}(t)-V_{bCH_4}(t)}{COD_{in}}$$

where $BMP_{exp}$ represents the experimental $CH_4$ yield; $V_{sCH_4}(t)$ and $V_{bCH_4}(t)$ are the accumulated volumes (in mL) of $CH_4$ in the substrate sample and the blank sample, respectively, at time $t$, at 0°C and 1 atm; $W_{vs}$ (in g) is the initial substrate VS added to each of the assay bottles; and $COD_{in}$ (in g) is the initial substrate chemical oxygen demand COD added to each of the assay bottles.

Biodegradability index (BDI) is a measure of a substrate's anaerobic biodegradability, or how easily a substrate degrades in an anaerobic environment, given the experimental test conditions (Raposo et al., 2011). Ultimately, the BDI is calculated as the ratio of experimental to theoretical $CH_4$ yields, shown as follows (Yasim and Buyong, 2023):

$$BDI\,(\%)=\frac{BMP_{exp}}{BMP_{th}}\times 100$$

$$error\,(\%)=\frac{BMP_{exp}-BMP_{th}}{BMP_{exp}}\times 100$$

### 2.1.1 Kinetics of methane production

Several empirical and mathematical models have been developed over the years to understand the kinetics of $CH_4$ production. Assessment of BMP kinetics is imperative to understand the scalability of biogas production from a laboratory-scale study to an industry-level implementation. However, there is still a lack of a universally accepted standard model that precisely fits every type of dataset, and it is common practice to use the model that best fits the experimental values observed (Donatelli and Chang, 2024). Some of the most common kinetic models used throughout the globe for single-stage digestion are listed below:

a. **First-order kinetic model:**

$$BMP(t)=\begin{cases}BMP_{\infty}\left[1-e^{-k(t-t_{lag})}\right] & t>t_{lag}\\ 0 & t\le t_{lag}\end{cases}$$

where $BMP(t)$ indicates the cumulative biomethane yield (in mL $CH_4$/g COD) after time $t$ (in days), $BMP_{\infty}$ indicates the maximum yield (in mL $CH_4$/g COD) observed at the completion of the experiment and is also referred to as the biomethane potential, $t_{lag}$ (in days) represents the lag phase, and $k$ (in day$^{-1}$) is the first kinetic coefficient. The lag phase is the time taken by the microbes to acclimatize in the medium, and no significant $CH_4$ yield is observed during this time (Donatelli and Chang, 2024). Typically, this model is used in cases where no lag phase is observed (i.e., $t_{lag}=$ 0 days).

**b. Zwietering–Gompertz model:**

$$BMP(t) = BMP_{\infty} e^{-e^{-\left(e\frac{R_{max}}{BMP_{\infty}}\right)(t-t_{lag})+1}}$$

where $BMP(t)$ indicates the cumulative biomethane yield (in mL $CH_4$/g COD) after time $t$ (in days), $BMP_{\infty}$ indicates the maximum yield (in mL $CH_4$/g COD) observed at the completion of the experiment, $t_{lag}$ (in days) represents the lag phase, and $R_{max}$ represents the maximum specific $CH_4$ production potential (in mL $CH_4$/g COD/day) (Filer et al., 2019).

**c. France model:**

$$BMP(t) = \begin{cases} BMP_{\infty}\left[1-e^{k_1(t_{lag}-t)+k_2\left(\sqrt{t_{lag}}-\sqrt{t}\right)}\right] & t \geq t_{lag} \\ 0 & t < t_{lag} \end{cases}$$

where $BMP(t)$ indicates the cumulative biomethane yield (in mL $CH_4$/g COD) after time $t$ (in days), $BMP_{\infty}$ indicates the maximum yield (in mL $CH_4$/g COD) observed at the completion of the experiment and is also referred to as the biomethane potential, $t_{lag}$ (in days) represents the lag phase, and $k_1$ *and* $k_2$ (in $day^{-1}$) are kinetic coefficients. This model is an example of $CH_4$ production being described by two nonfirst-order reactions (Donatelli and Chang, 2024).

### 2.1.2 Methane yield curve

The cumulative $CH_4$ yields for different test substances and under different operational conditions, as observed from BMP tests, can be plotted as a function of time (in days), as shown in Fig. 9.2. Studies indicate four possible types of fitted curves, as illustrated in Fig. 9.2A (Filer et al., 2019). Curve-1(A) shows the yield of a material that is readily biodegradable, as indicated by the immediate production of biogas. Curve-2 is identical to Curve-1(A), except that it shows an initial lag phase, indicating the hydrolysis phase of AD as the rate-determining step. Curve-3 initially shows a negative yield during the incubation period, indicating the presence of inhibitory toxins that cause the $CH_4$ yield in the sample to drop lower than that of the blank. Curve-4 shows a negative yield throughout, implying the presence of toxins that inhibit the methanogens. The blank continues to slowly produce biogas, while the sample produces virtually no $CH_4$, which is why their difference gives an increasingly negative plot (Filer et al., 2019). Fig. 9.2B shows the $CH_4$ yield curves for the same material, without (Curve-1(A)) vs. with (Curve-1(B)) pretreatment (Atelge et al., 2020). This highlights just how important feedstock pretreatment is, as discussed in a later section.

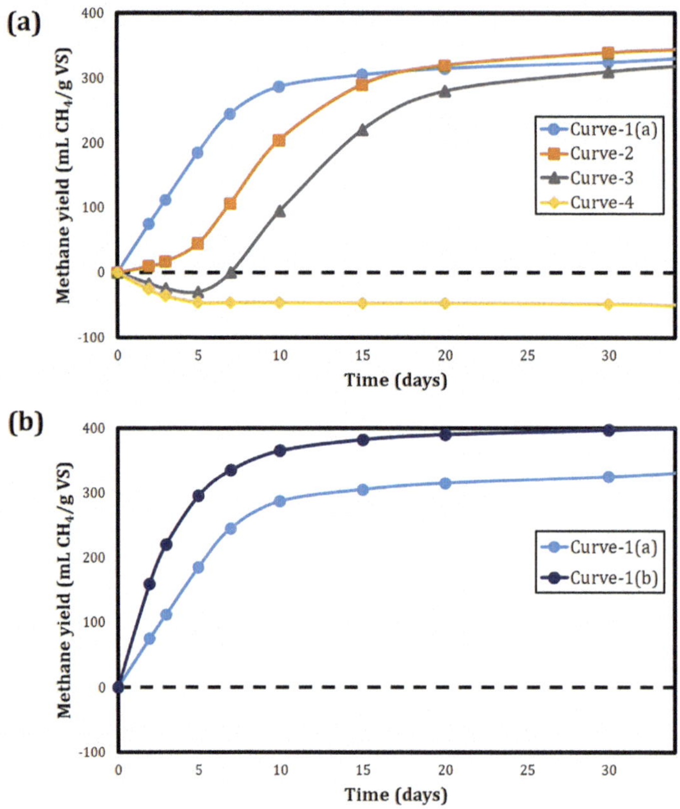


**FIG. 9.2 Examples of possible $CH_4$ yield curves.**

The curves represent the cumulative $CH_4$ yields for different substrates as a function of time, under different operational conditions: (A) for different test materials, and (B) for the same test material before (Curve-1(A)) and after (Curve-1(B)) pretreatment.

## 2.2 Feedstocks for AD

Crop residues, food waste, and agricultural waste are most commonly used as feedstock owing to their easy availability and are characterized by their lignocellulosic composition (Cheng and Brewer, 2021; Kucharska et al., 2018; Mudhoo, 2012). Lignocellulosic biomass (LCB) is stoichiometrically composed of polysaccharides and biopolymers such as cellulose $[C_6H_{10}O_5]_n$, hemicellulose $[C_5H_{10}O_5]_m$, and lignin $[C_{10}H_{11}O_{3.5}]$ (Basak et al., 2023; Ojo, 2023; Usmani et al., 2020). Fig. 9.3A highlights the approximate relative fractions of lignin, cellulose, and hemicellulose in the VS fractions of some commonly used lignocellulosic substrates (Cai et al., 2017). Animal waste and manure are rich in proteins, lipids, and carbohydrates from unprocessed tissues, blood, and by-products like fur, feathers, and hide (Habchi et al., 2024). Fig. 9.3B shows the approximate chemical composition of some slaughterhouse wastes in terms of their TS fractions (Habchi et al., 2024).

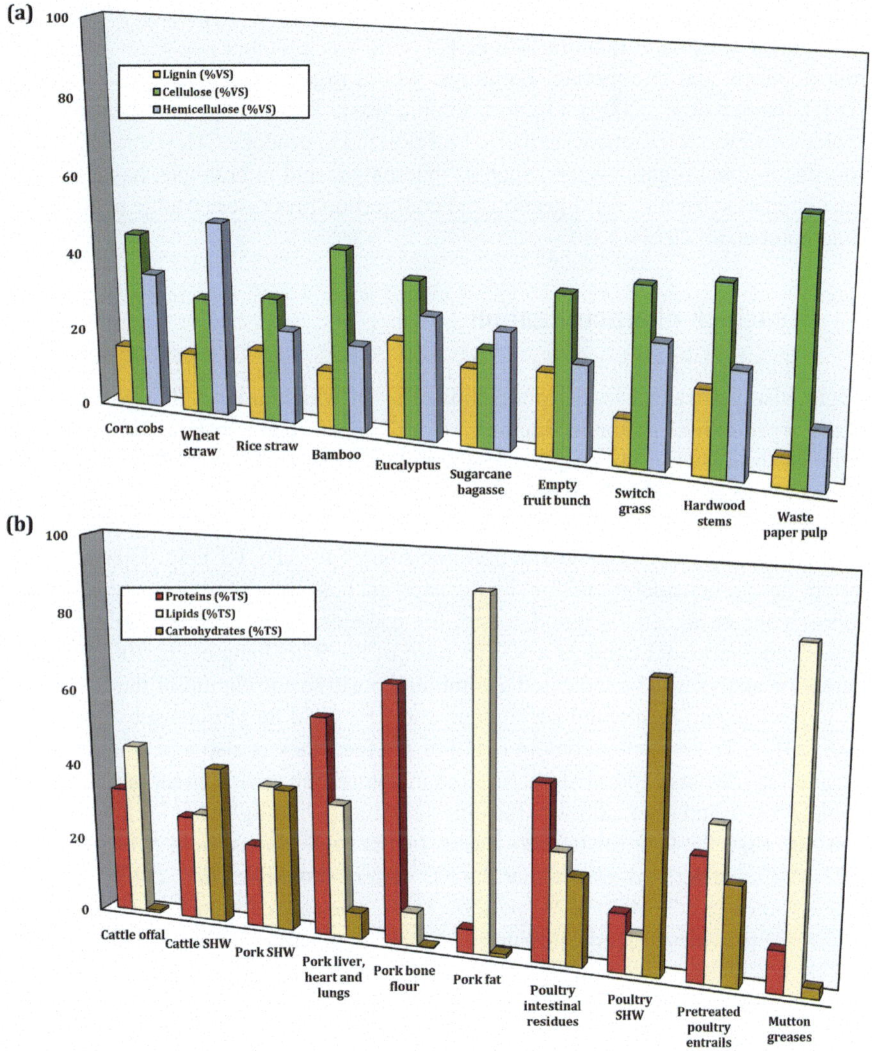


**FIG. 9.3 Approximate composition of some commonly used substrates.**

The figure shows the major chemical constituents of: (A) lignocellulosic biomass, (B) animal wastes. For LCB, higher percentages of hemicellulose and cellulose are preferred due to their readily degradable nature. The fiber-rich nature of lignin is the primary reason for the recalcitrant nature of LCB. Similarly, a greater concentration of energy-dense compounds like lipids is favored in animal wastes, while high amounts of proteins often lead to the formation of inhibitory compounds like $NH_3$ in the reactor.

Other substrates used include effluents from industries, wastewater sludge, and even animal waste and manure. Municipal solid waste comprises waste generated from domestic and commercial buildings, whose organic fraction is also used for AD (Dalbanjan et al., 2025). Organic soluble waste from sludge stabilization (i.e., aerobic wastewater treatment) can also be valorized to produce $CH_4$ (Gerardi, 2003). Nonconventional organic waste, such as macroalgae and microalgae, has also been considered as substrates for digestion and codigestion processes (Atelge et al., 2020; Khanthong et al., 2023).

## 2.3 Feedstock characterization

The organic feedstock chosen for an AD process is typically characterized by its dry weight, also referred to as the total solids (TS), of which around 70%–90% should be the organic biodegradable fraction, or the VS, for it to substantially contribute to the AD process (Orhorhoro, 2017; Vögeli et al., 2014). TS and VS are important parameters to consider when assessing operation specifications, loading rates, and digester performance (Chynoweth et al., 1987). A higher value of VS implies better yields, since only the organic fraction contributes to AD. TS help determine important design considerations for the digester, such as Bingham viscosity and yield stress (Nsair et al., 2020). Food waste, for instance, is an excellent substrate for biogas generation and exhibits a VS/TS ratio of 80–97 (Zhang et al., 2007). Linear regression analyses have indicated a combined positive correlation of the VS/TS ratio, lignin content, hemicellulose content, and $CH_4$ yield for various substrates (Yan et al., 2017). The moisture content (MC) of the feedstock is also a parameter worth considering (Avinash and Mishra, 2023). A moisture-rich environment facilitates microbial growth and interactions in the digester. Studies highlight that food waste with high MC (like 93.43% for tomato waste) favors high biogas yield (Mrosso et al., 2023). MC also determines whether the AD system should be "dry" ($MC < 85\%$, or $TS > 15\%$) or "wet" ($MC > 85\%$, or $TS < 15\%$). The calculation of the $CH_4$ yield of a substrate provides a good estimate of how economically viable its choice for AD will be. While feedstock with high BDI is undeniably preferable, substrates with low BDI can still be considered for biogas generation by considering pretreatment and codigestion options.

A typical macronutrient ratio of C/N/P/S of 600:15:5:1 is optimal for sustainable microbial growth (Nsair et al., 2020). In general, a C/N ratio of around 20–30, determined based on the total organic carbon and the total nitrogen in the biodegradable fraction of the biomass, is preferred for it to be considered for AD (Forster-Carneiro et al., 2012). A higher value of the ratio might indicate that the stoichiometric requirement of nitrogen by the bacteria is not being met, thus hindering their growth. This lowers the yield of volatile fatty acids (VFAs) and, consequently, biogas in the system. On the contrary, a substantially low value of the ratio might imply the accumulation of $NH_3$ in the system, which acts as an inhibitor to the efficiency of the methanogens (Kondusamy and Kalamdhad, 2014). Table 9.1 shows approximate values of the C/N ratio for some commonly used substrates. This ratio is, however,

**Table 9.1** C/N ratios for substrates commonly used in anaerobic digestion (Forster-Carneiro et al., 2012).

| Substrate | | C/N ratio |
|---|---|---|
| Sludge | Sewage sludge | 6:1 |
| | Digested sludge | 9–16:1 |
| Municipal solid waste | Organic fraction | 9–12:1 |
| Food waste | Sweet potato | 25:1 |
| | Sugarcane bagasse | 100:1 |
| | Restaurant waste | 14–38:1 |
| Lignocellulosic biomass | Corn straw | 60–120:1 |
| | Corn silage | 10:1 |
| | Shredded paper | 175:1 |
| | Cardboard waste | 350:1 |
| Animal waste | Cow manure | 20-35:1 |
| | Pig manure | 10–20:1 |

flexible in the sense that it can be optimized by adding carbon-rich or nitrogen-rich materials to the substrate for anaerobic codigestion.

# 3 Pretreatment of feedstock

## 3.1 Need for pretreatment

The polysaccharides present in LCB can be easily hydrolyzed into simple sugars. However, owing to the complex molecular structure of lignin that binds these molecules, even with the standard biogasification techniques, only around 20%–30% efficiency is achieved. The high degree of inter- and intramolecular hydrogen bonds in the hydroxyl groups in cellulose, due to its characteristic β-(1,4)-glycosidic bonds, significantly adds to the recalcitrance of LCB structures (Mankar et al., 2021). Animal waste with insoluble proteins resists degradation by proteolytic enzymes due to their interlinked hydrophobic polypeptide chains. Exocellular polymers in sludge limit its digestibility due to their binding mechanism. In the case of macroalgae, the thick cell walls of the cells need to be broken to ensure better liquefaction by releasing the intracellular substances into the solution for proper digestion by methanogenic microbes (Atelge et al., 2020).

Thus pretreatment of the feedstock becomes imperative, wherein various means and methods are employed with the common goal of simplifying the inherent molecular structure of the substrate, thereby enhancing the substrate's availability for microbial hydrolysis and, consequently, the entire AD process (Atelge et al., 2020; Nguyen et al., 2021; Mudhoo, 2012), as illustrated in Fig. 9.4. Pretreatment is often carried out in combination with the hydrolysis stage of AD, directly within the

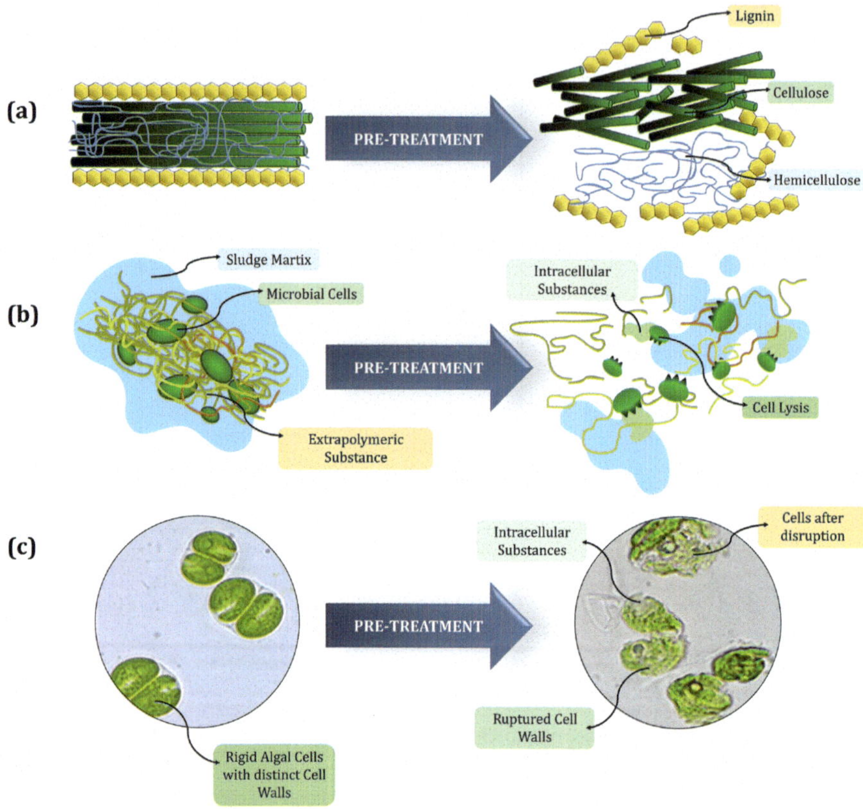


**FIG. 9.4 Simplified schematic diagram outlining the pretreatment of different types of feedstocks.**

The figure highlights the effect of pretreatment on: (A) lignocellulosic biomass, (B) wastewater sludge, and (C) algae.

digester. Pretreatment options can be broadly classified into physical, chemical, physicochemical, and biological methods. Each of these methods is described in the following sections.

## 3.2 Physical pretreatment

This form of pretreatment involves the breakdown of substrate by employing mechanical, thermal, or even ultrasonic agents, aiming to enhance the anaerobic biodegradability of the substrate and, in turn, increase the yield efficiency (Tiehm et al., 2001). However, these processes can often be power-intensive (Seidl and Goulart, 2016).

Mechanical pretreatment aims to physically reduce (using methods such as chipping, grinding, milling, shearing, and extrusion) the size of the substrate cells by rupturing their cell walls, decreasing the crystallinity of the complex constituent molecules, and increasing the surface area available to microbes for enzymatic hydrolysis (Arce and Kratky, 2022).

Thermal pretreatment involves substrate biodegradation using heat to destabilize hydrogen bonds in feedstock, partially solubilize the organic constituents (proteins, carbohydrates, and VFAs) and inorganic constituents (such as heavy metals, sulfur, and phosphorus) of the substrate, and even reduce the substrate volume (Appels et al., 2010; Kovačić et al., 2024). The most common thermal pretreatment options are hydrothermal pretreatment (HTP) at around 200°C–260°C and 700 psi, and low-temperature pyrolysis at around 200°C–300°C, with HTP being preferred for substrates with higher MC (Lee and Park, 2020; Yan et al., 2009).

Ultrasound-assisted pretreatment, on the contrary, uses high-frequency waves to improve the biodegradability of biomass before it undergoes hydrolysis. Waves of frequency greater than 20 kHz are mechanically transmitted through the substrate. This results in preexisting air microbubbles in the slurry medium producing local extremes of high temperature and pressure when they collapse (Kerrou et al., 2023; Pandiselvam et al., 2023). This phenomenon of acoustic cavitation leads to the chemical degradation of polymers and organic matter, as well as the formation of microtubes in the substrate tissue, which increases the surface area available for AD (Flores et al., 2021).

## 3.3 Chemical pretreatment

Chemical pretreatment involves the utilization of strong acids (the most common ones being sulfuric acid $H_2SO_4$ and phosphoric acid $H_3PO_4$), alkalis (such as solutions of caustic soda NaOH, caustic potash KOH, or slaked lime $Ca(OH)_2$), or oxidizing agents (such as ozone $O_3$, hydrogen peroxide $H_2O_2$, or even organic solvents such as ethanol $CH_3CH_2OH$) to treat biomass, either before or during the hydrolysis phase (Kucharska et al., 2018; Song et al., 2024). Fig. 9.5 shows the type of compounds that can be produced by the action of a reagent on LCB (Stamatelatou et al., 2012). These reagents work to enhance the biodegradability of organics in the solid phase and make them more susceptible to enzymatic attacks. The goal should be to maximize the production of simpler monomers and minimize that of the inhibitors.

Other pretreatment options include using aqueous ammonia ($NH_4^+$) to aid in delignification and cellulose modification, enhance its hydrolysis rate, and control the concentration of pathogens, such as *E. coli*, within it (Forster-Carneiro et al., 2012). Organic solvents (like solutions of 1-butyl-3-methylimidazolium halides) can also be utilized for these methods (Liu et al., 2024). Studies also indicate that the application of these reagents can significantly decrease the hydraulic retention time (HRT) of biomass in the AD process.

## 3.4 Physicochemical pretreatment

The conventional methods discussed above come with their own disadvantages, ranging from the physical methods often having high operational costs to the

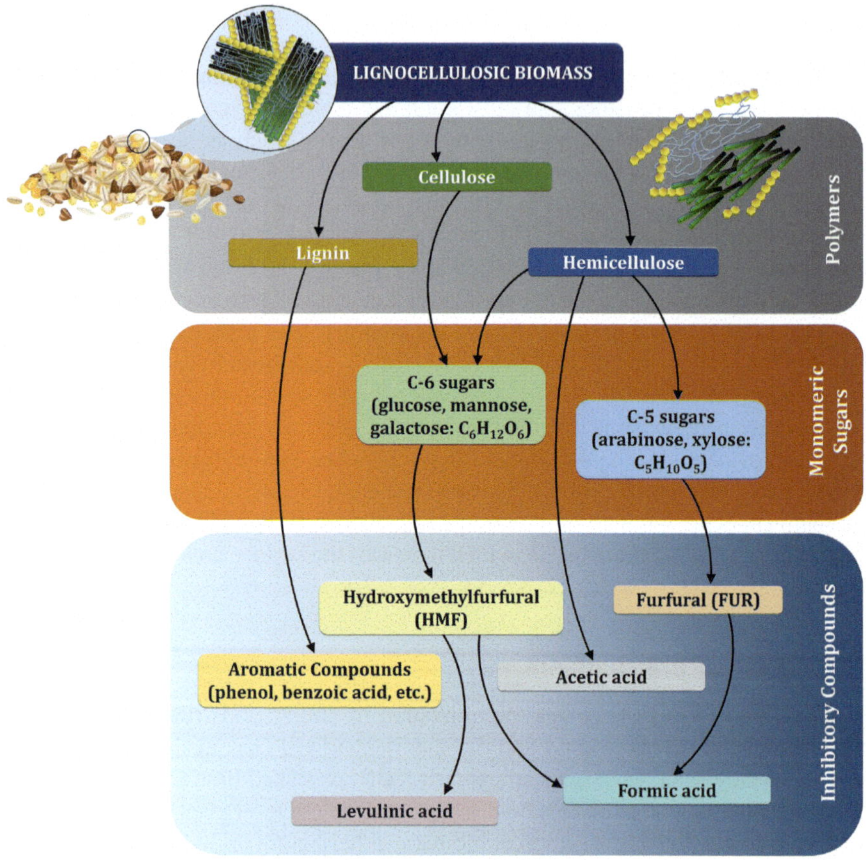


**FIG. 9.5 Effect of chemical pretreatment of lignocellulosic substrate.**

Three categories of compounds are formed: (A) simpler polymers (shown in the gray box), (B) monomeric sugars (shown in the orange box), and (C) inhibitory compounds (shown in the blue box).

chemical methods leading to the formation of toxic by-products and, in turn, causing equipment corrosion (Basak et al., 2023; Seidl and Goulart, 2016). Furthermore, the effectiveness of these mechanisms often increases when two or more of them are combined. These integrated approaches have, therefore, been intensively studied due to their economic and environmental merits.

A highly investigated pretreatment method, liquid hot water (LHW) pretreatment, is typically performed at around 150°C–260°C. LHW significantly alters complex polymeric structures, degrading them into soluble parts. Studies have observed an exponential increase in biogas yield after pretreatment with LHW, with the rate kinetics showing even better yields at elevated pressures and when aided by chemical catalysts (Chen et al., 2022; Rastogi and Shrivastava, 2017; Wan and Li, 2011). Another method that implements heat, pressure, and water for the pretreatment of substrate is the steam explosion method. This pretreatment method uses steam at high temperature and pressure to partially hydrolyze substrates. The feedstock, if dry, is first soaked in water to moisten it. Then, it is heated in a confined system at around 160°C–220°C, leading to a pressure build-up, which is then abruptly released after a short retention time. This causes a phase explosion, which leads the intracellular water in the biomass tissues to vaporize, improves its soluble dietary fiber content, and, in turn, enhances the biogas yield from the AD process (Cao et al., 2025; Ma et al., 2025).

$CO_2$ explosion can also enhance the digestibility of polymers such as lignin and hemicellulose. This method uses high-pressure $CO_2$ (up to around 28 MPa) heated to a specific temperature for a specified period of time. This supercritical $CO_2$ penetrates the biomass through small cellular pores. $CO_2$ also dissolves in water, forming carbonic acid, which further breaks down complex structures such as those of cellulose and hemicellulose (Kumar et al., 2009). When the pressure built up due to the $CO_2$ is explosively released, macromolecules deteriorate, increasing the surface area of the feedstock available for hydrolysis. Furthermore, the resulting increase in pore size aids the glucose yield from the biomass during the pretreatment process (Kucharska et al., 2018). This method also ensures that no inhibitory compounds are formed during the pretreatment process that would result in the loss of lignin or polysaccharide structures Olatunji et al. (2021). This method is, however, efficient for wet biomass only (Kucharska et al., 2018).

Pretreatment with ammonia under elevated temperature and pressure conditions has also yielded promising results. Ammonia fiber expansion employs anhydrous liquid ammonia (1:1 w/w) at around 60°C–170°C, under high pressures, for a short reaction time (5–60 min). This is followed by a sudden decompression of the system, resulting in the rapid evaporation of the ammonia and consequent cooling of the system (Kucharska et al., 2018; Mankar et al., 2021). Ammonia recovery process, on the contrary, makes use of an engineered column reactor where aqueous (5%–15%) ammonia solution is made to percolate through a packed column at elevated temperatures (150°C–180°C), at a particular flow rate, for a retention time of 10 min. The solution can be recovered and recycled, which makes the process economical (Kucharska et al., 2018).

## 3.5 Biological pretreatment

Microbes can be used to degrade complex macromolecules in feedstocks, ranging from selected bacterial strains, like *Sphingomonas paucimobilis* and *Bacillus circulans*, to several species of wood-decomposing fungi, such as white-rot, brown-rot, corrosive-rot, and soft-rot fungi (Forster-Carneiro et al., 2012; Atelge et al., 2020). Depending on the enzymatic requirements of the feedstock being used and the operational parameters being maintained, microbes are carefully selected for biological pretreatment.

A common method to increase $CH_4$ yield from feedstock is by carrying out the biogas production in two stages, with the first stage (hydrolysis and acidogenesis) being conducted in one reactor (pH 5.5–6.5), and the second stage (acetogenesis and methanogenesis) being conducted in another reactor (pH 6.8–7.2) (Atelge et al., 2020). While operational parameters such as pH, temperature, and HRT need to be meticulously engineered for both digesters, the two-stage AD (TSAD) process has shown significantly better yields of $CH_4$, with around 17% to 60% higher biogas yields. Sometimes, aerobic microbes can also be employed in the first reactor to increase the availability of organic matter for the anaerobic microbes in the subsequent stages. Sludge solubilization can also be enhanced by adding enzymes during the hydrolysis stage, such as cellulase, lipase, endoglucanases, exoglycanases, and dextranase, to the digester. The sugars thus produced act as a source of carbon for the fermenting microbial consortia that ultimately produce $CH_4$ (Atelge et al., 2020; Kucharska et al., 2018). While the energy and chemical requirements for these processes are minimal, the efficiencies of these pretreatment options are often lower than those previously discussed.

Ultimately, the efficiency of a pretreatment method is assessed in terms of how low its rate of production of inhibitory compounds is, how minimal its energy and chemical requirements are, and how low the cost of setting up and running the equipment is. Each of these methods has its own advantages and drawbacks, which is why industries often adopt hybrid methods of pretreatments that are a combination of two or more of the mechanisms discussed above, which, despite their complexities, have proven to be more efficient (Usmani et al., 2020).

# 4 Chemistry of AD

## 4.1 Process microbiology of AD

Most of our present understanding of AD stems from fermentation studies in the highly controlled environment of the rumen (Chynoweth et al., 1987). Recent advances in metagenomics have helped us better explore the microbial consortia and their metabolic interactions that drive the AD process (Kim et al., 2022). Our understanding of the microbial communities involved in AD mainly stems from the Microbial Database for Activated Sludge (MiDAS) project. A recent study has utilized the full-length 16S rRNA gene sequencing data of 285 full-scale AD plants from around the globe to develop the MiDAS 5 database (Dueholm et al., 2024). The AD process involves a cascade of complex biochemical reactions that break down biodegradable organic waste into biogas by utilizing microbes in the absence of oxygen. The complexity arises because the products produced by one group of microorganisms serve as the substrate for the next. As an insoluble substrate initially approaches a cell, an exoenzyme is secreted by the cell that coats the substrate outside of the cell. The enzyme then solubilizes the substrate, which then enters through the cell membrane and gets fermented by an endoenzyme within the protoplasm of the cell (Gerardi, 2003). The entire process of AD can be divided into four procedural stages, as depicted in

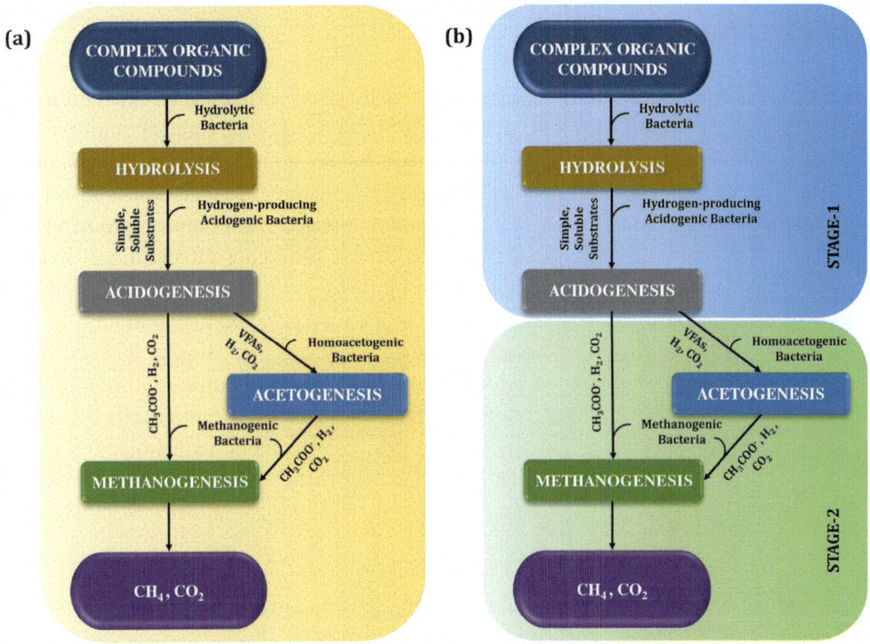


**FIG. 9.6 Schematic flowchart of the AD process.**

The figure highlights the intermediates formed in (A) a single-stage digester and (B) a two-stage digester.

Fig. 9.6. The different exoenzymes and endoenzymes that take part in the different stages in AD are exhaustively listed in Table 9.2, along with their functions.

**a. Hydrolysis** (Greek: *hydro*: water and *lysis*: unbind):

This step refers to the chemical breakdown of complex, insoluble organic compounds into simpler, soluble monomers and polymers by utilizing water. This is done by facultative hydrolytic bacteria that utilize extracellular hydrolases (EC 3 enzymes) to add hydrogen (H–) and hydroxide (OH–) groups (from water molecules) at the site of molecular cleavage of an H–O bond, facilitated by a change in the pH (a slightly acidic environment being optimal) of the medium. Hydrolysis is a relatively slow step, and even more so when solid waste is being digested, making it the rate-determining step for most AD processes (Anukam et al., 2019; Yadav et al., 2022). Fermentative bacteria of the genera *Enterobacter*, *Streptococcus*, *Lactobacillus*, *Bacteroides*, *Halocella*, *Trichococcus*, *Propionibacterium*, *Sporobacterium*, and *Bifidobacterium* are most commonly involved in feedstock hydrolysis. The general pathways for the

**Table 9.2** List of exoenzymes and endoenzymes produced by microbes at different stages of AD and their functions.

| AD stage | Enzyme type | Enzyme name | Enzyme class | Function | References |
|---|---|---|---|---|---|
| Hydrolysis | Exoenzyme | Cellulase | Hydrolase (glycosidase) | Cleaves β-(1,4)-glycosidic linkages in cellulose microfibrils in plant biomass, facilitating its conversion to monomers. | (Singh et al., 2021; Wang et al., 2020) |
| | | Hemicellulase | | Deconstructs hemicelluloses (like xylans, xyloglucans, and arabinoxylans) into monomers by targeting the glycosidic linkages in plant biomass. | (Agregán et al., 2022; Akram et al., 2024) |
| | | Amylase | | Targets α-(1,4)-glycosidic linkages in starch and degrades them into monosaccharides, such as glucose and maltose. | (Ali et al., 2025; Wang et al., 2020) |
| | | Pectinase | | Cleaves (1→4)-α-D-galacturonic linkages in pectins within plant cell walls through depolymerization and deesterification reactions. | (Yi, 2021) |
| | | Protease | Hydrolase (peptidase) | Catalyzes the solubilization of proteins by cleaving peptide bonds and further breaking peptidoglycans into amino acids. | (Pang et al., 2023) |
| | | Lipase | Hydrolase (esterase) | Hydrolyzes crude lipids and triacylglycerols to glycerol and fatty acids at the water–lipid interface by cleaving ester bonds, minimizing the inhibition caused by lipid accumulation. | (Affes et al., 2017; Meng et al., 2017) |

| | | | | | |
|---|---|---|---|---|---|
| Acidogenesis | Endoenzyme | Hydrogenase | Oxidoreductase | Catalyzes the reversible oxidation of $H_2$, playing a crucial role in interspecies electron transfer during the formation of VFAs. | (Hendrickson and Leigh, 2008) |
| | | Alcohol dehydrogenase | | Oxidizes alcohols to aldehydes and ketones (like ethanol to acetaldehyde and propanol to propanal) and vice versa, facilitating the recycling of $NAD^+$ and NADH during WLP. | (Torres and Ayala, 2013) |
| | | Lactate dehydrogenase | | Maintains the glycolytic flux in the system by oxidizing lactate to pyruvate, which can be converted into acetyl-CoA and then into acetate. | (Rosenbaum and Müller, 2025) |
| | | Aldehyde dehydrogenase | | Oxidizes aldehydes to carboxylic acids, thereby contributing to the VFA pool for short-chain fatty acids. | (Holbrook et al., 2022) |
| | | Formate lyase | Transferase | Cleaves pyruvate into acetyl-CoA and subsequently into acetate. | (Crain and Broderick, 2014) |
| | | Acetate kinase | | Catalyzes acetyl phosphate and ADP into acetate and ATP and is essential for the conservation of energy during acidogenesis. | (Gheshlaghi et al., 2009) |

*Continued*

**Table 9.2** List of exoenzymes and endoenzymes produced by microbes at different stages of AD and their functions—cont'd

| AD stage | Enzyme type | Enzyme name | Enzyme class | Function | References |
|---|---|---|---|---|---|
| Acetogenesis | Endoenzyme | Carbon monoxide dehydrogenase | Oxidoreductase | Catalyzes the oxidation of CO to $CO_2$ (or vice versa) and enables carbon fixation by the carboxyl branch of the WLP. | (Pezacka and Wood, 1984) |
| | | Formate dehydrogenase | | Catalyzes the reduction of $CO_2$ to formate (or vice versa) and enables carbon fixation by the methyl branch of the WLP. | (Zhou et al., 2018) |
| | | Hydrogenase | | Catalyzes the production and consumption of $H_2$ during the syntrophic interactions between the acetogenesis and methanogenesis phases. | (Harirchi et al., 2022) |
| | | Acetyl-CoA synthase | Transferase | Catalyzes the formation of acetyl-CoA at the pivot of the WLP by coupling CO, CoA, and a methyl group. | (Karekar et al., 2022) |
| | | Formyl tetrahydrofolate synthetase | Ligase | Catalyzes the transfer of a one-carbon (formyl) group from 10-formyltetrahydrofolate during the enzymatic reduction of $CO_2$ to acetate. | (Singh et al., 2021) |
| Methanogenesis | Endoenzyme | Formylmethanofuran dehydrogenase | Oxidoreductase | Catalyzes the reversible conversion of $CO_2$ to formylmethanofuran and also the production of $CH_4$ from $CO_2$ and $H_2$ during methanogenesis. | (Goswami et al., 2016; Wintsche et al., 2018) |
| | | Methylene-H4MPT dehydrogenase | | Catalyzes the oxidation of methylene-$H_4$MPT to methenyl-$H_4$MPT using $NADP^+$ (and vice versa) during the conversion of $CO_2$ into $CH_4$. | (Grabarse et al., 1999) |
| | | F420-reducing hydrogenase | | Uses $H_2$ to reduce the coenzyme $F_{420}$, which is a crucial electron donor in hydrogenotrophic methanogenesis. | (Hendrickson and Leigh, 2008) |
| | | Coenzyme F420-dependent Methylene-$H_4$MPT reductase | | Catalyzes the reduction of methylene-$H_4$MPT to $N^5$-methyl-$H_4$MPT during hydrogenotrophic methanogenesis. | (Aufhammer et al., 2005; Shima et al., 2000) |
| | | Methyl-CoM reductase | Transferase | Uses its coenzyme $F_{430}$ to reduce CoM to $CH_4$ during methanogenesis. | (Chen et al., 2020) |
| | | Methenyl-H4MPT cyclohydrolase | Hydrolase | Catalyzes the reversible conversion of methenyl-$H_4MPT^+$ from formyl-$H_4$MPT during the one-carbon metabolic reduction of $CO_2$ to $CH_4$. | (Catenacci et al., 2022) |

hydrolysis of common substrates, along with the first-order hydrolysis rate constants for their degradation kinetics, are listed below (Cesaro and Belgiorno, 2014):

$$(-C_6H_{10}O_5-)_n + nH_2O \rightarrow nC_6H_{12}O_6 \left[k = 0.5 - 0.2\,days^{-1}\right]$$
$$Polysaccharide + Water \rightarrow Monosaccharides$$

$$(-NH-CHR-CO-)_n + nH_2O \rightarrow nH_2N-CHR-COOH \left[k = 0.25 - 0.8\,days^{-1}\right]$$
$$Protein + Water \rightarrow Aminoacids$$

$$C_3H_5(OOCR)_n + nH_2O \rightarrow C_3H_5(OH)_n + nRCOOH \left[k = 0.1 - 0.7\,days^{-1}\right]$$
$$Lipid + Water \rightarrow Glycerol + Fattyacids$$

Complex carbohydrates (polysaccharides) are broken down into simpler carbohydrates (monosaccharides), like glucose, by saccharolytic hydrolases (Gerardi, 2003). Gram-positive hydrolytic bacteria of the genus *Cellulomonas* produce the enzyme cellulase, which degrades cellulose (Gerardi, 2003), while amylase, produced by certain strains of bacteria like *Bacillus subtilis*, *Bacillus halotolerans*, *Pseudomonas nitroreducens*, and *Streptococcus lutetiensis,* targets starch (Sanjaya et al., 2024). The proteolytic hydrolase, protease, breaks down proteins, first into polypeptides and then into amino acids by breaking the peptide bonds (Gerardi, 2003). Similarly, lipase, a lipolytic exoenzyme, cleaves the ester bonds in fats (lipids) to produce fatty acids and glycerol.

**b. Acidogenesis** (Latin: *acidus*: sour/acid and Greek: *genesis*: formation):

This is the second step of the AD process, where facultative and obligate anaerobes metabolize the simple monomers formed during hydrolysis into alcohols such as ethanol (favored at $pH < 5$), VFAs (favored at $pH > 5$) (Anukam et al., 2019). Only acetic acid ($CH_3COOH$), $CO_2$, and $H_2$ can be directly converted into $CH_4$ by methanogens, while the other VFAs (like formic, propionic, butyric, and valeric acids) need to go through the acetogenesis stage first. Some methanogens can also utilize formic acid (HCOOH). Fermentation of amino acids can also lead to the formation of $NH_3$ gas in the reactor. Typical governing reactions (for the fermentation of glucose) are listed as follows (Anukam et al., 2019):

$$C_6H_{12}O_6 \rightleftarrows 2CH_3CH_2OH + 2CO_2$$

$$C_6H_{12}O_6 + 2H_2 \rightleftarrows 2CH_3CH_2COOH + 2H_2O$$

$$C_6H_{12}O_6 \rightarrow 3CH_3COOH$$

Among the nonmethanogenic bacteria that have been isolated as acidogens are *Clostridium* spp., *Peptococcus anaerobus*, *Ruminococcus* spp., *Bifidobacterium* spp., *Desulfovibrio* spp., *Corynebacterium* spp., *Lactobacillus*, *Actinomycetes*, *Eubacterium* spp., and *Escherichia coli* (Harirchi et al., 2022; Yadav et al., 2022). Fermentative fungi of the genera Aspergillus, Humicola, and Penicillium can also be

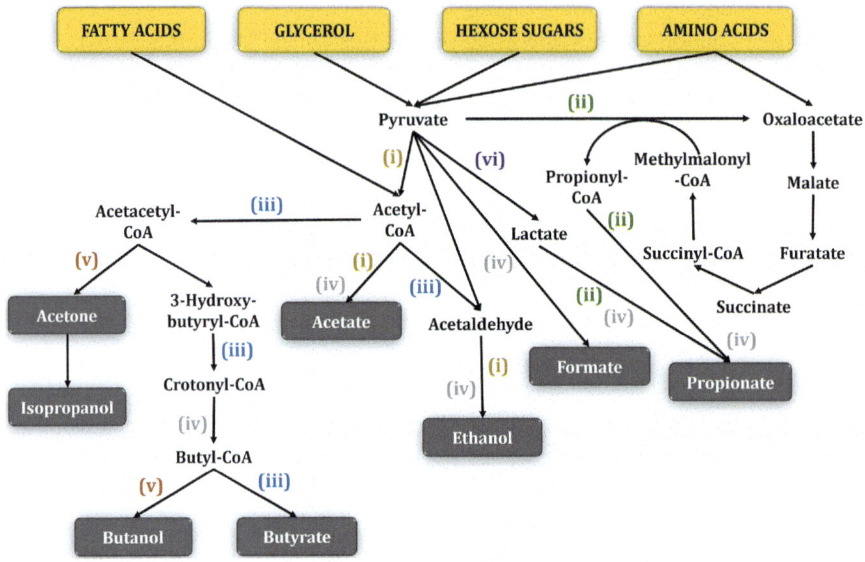


**FIG. 9.7 Flowchart outlining the metabolic pathways for the production of VFAs.**

The pathways demonstrated are: (i) acetate–ethanol fermentation, (ii) propionate fermentation, (iii) butyrate fermentation, (iv) mixed-acid fermentation, (v) acetone–butanol fermentation, and (vi) lactate-type fermentation. The compounds in the yellow boxes show the different products formed after the hydrolytic catabolism of the substrate in the digester, and the compounds in the gray boxes show the major products formed during acidogenesis through the pathway(s) represented by the numbers.

employed in this stage (Harirchi et al., 2022). The pathways for the formation and distribution of different VFAs are shown in Fig. 9.7 (Yadav et al., 2022; Zhou et al., 2018). Pyruvate proves to be a pivotal point in this network, and one that produces a plethora of products. Fatty acids undergo β-oxidation and break down into Acetyl-CoA units, as shown below (Louis and Flint, 2009).

**c. Acetogenesis** (Latin: *acetum*: vinegar and Greek: *genesis*: formation):

In this step, the long-chain VFAs and alcohols formed during acidogenesis are converted into $CH_3COO^-$, $CO_2$, and $H_2$ by syntrophic acetogens (Yadav et al., 2022). Acetogens are not merely bacteria that produce acetate as a product. These microbes characteristically follow the Wood–Ljungdahl pathway (WLP) and produce acetyl-CoA starting from two molecules of one-carbon source molecules (like $CO_2$, CO, and $HCOO^-$) (Schuchmann and Müller, 2016). The overall reaction for WLP can be represented as follows:

$$4H_2 + 2CO_2 \rightarrow CH_3COOH + 2H_2O\,[\Delta G_0' = -95.0 kJ/mol]$$

Examples include bacteria from genera like *Acetobacterium*, *Syntrophomonas*, *Acetanaerobacterium*, *Anaerobacterium*, and *Candidatus*. The $H_2$ gas produced by syntrophic acetate-oxidizing bacteria (Yadav et al., 2022) is responsible for the dehydrating nature of this stage (Anukam et al., 2019). This effect can be inhibitory to the growth of acetogens, but some hydrogen-scavenging methanogenic bacteria utilize this hydrogen to produce $CH_4$ (Anukam et al., 2019).

$$CH_3CH_2COO^- + 3H_2O \rightarrow CH_3COO^- + HCO_3^- + H^+ + 3H_2\,[\Delta G_0' = +76.1 kJ/mol]$$

$$CH_3CHOHCOO^- + 2H_2O \rightarrow CH_3COO^- + HCO_3^- + H^+ + 2H_2\,[\Delta G_0' = -4.2 kJ/mol]$$

$$CH_3CH_2OH + H_2O \rightarrow CH_3COO^- + H^+ + 2H_2\,[\Delta G_0' = +9.6 kJ/mol]$$

$$CH_3OH + 2H_2O \rightarrow HCO_3^- + H^+ + 3H_2\,[\Delta G_0' = +23.5 kJ/mol]$$

Since around 70% of the total $CH_4$ is produced from the action of methanogens on the acetate produced in this step, acetogenesis effectively determines the overall efficiency of the AD process (Anukam et al., 2019; Yadav et al., 2022).

**d. Methanogenesis** (Greek: *methy* wine/alcohol and *genesis* formation):

This is the final stage of the AD process, driven by strictly anaerobic archaea acting on the products formed in the previous steps, as shown below. The principal genera of methanogens are rod-shaped (e.g., *Methanosaeta* spp., *Methanobacterium* spp., and *Methanobacillus* spp.) or spherical (e.g., *Methanolobus* spp., *Methanococcoides* spp., *Methanosarcina* spp., and *Methanococcus* spp.). As already discussed, methanogenesis occurs in close symbiosis with acetogenesis due to the production of $H_2$ by acetogens and its simultaneous utilization by methanogens. Archaea of the genera *Methanoplasmatales*, *Methanobacteriales*, *Methanomicrobiales*, and *Methanosarcinales* are most commonly used in AD (Harirchi et al., 2022).

$$4H_2 + HCO_3^- + H^+ \rightarrow CH_4 + 3H_2O[\Delta G_0' = -135.6 kJ/mol]$$

$$CH_3COO^- + H_2O \rightarrow CH_4 + HCO_3^-\,[\Delta G_0' = -31.0 kJ/mol]$$

$$CH_3OH \rightarrow \frac{3}{4}CH_4 + \frac{1}{4}HCO_3^- + \frac{1}{4}H^+ + \frac{1}{4}H_2O[\Delta G_0' = -78.2 kJ/mol]$$

Acetophilic methanogens convert $CH_3COO^-$ into $CH_4$ by decarboxylation, while hydrogenophilic methanogens reduce $CO_2$ and $H_2$ into $CH_4$ (Anukam et al., 2019). Methylotrophic methanogenesis utilizes methyl alcohol ($CH_3OH$) to produce $CH_4$ (Gerardi, 2003). There are six known major pathways for methanogens to utilize a reagent to produce $CH_4$, namely, $CH_3COOH$, $HCOOH$, $CO_2$, $(CH_3)_2SO_4$, $CH_3OH$, and $CH_3NH_2$ (Anukam et al., 2019). Furthermore, the efficiencies of these pathways are substantially affected by the nature of the substrate used and the operational parameters employed for the AD process. $CH_4$, being sparingly soluble in water, leads to the product naturally separating in the digester, thus limiting any downstream processing requirements (Hunter et al., 2021). The presence of sulfate-reducing bacteria

(SRB), like *Desulfobulbus* spp. and *Desulfomonas* spp., competes with methanogens for $H^+$ and $CH_3COO^-$, and produces inhibitory compounds like $H_2S$, which hinder the efficiency of the archaea (Harirchi et al., 2022).

## 4.2 Syntrophy in AD

Syntrophic interactions among the different microbial communities are imperative for the efficient functioning of an anaerobic system. For instance, the thermodynamic dependence between acetogens and methanogens has been demonstrated in the following example of two strains isolated from a culture of *Methanobacillus omelianskii* (Zhang et al., 2023).

$$\textbf{StrainS}: 2CH_3CH_2OH + 2H_2O \rightarrow 2CH_3COO^- + 2H^+ + 4H_2\,[\Delta G_0' = +9.6 kJ/mol]$$

$$\textbf{StrainMoH}: 4H_2 + CO_2 \rightarrow CH_4 + 2H_2O[\Delta G_0' = -131 kJ/mol]$$

$$\textbf{Co-culture}: 2CH_3CH_2OH + CO_2 \rightarrow 2CH_3COO^- + 2H^+ + CH_4\,[\Delta G_0' = -122 kJ/mol]$$

It is evident from the above reactions that the production of acetate by Strain S (acetogens) has a positive Gibbs free energy under standard conditions, making it thermodynamically impractical. However, in the coculture, not only is the overall reaction thermodynamically feasible, but the utilization of $H_2$ by Strain MoH (methanogens) as a substrate to produce $CH_4$ further promotes the activity of the acetogens to produce more (Zhang et al., 2023). Furthermore, the expression for Gibbs free energy ($\Delta G_0$) for the coculture system can be given as:

$$\Delta G_0 = \Delta G_0' + RT\ln\left(\frac{[CH_3COO^-][H_2]^4}{[CH_3CH_2OH]^2}\right)$$

A negative value for Gibbs free energy is observed only when the simultaneous production and utilization of $H_2$ in the acetogenic and methanogenic phases maintain the hydrogen partial pressure (HPP) at $<10^{-4}$ atm. This also ensures that fermentation products other than $CO_2$ and $CH_4$ do not accumulate in the system, while promoting the further production of $CH_3COO^-$ and $H^+$ (Gerardi, 2003; Zhang et al., 2023).

This example clearly explains the obligate dependence of acetogens and methanogens in an anaerobic digester. It justifies why the accumulation of VFAs, without the syntrophy between the hydrogenotrophic methanogens and VFA-oxidizing acetogens, in the reactor deteriorates the $CH_4$ yield (Yue et al., 2021; Zhang et al., 2023). Research has also been conducted to expand the concept of syntrophy beyond acetogenesis to the broader microbial consortia involved in AD, particularly to the oxidation of VFAs such as propionate and butyrate, which also rely heavily on methanogens and $H_2$ concentrations of $10^{-5}$–$10^{-3}$ atm. Process parameters like temperature and pH also play a role in these thermodynamic relationships and are imperative to the design of an AD plant.

# 5 Process description of AD

## 5.1 Design considerations and operational parameters

The growth patterns of the microbes, their enzymatic actions on substrates, and their syntrophic interactions with one another in AD systems are heavily reliant on the environment within the reactors. Any change in the digester environment needs to be gradually introduced to account for the acclimatization of the microbes within it. In general, fermentative bacteria are more tolerant to change than methanogenic archaea. Thus overloading the digester with substrate can hinder the metabolism of methanogens, while acidogens continue producing VFAs. This can lead to acidification or "souring" of the system, which affects digester performance (Vögeli et al., 2014). Alongside a judicious choice of the feedstock to be used, there are several design considerations and operational parameters that need to be meticulously engineered for efficient digester performance (Meegoda et al., 2018; Nsair et al., 2020).

### *5.1.1 Phase separation*

Separating the acid-forming and methane-forming phases in AD, as opposed to conducting the whole process in a single digester, yields better results in terms of VS removal, the quantity of $CH_4$ produced, and even its quality (Dareioti et al., 2021; Mazzanti et al., 2025). Performing AD in multiple stages facilitates optimization of variables in each reactor, thus generating better overall results (Dareioti et al., 2021). A comparison of the codigestion of food waste and activated sludge in single-stage and two-stage processes reported an increase of 26% in biogas production, along with a greater percentage of $CH_4$, in the latter (Baldi et al., 2019). Fig. 9.8 shows the schematic design of the equipment used in single-stage AD and TSAD.

### *5.1.2 Feeding mode*

Depending on the way the feedstock is introduced into the digester, AD technologies can also be broadly classified into batch mode and continuous mode. In a batch mode process, the feedstock is introduced into the digester all at once. The digester is then appropriately closed for a defined retention time before being opened again. Batch digesters prove to be an economically viable option for many developing nations due to their simple design and low implementation cost (Mir et al., 2016). In the continuous mode of feeding, the feedstock is introduced into the digester at a constant flow rate, while an equivalent volume of slurry exits it (Vögeli et al., 2014). Researchers attest to the enhanced biogas yields by continuous mode processes, reporting a productivity of 2.78 $CH_4$/L/day at 8.6 kg COD/$m^3$/day, using diluted food waste as feedstock (Park et al., 2018). Newer technologies that incorporate the advantages of both these methods, while optimizing the economic feasibility of the overall AD process, are also being studied.

### *5.1.3 Temperature*

Temperature significantly influences the growth rates and chemical kinetics of different microbes within the reactor. Methanogens are the most temperature-sensitive,

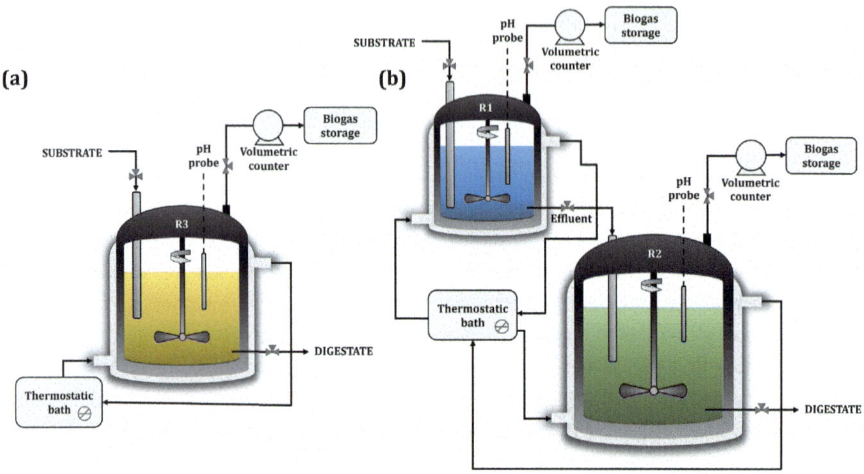


**FIG. 9.8 Schematic design of the AD process, carried out in: (A) a single-stage digester, (B) a two-stage digester.**

A single-stage digester (R3) optimally operates in a single stage at a pH of around 6.8 to 7.2 and at 35°C–37°C. For a two-stage digester, the fermentation stage (R1) typically operates at around 55°C and within an acidic pH range of 5.5–6.5 to enhance the hydrolysis phase. The methane-forming stage (R2) operates at around 35°C and within a near-neutral pH regime of 6.8–7.2 to stabilize the methanogenesis phase. The temperature of each reactor is maintained by circulating water in a thermostatic vest, as shown.

exhibiting optimum growth rates under mesophilic conditions (25°C–40°C). 37°C is optimal for most mesophilic enzymes (Nsair et al., 2020). Some studies also argue that thermophilic conditions (50°C–70°C) work better for substrates that are easily biodegradable, requiring shorter start-up times and enhanced reaction rates (Nsair et al., 2020). Ultimately, there is no universal temperature range that works for every type of feedstock, and it needs to be tailored according to the process requirements. Beyond these ranges, $CH_4$ yields drop dramatically, and the formation of inhibitory compounds is also possible. Fluctuations beyond ± 3°C/day for mesophilic systems and ± 1°C/day for thermophilic systems should be avoided (Nsair et al., 2020). Reactors are typically built underground to utilize the temperature buffer capacity of the soil (Vögeli et al., 2014). Another approach is to use a thermostatic jacket that utilizes hot air circulation to maintain temperatures inside the reactor, as shown in Figure 12.8.

### 5.1.4 pH

Different microbes have different pH ranges optimal for their growth. Acidogens typically prefer a pH of around 5.0–6.0, while methanogens are better suited to a pH of around 6.5–8.0 (Nsair et al., 2020). Monitoring and optimizing pH becomes easier in TSAD, as it becomes easier to cater to the specific requirements of the fermenting and methane-producing groups of microbes. The pH of the reactor also plays an important role in determining the metabolic routes of production and distribution of VFAs during acidogenesis. An increase in alkalinity can lead to a decrease in pH and thus a decrease in acid production. The concentrations of VFAs and inhibitory compounds like $NH_3$ and $H_2S$ also affect the pH within the reactors (Nsair et al., 2020). Digesters typically operate at a pH of 6.5–8.4, and an alkalinity of around 3000 mg/L as $CaCO_3$ is ideally maintained at all times to provide sufficient buffering capacity (Nsair et al., 2020; Vögeli et al., 2014). Reagents like lime (CaO) and sodium bicarbonate ($NaHCO_3$) are used to control the reactor pH when it becomes too acidic (Vögeli et al., 2014).

### 5.1.5 Organic loading rate

The total capacity of a digester should include the total volume of the slurry and the volume of the gas produced, designed on the basis of a fixed retention period. Furthermore, the total volume should also account for the slurry rise in the digester. Organic loading rate (OLR) is an important parameter to design a digester that economically yields biogas. OLR refers to the amount of substrate introduced per unit volume of the reactor at a given time (Nsair et al., 2020; Vögeli et al., 2014). OLR can be calculated as:

$$OLR\left(kgoDM/m^3day\right) = \frac{m \times c}{V_R \times 100}$$

where m (in kg $day^{-1}$) refers to the amount of substrate fed per unit time, c (% oDM) is the concentration of organic dry matter (oDM), and $V_R$ (in $m^3$) is the total volume of the reactor. A low OLR leads to a lower $CH_4$ yield efficiency, while a very high OLR might lead to the accumulation of VFAs in the system (Nsair et al., 2020). OLR is typically in the range of 4–8 kg VS/$m^3$ day for continuous stirred tank reactors (CSTRs) and below 2 kg VS/$m^3$ day for nonstirred systems (Vögeli et al., 2014).

### 5.1.6 Hydraulic retention time

HRT, or θ, refers to the time the feedstock remains in the digester. HRT is an important design parameter for AD, since an increased HRT implies a lowered OLR. HRT can be calculated as follows:

$$\theta\left(day\right) = \frac{V_R}{Q}$$

where $V_R$ (in $m^3$) is the digester volume and Q is the volumetric flow rate of the digester.

Shorter HRTs can be associated with the accumulation of VFAs, leading to inhibitory effects (Meegoda et al., 2018). HRTs below the range of 10–25 days may lead to the washout of microbes from the reactor and, thus, lower biogas yields. For a typical digester, the HRT should optimally range from 16 to 60 days (Nsair et al., 2020). For the purpose of AD, HRT is often used interchangeably with solid retention time.

### 5.1.7 Mixing

Mixing in the reactor helps maintain the homogeneity in the substrate medium by continuously inoculating new substrate with the microbes, thus aiding microbial metabolism (Mazzanti et al., 2025; Nsair et al., 2020). It reduces the stagnation of sediments at the bottom layer and the accumulation of foam and scum at the upper layer of the slurry. It also prevents stratification due to gradients of temperature or pH, enhances process stability, and aids the upward flow of bubbles through the medium (Nsair et al., 2020).

Based on the design considerations discussed above, several reactor models have been developed to maximize $CH_4$ yields while maintaining the economic feasibility of the AD process. Some of the commonly used reactor models are discussed below. The quality of feedstock used, its TS percentage, and the operational parameters are important considerations in deciding the type of reactor to be used (Kiani et al., 2022).

a. **Continuous stirred tank reactor:**

A CSTR consists of a well-mixed tank where microbes and the substrate are continuously stirred to ensure homogeneity (Ahmad et al., 2016). Continuous feeding maintains steady biomass and gas production. CSTRs typically operate at mesophilic or thermophilic temperatures and a neutral pH. They are mostly used for slurries with moderate solid concentrations (5%–10% TS) (Kiani et al., 2022).

b. **Upflow anaerobic sludge blanket (UASB):**

This reactor is a high-rate digester where the slurry is made to flow upward through a dense sludge bed of anaerobic granules (Ajmal et al., 2025; España-Gamboa et al., 2012; Kiani et al., 2022). The upward fluid flow facilitates natural mixing within the reactor. A three-phase separator at the top of the reactor retains the biomass and forces heavy sludge to settle at the bed, while the biogas and the effluent leave (Kiani et al., 2022). Above the granular bed, a lighter sludge blanket forms with a lower solid concentration. UASBs are widely used for municipal sewage and industrial wastewater, especially in warm climates. High $CH_4$ yields at lower HRTs are a major advantage of these reactors (Kiani et al., 2022).

c. **Plug-flow reactor (PFR):**

Anaerobic plug-flow digesters are long, horizontal, tubular tanks with no mixing along their length. PFRs are often fed with high-solids slurries (10%–15% TS) and can be operated under mesophilic or thermophilic conditions. PFRs do not require additional mixing. For comparable HRT, PFRs tend to achieve similar or higher VS conversion than CSTRs because plug-flow hydraulics and high solids provide longer contact times (Kiani et al., 2022). PFRs are typically used for dry digestion processes, utilizing feedstocks like manure slurries and municipal wastes.

d. **Fixed-bed reactor (FBR):**

An anaerobic filter or an FBR packs an inert supporting medium (like gravel, granular activated carbon (GAC), polyvinyl chloride (PVC), plastic, foam) through which wastewater flows, allowing microbes to attach and form biofilms. Common fixed-bed designs include upflow filters, anaerobic fluidized beds, and structured or randomly packed beds. Because the biomass is retained on the media, HRT can be much shorter than the solids retention time, making these systems high-rate. Furthermore, no mixing is required, which is an added advantage (Kiani et al., 2022). An average COD removal efficiency of 70% was observed in the treatment of domestic sewage in an FBR, and the system maintained its stability and consistently operated over a remarkably long period (Silva et al., 2024).

## 5.2 Inhibitors in AD

Depending on their concentration in the reactor, certain compounds can be toxic to the microbial consortia involved (Vögeli et al., 2014). The most common inhibitors reported in the literature include $O_2$, $H_2S$, $NH_3$, metal ions, and certain organics. Methanogens are typically the most severely affected by these compounds (Nsair et al., 2020; Vögeli et al., 2014). The levels at which these species hinder $CH_4$ yields vary depending on the operational parameters adopted.

### 5.2.1 Oxygen

Obligate anaerobes show little to no tolerance to $O_2$. Studies have reported damage to chromosomal DNA and complete cell lysis in these microbes in an $O_2$-rich environment. Further, facultative anaerobes, in the presence of $O_2$, can metabolize through aerobic respiration, which can decrease $CH_4$ yields in the digester. However, micro-aeration with $O_2$ has proven to improve hydrolysis rates and decrease the concentration of $H_2S$ in the produced biogas (Botheju, 2011; Nsair et al., 2020). Limited aeration can also provide some buffer to pH instabilities and VFA accumulation in the digester (Botheju, 2011). Generally, an $O_2$ level of $<0.1$ ppm is tolerable by methanogens (Nsair et al., 2020).

### 5.2.2 Sulfide

Sulfate ($SO_4^{2-}$) in substrates like industrial wastewater can be easily reduced to sulfide ($S^{2-}$) by SRB. The reduction of $SO_4^{2-}$ using $H_2$ and $CH_3COO^-$ is thermodynamically more favored than their utilization by methanogens to produce $CH_4$. SRBs can utilize a variety of compounds, making them potential competitors to even the fermentative bacteria in the reactor (Chen et al., 2008). SRBs generate $H_2S$, which is particularly inhibitory to acetotrophic methanogens, with an $IC_{50}$ of 160 ppm (Chen et al., 2014). These bacteria also act as alkalinity sources and decrease the $CH_4$ yield (Nsair et al., 2020). $S^{2-}$ above 50 ppm has been reported to be toxic to unacclimatized methanogens (Chen et al., 2014).

### 5.2.3 Ammonia

$NH_3$ is mainly formed by the degradation of nitrogenous substrates, like proteins and urea. Optimal concentrations of $NH_3$ and $NH_4^+$ help maintain adequate buffering capacity in the digester (Nsair et al., 2020). However, excessive amounts of these species can act as inhibitors in the medium, with methanogens being the most susceptible (Chen et al., 2008). Free $NH_3$ can diffuse passively into cells and cause a proton imbalance. Total ammonia nitrogen levels above 1400–17,000 ppm have been found to be inhibitory (Nsair et al., 2020). Engineered concentrations of ions like $Na^+$, $Ca^{2+}$, and $Mg^{2+}$ can be antagonistic to $NH_3$ inhibition and improve $CH_4$ yields (Chen et al., 2008).

#### 5.2.4 Organic compounds

Organic compounds of limited solubility and nonpolar nature can get adsorbed onto the sludge solids and accumulate in the digester. Other organics can degrade by pathways that introduce other toxic compounds into the reactor. These organics can hinder cell metabolism by disrupting ion gradients and, consequently, affect $CH_4$ yields (Chen et al., 2008). Some of these organics include alkyl and halide derivatives of benzenes, phenols, and alcohols; aldehydes and ketones; organic compounds containing nitrogen; and long-chain fatty acids (Nsair et al., 2020). The degradation of substrates like LCBs also produces several inhibitory organics, such as furfurals and their derivatives (Xia et al., 2022). Therefore addressing the accumulation of these compounds in the reactor and exploring viable degradation pathways that supplement $CH_4$ yields are of utmost importance.

## 6 Products of AD

### 6.1 Biogas

#### 6.1.1 Storage, purification, and utilization of biogas

The biogas produced is initially stored in facilities that are engineered to be resistant to UV, temperature, pressure, and other environmental conditions. Conventional units include floating drums, fixed domes, biogas storage bags, tanks, and cylinders, depending on the pressure at which it is being stored. Compression of biogas helps minimize the volume of the storage unit (Vögeli et al., 2014). Conventional techniques to clean and upgrade biogas are mentioned below:

**a. Absorption:**

Adsorption is typically performed using water under high pressure or with organic solvents such as N-methylpyrrolidone, methanol, and polyethylene glycol ether. The driving principle is that the solubility of gases like $CO_2$, $H_2S$, and $NH_3$ in these solvents is much higher than that of $CH_4$. The scrubbers can also be integrated with heat, pressure, and ultrasound systems to target specific gases, including traces of $N_2$ and $O_2$ present (Jepleting et al., 2025).

**b. Adsorption:**

Gases like $CO_2$, $H_2S$, and $NH_3$, as well as moisture, can be adsorbed into the bulk of porous materials such as kaolinite, zeolites, silica gel, activated carbon, iron-rich soils, and biochar. However, the efficiency of the mass transfer from the gaseous to the solid phase decreases as a function of time (Jepleting et al., 2025).

**c. Membrane separation:**

Membrane separation employs a semipermeable membrane to separate the gaseous impurities from the produced biogas. Membranes are typically made of inorganics like ceramics and metal-based materials, or polymers of different types. This method is not energy-intensive, but an effective upgrading of biogas requires several membranes to be used in series (Jepleting et al., 2025).

**d. Cryogenic separation:**

This method is based on the differential condensation points of different components of biogas. The raw biogas is cooled to extreme temperatures at above 80 bar, and the impurities like $CO_2$, $H_2S$, and $NH_3$ appropriately separate as a liquid or solid fraction. The $CH_4$ accumulates in the gas phase, with the effluent gas stream being 90%–99% pure. The set-up and running costs are, however, very high (Upadhyay et al., 2022). Currently, several biological methods of upgrading are also being studied. $CO_2$ can be converted to $CH_4$ by the chemoautotrophic or photoautotrophic action of hydrogenotrophic methanogens. Introduction of $H_2$ into the reactor is another option (Upadhyay et al., 2022). Biogas shows promise in terms of reducing greenhouse gas (GHG) emissions and replacing traditional fuel sources. Once upgraded, it can be easily used as a fuel source at domestic levels. Energy can be utilized from the direct combustion of gas for purposes like cooking and as lamps. Industrial applications include utilization in industrial burners and the production of electricity at off-grid levels (Paramasivam et al., 2025). Roughly 6 kWh of energy can be produced from 1 $m^3$ of biogas, of which 2 kWh can be utilized as useful electricity (Vögeli et al., 2014).

## 6.2 Digestate

### 6.2.1 Posttreatment and valorization of digestate

The carbon content of the feedstock is not completely exhausted by microbes at the end of the AD process. The digestate produced has a semifluid consistency, with the MC depending on the digestion process implemented (Mazzanti et al., 2025). The ever-expanding AD market implies an increasing production of residual digestate, which can be repurposed to promote a circular economy (Mazzanti et al., 2025). A study, for instance, estimated a yield of 0.20–0.47 tons of digestate per ton of food waste utilized (Ren et al., 2018). Separation of the solid and liquid phases of the effluent from AD is the most rudimentary form of posttreatment in practice. Commercially available precipitating agents or coagulants are added to the digestate to facilitate this phase separation (Chozhavendhan et al., 2023). The subsequent treatments of the solid and liquid fractions are usually purpose-specific.

**a. Liquid fraction processing (recovery of N and P):**

The abundance of nitrogen in the liquid fraction makes its stripping economical. This is done using air or water, and the gaseous $NH_3$ thus produced is countercurrently treated with acids like $H_2SO_4$, $H_3PO_4$, or $HNO_3$ to produce ammonium salts such as $(NH_4)_2SO_4$, $(NH_4)_3PO_4$, or $NH_4NO_3$, respectively, all of which are commercial fertilizers (Chozhavendhan et al., 2023). Other recovery techniques, such as ozonation, microwave irradiation, vacuum evaporation, and membrane technologies, have also been studied (Mazzanti et al., 2025). Precipitation reactions with select ionic salts can produce phosphorus-based slow-release fertilizers like struvite ($MgNH_4PO_4 \cdot 6H_2O$), vivianite ($Fe_3(PO_4)_2 \cdot 8H_2O$), hydroxyapatite ($Ca_5(PO_4)_3$), and calcium phosphate ($Ca_3(PO_4)_2$) (Khaita et al., 2025; Mazzanti et al., 2025).

**b. Solid fraction processing (production of char and other carbon-based compounds):**

Carbon constitutes 15%–55% of the solid fraction, making its exploitation through thermochemical processes promising. Pyrolysis at optimized temperatures can yield a range of products, including biochar, pyrogas, and biooil. Gasification with sub-stoichiometric $O_2$ levels can generate syngas, tar, and char. Hydrothermal carbonization at milder temperatures can also produce hydrochar. The solid residue from these processes can be utilized for soil amendment or to further produce products like bioethanol and biomethane (Wang et al., 2023). These novel integrated solid/liquid fractions valorization pathways further amplify the notability of AD in promoting a circular economy.

# 7 Environmental impacts of AD

## 7.1 Environmental impact assessment

A comprehensive understanding of the environmental impacts of AD is crucial to understanding its potential as a sustainable waste management practice. As opposed to conventional waste management practices like incineration or landfilling, AD prevents the release of $CO_2$ and $CH_4$ into the atmosphere. This makes it sustainable in terms of reducing GHG emissions. Improper disposal of organic wastes also potentially contaminates soil and groundwater resources through the leaching of pollutants like heavy metals, inorganic salts, and organic acids (Dagwar and Dutta, 2024; Dalbanjan et al., 2025). AD also addresses this issue by utilizing these wastes to produce bioenergy and other valuable products. Furthermore, the utilization of biogas to produce energy reduces the dependence on nonrenewable energy sources like coal, liquefied petroleum gas, and fossil fuels (Dalbanjan et al., 2025; Piadeh et al., 2024). Valorization of wastes to produce energy also reduces the stress on energy crops, and thus on land use. Consequently, arable land can be reserved for agricultural purposes.

However, the successful implementation of AD in waste management has certain prerequisites. A continuous supply of feedstock is imperative for biogas generation. Pretreatment methods have energy requirements, and the optimization parameters need to be carefully monitored for maximized $CH_4$ yields. Furthermore, leakages in digesters can release GHGs like $CH_4$ and $CO_2$ into the atmosphere, thus causing global warming. Public awareness and preparedness to switch from chemical fertilizers to biofertilizers are also important to successfully utilize the digestate (Dalbanjan et al., 2025).

## 7.2 Life-cycle assessment

A comprehensive cradle-to-grave analysis of sustainability metrics at each step involved in AD is essential for understanding its benefits and challenges. A thorough life-cycle assessment (LCA) of an AD plant should typically be a four-tier process, listed as follows (Dalbanjan et al., 2025):

1. Assessment of the sourcing of the feedstock to be used and, consequently, its pretreatment requirements,
2. Assessment of the AD process itself, in terms of the digestion efficiency,
3. Assessment of the upgrading and utilization of the biogas produced, and
4. Assessment of the posttreatment and valorization of the effluent digestate.

A positive energy balance for the process implies that more energy is produced by the process than is consumed by it. This is because the energy that can be generated from the $CH_4$ produced is greater than the net total of the direct and indirect energies utilized in all the processes associated with its production (Dalbanjan et al., 2025). The total GHG emissions from the production and utilization of the biogas are represented by the carbon footprint of the process. AD also has a low carbon footprint since it captures the $CH_4$ that would otherwise be released into the atmosphere (Dalbanjan et al., 2025). Water footprint refers to the net water consumption in each of these steps. It is typically low for AD since the MC of the feedstock itself is utilized. The water utilized in the collection of the feedstock, and the water saved from using the digestate as organic fertilizers, are also included in this calculation (Dalbanjan et al., 2025). LCA studies help researchers, engineers, and policymakers assess the environmental sustainability of large-scale implementations of digestion plants (Vasco-Correa et al., 2018).

# 8 Conclusion

AD continues to remain the most common in-plant sludge digestion method, and with good reason. Beyond enabling the recovery of a fraction of the feedstock's embedded chemical energy, it also produces a nutrient-rich digestate, which can be processed into several value-added products. This allows for a complete closed-loop valorization of organic wastes. Further, pretreatment options can be engineered based on the BMP of the feedstock to maximize $CH_4$ yields. The operational parameters adopted (such as the pH, temperature, OLR, and HRT), and even the design specifications of the reactor itself, impact the reaction kinetics of biogas production. Notably, minimizing sharp changes in pH and temperature and controlling the production of inhibitory compounds (like $NH_4^+$, $H_2S$, $O_2$, and certain organics) help maximize $CH_4$ production at a consistent rate. It is also crucial to understand the syntrophic interactions within the microbial communities inside the reactor. Cleaning and posttreatment of the products of AD are also imperative to enhance the calorific value of the biogas produced and the nutrient profile of the digestate formed. While the implementation of AD on an industrial scale often calls for a judicious trade-off between operational costs (like net energy consumption) and the biogas yield of the feedstock, environmental impact assessment and LCA studies attest to the viability of AD as a sustainable waste management technology. In addition, policies and regulations related to agriculture, energy management, and waste handling continue to evolve to encourage the effective incorporation of AD for waste valorization all around the globe.

## References

Affes, M., Aloui, F., Hadrich, F., Loukil, S., Sayadi, S., 2017. Effect of bacterial lipase on anaerobic co-digestion of slaughterhouse wastewater and grease in batch condition and continuous fixed-bed reactor. Lipids Health Dis. 1476-511X. 16 (1). doi:10.1186/s12944-017-0587-2.

Agregán, R., Lorenzo, J.M., Kumar, M., Shariati, M.A., Khan, M.U., Sarwar, A., Sultan, M., Rebezov, M., Usman, M., 2022. Anaerobic digestion of lignocellulose components: challenges and novel approaches. Energies 19961073. 15 (22). Available from: http://www.mdpi.com/journal/energies/. doi:10.3390/en15228413.

Ahmad, A., Buang, A., Bhat, A.H., 2016. Renewable and sustainable bioenergy production from microalgal co-cultivation with palm oil mill effluent (POME): a review. Renew. Sustain. Energy Rev. 13640321. 65, 214–234. Available from: http://www.journals.elsevier.com/renewable-and-sustainable-energy-reviews/. doi:10.1016/j.rser.2016.06.084.

Ajmal, Z et al., 2025. A comprehensive review on MXenes for various applications. Applied Energy 397, 126136. doi:10.1016/j.apenergy.2025.126136.

Akram, F., Fatima, T., Ibrar, R., Shabbir, I., Shah, F.I., Haq, I.U., 2024. Trends in the development and current perspective of thermostable bacterial hemicellulases with their industrial endeavors: a review. Int. J. Biol. Macromol. 18790003. 265. Available from: https://www.sciencedirect.com/science/journal/01418130. doi:10.1016/j.ijbiomac.2024.130993.

Ali, Z., Abdullah, M., Yasin, M.T., Amanat, K., Sultan, M., Rahim, A., Sarwar, F., 2025. Recent trends in production and potential applications of microbial amylases: a comprehensive review. Protein Expr. Purif. 10465928. 227, 106640. doi:10.1016/j.pep.2024.106640.

Anukam, A., Mohammadi, A., Naqvi, M., Granström, K., 2019. A review of the chemistry of anaerobic digestion: methods of accelerating and optimizing process efficiency. Processes. 2227-9717. 7 (8), 504. doi:10.3390/pr7080504.

Appels, L., Degrève, J., Van der Bruggen, B., Van Impe, J., Dewil, R., 2010. Influence of low temperature thermal pre-treatment on sludge solubilisation, heavy metal release and anaerobic digestion. Bioresour. Technol. 09608524. 101 (15), 5743–5748. doi:10.1016/j.biortech.2010.02.068.

Arce, C., Kratky, L., 2022. Mechanical pretreatment of lignocellulosic biomass toward enzymatic/fermentative valorization. iScience. 25890042. 25 (7), 104610. doi:10.1016/j.isci.2022.104610.

Atelge, M.R., Atabani, A.E., Banu, J.R., David Krisa, M.K., Eskicioglu, C., Kumar, G., Changsoo Lee, Y.Ş.Y., Unalan, S., Mohanasundaram, R., Duman, F., 2020. A critical review of pretreatment technologies to enhance anaerobic digestion and energy recovery. Fuel. 00162361. 270, 117494. doi:10.1016/j.fuel.2020.117494.

Aufhammer, S.W., Warkentin, E., Ermler, U., Hagemeier, C.H., Thauer, R.K., Shima, S., 2005. Crystal structure of methylenetetrahydromethanopterin reductase (Mer) in complex with coenzyme F420: architecture of the F420/FMN binding site of enzymes within the nonprolyl cis-peptide containing bacterial luciferase family. Protein Sci. 14 (7), 1840–1849. Available from: http://www.proteinscience.org/cgi/reprint/14/7/1840. doi:10.1110/ps.041289805Germany.

Avinash, L.S., Mishra, A., 2023. Enhancing biogas production in anaerobic digestion of MSW with addition of bio-solids and various moisture sources. Fuel. 00162361. 354, 129414. doi:10.1016/j.fuel.2023.129414.

Baldi, F., Pecorini, I., Iannelli, R., 2019. Comparison of single-stage and two-stage anaerobic co-digestion of food waste and activated sludge for hydrogen and methane production. Renew. Energy 18790682. 143, 1755–1765. Available from: http://www.journals.elsevier.com/renewable-and-sustainable-energy-reviews/. doi:10.1016/j.renene.2019.05.122.

Basak, B., Kumar, R., Bharadwaj, A.V.S.L.S., Kim, T.H., Kim, J.R., Jang, M., Oh, S.E., Roh, H.S., Jeon, B.H., 2023. Advances in physicochemical pretreatment strategies for lignocellulose biomass and their effectiveness in bioconversion for biofuel production. Bioresour. Technol. 18732976. 369. Available from: https://www.elsevier.com/locate/biortech. doi:10.1016/j.biortech.2022.128413.

Botheju, D., 2011. Oxygen effects in anaerobic digestion – a review. Open Waste Manag. J. 18764002. 4 (1), 1–19. doi:10.2174/1876400201104010001.

Cai, J., He, Y., Yu, X., Banks, S.W., Yang, Y., Zhang, X., Yu, Y., Liu, R., Bridgwater, A.V., 2017. Review of physicochemical properties and analytical characterization of lignocellulosic biomass. Renew. Sustain. Energy Rev. 18790690. 76, 309–322. Available from: https://www.journals.elsevier.com/renewable-and-sustainable-energy-reviews. doi:10.1016/j.rser.2017.03.072.

Cao, R., Gao, Y., Wang, L., Chaomin, L., Wang, Z., Yinta, L., Qiu, J., 2025. Mechanistic insights into starch behavior and starch-protein interactions in whole grain oat flour: structural and functional modifications by steam explosion treatment. Int. J. Biol. Macromol. 01418130. 307 (Pt 2), 142058. doi:10.1016/j.ijbiomac.2025.142058.

Catenacci, A., Santus, A., Malpei, F., Ferretti, G., 2022. Early prediction of BMP tests: a step response method for estimating first-order model parameters. Renew. Energy. 09601481. 188, 184–194. doi:10.1016/j.renene.2022.02.017.

Cesaro, A., Belgiorno, V., 2014. Pretreatment methods to improve anaerobic biodegradability of organic municipal solid waste fractions. Chem. Eng. J. 13858947. 240, 24–37. doi:10.1016/j.cej.2013.11.055.

Chen, H., Gan, Q., Fan, C., 2020. Methyl-coenzyme M reductase and its post-translational modifications. Front. Microbiol. 1664-302X. 11, 578356. Available from: https://www.frontiersin.org/article/10.3389/fmicb.2020.578356/full. doi:10.3389/fmicb.2020.578356.

Chen, J.L., Ortiz, R., Steele, T.W.J., Stuckey, D.C., 2014. Toxicants inhibiting anaerobic digestion: a review. Biotechnol. Adv. 07349750. 32 (8), 1523–1534. Available from: https://www.elsevier.com/inca/publications/store/5/2/5/4/5/5/index.htt. doi:10.1016/j.biotechadv.2014.10.005.

Chen, W.-H., Nižetić, S., Sirohi, R., Huang, Z., Luque, R., M.Papadopoulos, A., Sakthivel, R., Nguyen, X.P., Hoang, A.T., 2022. Liquid hot water as sustainable biomass pretreatment technique for bioenergy production: a review. Bioresour. Technol. 09608524. 344 (Pt A), 126207. doi:10.1016/j.biortech.2021.126207.

Chen, Y., Cheng, J.J., Creamer, K.S., 2008. Inhibition of anaerobic digestion process: a review. Bioresour. Technol. 09608524. 99 (10), 4044–4064. Available from: https://linkinghub.elsevier.com/retrieve/pii/S0960852407001563. doi:10.1016/j.biortech.2007.01.057.

Cheng, F., Brewer, C.E., 2021. Conversion of protein-rich lignocellulosic wastes to bio-energy: review and recommendations for hydrolysis + fermentation and anaerobic digestion. Renew. Sustain. Energy Rev. 13640321. 146, 111167. doi:10.1016/j.rser.2021.111167.

Chozhavendhan, S., Karthigadevi, G., Bharathiraja, B., Kumar, R.P., Lata Deso Abo, S.V.P., Balachandar, R., Jayakumar, M., 2023. Current and prognostic overview on the strategic exploitation of anaerobic digestion and digestate: a review. Environ. Res. 00139351. 216 (Pt 2), 114526. doi:10.1016/j.envres.2022.114526.

Chynoweth, D.P., Isaacson, R., Chynoweth, D.P., 1987. Anaerobic digestion of biomass.

Crain, A.V., Broderick, J.B., 2014. Pyruvate formate-lyase and its activation by pyruvate formate-lyase activating enzyme. J. Biol. Chem. 1083351X. 289 (9), 5723–5729. Available from: http://www.jbc.org/content/289/9/5723.full.pdf+html. doi: 10.1074/jbc.M113.496877

da Silvada Silva, J.A., Braga, A.F.M., Quartaroli, L., Fermoso, F.G., Zaiat, M., Silva, G.H.R.D., 2024. Long-Term anaerobic structured fixed-bed reactor operation for domestic sewage treatment: performance and metal dynamics. Methane. 2674-0389. 3 (3), 421–436. doi:10.3390/methane3030024.

Dagwar, P.P., Dutta, D., 2024. Landfill leachate a potential challenge towards sustainable environmental management. Sci. Total Environ. 00489697. 926, 171668. doi:10.1016/j.scitotenv.2024.171668.

Dalbanjan, N.P., Korgaonkar, K., Kadapure, A.J., Halladamani, S.B., Ramangouda, G., K, P.K.S., 2025. Green energy from waste: evaluating the sustainability of anaerobic biofuel technologies. The Microbe. 29501946. 7, 100410. doi:10.1016/j.microb.2025.100410.

Dareioti, M.A., Vavouraki, A.I., Tsigkou, K., Kornaros, M., 2021. Assessment of single- vs. Two-stage process for the anaerobic digestion of liquid cow manure and cheese whey. Energies 19961073. 14 (17). Available from: https://www.mdpi.com/1996-1073/14/17/5423/pdf. doi:10.3390/en14175423.

Donatelli, J.A., Chang, S., 2024. Biological methane potentials of food waste of different components: methane yields, production kinetics, and element balance. Bioresour. Technol. 09608524. 413, 131435. doi:10.1016/j.biortech.2024.131435.

Dueholm, M.K.D., Andersen, K.S., Korntved, A.K.C., Rudkjøbing, V., Alves, M., Bajón-Fernández, Y., Batstone, D., Butler, C., Cruz, M.C., Davidsson, Å., Erijman, L., Holliger, C., Koch, K., Kreuzinger, N., Lee, C., Lyberatos, G., Mutnuri, S., O'Flaherty, V., Oleskowicz-Popiel, P., Pokorna, D., Rajal, V., Recktenwald, M., Rodríguez, J., Saikaly, P.E., Tooker, N., Vierheilig, J., De Vrieze, J., Wurzbacher, C., Nielsen, P.H., 2024. MiDAS 5: global diversity of bacteria and archaea in anaerobic digesters. Nat. Commun. 20411723. 15 (1). Available from: https://www.nature.com/ncomms/. doi:10.1038/s41467-024-49641-y.

España-Gamboa, E.I., Mijangos-Cortés, J.O., Hernández-Zárate, G., Maldonado, J.A.D., Alzate-Gaviria, L.M., 2012. Methane production by treating vinasses from hydrous ethanol using a modified UASB reactor. Biotechnol. Biofuels. 17546834. 5. doi:10.1186/1754-6834-5-82.

Filer, J., Ding, H.H., Chang, S., 2019. Biochemical methane potential (BMP) assay method for anaerobic digestion research. Water. 2073-4441. 11 (5), 921. doi:10.3390/w11050921.

Flores, E.M.M., Cravotto, G., Bizzi, C.A., Santos, D., Iop, G.D., 2021. Ultrasound-assisted biomass valorization to industrial interesting products: state-of-the-art, perspectives and challenges. Ultrason. Sonochem. 18732828. 72. Available from: https://www.elsevier.com/inca/publications/store/5/2/5/4/5/1. doi:10.1016/j.ultsonch.2020.105455.

Forster-Carneiro, T., Isaac, R., Pérez, M., Schvartz, C., Mudhoo, A., 2012. Anaerobic Digestion: Pretreatments of Substrates Biogas Production, Vol. 1. Wiley, pp. 1–25. Available from: https://onlinelibrary.wiley.com/doi/10.1002/9781118404089.ch1.

Gerardi, M.H., 2003. The Microbiology of Anaerobic Digesters the Microbiology of Anaerobic Digesters. Wiley, pp. 1–177. Available from: https://onlinelibrary.wiley.com/doi/book/10.1002/0471468967. doi:10.1002/0471468967.

Gheshlaghi, R., Scharer, J.M., Moo-Young, M., Chou, C.P., 2009. Metabolic pathways of clostridia for producing butanol. Biotechnol. Adv. 07349750. 27 (6), 764–781. doi:10.1016/j.biotechadv.2009.06.002.

Goswami, R., Chattopadhyay, P., Shome, A., Banerjee, S.N., Chakraborty, A.K., Mathew, A.K., Chaudhury, S., 2016. An overview of physico-chemical mechanisms of biogas production by microbial communities: a step towards sustainable waste management. 3 Biotech 21905738. 6 (1). Available from: http://www.springerlink.com/content/2190-572x/. doi:10.1007/s13205-016-0395-9.

Grabarse, W., Vaupel, M., Vorholt, J.A., Shima, S., Thauer, R.K., Wittershagen, A., Bourenkov, G., Bartunik, H.D., Ermler, U., 1999. The crystal structure of methenyltetrahydromethanopterin cyclohydrolase from the hyperthermophilic archaeon *Methanopyrus kandleri*. Structure 09692126. 7 (10), 1257–1268. Available from: http://www.journals.elsevier.com/structure/. doi:10.1016/S0969-2126(00)80059-3.

Habchi, S., Pecha, J., Šánek, L., Karouach, F., Bari, H.E., 2024. Sustainable valorization of slaughterhouse waste through anaerobic digestion: a circular economy perspective. J. Environ. Manag. 03014797. 366, 121920. doi:10.1016/j.jenvman.2024.121920.

Harirchi, S., Wainaina, S., Sar, T., Nojoumi, S.A., Parchami, M., Parchami, M., Varjani, S., Khanal, S.K., Wong, J., Awasthi, M.K., Taherzadeh, M.J., 2022. Microbiological insights into anaerobic digestion for biogas, hydrogen or volatile fatty acids (VFAs): a review. Bioengineered 21655987. 13 (3), 6521–6557. Available from: http://www.tandfonline.com/loi/kbie20. doi:10.1080/21655979.2022.2035986.

Hendrickson, E.L., Leigh, J.A., 2008. Roles of coenzyme $F_{420}$-reducing hydrogenases and hydrogen- and $F_{420}$-dependent methylenetetrahydromethanopterin dehydrogenases in reduction of $F_{420}$ and production of hydrogen during methanogenesis. J. Bacteriol. 0021-9193 1098-5530. 190 (14), 4818–4821. Available from: https://journals.asm.org/doi/10.1128/JB.00255-08. doi:10.1128/JB.00255-08.

Hendrickson, E.L., Leigh, J.A., 2008. Roles of coenzyme F420-reducing hydrogenases and hydrogen- and F420-dependent methylenetetrahydromethanopterin dehydrogenases in reduction of F420 and production of hydrogen during methanogenesis. J. Bacteriol. 00219193. 190 (14), 4818–4821. doi:10.1128/JB.00255-08.

Holbrook, O.T., Molligoda, B., Bushell, K.N., Gobrogge, K.L., 2022. Behavioral consequences of the downstream products of ethanol metabolism involved in alcohol use disorder. Neurosci. Biobehav. Rev. 18737528. 133. Available from: https://www.elsevier.com/locate/neubiorev. doi:10.1016/j.neubiorev.2021.12.024.

Hunter, S.M., Blanco, E., Borrion, A., 2021. Expanding the anaerobic digestion map: a review of intermediates in the digestion of food waste. Sci. Total Environ. 0048-9697. 767, 144265. Available from: https://linkinghub.elsevier.com/retrieve/pii/S0048969720377962. doi:10.1016/j.scitotenv.2020.144265.

Jepleting, A., Mecha, A.C., Sombei, D., Moraa, D., Chollom, M.N., 2025. Potential of low-cost materials for biogas purification, a review of recent developments. Renew. Sustain. Energy Rev. 18790690. 210. Available from: https://www.sciencedirect.com/science/journal/13640321. doi:10.1016/j.rser.2024.115152.

Karekar, S., Stefanini, R., Ahring, B., 2022. Homo-acetogens: their metabolism and competitive relationship with hydrogenotrophic methanogens. Microorganisms 2076-2607. 10 (2), 397. Available from: https://www.mdpi.com/2076-2607/10/2/397. doi:10.3390/microorganisms10020397.

Kerrou, O., Lahboubi, N., Bakraoui, M., Bari, H.E., 2023. Improving methane production from palm leaflets waste with thermal and ultrasound pre-treatment. Phys. Chem. Earth, Parts A/B/C. 14747065. 132, 103482. doi:10.1016/j.pce.2023.103482.

Khaita, C., Chinwetkitvanich, S., Polprasert, S., Polprasert, C., Patthanaissaranukool, W., 2025. Phosphorus and bioenergy recovery from anaerobic co-digestion of food waste and corn residues with digested sludge: effect of HRT and pre-treatment methods. Clean Technol. Environ. Policy. 1618-954X. 27 (5), 2007–2027. doi:10.1007/s10098-024-02941-9.

Khanthong, K., Kadam, R., Kim, T., Park, J., 2023. Synergetic effects of anaerobic co-digestion of food waste and algae on biogas production. Bioresour. Technol. 09608524. 382, 129208. doi:10.1016/j.biortech.2023.129208.

Kiani, M.K.D., Parsaee, M., Ardebili, S.M.S., Reyes, I.P., Fuess, L.T., Karimi, K., 2022. Different bioreactor configurations for biogas production from sugarcane vinasse: a comprehensive review. Biomass Bioenergy 18732909. 161. Available from: http://www.journals.elsevier.com/biomass-and-bioenergy/. doi:10.1016/j.biombioe.2022.106446.

Kim, N.K., Lee, S.H., Kim, Y., Park, H.D., 2022. Current understanding and perspectives in anaerobic digestion based on genome-resolved metagenomic approaches. Bioresour. Technol. 18732976. 344. Available from: https://www.elsevier.com/locate/biortech. doi:10.1016/j.biortech.2021.126350.

Kondusamy, D., Kalamdhad, A.S., 2014. Pre-treatment and anaerobic digestion of food waste for high rate methane production – a review. J. Environ. Chem. Eng. 22133437. 2 (3), 1821–1830. doi:10.1016/j.jece.2014.07.024.

Kovačić, Đ., Radočaj, D., Samac, D., Jurišić, M., 2024. Influence of thermal pretreatment on lignin destabilization in harvest residues: an ensemble machine learning approach. AgriEngineering. 2624-7402. 6 (1), 171–184. doi:10.3390/agriengineering6010011.

Kucharska, K., Rybarczyk, P., Hołowacz, I., Łukajtis, R., Glinka, M., Kamiński, M., 2018. Pretreatment of lignocellulosic materials as substrates for fermentation processes. Molecules. 1420-3049. 23 (11), 2937. doi:10.3390/molecules23112937.

Kumar, P., Barrett, D.M., Delwiche, M.J., Stroeve, P., 2009. Methods for pretreatment of lignocellulosic biomass for efficient hydrolysis and biofuel production. Ind. Eng. Chem. Res. 08885885. 48 (8), 3713–3729. Available from: http://pubs.acs.org/doi/pdfplus/10.1021/ie801542g. doi: 10.1021/ie801542g.

Lee, J., Park, K.Y., 2020. Impact of hydrothermal pretreatment on anaerobic digestion efficiency for lignocellulosic biomass: influence of pretreatment temperature on the formation of biomass-degrading byproducts. Chemosphere. 00456535. 256, 127116. doi:10.1016/j.chemosphere.2020.127116.

Liu, L., Peng, H., Shen, Y., Shen, L., Guiming, F., Wan, Y., Zhaodi, X., 2024. 1-Butyl-3-methylimidazolium halide based ionic liquids as solvents for the pretreatment of bamboo slices: effect of electronegativity of anions. Ind. Crops Prod. 09266690. 217, 118877. doi:10.1016/j.indcrop.2024.118877.

Louis, P., Flint, H.J., 2009. Diversity, metabolism and microbial ecology of butyrate-producing bacteria from the human large intestine. FEMS Microbiol. Lett. 15746968. 294 (1), 1–8. doi:10.1111/j.1574-6968.2009.01514.x.

Ma, X., Zhang, Y., Chu, X., Wei, L., Li, P., Yu, K., 2025. Effects of steam explosion on the structural, physicochemical, and functional characteristics of dietary fiber in sorghum grains. LWT. 00236438. 218, 117522. doi:10.1016/j.lwt.2025.117522.

Mankar, A.R., Pandey, A., Arindam Modak, K.K.P., 2021. Pretreatment of lignocellulosic biomass: a review on recent advances. Bioresour. Technol. 09608524. 334, 125235. doi:10.1016/j.biortech.2021.125235.

Mazzanti, G., Demichelis, F., Fino, D., Tommasi, T., 2025. A closed-loop valorization of the waste biomass through two-stage anaerobic digestion and digestate exploitation. Renew. Sustain. Energy Rev. 13640321. 207, 114938. doi:10.1016/j.rser.2024.114938.

Meegoda, J.N., Li, B., Patel, K., Wang, L.B., 2018. A review of the processes, parameters, and optimization of anaerobic digestion. Int. J. Environ. Res. Public Health 16604601. 15 (10). Available from: https://www.mdpi.com/1660-4601/15/10/2224/pdf. doi:10.3390/ijerph15102224.

Meng, Y., Luan, F., Yuan, H., Chen, X., Xiujin, L., 2017. Enhancing anaerobic digestion performance of crude lipid in food waste by enzymatic pretreatment. Bioresour. Technol. 09608524. 224, 48–55. doi:10.1016/j.biortech.2016.10.052.

Mir, M.A., Hussain, A., Verma, C., Dubey, S., 2016. Design considerations and operational performance of anaerobic digester: a review. Cogent Eng. 2331-1916. 3 (1), 1181696. doi:10.1080/23311916.2016.1181696.

Mrosso, R., Mecha, A.C., Kiplagat, J., 2023. Characterization of kitchen and municipal organic waste for biogas production: effect of parameters. Heliyon 24058440. 9 (5), e16360. Available from: https://linkinghub.elsevier.com/retrieve/pii/S2405844023035673. doi:10.1016/j.heliyon.2023.e16360.

Mudhoo, A., 2012. Biogas Production: Pretreatment Methods in Anaerobic Digestion. Wiley. doi:10.1002/9781118404089.

Nguyen, V.K., Chaudhary, D.K., Dahal, R.H., Trinh, N.H., Kim, J., Chang, S.W., Hong, Y., La, D.D., Nguyen, X.C., Ngo, H.H., Chung, W.J., Nguyen, D.D., 2021. Review on pretreatment techniques to improve anaerobic digestion of sewage sludge. Fuel 00162361. 285. Available from: http://www.journals.elsevier.com/fuel/. doi:10.1016/j.fuel.2020.119105.

Nsair, A., Cinar, S.O., Alassali, A., Qdais, H.A., Kuchta, K., 2020. Operational parameters of biogas plants: a review and evaluation study. Energies. 1996-1073. 13 (15), 3761. doi:10.3390/en13153761.

Ojo, A., 2023. An overview of lignocellulose and its biotechnological importance in high-value product production. Fermentation. 2311-5637. 9 (11), 990. doi:10.3390/fermentation9110990.

Olatunji, K.O., Ahmed, N.A., Ogunkunle, O., 2021. Optimization of biogas yield from lignocellulosic materials with different pretreatment methods: a review. Biotechnol Biofuels 14 (1). doi:10.1186/s13068-021-02012-x.

Orhorhoro, E.K., 2017. Experimental determination of effect of total solid (TS) and volatile solid (VS) on biogas yield. Am. J. Mod. Energy. 2575-3908. 3 (6), 131. doi:10.11648/j.ajme.20170306.13.

Pandiselvam, R., Yüksel Aydar, A., Kutlu, N., Aslam, R., Sahni, P., Mitharwal, S., Gavahian, M., Kumar, M., Raposo, A., Yoo, S., Han, H., Kothakota, A., 2023. Individual and interactive effect of ultrasound pre-treatment on drying kinetics and biochemical qualities of food: a critical review. Ultrason. Sonochem. 13504177. 92, 106261. doi:10.1016/j.ultsonch.2022.106261.

Pang, H., Yumeng, X., Ren, R., Junguo, H., Pan, X., Wang, L., 2023. Enhanced anaerobic digestion of waste activated sludge by alkaline protease-catalyzing hydrolysis: role and significance of initial pH adjustment. Chem. Eng. J. 13858947. 467, 143323. doi:10.1016/j.cej.2023.143323.

Paramasivam, P., Alruqi, M., Ağbulut, Ü., 2025. Waste to energy: enhancing biogas utilization in dual-fuel engines using machine learning based prognostic analysis. Fuel 00162361. 381, 133093. Available from: https://linkinghub.elsevier.com/retrieve/pii/S0016236124022427. doi:10.1016/j.fuel.2024.133093.

Park, J.H., Kumar, G., Yun, Y.M., Kwon, J.C., Kim, S.H., 2018. Effect of feeding mode and dilution on the performance and microbial community population in anaerobic digestion of food waste. Bioresour. Technol. 18732976. 248, 134–140. Available from: https://www.elsevier.com/locate/biortech. doi:10.1016/j.biortech.2017.07.025.

Pezacka, E., Wood, H.G., 1984. Role of carbon monoxide dehydrogenase in the autotrophic pathway used by acetogenic bacteria. Proc. Natl. Acad. Sci. U. S. A. 0027-8424. 81 (20), 6261–6265. doi:10.1073/pnas.81.20.6261.

Piadeh, F., Offie, I., Behzadian, K., Rizzuto, J.P., Bywater, A., Córdoba-Pachón, J.R., Walker, M., 2024. A critical review for the impact of anaerobic digestion on the sustainable development goals. J. Environ. Manag. 10958630. 349. Available from: https://www.sciencedirect.com/journal/journal-of-environmental-management. doi:10.1016/j.jenvman.2023.119458.

Raposo, F., Fernández-Cegrí, V., De la Rubia, M.A., Borja, R., Béline, F., Cavinato, C., Demirer, G., Fernández, B., Fernández-Polanco, M., Frigon, J.C., Ganesh, R., Kaparaju, P., Koubova, J., Méndez, R., Menin, G., Peene, A., Scherer, P., Torrijos, M., Uellendahl, H., Wierinck, I., de Wilde, V., 2011. Biochemical methane potential (BMP) of solid organic substrates: evaluation of anaerobic biodegradability using data from an international interlaboratory study. J. Chem. Technol. Biotechnol. 0268-2575. 86 (8), 1088–1098. doi:10.1002/jctb.2622.

Rastogi, M., Shrivastava, S., 2017. Recent advances in second generation bioethanol production: an insight to pretreatment, saccharification and fermentation processes. Renew. Sustain. Energy Rev. 13640321. 80, 330–340. doi:10.1016/j.rser.2017.05.225.

Ren, Y., Miao, Y., Chuanfu, W., Wang, Q., Gao, M., Huang, Q., Liu, Y., 2018. A comprehensive review on food waste anaerobic digestion: research updates and tendencies. Bioresour. Technol. 09608524. 247, 1069–1076. doi:10.1016/j.biortech.2017.09.109.

Rosenbaum, F.P., Müller, V., 2025. Purification and characterization of a thermophilic NAD+-dependent lactate dehydrogenase from *Moorella thermoacetica*. FEBS Open Bio 22115463. 15 (5), 714–725. Available from: http://www.elsevier.com/wps/find/journaldescription.cws_home/726807/description#description. doi:10.1002/2211-5463.13964.

Sanjaya, E.H., Suharti, S., Alvionita, M., Telussa, I., Febriana, S., Clevanota, H., 2024. Isolation and Characterization of Amylase Enzyme Produced by Indigenous Bacteria from Sugar Factory Waste. TOBIOTJ,18 (1). doi:10.2174/0118740707296261240418114958.

Schuchmann, K., Müller, V., 2016. Energetics and application of heterotrophy in acetogenic bacteria. Appl. Environ. Microbiol. 10985336. 82 (14), 4056–4069. Available from: http://aem.asm.org/content/82/14/4056.full.pdf. doi:10.1128/AEM.00882-16.

Seidl, P.R., Goulart, A.K., 2016. Pretreatment processes for lignocellulosic biomass conversion to biofuels and bioproducts. Curr. Opin. Green Sustain. Chem. 24522236. 2,

48–53. Available from: http://www.journals.elsevier.com/current-opinion-in-green-and-sustainable-chemistry. doi:10.1016/j.cogsc.2016.09.003.

Shima, S., Warkentin, E., Grabarse, W., Sordel, M., Wicke, M., Thauer, R.K., Ermler, U., 2000. Structure of coenzyme F420 dependent methylenetetrahydromethanopterin reductase from two methanogenic archaea. J. Mol. Biol. 00222836. 300 (4), 935–950. doi:10.1006/jmbi.2000.3909.

Singh, A., Bajar, S., Devi, A., 2021. Deepak Pant, An overview on the recent developments in fungal cellulase production and their industrial applications. Bioresour. Technol. Rep. 2589014X. 14, 100652. doi:10.1016/j.biteb.2021.100652.

Song, C., Zhang, Z., Wang, X., Xuejun, H., Chen, C., Liu, G., 2024. Machine learning-aided unveiling the relationship between chemical pretreatment and methane production of lignocellulosic waste. Waste Manag. 0956053X. 187, 235–243. doi:10.1016/j.wasman.2024.07.004.

Stamatelatou, K., Antonopoulou, G., Ntaikou, I., Lyberatos, G., Mudhoo, A., 2012. The Effect of Physical, Chemical, and Biological Pretreatments of Biomass on Its Anaerobic Digestibility and Biogas Production Biogas Production, Vol. 1. Wiley, pp. 55–90. Available from: https://onlinelibrary.wiley.com/doi/10.1002/9781118404089.ch3.

Tarvin, D., Buswell, A.M., 1934. The methane fermentation of organic acids and carbohydrates 1,2. J. Am. Chem. Soc. 0002-7863. 56 (8), 1751–1755. doi:10.1021/ja01323a030.

Tiehm, A., Nickel, K., Zellhorn, M., Neis, U., 2001. Ultrasonic waste activated sludge disintegration for improving anaerobic stabilization. Water Res. 00431354. 35 (8), 2003–2009. doi:10.1016/s0043-1354(00)00468-1.

Torres, E., Ayala, M., 2013. Biocatalysis by Metalloenzymes, Vol. 6. Elsevier BV, pp. 685–735. doi:10.1016/b978-0-08-097774-4.00625-2.

Upadhyay, A., Kovalev, A.A., Zhuravleva, E.A., Kovalev, D.A., Litti, Y.V., Masakapalli, S.K., Pareek, N., Vivekanand, V., 2022. Recent development in physical, chemical, biological and hybrid biogas upgradation techniques. Sustainability 2071-1050. 15 (1), 476. Available from: https://www.mdpi.com/2071-1050/15/1/476. doi:10.3390/su15010476.

Usmani, Z., Sharma, M., Gupta, P., Karpichev, Y., Gathergood, N., Bhat, R., Gupta, V.K., 2020. Ionic liquid based pretreatment of lignocellulosic biomass for enhanced bioconversion. Bioresour. Technol. 09608524. 304, 123003. doi:10.1016/j.biortech.2020.123003.

Vasco-Correa, J., Khanal, S., Manandhar, A., Shah, A., 2018. Anaerobic digestion for bioenergy production: global status, environmental and techno-economic implications, and government policies. Bioresour. Technol. 09608524. 247, 1015–1026. doi:10.1016/j.biortech.2017.09.004.

Vögeli, Y., Lohri, C.R., Gallardo, A., Diener, S., Zurbrügg, C., 2014. Anaerobic digestion of biowaste in developing countries - practical information and case studies. Eawag. Available from:. doi:10.13140/2.1.2663.1045.

Wan, C., Li, Y., 2011. Effect of hot water extraction and liquid hot water pretreatment on the fungal degradation of biomass feedstocks. Bioresour. Technol. 09608524. 102 (20), 9788–9793. doi:10.1016/j.biortech.2011.08.004.

Wang, W., Chang, J.S., Lee, D.J., 2023. Anaerobic digestate valorization beyond agricultural application: current status and prospects. Bioresour. Technol. 18732976. 373. Available from: https://www.elsevier.com/locate/biortech. doi:10.1016/j.biortech.2023.128742.

Wang, X., Cheng, S., Zifu, L., Men, Y., Jiajun, W., 2020. Impacts of cellulase and amylase on enzymatic hydrolysis and methane production in the anaerobic digestion of corn straw. Sustainability. 2071-1050. 12 (13), 5453. doi:10.3390/su12135453.

Wintsche, B., Jehmlich, N., Popp, D., Harms, H., Kleinsteuber, S., 2018. Metabolic adaptation of methanogens in anaerobic digesters upon trace element limitation. Front. Microbiol. 1664-302X. 9. doi:10.3389/fmicb.2018.00405.

Xia, A., Feng, D., Huang, Y., Zhu, X., Wang, Z., Zhu, X., Liao, Q., 2022. Activated carbon facilitates anaerobic digestion of furfural wastewater: effect of direct interspecies electron transfer. ACS Sustain. Chem. Eng. 2168-0485. 10 (25), 8206–8215. doi:10.1021/acssuschemeng.2c01907.

Yadav, M., Joshi, C., Paritosh, K., Thakur, J., Pareek, N., Masakapalli, S.K., Vivekanand, V., 2022. Organic waste conversion through anaerobic digestion: a critical insight into the metabolic pathways and microbial interactions. Metab. Eng. 10967184. 69, 323–337. Available from: http://www.elsevier.com/inca/publications/store/6/2/2/9/1/3/index.htt. doi:10.1016/j.ymben.2021.11.014.

Yan, H., Zhao, C., Zhang, J., Zhang, R., Xue, C., Liu, G., Chen, C., 2017. Study on biomethane production and biodegradability of different leafy vegetables in anaerobic digestion. AMB Express. 2191-0855. 7 (1). doi:10.1186/s13568-017-0325-1.

Yan, W., Acharjee, T.C., Coronella, C.J., Vásquez, V.R., 2009. Thermal pretreatment of lignocellulosic biomass. Environ. Prog. Sustain. Energy 19447442. 28 (3), 435–440. Available from: http://www3.interscience.wiley.com/cgi-bin/fulltext/122539482/PDF-START. doi: 10.1002/ep.10385.

Yasim, N.S.E.M., Buyong, F., 2023. Comparative of experimental and theoretical biochemical methane potential generated by municipal solid waste. Environ. Adv. 26667657. 11. Available from: https://www.journals.elsevier.com/environmental-advances. doi:10.1016/j.envadv.2023.100345.

Yi, Y., 2021. Tiny Bugs Play Big Role Advances in 2nd Generation of Bioethanol Production. Elsevier, China, pp. 113–136. Available from: https://www.sciencedirect.com/book/9780128188620. doi:10.1016/B978-0-12-818862-0.00007-8.

Yue, Y., Wang, J., Xiayuan, W., Zhang, J., Chen, Z., Kang, X., Zuopeng, L., 2021. The fate of anaerobic syntrophy in anaerobic digestion facing propionate and acetate accumulation. Waste Manag. 0956053X. 124, 128–135. doi:10.1016/j.wasman.2021.01.038.

Zhang, R., Elmashad, H., Hartman, K., Wang, F., Liu, G., Choate, C., Gamble, P., 2007. Characterization of food waste as feedstock for anaerobic digestion. Bioresour. Technol. 09608524. 98 (4), 929–935. doi:10.1016/j.biortech.2006.02.039.

Zhang, Y., Chunxing, L., Yuan, Z., Wang, R., Angelidaki, I., Zhu, G., 2023. Syntrophy mechanism, microbial population, and process optimization for volatile fatty acids metabolism in anaerobic digestion. Chem. Eng. J. 13858947. 452, 139137. doi:10.1016/j.cej.2022.139137.

Zhou, M., Yan, B., Wong, J.W.C., Zhang, Y., 2018. Enhanced volatile fatty acids production from anaerobic fermentation of food waste: a mini-review focusing on acidogenic metabolic pathways. Bioresour. Technol. 18732976. 248, 68–78. Available from: https://www.elsevier.com/locate/biortech. doi:10.1016/j.biortech.2017.06.121.

CHAPTER

# Microbial conversion processes for resource recovery and circular economy 10

**Pradeep Kumar Ramteke and Ajit P. Rathod**
*Department of Chemical Engineering, Visvesvaraya National Institute of Technology (VNIT), Nagpur, India*

## 1 Introduction

Microbial biotechnology is essential for a circular economy, particularly in waste management and resource recovery. Microorganisms are crucial for converting biological waste into valuable products such as biogas, bioplastics, and chemicals (O'Connor, 2021). Microbial processes integrate resource recovery, energy production, and fresh water in wastewater treatment. Agroindustrial biowastes are a promising carbon source for microbial fermentation, which can be used to transform waste into value-added products such as bioethanol and biofuels (Kee et al., 2021). The bioconversion of lignocellulose waste through microbial dynamics offers a cost-effective solution for bioenergy and product development (Sarsaiya et al., 2019). These microbial processes contribute to sustainable waste management, promote a low-carbon economy, and support the United Nations (UN) Sustainable Development Goals (Nielsen, 2017). This chapter explores the potential of biowaste to produce bioenergy and bioproducts within a circular economy framework. It emphasizes the significance of research in this field and its impact on various sectors (Jain et al., 2022). In addition, it discusses the potential for transforming biowaste into value-added products using microbes to enhance soil health and crop production, thereby supporting a circular economy (Upadhyay et al., 2024). It also discusses the potential of microbial gas fermentation to absorb and utilize waste carbon, establishing a circular economy. It provides an overview of systems-level metabolism, key metabolic pathways, opportunities for scale-up commercial success, and the latest technological advancements in strain and process engineering for both anaerobic acetogens and aerobic hydrogenotrophs (Pavan et al., 2022). Furthermore, it reviews the use of microbes and microbial genes to facilitate a circular and sustainable biobased economy for polyethylene terephthalate (PET), a widely used plastic (Salvador et al., 2019).

Microbial Technologies for Sustainable Waste Management. https://doi.org/10.1016/B978-0-443-44093-9.00010-4

## 1.1 Global challenges: Waste, resources, and environment

The challenges of global waste generation, resource depletion, and environmental degradation are driving a shift toward circular economy approaches in microbial conversion processes for resource recovery. Proper waste management and moving toward a circular economy are essential for sustainable growth and addressing these challenges (Najar et al., 2024). These approaches contribute to waste valorization, promote a low-carbon economy, and encourage the world's shift to a circular economy and sustainable resource management. Microbial biotechnology is essential for establishing a circular economy to clean wastewater by combining electricity generation and resource recovery to produce clean water. Further research is required to better understand the critical microorganisms involved (Islam et al., 2024; Nielsen, 2017). Anaerobic and aerobic digestion methods can also be used to treat organic solid waste in the context of a circular economy and resource recovery (Wainaina et al., 2020). Microbial fuel cells (MFCs), including microbial electrolysis cells (MECs), microbial recycling cells, and microbial electrochemical technologies (METs), enable the simultaneous generation of electricity and recovery of nutrients from wastewater, thus advancing a circular economy and supporting carbon neutrality goals (Deng et al., 2023). This chapter explores the repurposing of food and agricultural waste into valuable biobased products, including physiologically active molecules for potential antimicrobial medication, as well as the microbial synthesis of biopolymers and biosurfactants. This approach offers a promising solution to address environmental concerns and progress toward sustainable development goals (Iacono et al., 2024). Fig. 10.1 shows the different types of sustainable environmental technologies.

## 1.2 Role of microbial processes in converting waste into useful properties

Microbial and enzymatic methods are essential for converting organic waste into useful bioproducts and addressing waste management issues while producing sustainable resources. These bioconversion processes can yield many different products, such as enzymes, biofuels, bioplastics, and bioactive compounds (Chavan et al., 2022). MFCs and other microbial electrochemical systems (MESs) offer promising applications in waste-to-energy conversion and chemical production (Logan and Rabaey, 2012). Recent advancements in bioprocess engineering and microbial biotechnology have led to innovations in hybrid electrosynthesis, plastic degradation, and fermentation technology (Rene et al., 2023). Despite advancements in refining these processes at the laboratory level, further investigation is required to address challenges in efficiency, scalability, and system reliability for commercial implementation. The integration of these technologies facilitates the shift to a circular economy and sustainable resource management (Torres-León et al., 2021). Recent developments in the microbial transformation of biowastes into valuable products have highlighted the circular economic approach to waste management and resource

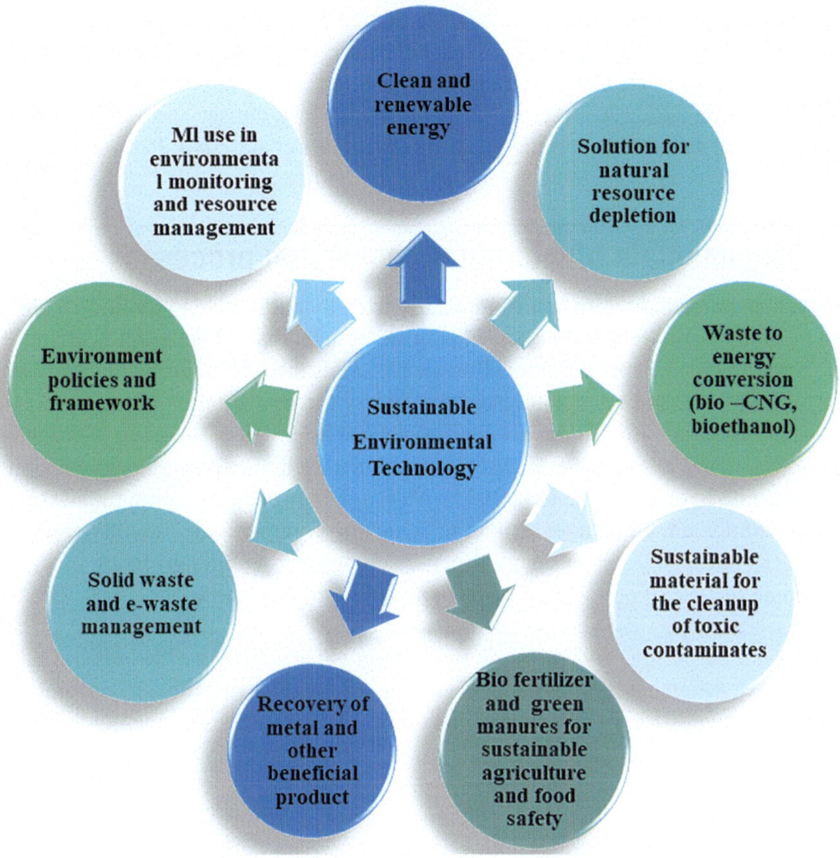


**FIG. 10.1**

**Various sources of sustainable environmental technology**, which include renewable energy (solar, wind, bioenergy), waste management (recycling, composting, waste-to-energy), water purification, and sustainable farming practices.

utilization (Saravanan et al., 2023; Kee et al., 2021). This chapter explores microbial waste valorization processes applicable to a variety of waste streams, including carbohydrate-rich, lipid-rich, fruit/vegetable, plant/animal, and plastic waste (Wackett, 2023). It discusses the diverse sources of food waste and the limitations of traditional disposal methods. In addition, it highlights how microorganisms can serve as environmentally beneficial and sustainable solutions for valorizing food waste by producing useful products such as biofuels, electrical energy, biosurfactants, bioplastics, and biofertilizers (Sharma et al., 2020).

Microorganisms are essential in both the biological and technical aspects of the circular economy, contributing to the transformation of biological waste into useful products like biogas, bioplastics, and compost. Anaerobic digestion (AD) and aerobic digestion technologies are key to resource recovery from organic solid waste, producing biogas and high-value chemicals, whereas composting enables nutrient recovery. These microbial-based approaches align with the principles of reducing dependence on finite resources in the circular economy, conserving ecosystems, and promoting sustainable economic practices (Wainaina et al., 2020; Papa et al., 2023).

## 2 Fundamentals of microbial conversion processes

Microbial conversion is an environmentally friendly alternative to traditional chemical methods for producing fuels and chemicals from biomass. These processes involve the breakdown of complex lignocellulosic materials through microbial hydrolysis of lignin, cellulose, and hemicellulose, followed by the fermentation of sugars to ethanol (Dionisi et al., 2015). Various feedstocks can be utilized, including solid waste, wastewater, and gaseous waste. MFCs have been used to explore the conversion of biomass into hydrogen (Wackett, 2022). Although the potential for microbial conversion is significant, challenges remain in achieving commercially attractive rates and yields (Kennes, 2018). This chapter focuses on optimizing bioconversion processes, including the use of engineered microorganisms and enzyme applications. The field of microbial energy conversion encompasses diverse areas, such as biomass production, waste recycling, and methane production (Yuan, 2018), with ongoing efforts to improve efficiency and economic viability. Microbial metabolism encompasses diverse strategies for energy acquisition and carbon assimilation, including heterotrophy, chemoautotrophy, and photoautotrophy. Prokaryotes exhibit a wider range of metabolic types than eukaryotes and utilize various substrates and energy generation processes (Oren, 2010). Aerobic and anaerobic pathways, such as glycolysis, fermentation, the TCA cycle, and oxidative phosphorylation, are fundamental to microbial metabolism (Fuhrmann, 2021). Understanding microbial metabolism as a system requires the combination of in vivo, in vitro, and in silico methods (Downs, 2006). Heterocyclic aromatic compounds biodegrade in both aerobic and anaerobic environments, with different organisms using distinct pathways for biotransformation. The substituents in pyridine derivatives determine the rate of transformation, with the highest rates observed in pyridine carboxylic acids. Microorganisms play crucial roles in biogeochemical cycles by transforming carbon, nitrogen, sulfur, and other elements (Kaiser et al., 1996). Fig. 10.2 illustrates the applications of microbial conversion.

Microorganisms play a crucial role in resource recovery and waste treatment. Bacteria, fungi, archaea, and algae are key players in these processes. Microalgae (MA)–bacterial consortia have shown promise in treating food industry wastewater, achieving high chemical oxygen demand and nutrient removal efficiencies. These consortia also demonstrate the potential for carbon fixation and resource recovery (Sirohi et al., 2022). Bacteria, fungi, and archaea are effective in bioextracting metals from various industrial wastes, including metallurgical by-products, e-waste, and municipal incinerator ash (Lee and Pandey, 2012). Acidophilic microorganisms,

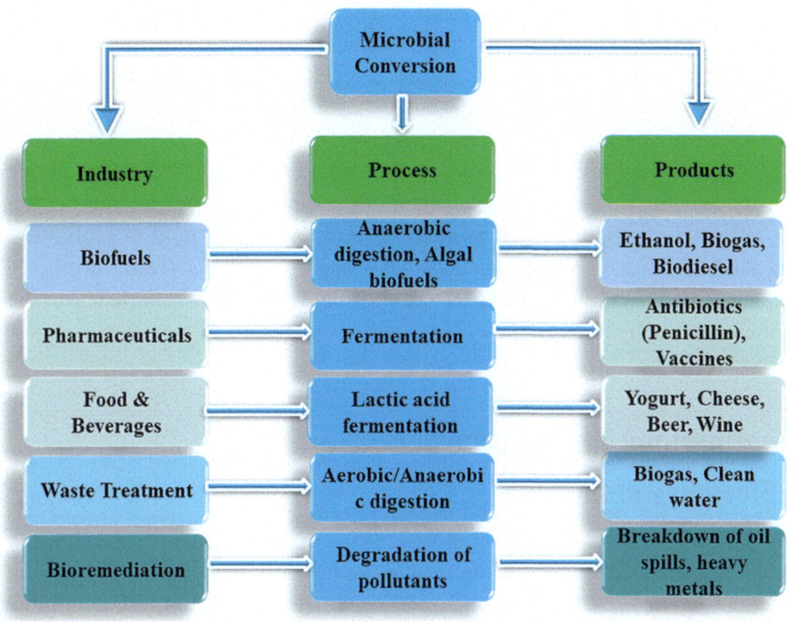


**FIG. 10.2**

**Applications of microbial conversion**, which refer to the use of microorganisms to transform organic materials into valuable products, such as biofuels, bioplastics, and fertilizers, as well as to break down waste into simpler, environmentally friendly compounds.

such as *Acidithiobacillus ferrooxidans* and *Aspergillus niger*, are particularly useful for recovering heavy metals from industrial wastes, such as spent catalysts and electronic scraps (Mishra, 2010). These microbial methods offer eco-friendly and cost-effective alternatives to conventional physical and chemical treatments, thereby addressing both resource recovery and pollution mitigation challenges. Various types of substrates can be used for sustainable waste management and energy creation. Agricultural and agroindustrial wastes serve as low-cost substrates for fermentation processes, yielding valuable microbial products, such as enzymes, pigments, and biosurfactants (Astudillo et al., 2023). The agrifood industry provides diverse substrates for recycling, including plant and animal residues, which can be used to produce bioactive substances and construction materials (Yildiz et al., 2023). Organic waste AD, including municipal solid waste, animal manure, and industrial waste, produces biogas with varying methane yields depending on the substrate composition. MFCs can convert chemical energy from diverse wastewater sources, such as industrial, agricultural, domestic, and pharmaceutical sources, into electricity while simultaneously treating waste (Nwokolo et al., 2020). These technologies offer promising solutions for waste management, energy production, and environmental sustainability.

Microbial efficiency in soil and bioelectrochemical systems (BESs) is influenced by several factors. pH plays a crucial role, with a threshold of approximately 5.5, below which efficiency declines, coinciding with increased aluminum solubility (Jones et al., 2019). Temperature, nutrient availability, and substrate composition also significantly affect microbial carbon use efficiency (CUE) and degradation processes. In constructed wetlands, organic carbon availability, redox conditions, and media characteristics are the key parameters affecting microbial activity (Meng et al., 2014). For BESs, conductivity, inoculum selection, and reactor design parameters, such as electrode material and potential, are important considerations (Gupta, 2017). While some factors, such as external nutrient inputs, tend to promote CUE, others, such as climate warming, may have negative effects (Dang et al., 2024). Understanding these factors is crucial for optimizing microbial efficiency in various environmental applications.

# 3 Applications in resource recovery

## 3.1 Bioenergy production

AD is an effective renewable energy approach that transforms organic waste into methane-rich biogas, offering a sustainable solution for energy generation (Stamatelatou et al., 2011). Various microorganisms, including methanogenic archaea, carry out the four phases of this process: hydrolysis, acidogenesis, acetogenesis, and methanogenesis (Torres-Sebastián et al., 2022). AD can utilize diverse feedstocks, such as farming waste, manure, and municipal solid waste, to produce biogas with 50%–84% methane content. This technology offers multiple benefits, including waste reduction, improved sanitation, and decreased greenhouse gas emissions (Jame et al., 2021). Biogas can be upgraded to >90% methane, making it comparable to natural gas for various applications in the industry. It serves as an alternative energy source for both urban and rural areas, generating electricity, heat, and mechanical energy. However, challenges such as long startup times and the pH sensitivity of microorganisms must be addressed for optimal production (das Neves et al., 2009). MECs are promising BESs for producing bioelectricity and biohydrogen from organic waste streams (Zhou et al., 2013). MECs offer several benefits over traditional techniques, including higher yields and lower energy requirements. However, challenges remain in improving efficiency and reducing costs for large-scale applications (Yang et al., 2021). Recent research has focused on developing self-sustainable MECs by integrating carbon-neutral technologies such as solar and osmotic power. Other strategies to enhance performance include the inhibition of methanogens, the development of novel cathode catalysts, and the optimization of reactor design (Cheng et al., 2022). MECs also have the potential to extract fertilizers and heavy metals from sewage streams, thereby generating multiple sources of revenue. Despite progress, more power output enhancements, cost savings, and scaling up are necessary for the practical implementation of MEC technology (Jensen et al., 2022). Fig. 10.3 shows the bioenergy sources.

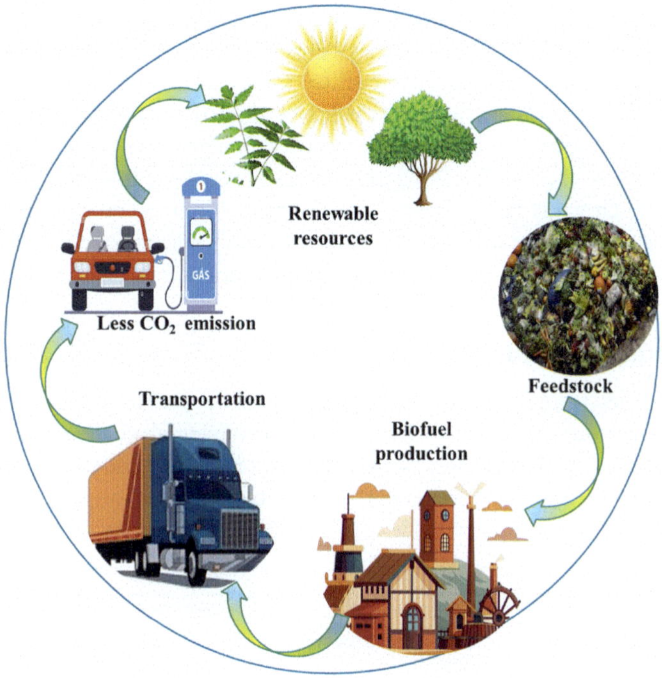


**FIG. 10.3**

**Bioenergy production sources**, which refer to organic materials, such as agricultural waste, algae, and food scraps, that are converted by biological processes (like fermentation or anaerobic digestion) into renewable energy forms, including biogas, bioethanol, and biodiesel.

## 3.2 Nutrient recovery

Nutrient recovery from wastewater, particularly nitrogen (N) and phosphorus (P), is crucial for sustainable development and addressing environmental concerns. BESs show great potential for recovering ammonium nitrogen from waste streams with high nitrogen content (Kelly and He, 2014). Chemical precipitation can remove 80%–99% of P, whereas air stripping can recover 70%–92% of N. Membrane separation achieves 99%–100% N recovery without secondary pollutants. Biological processes in constructed wetlands can remove both N (83%–87%) and P (70%–85%) (Sengupta et al., 2015; Ramteke et al., 2025). Recent advancements have focused on reducing energy requirements and improving recovery efficiency (Nancharaiah et al., 2016). Microbial biofertilizers are an eco-friendly alternative to chemical fertilizers for sustainable agriculture and nutrient recovery. Biofertilizers, including MA and purple phototrophic bacteria, can recover nutrients from agricultural waste and improve soil fertility. Through several procedures, including the generation of phytohormones, nitrogen fixation, and nutrient solubilization, they promote plant growth (Kour et al., 2020). MA can act as a slow-release fertilizer, improving nitrogen retention and cycling in the soil. Biofertilizers also promote beneficial rhizosphere bacteria and soil carbon cycling (Zarezadeh et al., 2019). Recent developments in biofertilizer formulations include microencapsulation and nanoimmobilization techniques (Kumar et al., 2022b). Integrating nutrient recovery into biorefinery design can produce high-value fertilizer products alongside biofuels and chemicals, thereby promoting sustainability and closing anthropogenic nutrient cycles (Carey et al., 2016). Fig. 10.4 illustrates the importance of the nutrient recovery process.

## 3.3 Bioproduct synthesis

Bioplastics made from renewable resources provide a sustainable substitute for plastics made from petrochemicals, addressing environmental concerns and resource depletion (Ali et al., 2024). Polyhydroxyalkanoates (PHA) and polylactic acid are prominent bioplastics with properties comparable to those of synthetic plastics.

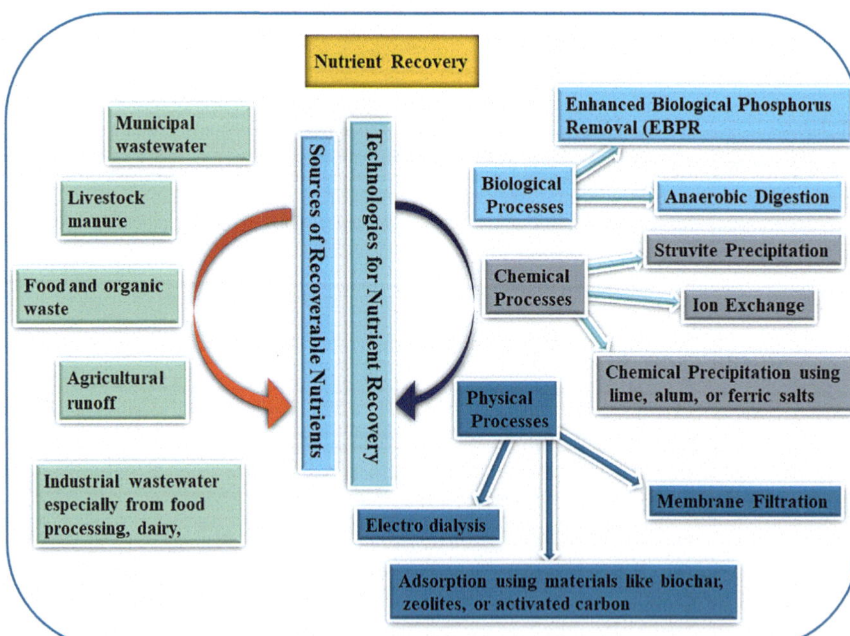


**FIG. 10.4**

**Importance of nutrient recovery process**, which lies in its ability to extract valuable nutrients, such as nitrogen, phosphorus, and potassium, from waste streams (like sewage or agricultural runoff), allowing for their reuse in agriculture and reducing the environmental impact of nutrient pollution.

Organic waste from different sources, particularly PHA, serves as a cost-effective feedstock for bioplastic production through microbial fermentation. This method not only reduces waste and environmental pollution but also lowers production costs (Kuang et al., 2023). Genetically engineered bacteria, such as *Escherichia coli*, can produce PHA from waste carbon sources (Sushmitha, 2016). Despite their potential, challenges remain in scaling up production and improving mechanical properties. The optimization of production techniques should be the main focus of future studies, enhancing material properties and assessing the environmental impacts of bioplastic degradation products (Thomas et al., 2023). Enzymes are essential components in the synthesis of bioactive compounds with diverse applications. Lipases are particularly suitable for producing bioactive esters with cosmeceutical properties through esterification or transesterification, thus offering a more sustainable alternative to traditional chemical catalysts (Alvarez et al., 2021). Lactobionic acid, a versatile polyhydroxy acid, has gained attention as a bioactive compound with applications in industries such as medicine, food, and cosmetics. Its biotechnological production through microbial or enzymatic approaches has been explored (Alonso et al., 2013). Tartaric acid and its derivatives are important synthons for the synthesis of bioactive molecules (Ghosh et al., 2001). Biocatalysis, which utilizes enzymes for organic synthesis, has numerous applications in the pharmaceutical, flavor, and fragrance industries, as well as in the production of specialty and bulk chemicals. Enzymatic synthesis offers opportunities for scalable chemical production, although some limitations exist (Wu et al., 2021).

## 3.4 Bioremediation

Microbial consortia, particularly fungal-bacterial communities, offer a promising approach for the bioremediation of petroleum hydrocarbons and heavy metals in contaminated soils (Haque et al., 2022). These consortia exhibit synergistic mechanisms, including enhanced enzyme pollutant production and cometabolism, leading to more effective degradation compared with single-strain approaches (Zhang and Zhang, 2022). Microorganisms naturally biodegrade pollutants through enzymatic catalytic activities, making bioremediation an economical and environmentally sustainable solution (Varjani, 2017). Recent advancements in tools for gene editing and synthetic biology have enabled the design of resilient and stable systems of artificial microbial consortia. However, optimizing environmental factors and addressing the potential antagonistic interactions between consortium members remain a challenge. The addition of surfactants can further improve degradation efficiency (Rezaei and Moghimi, 2024). Overall, microbial consortia-based bioremediation has great potential for addressing soil pollution issues. The concept of closing material loops in industrial, agricultural, and urban systems has been a focus of research for decades. Waste valorization, which involves treating waste for useful purposes, such as energy or raw materials, is a key strategy in loop-closing efforts (Nzihou and Lifset, 2010). This approach extends beyond household recycling to include industrial symbiosis and by-product exchange. Integrating agricultural systems with sanitary systems for nutrient recycling is a promising method for closing loops in urban and periurban areas (Bahri, 2001). The circular economy model aims to minimize waste and emissions through resource-efficient designs and closed-loop systems (Camilleri, 2018). Contrary to common belief, historical evidence suggests that industrial resource recovery and loop-closing practices were widespread in earlier eras, with animal by-product recovery being a notable example from the Neolithic to the mid-20th century (Desrochers, 2000). These findings highlight the importance of resource efficiency in various sectors (Ramteke et al., 2025b).

## 4 Case studies: Microbial conversion processes for resource recovery

More and more people are considering municipal wastewater treatment facilities to be centers for resource recovery, shifting from traditional treatment to circular economic models. These facilities can recover valuable resources, such as energy, nutrients, and water (Mo and Zhang, 2013). Advanced technologies, such as AD, aerobic granular sludge, and membrane bioreactors, enable the recovery of biogas, freshwater, and nutrients (Ali et al., 2022). Mathematical modeling tools can evaluate and optimize resource recovery options, particularly for energy production through AD. Case studies, such as the Palermo University project, demonstrate the potential for resource recovery and circular economy implementation in wastewater treatment (Ramin et al., 2021). However, challenges remain in integrating multiple recovery methods and assessing their combined benefits and trade-offs from a life cycle perspective. Overall, converting wastewater treatment plants into facilities for resource recovery offers promising opportunities for sustainability and economic benefits in the water sector (Mannina et al., 2021). Various biotransformation methods, including fermentation, composting, and AD, can convert food waste into biofuels, bioplastics, enzymes, and biofertilizers (Sharma et al., 2020; Kumar et al., 2022a). These processes address the water-energy-food nexus by recovering resources from waste streams in food production plants. However, implementing resource recovery solutions in industrial settings requires careful consideration of the effects on the environment, the readiness of the technology, and economic viability. Despite these challenges, methods used by microbes to biotransform food waste into resources offer promising approaches to waste management, promoting sustainability and aligning with the sustainable development goals of the UN (Udugama et al., 2020).

The implementation of sustainable waste management practices offers significant environmental and economic benefits. Codigestion of biosolids and organic materials using high-solid anaerobic waste demonstrates the lowest environmental impacts and life cycle costs compared to alternatives such as incineration and landfilling (Lee et al., 2020). In-vessel composting systems at universities can reduce greenhouse gas emissions, fossil fuel use, and eutrophication while generating profit through vegetable sales (Mu et al., 2017). Adopting reverse logistics to recycle construction and demolition waste provides economic advantages, with companies reporting substantial profits from selling recycled materials (Neto and Correia, 2019). Lean construction techniques, such as Building Information Modeling and Pull Planning, contribute to waste reduction at the source, resulting in material and labor savings while decreasing greenhouse gas emissions (Ghosh et al., 2020). This chapter highlights the potential of waste management strategies to minimize environmental impacts, reduce landfill use, and yield cost savings across various sectors.

## 5 Challenges, limitations, and future directions

MA mass production faces challenges, including contamination, low productivity, and high costs in open ponds, whereas photobioreactors offer better control but are expensive (Shen et al., 2009). Enzyme immobilization can improve industrial processes but requires a multidisciplinary approach for successful scale-up (Chapman et al., 2018). Microbial lactic acid production encounters obstacles such as substrate inhibition, end-product inhibition, and contamination, which can be addressed using fermentation technology and metabolic engineering (Abdel-Rahman and Sonomoto, 2016). Plant cell culture systems, which are used for synthesizing biomolecules, have low yields and production variability. Approaches to address these challenges include elicitation, in situ product recovery, and metabolic engineering. Although plant cell cultures offer advantages for protein production, such as proper glycosylation and reduced contamination risk, high costs remain a significant barrier to commercialization (Wilson and Roberts, 2012).

Microbial conversion processes offer promising opportunities for resource recovery from wastewater, addressing both environmental and economic concerns. These technologies can recover valuable products, such as bioenergy, biopolymers, and nutrients (Puyol et al., 2017). BESs have emerged as a versatile approach to metal recovery and energy generation from waste streams (Christgen et al., 2019). Various bioprocesses have been developed to extract and recover valuable molecules, including biofuels, organic acids, and bioplastics, from wastewater and sludge (Gudiukaite et al., 2021). While microbial waste-to-resource technologies promise sustainability, high costs and market barriers hinder scalability. Research focuses on efficiency gains to enable viable circular bioeconomy solutions (Makepa and Chihobo, 2024).

Resource recovery from waste faces regulatory and social challenges when transitioning to a circular economy. Policy gaps exist between short-term economic concerns and long-term sustainability goals, with conflicts among regulations impeding progress (Purnell et al., 2019). Public perception and market uncertainties regarding feedstock quality and availability hinder investment in innovative conversion processes (Khan and Evans, 1997). To address these issues, a novel framework has been proposed for risk assessment and quality monitoring of products for the recovery of wastewater resources, encompassing quality, health, and environmental aspects (Ardiyanti et al., 2024). This framework aims to bridge regulatory gaps and ensure comprehensive oversight across various recovery scenarios. Advancements in microbial-based technologies have enabled simultaneous resource recovery and wastewater treatment; consequently, bioenergy and value-added products are generated. However, procedural limitations still exist, necessitating further studies to support more sustainable and effective waste utilization routes (Gudiukaite et al., 2021). MESs are emerging technologies that use microorganisms to convert chemical energy into electricity and valuable chemicals (Wang and Ren, 2013). These versatile platforms can be applied in various configurations, including MFCs, electrosynthesis, and desalination cells (Zhang and Tremblay, 2016). Recent developments in technologies for RNA-based device systems, biosensors, and CRISPR-C, which are examples of programmable synthetic biology, have opened up new opportunities for creating highly efficient microbial cell factories (Han et al., 2023).

Recent studies highlight the growing role of artificial intelligence (AI) and machine learning (ML) in enhancing microbiological processes. These technological advancements are being utilized to improve productivity, yield, and process control in various bioprocesses, including wastewater treatment, MA cultivation, and biohydrogen production (Sahu et al., 2023). AI/ML algorithms, such as support vector machines, genetic algorithms, and artificial neural networks, are employed for species identification, real-time monitoring, growth condition optimization, and prediction of process performance. These tools can analyze complex, high-dimensional data from bioprocesses, enabling precise control and improved efficiency (Ahmed and Asadullah, 2020). In biohydrogen production, ML has been applied to model nonlinear relationships between operational parameters and predict microbial population dynamics (Pandey et al., 2023). Despite challenges such as data availability and model complexity, AI/ML technologies offer significant potential for enhancing system performance, reducing costs, and minimizing environmental impact in microbial conversion processes (Imamoglu, 2024).

## 6 Conclusion

Microbial biotechnology has emerged as a cornerstone as the circular economy is being implemented, particularly in revolutionizing waste management and optimizing resource recovery. By harnessing the metabolic prowess of microorganisms, we can transform an array of biowastes into valuable resources, including bioenergy sources such as biogas and biohydrogen, biodegradable bioplastics, and sustainable biofertilizers. This approach not only mitigates the escalating global challenges of environmental deterioration, resource depletion, and waste production, but also supports the UN Sustainable Development Goals by promoting a low-carbon economy. The applications of microbial processes are diverse, ranging from wastewater treatment that integrates energy production and clean water generation to the valorization of agro-industrial waste into bioethanol and other biofuels. Microbial gas fermentation offers innovative solutions for capturing and utilizing waste carbon, while microbial degradation of plastics contributes to a circular and sustainable biobased economy for materials such as PET. However, the widespread adoption of these technologies hinges on overcoming existing challenges, such as optimizing process stability, improving efficiency and scalability, and ensuring economic viability. Future research and development efforts should focus on strain engineering, bioprocess optimization, and system integration to unlock the full potential of microbial biotechnology. Furthermore, understanding and manipulating variables such as substrate type, temperature, pH, and nutrient availability are crucial for maximizing microbial efficiency in various environmental applications. By embracing a holistic approach that considers the entire product lifecycle and fosters interdisciplinary collaboration, we can fully harness the transformative power of microorganisms to build a sustainable, resource-efficient, and circular future.

## References

Abdel-Rahman, M.A., Sonomoto, K., 2016. Opportunities to overcome the current limitations and challenges for efficient microbial production of optically pure lactic acid. J. Biotechnol. 18734863. 236, 176–192. Available from: https://www.elsevier.com/locate/jbiotec. doi:10.1016/j.jbiotec.2016.08.008.

Ahmed, A.A.A., Asadullah, A.B.M., 2020. Artificial intelligence and machine learning in waste management and recycling. Eng. Int. 2409-3629. 8 (1), 43–52. doi:10.18034/ei.v8i1.498.

Ali, M., Hong, P.Y., Mishra, H., Vrouwenvelder, J., Saikaly, P.E., 2022. Adopting the circular model: opportunities and challenges of transforming wastewater treatment plants into resource recovery factories in Saudi Arabia. Water Reuse 27096106. 12 (3). Available from: https://watermark.silverchair.com/jwrd0120346.pdf?. doi:10.2166/wrd.2022.038.

Ali, Z., Abdullah, M., Yasin, M.T., Amanat, K., Ahmad, K., Ahmed, I., Qaisrani, M.M., Khan, J., 2024. Organic waste-to-bioplastics: conversion with eco-friendly technologies and approaches for sustainable environment. Environ. Res. 00139351. 244, 117949. doi:10.1016/j.envres.2023.117949.

Alonso, S., Rendueles, M., Díaz, M., 2013. Bio-production of lactobionic acid: current status, applications and future prospects. Biotechnol. Adv. 07349750. 31 (8), 1275–1291. doi:10.1016/j.biotechadv.2013.04.010.

Alvarez, E., Villa, R., Nieto, S., Donaire, A., García-Verdugo, E., Luis, S.V., Lozano, P., 2021. The suitability of lipases for the synthesis of bioactive compounds with cosmeceutical applications. Mini-Rev. Org. Chem. 18756298. 18 (4), 515–528. Available from: www.eurekaselect.com/622/journal/mini-reviews-organic-chemistry. doi:10.2174/1570193X17999200805215623.

Ardiyanti, R., Azrague, K., Medema, G., Hallé, C., 2024. Navigating quality, health and environmental risk: a novel framework for wastewater resource recovery products. J. Clean. Prod. 09596526. 434, 140260. doi:10.1016/j.jclepro.2023.140260.

Astudillo, Á., Rubilar, O., Briceño, G., Cristina Diez, M., Schalchli, H., 2023. Advances in agroindustrial waste as a substrate for obtaining eco-friendly microbial products. Sustainability. 2071-1050. 15 (4), 3467. doi:10.3390/su15043467.

Bahri, A., 2001. Urban and peri-urban water-related relationships: closing the loops. Environ. Manag. Health. 0956-6163. 12 (4), 364–376. doi:10.1108/eum0000000005704.

Camilleri, M., 2018. Closing the loop for resource efficiency, sustainable consumption and production: a critical review of the circular economy. Int. J. Sustain. Dev. 0960-1406. 1 (1), 1. doi:10.1504/IJSD.2018.10012310.

Carey, D.E., Yang, Y., McNamara, P.J., Mayer, B.K., 2016. Recovery of agricultural nutrients from biorefineries. Bioresour. Technol. 18732976. 215, 186–198. Available from: https://www.elsevier.com/locate/biortech. doi:10.1016/j.biortech.2016.02.093.

Chapman, J., Ismail, A.E., Dinu, C.Z., 2018. Industrial applications of enzymes: recent advances, techniques, and outlooks. Catalysts 20734344. 8 (6). Available from: http://www.mdpi.com/2073-4344/8/6/238/pdf. doi:10.3390/catal8060238.

Chavan, S., Yadav, B., Atmakuri, A., Tyagi, R.D., Wong, J.W.C., Drogui, P., 2022. Bioconversion of organic wastes into value-added products: a review. Bioresour. Technol. 18732976. 344. Available from: https://www.elsevier.com/locate/biortech. doi:10.1016/j.biortech.2021.126398.

Cheng, D., Ngo, H.H., Guo, W., Chang, S.W., Nguyen, D.D., Zhang, S., Deng, S., Ding, A., Hoang, N.B., 2022. Impact factors and novel strategies for improving biohydrogen production in microbial electrolysis cells. Bioresour. Technol. 09608524. 346, 126588. doi:10.1016/j.biortech.2021.126588.

Christgen, B., Suarez, A., Milner, E., Boghani, H., Sadhukhan, J., Shemfe, M., Gadkari, S., Kimber, R.L., Lloyd, J.R., Rabaey, K., Feng, Y., Premier, G.C., Curtis, T., Scott, K., Yu, E., Head, I.M., 2019. Chapter 4. Metal Recovery Using Microbial Electrochemical Technologies. Royal Society of Chemistry (RSC), pp. 87–112. doi:10.1039/9781788016353-00087.

Dang, R., Liu, J., Lichtfouse, E., Zhou, L., Zhou, M., Xiao, L., 2024. Soil microbial carbon use efficiency and the constraints. Ann. Microbiol. 18692044. 74 (1). Available from: https://www.springer.com/journal/13213. doi:10.1186/s13213-024-01780-9.

das Neves, L.C.M., Converti, A., Penna, T.C.V., 2009. Biogas production: new trends for alternative energy sources in rural and urban zones. Chem. Eng. Technol. 0930-7516. 32 (8), 1147–1153. doi:10.1002/ceat.200900051.

Deng, S., Wang, C., Ngo, H.H., Guo, W., You, N., Tang, H., Hongbin, Y., Tang, L., Han, J., 2023. Comparative review on microbial electrochemical technologies for resource recovery from wastewater towards circular economy and carbon neutrality. Bioresour. Technol. 09608524. 376, 128906. doi:10.1016/j.biortech.2023.128906.

Desrochers, P., 2000. Market processes and the closing of 'industrial loops'. J. Ind. Ecol. 1088-1980. 4 (1), 29–43. doi:10.1162/108819800569276.

Dionisi, D., Anderson, J.A., Aulenta, F., Mccue, A., Paton, G., 2015. The potential of microbial processes for lignocellulosic biomass conversion to ethanol: a review. J. Chem. Technol. Biotechnol. 10974660. 90 (3), 366–383. Available from: http://onlinelibrary.wiley.com/journal/10.1002/(ISSN)1097-4660. doi:10.1002/jctb.4544.

Downs, D.M., 2006. Understanding microbial metabolism. Annu. Rev. Microbiol. 00664227. 60, 533–559. doi:10.1146/annurev.micro.60.080805.142308.

Fuhrmann, J.J., 2021. Microbial Metabolism Principles and Applications of Soil Microbiology, third ed. Elsevier, United States, pp. 57–87. Available from: https://www.sciencedirect.com/book/9780128202029. doi:10.1016/B978-0-12-820202-9.00003-4.

Ghosh, A.K., Koltun, E.S., Bilcer, G., 2001. Tartaric acid and tartrates in the synthesis of bioactive molecules. Synthesis 00397881. 9, 1281–1301. Available from: https://www.thieme-connect.de/products/ejournals/journal/10.1055/s-00000084. doi:10.1055/s-2001-15217.

Ghosh, R.S., Bhattacharjee, S., Ganapathy, 2020. A case study to examine environmental benefits of lean construction. Sustain. Green Lean 1, 133–144.

Gudiukaite, R., Nadda, A.K., Gricajeva, A., Shanmugam, S., Nguyen, D.D., Lam, S.S., 2021. Bioprocesses for the recovery of bioenergy and value-added products from wastewater: a review. J. Environ. Manag. 10958630. 300. Available from: https://www.sciencedirect.com/journal/journal-of-environmental-management. doi:10.1016/j.jenvman.2021.113831.

Gupta, N.M., 2017. Factors affecting the efficiency of a water splitting photocatalyst: a perspective. Renew. Sustain. Energy Rev. 18790690. 71, 585–601. Available from: https://www.journals.elsevier.com/renewable-and-sustainable-energy-reviews. doi:10.1016/j.rser.2016.12.086.

Han, Y.H., Kim, G., Seo, S.W., 2023. Programmable synthetic biology tools for developing microbial cell factories. Curr. Opin. Biotechnol. 18790429. 79. Available from: http://www.elsevier.com/locate/copbio. doi:10.1016/j.copbio.2022.102874.

Haque, S., Srivastava, N., Pal, D.B., Alkhanani, M.F., Almalki, A.H., Areeshi, M.Y., Naidu, R., Gupta, V.K., 2022. Functional microbiome strategies for the bioremediation of petroleum-hydrocarbon and heavy metal contaminated soils: a review. Sci. Total Environ. 00489697. 833, 155222. doi:10.1016/j.scitotenv.2022.155222.

Iacono, E., Sciandrone, G., Filipponi, C., Freer, G., Lai, M., Pistello, M., 2024. Converting biomass waste into valuable biomaterials and bioactive compounds: an overview of antimicrobial activities in the pursuit of global sustainability and health. New Microbiologica 11217138. 47 (2), 123–136. Available from: https://newmicrobiologica.org/latest-issue/.

Imamoglu, E., 2024. Artificial intelligence and/or machine learning algorithms in microalgae bioprocesses. Bioengineering. 2306-5354. 11 (11), 1143. doi:10.3390/bioengineering11111143.

Islam, N.F., Gogoi, B., Saikia, R., Yousaf, B., Narayan, M., Sarma, H., 2024. Encouraging circular economy and sustainable environmental practices by addressing waste management and biomass energy production. Reg. Sustain. 2666660X. 5 (4). Available from: https://www.sciencedirect.com/science/journal/2666660X. doi:10.1016/j.regsus.2024.100174.

Jain, A., Sarsaiya, S., Awasthi, M.K., Singh, R., Rajput, R., Mishra, U.C., Chen, J., Shi, J., 2022. Bioenergy and bio-products from bio-waste and its associated modern circular economy: current research trends, challenges, and future outlooks. Fuel. 00162361. 307, 121859. doi:10.1016/j.fuel.2021.121859.

Jame, R., Alaysuy, O., Omer, N., Alatawi, N.M., Elsaim, M.H., 2021. Production of bio-methane gas as renewable energy source - a review. J. Energy Environ. Chem. Eng. 2637-4331. 6 (4), 124. doi:10.11648/j.jeece.20210604.14.

Jensen, L.S., Kaul, C., Juncker, N.B., Thomsen, M.H., Chaturvedi, T., 2022. Biohydrogen production in microbial electrolysis cells utilizing organic residue feedstock: a review. Energies 19961073. 15 (22). Available from: http://www.mdpi.com/journal/energies/. doi:10.3390/en15228396.

Jones, D.L., Cooledge, E.C., Hoyle, F.C., Griffiths, R.I., Murphy, D.V., 2019. pH and exchangeable aluminum are major regulators of microbial energy flow and carbon use efficiency in soil microbial communities. Soil Biol. Biochem. 00380717. 138. Available from: https://www.elsevier.com/inca/publications/store/3/3/2. doi:10.1016/j.soilbio.2019.107584.

Kaiser, J.P., Feng, Y., Bollag, J.M., 1996. Microbial metabolism of pyridine, quinoline, acridine, and their derivatives under aerobic and anaerobic conditions. Microbiol. Rev. 01460749. 60 (3), 483–498. doi:10.1128/mmbr.60.3.483-498.1996.

Kee, S.H., Chiongson, J.B.V., Saludes, J.P., Vigneswari, S., Ramakrishna, S., Bhubalan, K., 2021. Bioconversion of agro-industry sourced biowaste into biomaterials via microbial factories – a viable domain of circular economy. Environ. Pollut. 18736424. 271. Available from: https://www.journals.elsevier.com/environmental-pollution. doi:10.1016/j.envpol.2020.116311.

Kelly, P.T., He, Z., 2014. Nutrients removal and recovery in bioelectrochemical systems: a review. Bioresour. Technol. 18732976. 153, 351–360. Available from: https://www.elsevier.com/locate/biortech. doi:10.1016/j.biortech.2013.12.046.

Kennes, C., 2018. Bioconversion processes. Fermentation 23115637. 4 (2). Available from: https://www.mdpi.com/2311-5637/4/2/21/pdf. doi:10.3390/fermentation4020021.

Khan, M.R., Evans, R.J., 1997. Conversion and utilization of waste materials. J. Sol. Energy Eng. 0199-6231. 119 (3), 263–263. doi:10.1115/1.2888030.

Kour, D., Rana, K.L., Yadav, A.N., Yadav, N., Kumar, M., Kumar, V., Vyas, P., Dhaliwal, H.S., Saxena, A.K., 2020. Microbial biofertilizers: bioresources and eco-friendly technologies for agricultural and environmental sustainability. Biocatal. Agric. Biotechnol. 18788181. 23, 101487. doi:10.1016/j.bcab.2019.101487.

Kuang, Z.Y., Yang, H., Shen, S.W., Lin, Y.N., Sun, S.W., Neureiter, M., Yue, H.T., Ye, J.W., 2023. Bio-conversion of organic wastes towards polyhydroxyalkanoates. Biotechnol. Notes 26659069. 4, 118–126. Available from: https://www.sciencedirect.com/science/journal/26659069. doi:10.1016/j.biotno.2023.11.006.

Kumar, A., Kumari, S., Dindhoria, K., Manyapu, V., Kumar, R., 2022a. Efficient Utilization and Bioprocessing of Agro-Industrial Waste. Springer Science and Business Media LLC, Springer Science and Business Media LLC, pp. 1–37. doi:10.1007/978-3-030-84405-9_1.

Kumar, S., Diksha, S.S.S., Kumar, R., 2022b. Biofertilizers: an ecofriendly technology for nutrient recycling and environmental sustainability. Curr. Res. Microb. Sci. 26665174. 3, 100094. doi:10.1016/j.crmicr.2021.100094.

Lee, E., Oliveira, D.S.B.L., Oliveira, L.S.B.L., Jimenez, E., Kim, Y., Wang, M., Ergas, S.J., Zhang, Q., 2020. Comparative environmental and economic life cycle assessment of high solids anaerobic co-digestion for biosolids and organic waste management. Water Res. 00431354. 171, 115443. doi:10.1016/j.watres.2019.115443.

Lee, J.C., Pandey, B.D., 2012. Bio-processing of solid wastes and secondary resources for metal extraction - A review. Waste Manag. 18792456. 32 (1), 3–18. doi:10.1016/j.wasman.2011.08.010.

Logan, B.E., Rabaey, K., 2012. Conversion of wastes into bioelectricity and chemicals by using microbial electrochemical technologies. Science 10959203. 337 (6095), 686–690. Available from: http://www.sciencemag.org/content/337/6095/686.full.pdf. doi:10.1126/science.1217412.

Makepa, D.C., Chihobo, C.H., 2024. Barriers to commercial deployment of biorefineries: a multi-faceted review of obstacles across the innovation chain. Heliyon 24058440. 10 (12). Available from: https://www.sciencedirect.com/science/journal/24058440. doi:10.1016/j.heliyon.2024.e32649.

Mannina, G., Alduina, R., Badalucco, L., Barbara, L., Capri, F.C., Cosenza, A., Di Trapani, D., Gallo, G., Laudicina, V.A., Muscarella, S.M., Presti, D., 2021. Water resource recovery facilities (WRRFs): the Case Study of Palermo University (Italy). Water. 2073-4441. 13 (23), 3413. doi:10.3390/w13233413.

Meng, P., Pei, H., Hu, W., Shao, Y., Li, Z., 2014. How to increase microbial degradation in constructed wetlands: influencing factors and improvement measures. Bioresour. Technol. 18732976. 157, 316–326. Available from: https://www.elsevier.com/locate/biortech. doi:10.1016/j.biortech.2014.01.095.

Mishra, Y.H.R., et al., 2010. Current research trends of microbiological leaching for metal recovery from industrial wastes. Current Research, Technology and Education Topics in Applied Microbiology and Microbial Biotechnology, vol 1. 1289–1292.

Mo, W., Zhang, Q., 2013. Energy-nutrients-water nexus: integrated resource recovery in municipal wastewater treatment plants. J. Environ. Manag. 10958630. 127, 255–267. Available from: https://www.sciencedirect.com/journal/journal-of-environmental-management. doi:10.1016/j.jenvman.2013.05.007.

Mu, D., Horowitz, N., Casey, M., Jones, K., 2017. Environmental and economic analysis of an in-vessel food waste composting system at Kean University in the U.S. Waste Manag. 18792456. 59, 476–486. Available from: https://www.elsevier.com/locate/wasman. doi:10.1016/j.wasman.2016.10.026.

Najar, I.N., Sharma, P., Das, R., Tamang, S., Mondal, K., Thakur, N., Gandhi, S.G., Kumar, V., 2024. From waste management to circular economy: leveraging thermophiles for sustainable growth and global resource optimization. J. Environ. Manag. 10958630. 360. Available from: https://www.sciencedirect.com/science/journal/03014797. doi:10.1016/j.jenvman.2024.121136.

Nancharaiah, Y.V., Mohan, S.V., Lens, P.N.L., 2016. Recent advances in nutrient removal and recovery in biological and bioelectrochemical systems. Bioresour. Technol. 18732976. 215, 173–185. Available from: https://www.elsevier.com/locate/biortech. doi:10.1016/j.biortech.2016.03.129.

Neto, G.C.O., Correia, J.M.F., 2019. Environmental and economic advantages of adopting reverse logistics for recycling construction and demolition waste: a case study of Brazilian construction and recycling companies. Waste Manag. Res. 10963669. 37 (2), 176–185. Available from: https://wmr.sagepub.com/. doi:10.1177/0734242X18816790.

Nielsen, P.H., 2017. Microbial biotechnology and circular economy in wastewater treatment. Microb. Biotechnol. 17517915. 10 (5), 1102–1105. Available from: http://onlinelibrary.wiley.com/journal/10.1111/(ISSN)1751-7915. doi:10.1111/1751-7915.12821.

Nwokolo, N., Mukumba, P., Obileke, K., Enebe, M., 2020. Waste to energy: a focus on the impact of substrate type in biogas production. Processes 22279717. 8 (10), 1–21. Available from: https://www.mdpi.com/2227-9717/8/10/1224/pdf. doi:10.3390/pr8101224.

Nzihou, A., Lifset, R., 2010. Waste valorization, loop-closing, and industrial ecology. J. Ind. Ecol. 15309290. 14 (2), 196–199. doi:10.1111/j.1530-9290.2010.00242.x.

O'Connor, K.E., 2021. Microbiology challenges and opportunities in the circular economy. Microbiology. 1350-0872. 167 (1). doi:10.1099/mic.0.001026.

Oren, A., 2010. Microbial Metabolism: Importance for Environmental Biotechnology. Springer Science and Business Media LLC, Springer Science and Business Media LLC, pp. 193–255. doi:10.1007/978-1-60327-140-0_5.

Pandey, A.K., Park, J., Jeun, K., Joo, H.-H., Raj, T., Singh, L.K., Singh, N., Kim, S.-H., 2023. Machine learning in fermentative biohydrogen production: advantages, challenges, and applications. Bioresour. Technol. 09608524. 370, 128502. doi:10.1016/j.biortech.2022.128502.

Papa, G., Cucina, M., Echchouki, K., De Nisi, P., Adani, F., 2023. Anaerobic digestion of organic waste allows recovering energy and enhancing the subsequent bioplastic degradation in soil. Resour. Conserv. Recycl. 09213449. 188, 106694. doi:10.1016/j.resconrec.2022.106694.

Pavan, M., Reinmets, K., Garg, S., Mueller, A.P., Marcellin, E., Köpke, M., Valgepea, K., 2022. Advances in systems metabolic engineering of autotrophic carbon oxide-fixing biocatalysts towards a circular economy. Metab. Eng. 10967184. 71, 117–141. Available from: http://www.elsevier.com/inca/publications/store/6/2/2/9/1/3/index.htt. doi:10.1016/j.ymben.2022.01.015.

Purnell, P., Velenturf, A.P.M., Marshall, R., 2019. Chapter 16. New Governance for Circular Economy: Policy, Regulation and Market Contexts for Resource Recovery from Waste. Royal Society of Chemistry (RSC, pp. 395–422. doi:10.1039/9781788016353-00395.

Puyol, D., Batstone, D.J., Hülsen, T., Astals, S., Peces, M., Krömer, J.O., 2017. Resource recovery from wastewater by biological technologies: opportunities, challenges, and prospects. Front. Microbiol. 1664302X. 7. Available from: http://journal.frontiersin.org/article/10.3389/fmicb.2016.02106/full. doi:10.3389/fmicb.2016.02106.

Ramin, P., Ramin, E., Feldman, H., Flores-Alsina, X., Gernaey, K.V., 2021. Evaluating Resource Recovery Options in Wastewater Treatment Plants Using Mathematical Models. Informa UK Limited, Informa UK Limited, pp. 45–69. doi:10.1201/9781003055501-2.

Ramteke, P.K., Rathod, A.P., Kodape, S.M., Sonawane, S.S., 2025. Development and evaluation of environmentally sustainable cenosphere ceramic membrane for the efficient separation of methylene blue dye. Chem. Pap. 0366-6352. 79 (1), 313–332. doi:10.1007/s11696-024-03782-y.

Rene, E.R., Sarangi, P.K., I Nogué, V.S., Schnürer, A., Salvachúa, D., 2023. Current trends in waste valorization. Microb. Biotechnol. 17517915. 16 (2), 173–176. Available from: http://onlinelibrary.wiley.com/journal/10.1111/(ISSN)1751-7915. doi:10.1111/1751-7915.14198.

Rezaei, Z., Moghimi, H., 2024. Fungal-bacterial consortia: a promising strategy for the removal of petroleum hydrocarbons. Ecotoxicol. Environ. Saf. 01476513. 280, 116543. doi:10.1016/j.ecoenv.2024.116543.

Sahu, S., Kaur, A., Singh, G., Arya, S.K., 2023. Harnessing the potential of microalgae-bacteria interaction for eco-friendly wastewater treatment: a review on new strategies involving machine learning and artificial intelligence. J. Environ. Manag. 03014797. 346, 119004. doi:10.1016/j.jenvman.2023.119004.

Salvador, M., Abdulmutalib, U., Gonzalez, J., Kim, J., Smith, A.A., Faulon, J.L., Wei, R., Zimmermann, W., Jimenez, J.I., 2019. Microbial genes for a circular and sustainable bio-PET economy. Genes 20734425. 10 (5). Available from: https://www.mdpi.com/2073-4425/10/5/373/pdf. doi:10.3390/genes10050373.

Saravanan, A., Karishma, S., Kumar, P.S., Rangasamy, G., 2023. A review on regeneration of biowaste into bio-products and bioenergy: life cycle assessment and circular economy. Fuel. 00162361. 338, 127221. doi:10.1016/j.fuel.2022.127221.

Sarsaiya, S., Jain, A., Awasthi, S.K., Duan, Y., Awasthi, M.K., Shi, J., 2019. Microbial dynamics for lignocellulosic waste bioconversion and its importance with modern circular economy, challenges and future perspectives. Bioresour. Technol. 18732976. 291. Available from: https://www.elsevier.com/locate/biortech. doi:10.1016/j.biortech.2019.121905.

Sengupta, S., Nawaz, T., Beaudry, J., 2015. Nitrogen and phosphorus recovery from wastewater. Curr. Pollut. Rep. 21986592. 1 (3), 155–166. Available from: https://springer.com/environment/pollution+and+remediation/journal/40726. doi:10.1007/s40726-015-0013-1.

Sharma, P., Gaur, V.K., Kim, S.H., Pandey, A., 2020. Microbial strategies for bio-transforming food waste into resources. Bioresour. Technol. 18732976. 299. Available from: https://www.elsevier.com/locate/biortech. doi:10.1016/j.biortech.2019.122580.

Shen, Y., Yuan, W., Pei, Z.J., Wu, Q., Mao, E., 2009. Microalgae mass production methods. Trans. ASABE. 2151-0040. 52 (4), 1275–1287. doi:10.13031/2013.27771.

Sirohi, R., Joun, J., Lee, J.Y., Yu, B.S., Sim, S.J., 2022. Waste mitigation and resource recovery from food industry wastewater employing microalgae-bacterial consortium. Bioresour.

Technol. 18732976. 352. Available from: https://www.elsevier.com/locate/biortech. doi:10.1016/j.biortech.2022.127129.

Stamatelatou, K., Antonopoulou, G., lyberatos, G., 2011. Production of Biogas via Anaerobic Digestion. Elsevier BV, Elsevier BV, pp. 266–304. doi:10.1533/9780857090492.2.266.

Sushmitha, D.B.E.R., 2016. Bioplastics - a review. IJMTER 3, 411–413.

Thomas, A.P., Kasa, V.P., Dubey, B.K., Sen, R., Sarmah, A.K., 2023. Synthesis and commercialization of bioplastics: organic waste as a sustainable feedstock. Sci. Total Environ. 18791026. 904. Available from: https://www.elsevier.com/locate/scitotenv. doi:10.1016/j.scitotenv.2023.167243.

Torres-León, C., Chávez-González, M.L., Hernández-Almanza, A., Martínez-Medina, G.A., Ramírez-Guzmán, N., Londoño-Hernández, L., Aguilar, C.N., 2021. Recent advances on the microbiological and enzymatic processing for conversion of food wastes to valuable bioproducts. Curr. Opin. Food Sci. 22147993. 38, 40–45. Available from: http://www.journals.elsevier.com/current-opinion-in-food-science/. doi:10.1016/j.cofs.2020.11.002.

Torres-Sebastián, M.J., Colli-Mull, J.G., Escobedo-Sánchez, L., Martínez-Fong, D., Rios-Solis, L., Gutiérrez-Castillo, M.E., López-Jiménez, G., Moreno-Rivera, M.L., Tovar-Gálvez, L.R., Espadas-Álvarez, A.J., 2022. Methane, a renewable biofuel: from organic waste to bioenergy. Biofuels 17597277. 13 (7), 907–917. Available from: http://www.tandfonline.com/loi/tbfu20?open=5&repitition=0#vol_5. doi:10.1080/17597269.2021.2016105.

Udugama, I.A., Petersen, L.A.H., Falco, F.C., Junicke, H., Mitic, A., Alsina, X.F., Mansouri, S.S., Gernaey, K.V., 2020. Resource recovery from waste streams in a water-energy-food nexus perspective: toward more sustainable food processing. Food Bioprod. Process. 09603085. 119, 133–147. Available from: http://www.elsevier.com/wps/find/journaldescription.cws_home/713883/description#description. doi:10.1016/j.fbp.2019.10.014.

Upadhyay, S.K., Singh, G., Rani, N., Rajput, V.D., Seth, C.S., Dwivedi, P., Minkina, T., Wong, M.H., Show, P.L., Khoo, K.S., 2024. Transforming bio-waste into value-added products mediated microbes for enhancing soil health and crop production: perspective views on circular economy. Environ. Technol. Innov. 23521864. 34. Available from: https://www.sciencedirect.com/science/journal/23521864. doi:10.1016/j.eti.2024.103573.

Varjani, S.J., 2017. Microbial degradation of petroleum hydrocarbons. Bioresour. Technol. 18732976. 223, 277–286. Available from: https://www.elsevier.com/locate/biortech. doi:10.1016/j.biortech.2016.10.037.

Wackett, L.P., 2023. Web alert: microbial waste valorization. Microb. Biotechnol. 1751-7915. 16 (2), 481–482. doi:10.1111/1751-7915.14223.

Wackett, L.P., alert, W., 2022. Microbial biomass conversion. Microb. Biotechnol. 17517915. 15 (10), 2683–2684. Available from: http://onlinelibrary.wiley.com/journal/10.1111/(ISSN)1751-7915. doi:10.1111/1751-7915.14149.

Wainaina, S., Awasthi, M.K., Sarsaiya, S., Chen, H., Singh, E., Aman Kumar, B.R., Awasthi, S.K., Liu, T., Duan, Y., Kumar, S., Zhang, Z., Taherzadeh, M.J., 2020. Resource recovery and circular economy from organic solid waste using aerobic and anaerobic digestion technologies. Bioresour. Technol. 09608524. 301, 122778. doi:10.1016/j.biortech.2020.122778.

Wang, H., Ren, Z.J., 2013. A comprehensive review of microbial electrochemical systems as a platform technology. Biotechnol. Adv. 07349750. 31 (8), 1796–1807. doi:10.1016/j.biotechadv.2013.10.001.

Wilson, S.A., Roberts, S.C., 2012. Recent advances towards development and commercialization of plant cell culture processes for the synthesis of biomolecules. Plant Biotechnol. J. 14677652. 10 (3), 249–268. doi:10.1111/j.1467-7652.2011.00664.x.

Wu, S., Snajdrova, R., Moore, J.C., Baldenius, K., Bornscheuer, U.T., 2021. Biocatalysis: enzymatic synthesis for industrial applications. Angew. Chem. Int. Ed. 15213773. 60 (1), 88–119. Available from: http://onlinelibrary.wiley.com/journal/10.1002/(ISSN)1521-3773. doi:10.1002/anie.202006648.

Yang, E., Mohamed, H.O., Sung-Gwan Park, M.O., Al-Qaradawi, S.Y., Castaño, P., Chon, K., Chae, K.-J., 2021. A review on self-sustainable microbial electrolysis cells for electro-biohydrogen production via coupling with carbon-neutral renewable energy technologies. Bioresour. Technol. 09608524. 320 (Pt B), 124363. doi:10.1016/j.biortech.2020.124363.

Yildiz, M.J., Kalinowska, M., Kalinowska-Wichrowska, K., Gołębiewska, E., Tarasewicz, P., Bobin, T., Tarapata, D., Szatyłowicz, E., Piekut, J., 2023. A short overview of the possibilities of using waste from the agri-food industry. Adv. Sci. Technol. Res. J. 2080-4075. 17 (2), 342–352. doi:10.12913/22998624/161840.

Yuan, Z., 2018. Microbial Energy Conversion Microbial Energy Conversion. De Gruyter, pp. 1–528. Available from: https://www.degruyter.com/document/doi/10.1515/9783110424737/html. doi:10.1515/9783110424737.

Zarezadeh, S., Moheimani, N.R., Jenkins, S.N., Hülsen, T., Riahi, H., Mickan, B.S., 2019. Microalgae and phototrophic purple bacteria for nutrient recovery from agri-industrial effluents: influences on plant growth, rhizosphere bacteria, and putative carbon- and nitrogen-cycling genes. Front. Plant Sci. 1664462X. 10. Available from: https://www.frontiersin.org/journals/plant-science. doi:10.3389/fpls.2019.01193.

Zhang, T., Tremblay, P.L., 2016. Editorial: current challenges and future perspectives on emerging bioelectrochemical technologies. Front. Microbiol. 1664302X. 7. Available from: http://journal.frontiersin.org/article/10.3389/fmicb.2016.00860. doi:10.3389/fmicb.2016.00860.

Zhang, T., Zhang, H., 2022. Microbial consortia are needed to degrade soil pollutants. Microorganisms. 2076-2607. 10 (2), 261. doi:10.3390/microorganisms10020261.

Zhou, M., Yang, J., Wang, H., Jin, T., Xu, D., Gu, T., 2013. Microbial fuel cells and microbial electrolysis cells for the production of bioelectricity and biomaterials. Environ. Technol. 1479487X. 34 (13-14), 1915–1928. Available from: http://www.tandf.co.uk/journals/titles/09593330.asp. doi:10.1080/09593330.2013.813951.

CHAPTER

# Microbial fuel cells: Harnessing energy from waste

11

**Abid Ullah[a], Muhammad Waqar[b], Muhammad Humayun[c], and Sumayya Shah[b]**

[a]*Advanced Energy and System Engineering, University of Science and Technology South Korea, Daejeon, South Korea*

[b]*Department of Chemistry, University of Malakand, Malakand, Pakistan*

[c]*Department of Basic Sciences, University of Engineering & Technology, Peshawar, Pakistan*

## 1 Introduction

The rising demand for sustainable and carbon-neutral energy sources has prompted significant global interest in renewable alternatives to fossil fuels (Janicek et al., 2014). These alternatives are not only vital for addressing the growing energy crisis but also crucial in mitigating the environmental degradation caused by conventional energy production (Ahmad et al., 2024). According to projections by the European Renewable Energy Council, nearly 50% of the world's energy could come from renewable sources by 2040 (Ahmed et al., 2023). In this context, various types of organic waste, widely available and rich in biodegradable content, present a substantial untapped resource for renewable energy recovery. Over the past few decades, waste management has evolved from a focus on disposal and treatment toward resource and energy recovery. Among the emerging technologies in this domain, bioelectrochemical systems (BESs) have attracted attention for their ability to efficiently treat waste while simultaneously recovering energy. A notable example of BESs is the microbial fuel cell (MFC), which enables the direct conversion of chemical energy in organic waste into electricity. This is achieved through a synergistic process involving microbial metabolism and a solid-state electron acceptor.

In an MFC, specialized microorganisms known as *exoelectrogens* oxidize organic substrates at the anode, releasing electrons and protons, as exhibited in Fig. 11.1. These electrons travel through an external circuit to the cathode, where they combine with protons and terminal electron acceptors, such as oxygen, an abundant and environmentally benign oxidant. Other electron acceptors, like ferricyanide, permanganate, dichromate, and persulfate, have also been studied, but oxygen remains the most practical due to its natural availability and cost-effectiveness. Compared to traditional anaerobic digestion (AD), MFCs offer several advantages. In AD, energy is recovered in the form of biogas (methane or hydrogen), but significant energy losses occur during gas combustion and electricity generation, often exceeding

Microbial Technologies for Sustainable Waste Management. https://doi.org/10.1016/B978-0-443-44093-9.00009-8

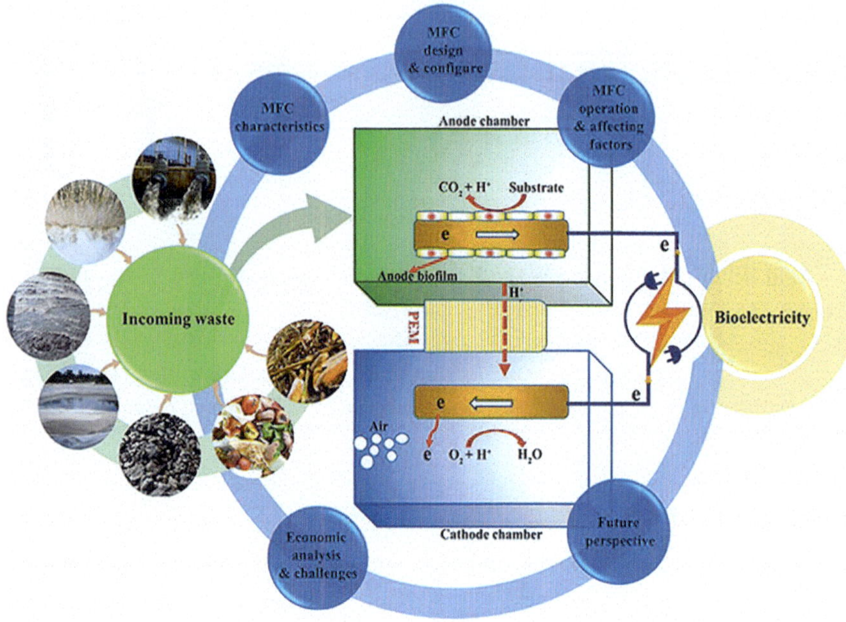


**FIG. 11.1**

Electricity generation from wastes by MFC.

65%. Moreover, biogas contains impurities such as hydrogen sulfide, necessitating additional purification steps. MFCs, in contrast, enable direct energy recovery with minimal losses and produce mainly carbon dioxide, which is biogenic and thus considered carbon-neutral. This significantly reduces greenhouse gas emissions, especially when replacing methane-based systems. In addition to environmental advantages, MFCs exhibit operational benefits over chemical fuel cells (Fig. 11.1). While chemical fuel cells often require expensive catalysts, high temperatures (500°C–1000°C), and extreme pH conditions, MFCs operate under mild, ambient conditions and can utilize a wide variety of low-grade organic wastes. This flexibility reduces operational costs and enhances safety and reliability, making MFCs a promising option for decentralized energy systems. Despite the promise of MFCs, their role as energy producers in real-world waste treatment applications remains underdeveloped. Much of the current research demonstrating high power outputs has been conducted using synthetic substrates and controlled laboratory conditions, rather than complex, heterogeneous real waste streams. In practical applications, MFCs are often optimized for organic removal rather than power generation, and their full potential for energy recovery is frequently overlooked. Realizing energy self-sufficient MFC systems requires a paradigm shift from merely achieving maximum theoretical power to developing strategies for effective energy harvesting and utilization (Saher, 2019b). Typically, MFCs are tested using external resistors to estimate their maximum power output; however, for real energy recovery, the external load must be replaced by energy storage devices or real-time applications. As a result, the operational conditions are often suboptimal, and actual energy extraction is lower than theoretical predictions.

This chapter presents a detailed examination of recent advancements in the application of MFCs for electricity generation from real waste streams, with a particular focus on the effective management and utilization of the harvested energy. It systematically analyzes the key factors that constrain MFC energy output and discusses potential strategies to enhance their long-term stability and performance. Furthermore, it explores innovative methods for energy storage and on-site utilization, offering insights into how MFC-generated power can be practically harnessed. By addressing these core challenges, the discussion aims to bridge the gap between experimental research and field-scale implementation, offering pathways for the realistic and scalable deployment of MFC technology in energy recovery systems.

## 2 Energy-generating capacity of MFC

To comprehensively evaluate the energy-generating potential of MFCs, several key performance metrics have been proposed. Among these, current density and power density are the most commonly used indicators, as they directly reflect the amount of electrical energy produced by the system. However, because MFCs often serve the dual purpose of electricity generation and organic waste treatment, it is essential to also consider parameters that relate energy output to waste degradation.

### 2.1 Current density and power density

Current density is one of the fundamental parameters used to evaluate the electricity-generating performance of MFCs. It expresses the electric current produced relative to either the electrode surface area or the reactor volume. Typically, the current is normalized to the geometric surface area of the anode, as this is where microbial oxidation of organic substrates occurs. However, when the cathodic reaction becomes the rate-limiting step, the cathode surface area may be used instead. The current density is calculated using the following expression (Harnisch and Schröder, 2010):

$$I_{AN} = \frac{I}{A_{AN}} \tag{1}$$

$$I_{Cat} = \frac{I}{A_{Cat}} \tag{2}$$

Likewise, when evaluating volumetric current density, the volume of the anodic chamber is used as the basis for normalization (Harnisch and Schröder, 2010):

$$I_V = \frac{I}{V_R} \tag{3}$$

where I is the current (A) and $I_{An}$ (A $m^{-2}$), $I_{Cat}$ (A $m^{-2}$), and $I_V$ (A $m^{-3}$) are current densities normalized by the anode surface area ($A_{An}$, $m^2$), cathode area ($A_{Cat}$, $m^2$), and anodic chamber volume ($V_R$, $m^3$), respectively. Power density is another widely used parameter to evaluate the power output of an MFC. Power density is expressed as the power (P, W) provided per unit surface area of the electrode or volume of the reactor. When the external resistance is equal to the internal resistance of an MFC, the maximum power density ($P_{max}$) can be achieved.

## 3 Energy harvesting from various wastes by MFC

Conventional waste treatment processes are often both energy-intensive and costly. In contrast, MFCs represent an innovative approach that enables the direct generation of electricity during waste treatment. MFCs can utilize a broad range of organic compounds—from simple molecules to complex polymers—as fuel sources, making them well-suited for energy recovery from diverse waste streams.

### 3.1 Domestic wastewater

Advancements in modern water management are driving the development of innovative domestic wastewater treatment technologies that emphasize reducing energy consumption while recovering energy, water, and other valuable resources. Traditional aerobic treatment combined with anaerobic sludge digestion often results in the loss of a significant portion of the energy stored in dissolved organic matter rather than recovering it. In contrast, MFCs enable the direct conversion of dissolved organic compounds into electricity, achieving energy capture with minimal additional energy input. MFCs offer particular advantages over AD when treating low-strength domestic wastewater. The potential for implementing energy self-sufficient MFC systems for household wastewater treatment has been demonstrated in liter-scale reactors. For example, a 200-L MFC stack (with an effective volume of 100 liters) treating domestic wastewater achieved a peak power output of 114 MW, enough to power a direct current (DC) pump. In addition, a 250-L stackable pilot-scale MFC generated a net power of 0.47 $Wm^{-3}$, with operational energy costs approximately half those of conventional aerobic treatment methods, as shown in Fig. 11.2.

### 3.2 Food wastes

Food processing wastes and food debris are promising feedstocks for bioenergy production due to their high moisture content, rich organic matter, and favorable carbon-to-nitrogen ratios that support efficient biodegradation. Various types of food waste,

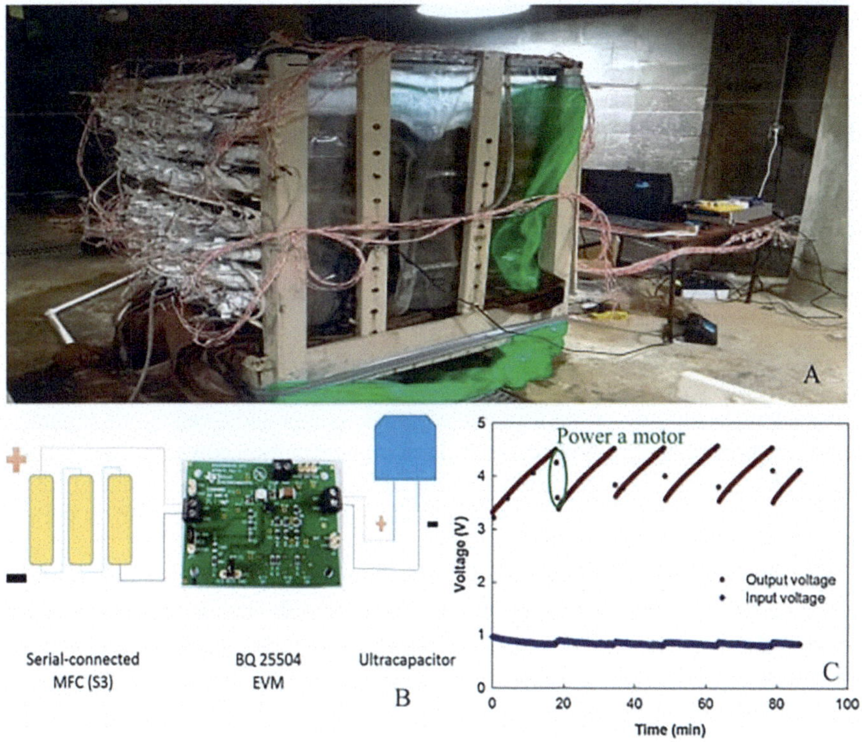


**FIG. 11.2**

(A) Prototype of a 200-L MFC stack fed with domestic wastewater; (B) schematic of charging/discharging circuit connection; and (C) charging and discharging of the ultracapacitors to drive a DC motor. When the voltage of the ultracapacitors reaches 4.5 V, they are discharged by powering the motor. When the voltage is lower than 3.5 V, the motor stops functioning.

including canteen scraps, molasses wastewater, starch processing effluent, brewery wastewater, palm oil mill wastewater, and dairy wastewater, have been explored as substrates for MFCs.

Wastewaters with high proportions of easily degradable carbohydrates, such as those from dairy, brewery, and molasses processing, generally yield better electricity generation than those dominated by more complex compounds like cellulose and lipids. For instance, an annular single-chamber MFC fueled with dairy wastewater achieved a power density of up to 20.2 $W.m^{-3}$ alongside a Coulombic efficiency (CE) of 26.9%. This highlights the strong potential of food waste-fed MFCs for energy self-sufficiency in larger-scale systems. A notable example includes a 100-L stackable pilot-scale reactor operating continuously on brewery wastewater, which generated 0.097 $kWh.m^{-3}$ of energy, enough to power a pumping system requiring 0.027 $kWh.m^{-3}$ for self-sustained feeding.

Despite the high energy content, a key challenge in using food wastes in MFCs is their generally low ion conductivity, which limits power generation. While MFC performance improves with solution conductivity up to about 20 mS/cm, most food wastewaters typically exhibit conductivities below 6 $mS.cm^{-1}$. For example, adding 100 mM NaCl to food waste leachate significantly increased maximum power output from 366 to 1000 $MW.m^{-3}$ by enhancing conductivity. In addition, because food wastes rich in soluble chemical oxygen demand (COD) can sometimes inhibit microbial activity, dilution is often necessary, making low-strength wastewaters, such as domestic wastewater, suitable as dilution media.

## 3.3 Landfill leachate

Landfill leachate, produced from the decomposition of municipal solid waste, contains a complex mixture of biodegradable organic compounds, xenobiotic organics, sulfides, ammonia, and heavy metals. Its high organic carbon content makes landfill leachate a potentially valuable substrate for MFCs. However, the presence of high COD levels, recalcitrant organics, and inhibitory substances often restricts the energy generation capacity of MFCs using this feedstock.

Typically, MFCs fueled with landfill leachate exhibit power densities below 1 $W/m^3$ and CEs under 20%. For example, an upflow air-cathode, membrane-less MFC treating landfill leachate produced electricity at a power density of 12.8 $W.m^{-3}$ but achieved a low CE of only 1.2%. While high COD concentrations may enhance power output, they can also drastically reduce CE. In one study, increasing the COD loading of young landfill leachate from 1 to 50 g/L caused the CE to drop sharply from 57% to just 1%.

## 3.4 Complex industrial wastes

Recalcitrant compounds often make up a larger portion of the total carbon pool compared to more readily biodegradable (labile) substances. A diverse range of these persistent chemicals, including petroleum hydrocarbons, chlorinated organics, nitrogen-containing compounds, heterocyclic molecules, and various polymers, have been investigated as substrates for MFCs. However, studies using actual field wastes remain limited. While MFCs have shown high COD removal efficiencies when treating complex wastewater, such as those from paper recycling and pharmaceutical industries, the associated power densities were relatively low, typically below 1 $MW.m^{-2}$. In contrast, significantly higher power outputs have been reported for more specialized industrial effluents, with dye wastewater generating power densities around 8 $W.m^{-3}$ and steroidal drug manufacturing effluents reaching up to 822.3 $W.m^{-3}$.

Using electrodes as electron acceptors directly in soils or sediments presents an appealing approach, as the contaminant-degrading microbes naturally colonize the anode surface, enabling efficient biodegradation. This setup offers a continuous, long-term electron sink that facilitates the breakdown of hazardous pollutants. For example, microbial electrochemical remediation of petroleum-contaminated soil using an MFC improved hydrocarbon degradation efficiency from just 2% in a control system to 24% under MFC operation, accompanied by a maximum power output of 2162 MW.$m^{-3}$. Similarly, an MFC applied to hexachlorobenzene-contaminated topsoil achieved a pesticide removal rate of 71.2%, with a peak power density of 77.5 MW.$m^{-2}$.

## 3.5 Sewage sludge

Sludge disposal costs can represent up to half of the total expenses in sewage treatment, posing a significant challenge for many municipal wastewater facilities. Importantly, a substantial portion of the energy contained in wastewater becomes concentrated in the sludge following aerobic treatment, making sludge a valuable resource for energy recovery. To date, MFCs utilizing sewage sludge as a substrate have achieved maximum power outputs of 4.2 W.$m^{-3}$ in abiotic cathode systems and up to 13.2 W.$m^{-3}$ when biocathodes are employed. Since sewage sludge primarily consists of insoluble particulate matter, while MFC microbes favor soluble, readily degradable organics, pretreatment methods such as ultrasonication, heating, alkalization, or prefermentation are often applied. These treatments help break down complex solids, thereby improving substrate bioavailability and enhancing energy recovery efficiency.

Although electricity generation from sludge-fed MFCs remains modest, the technology is attractive due to the simultaneous production of methane within the anodic chamber. For example, a two-stage MFC system designed for sludge treatment demonstrated a total energy output, including both electrical and biogas energy, of 23.22 kWh.$m^{-3}$ at a hydraulic retention time of 14 days, which is comparable to the energy yields from conventional AD.

## 3.6 Animal wastes

The rapid expansion of modern livestock farming has significantly increased the generation of animal waste. Both manure and urine are rich in organic matter, making them suitable substrates for MFCs. However, reported power densities for MFCs fueled by animal waste vary widely, ranging from several hundred milliwatts to several watts per cubic meter of reactor volume. For example, an MFC powered by dairy manure achieved a maximum power density of 15.1 W.$m^{-3}$ using a biocathode, and 16.3 W.$m^{-3}$ in a cassette-electrode configuration. In contrast, a 4-L loop-configured MFC fed with piggery wastewater produced a much lower power density of only 1.416 W.$m^{-3}$.

The power output of animal waste-fueled MFCs is strongly influenced by the solid and moisture content of the substrate. Increasing the solid content from 2% to 10% initially boosted power density, but higher solid concentrations subsequently caused a steep decline. In another study, MFCs treating animal wastes with moisture contents of 80%, 70%, and 60% achieved maximum power densities of 349 ± 39, 36 ± 9, and 12 ± 2 MW.$m^{-2}$, respectively.

A major limitation of using animal waste in MFCs is its high ammonia concentration, which can inhibit the activity of exoelectrogenic bacteria. In addition, ammonia can be converted to nitrate and nitrite, which compete with the electrode for electrons, thereby reducing overall energy recovery efficiency.

## 3.7 Plant wastes

The abundant and renewable nature of lignocellulosic materials derived from plant wastes makes them an attractive and cost-effective feedstock for bioenergy production. The potential of MFCs to utilize various agricultural residues, such as corn stover, wheat straw, rice straw, bean residue, and ground coffee, as well as aquatic plants like *Canna indica*, has been extensively studied (Ahmad et al., 2021).

Lignocellulosic biomass is primarily composed of cellulose, hemicellulose, and lignin. Since exoelectrogenic microorganisms cannot directly metabolize these complex polymers, the biomass must first be broken down into monosaccharides or other low-molecular-weight compounds through hydrolysis and fermentation before it can be effectively used for electricity generation. The relatively low biodegradability of lignocellulosic materials often limits the power output of MFCs.

As illustrated in Fig. 11.3, cellulose and hemicellulose polysaccharides are tightly bound within a lignin matrix, which hinders microbial access. Therefore pretreatment methods that disrupt the rigid lignocellulosic structure are essential to enhance microbial accessibility. Typically, such pretreatments yield carbohydrate-rich hydrolysates by hydrolyzing cellulose and hemicellulose. For example, using *Oscillatoria annae* to convert lignocellulose into glucose, a three-compartment MFC achieved high maximum power densities of 8.78 $W.m^{-3}$ and 6.73 $W.m^{-3}$ when using sugarcane bagasse and corn cob as substrates, respectively.

Comparing MFC performance across studies is challenging due to variations in operational conditions, reactor designs, electrode and membrane types, and microbial communities involved. Nonetheless, the potential of MFCs to recover electric energy from real wastes can be roughly estimated. As summarized in Table 11.1, energy-generating capacities vary widely depending on waste composition, strength, and solution chemistry.

Unlike experiments with pure substrates, microbial degradation of complex wastes involves energy-intensive processes with intricate networks of sequential and parallel degradation pathways. Consequently, wastes rich in easily biodegradable organics typically yield more energy than those containing complex substrates, refractory compounds, or insoluble materials. Moreover, energy recovery can be impaired by electron losses to competing electron acceptors such as nitrate, nitrite, and sulfate. Toxic compounds that inhibit exoelectrogenic activity also need to be removed or transformed, as they can significantly reduce MFC power output. Recognizing that many real wastes are not ideal for direct energy generation in MFCs, pretreatment offers promising opportunities to enhance biodegradability and remove inhibitory or competitive substances. Currently, the average normalized energy recovery (NER) for MFCs treating domestic wastewater is approximately 0.04 $kWh.m^{-3}$ of wastewater or 0.17 kWh/kg COD, while industrial wastewater yields about

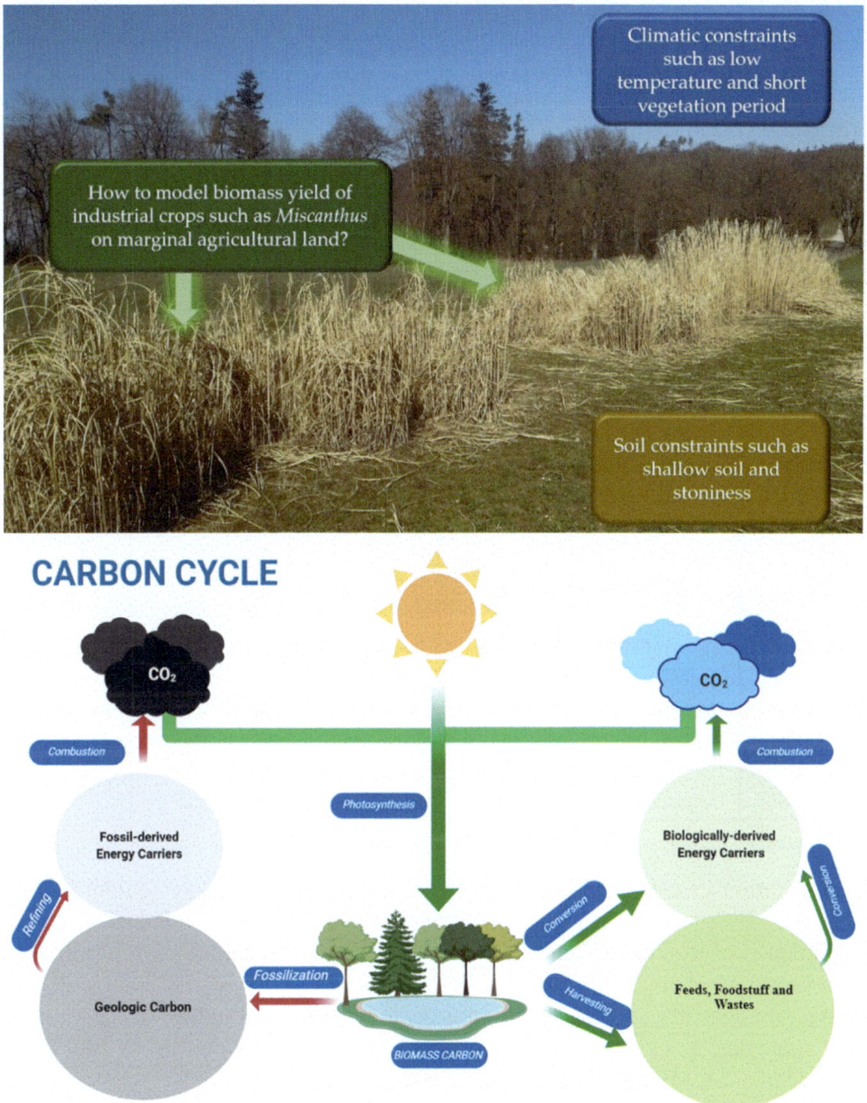


**FIG. 11.3**

Modeling biomass yield and carbon flow in miscanthus cultivation under climatic and soil constraints.

0.10 kWh.m$^{-3}$ or 0.04 kWh/kg COD. These values are considerably lower than those achieved by conventional AD, which typically recovers 0.34–0.49 kWh.m$^{-3}$ wastewater or 0.69–0.98 kWh/kg COD. However, some MFCs have reached NER values exceeding 2.0 kWh.m$^{-3}$ or 1.95 kWh/kg COD, indicating the potential for significant improvements after suitable pretreatment to enhance waste degradability.

**Table 11.1** Influence of waste characteristics on energy generation potential.

| Waste type | Composition (organic content) | Waste strength (COD/BOD) | Solution chemistry (pH, nutrients, salinity) | Suitable microbial technology | Energy output potential |
|---|---|---|---|---|---|
| Municipal solid waste (MSW) | Mixed organics, plastics | Medium | Neutral pH, variable nutrients | Anaerobic digestion | Moderate |
| Food waste | High biodegradable organics | High | Slightly acidic, nutrient-rich | Anaerobic digestion/fermentation | High |
| Agricultural waste | Lignocellulosic biomass | Medium–high | Variable pH, low nutrients | Biohydrogen/biogas production | Moderate–high |
| Industrial wastewater | Soluble organics, chemicals | High | Variable pH, possible toxins | Microbial fuel cells (MFCs) | Variable |
| Sewage sludge | High organic matter | High | Neutral pH, nutrient-rich | Anaerobic digestion | High |
| Landfill leachate | Complex organics, heavy metals | Very High | High salinity, toxic compounds | MFCs/hybrid systems | Low–moderate |
| Dairy wastewater | Proteins, fats, lactose | High | Slightly acidic | Anaerobic digestion | High |
| Textile wastewater | Dyes, low biodegradability | Low–Medium | Alkaline, toxic | MFCs/bioremediation | Low |

In addition to energy recovery efficiency, increasing the power output of real waste-fed MFCs remains critical. To compete with AD, MFCs should ideally achieve power densities around 1 $kW.m^{-3}$, yet most real waste-fed systems report power densities below 10 $W.m^{-3}$ (see Table 11.1). Nonetheless, there is potential to elevate these power densities to hundreds of watts per cubic meter through effective pretreatment strategies, as evidenced by the highest reported power density of 200 $W/m^3$ obtained in a 4-L MFC using acetate as a substrate.

## 4 Energy capture and storage with MFC

Currently, the direct power output from individual MFCs or MFC stacks is neither sufficient nor stable enough to continuously power standard electrical devices. Therefore effectively harnessing and utilizing the energy produced by MFCs in practical applications remains a significant challenge. To make MFC energy usable, it is essential to integrate systems capable of capturing, storing, and amplifying the power generated.

Recently, power converter-based energy storage systems have been investigated as alternatives to traditional external resistors used for extracting power from MFCs. Unlike external resistors, which dissipate energy as heat, these electronic circuits can simulate an equivalent load while simultaneously harvesting and storing the electrical energy produced by the MFC for later use.

## 4.1 Energy storage technologies for MFC

Currently, electrochemical capacitors are widely used to store energy generated by MFCs for powering low-energy devices. An electrochemical capacitor typically consists of two conductive electrodes separated by a dielectric material. The primary mechanism for charge storage in these capacitors is the electric double-layer effect, although pseudocapacitance also contributes to their capacity. Modern capacitors can rapidly absorb and release large amounts of energy over hundreds of thousands of cycles without generating heat or hazardous by-products. One key advantage of capacitors is their ability to smooth out high-frequency power fluctuations, thereby enhancing overall power quality. By repeatedly charging and discharging, capacitors can effectively multiply the current, voltage, and power output from MFCs. Since capacitors stop charging once their voltage reaches the MFC's open-circuit voltage, arrays of MFCs and capacitors are often combined to boost power. For instance, charging a bank of parallel-connected capacitors from four MFC units and then discharging them in series increased the output voltage from 0.7 V to 2.5 V, while peak power rose 2.6-fold with minimal energy loss in the circuit.

In some designs, a capacitive electrode is integrated directly with the MFC anode, allowing exoelectrogenic microbes to transfer electrons straight to the electrode for storage. This configuration enhances MFC power output, although its energy storage capacity still falls short compared to using an external capacitor. Another promising energy storage technology is the superconducting magnetic energy storage (SMES) system, which stores energy in the magnetic field generated by a persistent DC flowing through a superconducting coil. The current in the coil circulates indefinitely without decay, enabling long-term energy storage with minimal loss. SMES systems have been shown to harvest about 67% of the energy from MFCs. Despite the high initial cost, their low energy loss and excellent power quality make SMES an appealing option for storing energy produced by MFCs.

## 4.2 Capacitor-based power management system

Currently, capacitors cannot provide a continuous power output but are effective in delivering intermittent bursts of higher power. This intermittent supply suits applications like environmental monitoring sensors, which often operate intermittently. In such setups, a capacitor is typically paired with a charge pump, a boost converter, and the load, forming a power management system (Fig. 11.4). The charge pump draws a low current from the MFC to gradually charge the capacitor, while the boost converter raises the capacitor's output voltage to meet the load's requirements. This

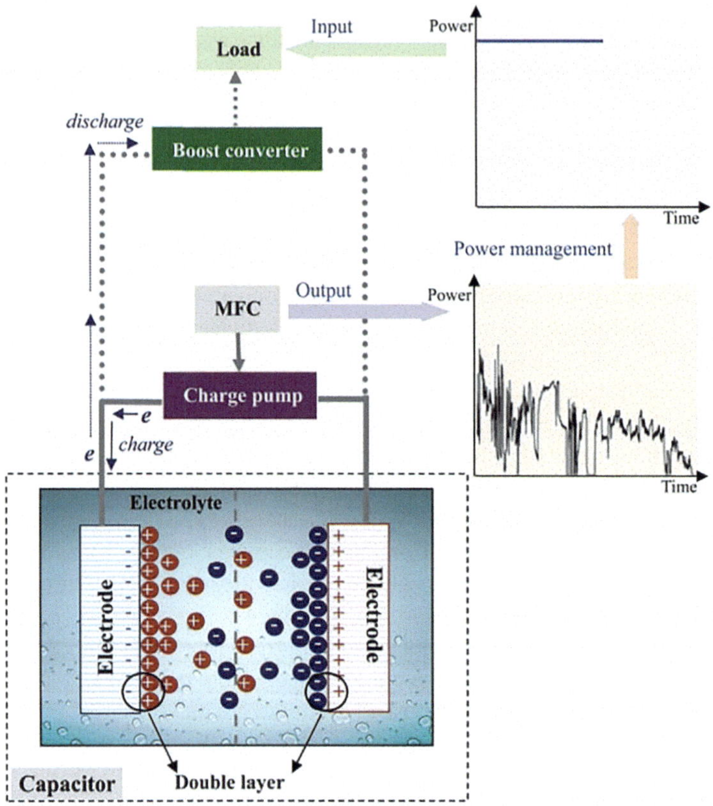


**FIG. 11.4**

Capacitor-based power management system for energy harvesting from MFC.

power management system has been successfully used to support benthic MFCs as long-term power sources for remote sensors.

Another practical application of this circuitry is in self-sustaining MFC stacks. For example, a stack consisting of 40 identical 20-mL single-chamber 3D-printed MFC units was developed to autonomously manage feeding, hydration, self-monitoring, and reporting tasks using only its own generated power. Electricity from this stack continuously powered a microcontroller to monitor and report stack voltage and ambient temperature every 10 min. It also charged a 12.5 F supercapacitor pack, which powered the feeding of anolyte every 48 h and catholyte hydration every 12 h.

Scaling up MFC stacks enables sufficient power output to operate electrical devices such as DC pumps. For instance, a 24-unit tubular MFC stack with a 2-L working volume charged 25 F ultracapacitors through a battery management evaluation module board (Fig. 11.2). The ultracapacitors maintained a stable output voltage between 3.5 and 4.5 V to drive a DC pump, despite the input voltage being only 1 V. Similarly, a 100-L brewery wastewater-fed MFC stack employing five capacitor-based circuits charged in parallel and discharged in series demonstrated a self-powered feeding cycle. The energy consumed by pumping was less than half of the total energy generated by the MFC, allowing surplus energy for extended operation.

Recently, a 6-L MFC treating synthetic wastewater with a COD of 1000 mg/L harvested 0.27 $kWh.m^{-3}$ of energy. Utilizing a circuit composed of 3.3 F capacitors and relays controlled by a programmable microcontroller, this system powered both the MFC's pumping (energy consumption of 0.014 $kWh.m^{-3}$) and an intermittent aeration system for a biological filter (energy consumption of 0.22 $kWh.m^{-3}$). Such power management systems pave the way for more stable and durable energy supply from large-scale MFC stacks to support electrical equipment in wastewater treatment plants.

To date, nearly all power management systems for MFC energy harvesting have focused on DC output for low-power electronics. However, most electrical appliances in waste treatment facilities require alternating current (AC), creating a need for energy management systems capable of DC-to-AC conversion at scale. Recently, a DC–AC converter with about 95% efficiency, capable of producing alternating voltage at any desired frequency, was developed. Yet, integrating this converter with capacitors in MFC systems remains a technical challenge. Traditional evaluation methods, involving an external resistor connected between the anode and cathode, measure continuous power supply potential but do not capture usable energy. When an MFC is connected to a capacitor for energy storage, these criteria are no longer appropriate. Instead, system performance should be assessed based on the energy harvested by the capacitor, considering parameters such as capacitor capacitance, the charging voltage that maximizes energy harvest, and the charging frequency required to achieve the desired voltage.

Research on MFC power management systems is still in its early stages. Various electrical energy storage technologies exist, including capacitor storage, superconducting magnetic storage, battery storage, flywheel storage, pumped hydro storage, and compressed air storage, each suitable for different power scales. Given the current power output levels of MFCs, capacitors remain the most practical storage option, although their effectiveness as long-term storage for large-scale MFCs requires further evaluation. Exploring and adapting alternative storage technologies will be essential to meet the evolving demands of MFC applications.

## 5 Direct power utilization from MFCs

Due to the limited and intermittent power output of individual MFCs, directly powering conventional electronic devices remains a challenge. Therefore strategies for in situ utilization of the electrical energy generated by MFCs must be explored. To date, various systems have been developed based on the MFC platform, incorporating diverse functionalities and structural configurations.

### 5.1 Microbial electrolysis cell

MFCs can be operated in a microbial electrolysis cell (MEC) mode, wherein the power generated at the anode is supplemented with an external voltage to drive thermodynamically unfavorable reactions at the cathode. A common application involves applying an external voltage exceeding 0.25 V alongside the MFC's inherent potential to initiate hydrogen ($H_2$) evolution at the cathode via proton reduction. This required voltage is significantly lower than that needed for conventional water electrolysis (typically 1.8–2.0 V). Notably, MFCs can be integrated with MECs in MEC–MFC hybrid systems to meet these additional energy requirements. In such configurations, the bioelectricity from both MFC and MEC anodes is synergistically utilized to overcome the energy barrier associated with proton-to-hydrogen conversion, thereby enabling efficient hydrogen production from organic waste (Khan et al., 2026).

Similar methodologies have been employed to synthesize various value-added chemicals at the cathode. This showcases the external voltages used to drive the cathodic production of different compounds reported in the literature. For instance, methane ($CH_4$), formic acid, and acetate have been produced through the electrochemical reduction of $CO_2$, powered by electrons originating from the anode (Hussain et al., 2026). In addition, the cathodic generation of hydrogen peroxide ($H_2O_2$) has been demonstrated using a two-electron oxygen reduction pathway, coupled with microbial oxidation of organics at the anode. At an external voltage of 0.5 V, the system achieved an $H_2O_2$ production efficiency of 83% from acetate. Given that $H_2O_2$ readily decomposes in aqueous environments, its in situ application for the degradation of biorefractory pollutants catalyzed by ferrous ions has been proposed.

Furthermore, the electrical output from MFCs can be harnessed for metal recovery from wastewater. Metals with high reduction potentials can be directly recovered at the cathode, whereas those with lower reduction potentials require additional power input. For example, the recovery of Cu, Pb, Cd, and Zn required external voltages of 0, 0.34, 0.51, and 1.7 V, respectively, corresponding to energy consumptions of 0, 3.8, 7.7, and 283.9 kWh per kg of metal (Fig. 11.5).

### 5.2 Microbial desalination cell

The concept of a microbial desalination cell (MDC) is based on harnessing the chemical energy stored in organic substrates to generate a potential gradient between the anode and cathode, thereby driving the desalination process. A typical MDC configuration comprises three chambers: an anode chamber, where organic matter is oxidized to produce electrons and protons; a central desalination chamber, where ionic separation occurs; and a cathode chamber, which completes the electrochemical circuit (Ullah et al., 2026a).

Unlike conventional desalination technologies that require external energy input, MDCs offer the dual benefit of extracting freshwater from saline water while simultaneously recovering energy from wastewater. For instance, a liter-scale up flow MDC demonstrated the capacity to generate 1.8 kWh of energy while achieving a

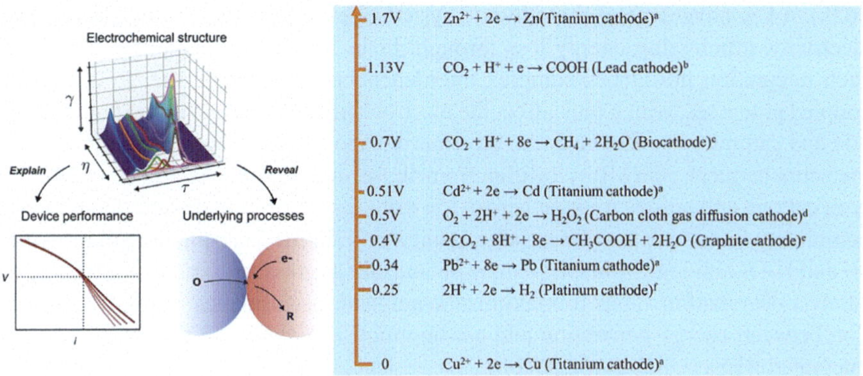


**FIG. 11.5**

**External voltages applied for the production of various chemicals.** This figure illustrates the externally applied voltages required to drive different electrochemical reactions, highlighting the energy thresholds associated with the production of specific chemicals.

90% reduction in salinity from 1 $m^3$ of seawater. In contrast, a typical reverse osmosis system requires approximately 2.2 kWh of energy to recover just 50% of water, highlighting the superior energy efficiency of MDC technology.

## 5.3 MFC-assisted photoelectrocatalytic system and biophotoelectrochemical cell

Photocatalytic oxidation is a promising technique for the degradation of organic pollutants; however, its practical efficiency is often limited by the rapid recombination of photogenerated electron–hole pairs. This significant drawback can be mitigated by coupling the photocatalytic system with an MFC to provide an external anodic bias. Remarkably, the pollutant degradation rate in such an integrated system was found to be twice the combined rate of standalone photocatalytic and electrochemical processes, indicating a strong synergistic enhancement between the MFC and photocatalytic components.

Beyond pollutant removal, bioelectricity produced at the MFC anode has also been applied to facilitate hydrogen ($H_2$) evolution at a photocathode. For example, a self-biased biophotoelectrochemical cell (BPEC) was developed using an $MoS_3$-modified silicon nanowire photocathode, enabling simultaneous spontaneous $H_2$ production and electricity generation under visible light. In this system, photogenerated holes in the semiconductor's valence band were neutralized by electrons from the MFC bioanode, while photo-excited conduction band electrons combined with protons to form $H_2$. This process effectively suppressed electron–hole recombination, thereby enhancing overall hydrogen production efficiency. These examples reflect the versatility of MFCs, where the microbial oxidation at the anode is leveraged to direct electron flow toward a variety of applications (Aziz et al., 2025; Ullah et al.,

2025). A key advantage of these hybrid systems lies in the in situ utilization of bioelectricity, minimizing energy loss through direct internal consumption. However, such integration introduces complex interdependencies between the MFC and the coupled processes, which may affect the electrochemical environment at both the anode and cathode. For instance, the electron transport dynamics in a microbial photoelectrochemical cell(MPEC) differ from those in a standalone MFC. In MDCs, both cations and anions migrate across ion exchange membranes, introducing ionic complexity (Fatima et al., 2024; Saher et al., 2019). In MECs, cathodic products like $H_2$ and $H_2O_2$ although valuable, can be chemically aggressive, potentially leading to catalyst degradation. Thus to maximize synergistic performance, careful coordination between energy generation and consumption is essential in the design of such integrated BESs (Ullah et al., 2026b).

## 6 Challenges and perspectives

MFCs have long been proposed as a means to achieve energy-neutral or even energy-positive wastewater treatment, contrasting with the traditionally energy-intensive nature of conventional treatment technologies. The primary energy demands in MFC operation stem from pumping, mixing, and recirculation processes. According to current best practices, treating domestic wastewater with MFCs consumes approximately 0.024 kWh per cubic meter for reactor feeding and mixing, while generating around 0.026 kWh per cubic meter of electricity. In the case of food waste treatment, such as brewery wastewater, the energy required to operate the pump system was reported at 0.027 kWh per cubic meter, which represented only 27.8% of the total energy generated by the MFC (Hussain et al., 2020). These data suggest that a net-positive energy balance is achievable if the energy content in wastewater is more effectively utilized. To move toward this goal, strategies are needed to minimize internal energy losses, employ suitable pretreatment methods to release biodegradable substrates, and integrate energy management systems (EMS) to enhance and stabilize MFC power generation (Alam et al., 2021).

The theoretical energy content of various wastewaters ranges from 4.92 to 7.97 kWh per kg of COD, yet MFCs currently recover <1.0 kWh per kg COD in real wastewater applications. To bridge this gap, advancements are required in multiple areas, including genetic engineering of exoelectrogenic microbes to enhance electron transfer, optimizing biofilm thickness for efficient electron conduction and substrate access, and developing novel electrode architectures such as three-dimensional macroporous structures that support robust microbial colonization while minimizing clogging. Improving cathodic oxygen reduction is also critical and can be addressed through the development of high-performance catalysts, such as metal alloys or biomimetic materials with high selectivity, durability, and activity. Likewise, the implementation of advanced membranes, especially composite or forward osmosis types with high ionic conductivity and resistance to fouling, will be essential. Despite technical progress, the high cost of MFC components remains a major barrier to scalability, with current systems estimated to be up to 800 times more expensive than anaerobic digesters, primarily due to the high costs of electrodes and separators (Shahbaz et al., 2021; Ullah et al., 2024d; Khan et al., 2024). Thus low-cost alternatives must be developed to make MFCs economically viable for widespread wastewater treatment.

While significant efforts have been made to optimize MFC design and operation, the composition of the feedstock plays an even more critical role in determining performance. Real wastewater typically contains complex organic compounds that require longer degradation pathways, leading to greater energy losses and reduced electrical output compared to simpler substrates. In addition, the frequent presence of competitive electron acceptors, such as nitrates, further reduces energy recovery. Therefore MFC systems should be designed with strategies to convert complex carbon-rich waste into substrates that are more favorable for exoelectrogenic metabolism and to mitigate losses caused by competing redox processes. This can be achieved through multistage treatment schemes, such as anaerobic pretreatment or series-connected MFC reactors, which improve substrate accessibility and reduce the impact of competing electron acceptors. However, pretreatment approaches must be carefully managed (Hussain et al., 2024; Ullah et al., 2022). Although physicochemical treatments, such as ultrasonic, ozone, or thermal methods, can increase biodegradability, they often require significant energy inputs ranging from 2.60–2.80, 4.49–5.13, and up to 45.52 kWh per kg of total suspended solids (TSS), respectively, thereby reducing the net energy gain. Pretreatment methods must therefore be selected based on a balance between enhancing microbial accessibility and minimizing energy and cost inputs. For instance, a 3.6-L two-stage MFC treating primary sludge achieved a total energy output of 23.22 kWh per cubic meter of anode liquid volume over 120 days, translating to a net energy recovery of 2.71 kWh per cubic meter of sludge, or 0.05–0.11 kWh per kg TSS depending on sludge concentration (Ullah et al., 2021; Ahmad et al., 2019).

Despite the ability of MFCs to generate electricity from actual waste streams, one of the most pressing challenges remains the practical utilization of that energy. Currently, the low and fluctuating power output of individual MFCs is insufficient to continuously operate standard electronic devices. Even with the use of MFC stacks, issues such as voltage reversal, biofouling, and system instability over extended operation periods limit large-scale applicability (Ajmal et al., 2025). To overcome these limitations, the development of customized energy management systems is essential. These systems should be capable of storing and stabilizing the low-voltage DC output from MFCs, converting it into AC when necessary, and managing energy flow to meet real-time demand (Saher, 2019a; Hussain et al., 2025). Supercapacitors have emerged as promising candidates for MFC energy storage due to their high energy capacity and excellent buffering ability. However, current systems rely on manual selection of charging and discharging parameters, which limits their efficiency. Future EMS designs must incorporate smart regulation technologies that automatically adjust to the fluctuating power profiles of MFCs to ensure reliable energy storage and discharge.

Given that field-scale MFCs alone cannot yet meet substantial energy demands, integration into hybrid renewable energy systems presents a promising pathway. Such systems, which combine MFCs with other energy sources like solar or wind, can offset the limitations of each individual source while improving overall energy reliability and reducing dependence on fossil fuels. Hybrid systems have already shown practical success in real-world applications, and similar integration involving MFCs is increasingly being explored. For example, a multisource system combining MFCs with acoustic piezoelectric harvesters was developed to provide continuous power for underwater wireless sensor networks. In another application, a hybrid energy harvester combining MFCs with a dielectric elastomer generator was employed to power an autonomous robot (EcoBot; Ullah et al., 2024a, 2024b, 2024c). The robot's operation was managed through energy storage during dormant periods, followed by bursts of activity using the stored electricity. Such examples underscore the potential of MFC-based hybrid energy systems to address low-power applications where continuous, high-output electricity is not required (Sadiq et al., 2021).

Despite the ability to increase overall power output through energy storage devices, energy losses inevitably occur during each charging and discharging cycle. In contrast, in situ utilization strategies, where the electricity generated by MFCs is immediately consumed within the system, offer a more energy-efficient approach. Studies from our laboratory, including MECs, MPECs, and BPECs, have demonstrated multiple ways to directly harness electron flow from MFCs to drive various reduction-based processes. Of particular interest is microbial electrosynthesis using MECs, which enables the conversion of organic waste at the anode into value-added chemicals at the cathode. One major advantage of this approach is that the instability of the MFC-generated power can be compensated for by adjusting the intensity of the external voltage supplied to the cathode. However, achieving high product specificity remains a challenge, requiring further research into biocatalyst development and system optimization. Ultimately, the bioelectrochemical reactions enabled by MFCs present a powerful platform for converting chemical energy in waste into practical applications. As the field evolves, greater emphasis should be placed on in situ power utilization strategies and the integration of MFCs with other low-energy-demand technologies to fully realize the potential of this sustainable energy device.

## References

Ahmad, S., Nawaz, T., Ullah, A., Ahmed, M., Khan, M.O., Saher, S., Qamar, A., Sikandar, M.A., 2021. Thermal optimization of manganese dioxide nanorods with enhanced ORR activity for alkaline membrane fuel cell. Electrochem. Sci. Adv. 26985977. 1 (3). Available from: https://chemistry-europe.onlinelibrary.wiley.com/journal/26985977. doi:10.1002/elsa.202000032.

Ahmad, S., Saher, S., Ahmed, M., Ullah, A., 2019. Performance evaluation of nickle platinum novel electrocatalysts for proton exchange membrane fuel cell. Int J Eng Work 6 (01), 39–43.

Ahmad, S., Ullah, A., Samreen, A., Qasim, M., Nawaz, K., Ahmad, W., Alnaser, A., Kannan, A.M., Egilmez, M., 2024. Hydrogen production, storage, transportation and utilization for energy sector: a current status review. J. Energy Storage. 2352152X. 101, 113733. doi:10.1016/j.est.2024.113733.

Ahmed, S., Kazmi, W.W., Hussain, A., Khan, M.Z., Bibi, S., Saleem, M., Song, R.H., Sajid, Z., Ullah, A., Khan, M.K., 2023. Facile and low-temperature synthesis approach to fabricate Sm0.5Sr0.5CoO3−δ cathode material for solid oxide fuel cell. J. Korean Ceram. Soc. 12297801. 60 (2), 272–282. Available from: https://www.springer.com/journal/43207. doi:10.1007/s43207-022-00261-6.

Ajmal, Z., Qadeer, A., Singh, K., Mahmud, A.A., Lakhan, M.N., Pradeep, H., Hussain, B., Ullah, A., Khan, H.R., Ibrahim, E.H., Kewate, O.J., Cui, J., Rosaiah, P., Akkinepally, B., Hussain, I., Shuhang, W., 2025. A comprehensive review on MXenes for various applications. Appl. Energy. 03062619. 397, 126136. doi:10.1016/j.apenergy.2025.126136.

Alam, K., Khan, K.I., Ullah, A., Ullah, A., Ali, S., Ullah, S., Ali, A., Hussain, S., 2021. Fabrication of superhydrophillic and graded index antireflective double layer coating for solar photovoltaics module using aerosol impact deposition assembly. Thin Solid Films. 00406090. 721, 138518. doi:10.1016/j.tsf.2021.138518.

Aziz, D., Qamar, S., Fatima, H., Ullah, A., Hussain, B., 2025. System Design, Sustainability, and Efficiency in Direct Liquid Fuel Cells. Informa UK Limited, pp. 27–49. doi:10.1201/9781003512592-2.

Fatima, H., Qamar, S., Choudhury, R., Shamsi, H., Ullah, A., 2024. Green Materials for Energy Storage Applications Encyclopedia of Green Materials. Springer Nature, Pakistan, pp. 845–856. Available from: https://link.springer.com/referencework/10.1007/978-981-97-4618-7. doi:10.1007/978-981-97-4618-7_205.

Harnisch, F., Schröder, U., 2010. From MFC to MXC: chemical and biological cathodes and their potential for microbial bioelectrochemical systems. Chem. Soc. Rev. 14604744. 39 (11), 4433–4448. doi:10.1039/c003068f.

Hussain, B., Ullah, A., Abbas, W., Ahmad, S., Egilmez, M., Rosaiah, P., Usmani, Y.S., Tamang, T.L., Hussain, I., 2025. Colloidal $TiO_2$ solid spheres as high-performance anodes for Lithium-ion batteries: synthesis, characterization, and optimization. Mater. Res. Bull. 00255408. 183, 113221. doi:10.1016/j.materresbull.2024.113221.

Hussain, B., Ullah, A., Abbas, W., Hussain, S.S., 2024. Biomass-Derived Materials for Lithium Secondary Batteries Encyclopedia of Green Materials. Springer Nature, South Korea, pp. 288–295. Available from: https://link.springer.com/referencework/10.1007/978-981-97-4618-7. doi:10.1007/978-981-97-4618-7_213.

Hussain, B., Ullah, A., Ali, I., Haider, S., Han, H.J., Jin, G., 2026. Advances in semiconductor materials and device architectures for biomedical systems: a mini review. Biomed. Eng. Lett. 16 (3), 663–680. doi:10.1007/s13534-026-00573-0.

Hussain, K., Ali, I., Hasnain, S., Hussain, S.S., Hussain, B., Khan, M.S., Ammar, S.M., Hussain, B., Hussain, A., Javed, M.A., Ullah, A., Al Arni, S., Elboughdiri, N., Abbas, N., 2020. Reagents assisted Mg-doped $CeO_2$ for high-performance energy-storage applications. J. Electroanal. Chem. 15726657. 873. Available from: https://www.journals.elsevier.com/journal-of-electroanalytical-chemistry. doi:10.1016/j.jelechem.2020.114401.

Janicek, A., Fan, Y., Liu, H., 2014. Design of microbial fuel cells for practical application: a review and analysis of scale-up studies. Biofuels. 1759-7269. 5 (1), 79–92. doi:10.4155/bfs.13.69.

Khan, M.R., Jarząbek, B., Ullah, A., 2026. Role of solid additives in morphological and structural optimization of bulk heterojunction organic solar cells. Materials 19 (7), 1387. doi:10.3390/ma19071387.

Khan, Z.U., Khan, A.D., Khan, K., AlKhatib, S.A.K., Khan, S., Khan, M.Q., Ullah, A., 2024. A review of degradation and reliability analysis of a solar PV module. IEEE Access. 2169-3536. 1. doi:10.1109/access.2024.3432394.

Sadiq, M., Arif, M., Ullah, A., Naveed, A., Afridi, S., Humayun, M., Khan, M.N., Asif, M., 2021. Synthesis of $MnO_2$ carbon nanotubes catalyst with enhanced oxygen reduction reaction for polymer electrolyte membrane fuel cell. J. Pak. Inst. Chem. Eng. 1813-4092. 49 (2), 19–27. doi:10.54693/piche.04923.

Saher, S., 2019a. Influence of dual layer silica nanoparticles coating on the performance enhancement of solar PV modules. J. Mech. Contin. Math. Sci. 09738975. 14 (1). doi:10.26782/jmcms.2019.02.00025.

Saher, S., 2019b. Ultra stable and highly efficient nickel nanotube catalyst for PEMFC electrochemical oxygen reduction reaction. J. Mech. Contin. Math. Sci. 09738975. 14 (1). doi:10.26782/jmcms.2019.02.00024.

Saher, S., Alam, K., Qamar, A., Ullah, A., Iqbal, J., 2019. Performance enhancement of intermediate temperature SOFC cathode by nano-composite coating. J. Mech. Cont. Math. Sci. 14 (2454–7190), 340–349.

Shahbaz, A., Afaf, A., Tahir, N., Ullah, A., Saim, S., 2021. Non precious metal catalysts: a fuel cell and ORR study of thermally synthesized nickel and platinum mixed nickel nanotubes for PEMFC. Key Eng. Mater. 1662-9795. 875, 193–199. doi:10.4028/www.scientific.net/kem.875.193.

Ullah, A., Ahmad, S., Ali, T., Samreen, A., Saher, S., Arif, M.Z., Kannan, A.M., 2026a. Advanced fabrication strategies for durable, high-efficiency, and scalable solid oxide fuel cells. Fuel 407, 137515. doi:10.1016/j.fuel.2025.137515.

Ullah, A., Hussain, J., Waqar, M., Humayun, M., Khan, M.R., 2026b. Optimizing oxygen reduction efficiency with a robust nickel nitrogen-doped carbon nanotube catalyst for polymer electrolyte membrane fuel cells. Results Chem. 19, 102918. doi:10.1016/j.rechem.2025.102918.

Ullah, A., Ahmad, S., Arif, M.Z., Egilmez, M., Hussain, B., Khan, M.R., Hussain, A., Samreen, A., 2024a. Alternative electrocatalysts: exploring quantum dots and transition metal MXenes. RSC Catal. Ser. 47, 447–471. Available from: http://www.rsc.org/shop/books/series.asp?seriesid=83. doi:10.1039/9781837674664-00447.

Ullah, A., Hussain, A., Qamar, S., Ali, M.M., Rehman, S.U., 2024c. Green Materials and Methods for Wastewater Treatment and Remediation. Springer Science and Business Media LLC, pp. 1–10. doi:10.1007/978-981-16-4921-9_52-2.

Ullah, A., Hussain, B., Humayun, M., Khan, M.R., Qamar, S., 2025. Advances in Membrane and Catalyst in Direct Formic Acid Fuel Cells (Dfafcs). Informa UK Limited, pp. 122–147. doi:10.1201/9781003512592-5.

Ullah, A., Hussain, B., Hussain, S., Humayun, M., 2021. Synthesizing and optimizing Manganese dioxide nanorods and its behavior toward oxygen reduction reaction. In: AIChE Annual Meeting, Conference Proceedings. American Institute of Chemical Engineers, Pakistan. 9781713852834. https://www.aiche.org/conferences/.

Ullah, A., Hussain, B., Khan, M.R., Alam, K., Humayun, M., Arif, M., 2024b. Exploring the potential of $\alpha$-$MnO_2$ carbon nanotubes for improved oxygen reduction reaction performance at the cathode of alkaline fuel cells. Adv. Mater. Res. 1662-8985. 1179, 111–122. doi:10.4028/p-1ljocn.

Ullah, A., Hussain, B., Waqar, M., Khan, M.R., Jehanzeb, J., Alam, K., Rosaiah, P., Ali, M., Annu, B.A., Hussain, I., 2024d. Synergistic effects of carbon nanotubes on α-$MnO_2$ electrocatalysts for improved oxygen reduction in alkaline fuel cells. Diam. Relat. Mater. 09259635. 149, 111611. doi:10.1016/j.diamond.2024.111611.

Ullah, A., Hussain, S.S., Hussain, B., 2022. Ruthenium doped lscf based cathode for enhanced performance of solid oxide fuel cells. ECS Meet. Abstr. 2151-2043. MA2022-02 (23), 965. doi:10.1149/ma2022-0223965mtgabs.

CHAPTER

# Microbial biotechnologies in industrial applications: Fundamentals, innovations, and future prospects

12

**Deepchandra Joshi[a,b], Mustari Abrar[a,c], Babu Alappat[c], and T.R. Sreekrishnan[a]**

[a]*Department of Biochemical Engineering and Biotechnology, Indian Institute of Technology Delhi, New Delhi, Delhi, India*

[b]*Institute of Energy and Environment, Pennsylvania State University, State College, University Park, PA, United States*

[c]*Department of Civil Engineering, Indian Institute of Technology Delhi, New Delhi, Delhi, India*

## 1 Introduction

The unprecedented growth of industrial activity over the past decades has profoundly reshaped both economies and ecosystems. As per the report by the International Energy Agency, at present, industries are responsible for nearly 37% of total global energy consumption, estimated at approximately $157 \times 10^{18}$ joules per year (Tracking SDG7: The Energy Progress Report, 2024), which, according to the Intergovernmental Panel on Climate Change (IPCC), contributes over 9.2 gigatons of $CO_2$ equivalent emissions annually, accelerating anthropogenic climate change (IPCC, 2023). Simultaneously, unsustainable resource extraction and linear production models have increased wastewater pollution. Every year, global industries discharge nearly $450 \times 10^{12}$ L of wastewater, much of which is rich in recalcitrant organics, nutrients, metals, and emerging contaminants (ECs) (Amdeha, 2024). This reality highlights the urgent need to transform industrial systems to operate within planetary boundaries while maintaining economic resilience. At the heart of this transformation, microbial technologies have emerged as powerful, adaptable tools to advance a circular bioeconomy. Industries can close resource loops, reduce carbon footprints, and transform waste into value-added products by harnessing the catalytic and metabolic potential of microorganisms and their enzymes. Unlike conventional thermochemical processes that typically demand harsh conditions (temperatures > 200°C, corrosive chemicals, high pressures) (Jankovic and Wilkinson, 2025), microbial processes operate under relatively mild and energy-efficient regimes, typically pH 4–9 and temperatures between 20°C and 60°C, which significantly lowers operational energy requirements and greenhouse gas emissions (Prasad et al., 2021; Selvasembian et al.,

Microbial Technologies for Sustainable Waste Management. https://doi.org/10.1016/B978-0-443-44093-9.00015-3

2022). The specificity of microbial enzymes also enables targeted transformation of complex feedstocks, while microbial communities can adapt dynamically to fluctuating industrial waste streams, offering flexibility and robustness rarely matched by chemical catalysts. While much of the early focus has been on microbial production of lignocellulosic enzymes like xylanases to process fermentable sugars from plant biomass with thermostable variants demonstrating specific activities exceeding 1200 U/mg, operational stability up to 110°C, and activity across pH 4–10 (Ajeje et al., 2021; Fang et al., 2019; Juturu and Wu, 2014), the industrial applications of microbial technologies extend far beyond biomass valorization. Modern biotechnology now supports the large-scale production of biopolymers, biobased surfactants, biochemicals, and specialty enzymes used in food processing, textiles, detergents, and pharmaceuticals. The major areas of biotechnology currently being used at the industrial level are presented in Fig. 12.1.

With microbial production and degradation, parallel advances are also being achieved in bioelectrochemical systems (BESs), particularly microbial fuel cells (MFCs), which have enabled the direct conversion of organic pollutants in wastewater into electrical energy. Laboratory- and pilot-scale MFCs have achieved power densities of up to 1000 MW/m$^2$ and Coulombic efficiencies (CEs) reaching 80% (Baskar et al., 2025), while simultaneously removing 30%–50% of chemical oxygen demand (COD) from industrial effluents. Such systems exemplify the integrated recovery of energy and water, aligning closely with circular economy principles. Microbial platforms' critical contribution lies in sustainable industrial processes by facilitating the biodegradation of recalcitrant pollutants (Joshi et al., 2022), bioremediation of heavy metals and xenobiotics (El Hachimi et al., 2025), and the production of high-value compounds such as amino acids, vitamins, and organic acids through fermentation (Garavand et al., 2023). The decentralized and modular nature of many microbial technologies makes them particularly suited for small and medium enterprises, off-grid industrial units, and emerging economies, thus supporting broader goals of inclusive and sustainable industrialization as outlined in Sustainable

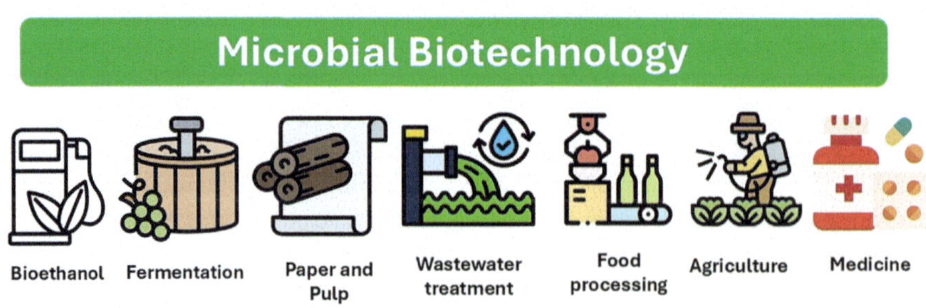


**FIG. 12.1**

Biotechnology industries. A major industrial sector with microbial biotechnology applications.

Development Goal 9 (SDG 9), which focuses on building resilient infrastructure, promoting inclusive and sustainable industrialization, and fostering innovation. Considering microbial technologies' importance and broad applications, this chapter discusses microbial technologies not as isolated innovations but as central pillars of a circular and low-carbon industrial future. It traces the evolution from enzyme-based biocatalysis to integrated BESs, highlighting how advances in synthetic biology, process engineering, and materials science have expanded their scalability and economic viability. The discussion integrates the scientific foundations of microbial catalysis and metabolism with real-world industrial practice, emphasizing the correlation between microbial process design and sustainability metrics such as energy intensity, carbon emissions, and resource recovery efficiency.

## 2 Fundamentals of microbial technologies

The microbial technologies have the ability to perform highly selective and efficient transformations under mild conditions, replacing energy-intensive chemical pathways and enabling new routes to valorize complex feedstocks, which are highly required for industrial applications (Prasad et al., 2021). At the core of these technologies lie three fundamental microbial processes: biotransformations (Branduardi, 2021), fermentation (Cruz-Casas et al., 2021), and enzymatic catalysis (Saravanan et al., 2021), each contributing distinctively to product formation. Biotransformations harness the metabolic versatility of microorganisms to convert complex substrates or pollutants into value-added products. For instance, microbial communities have evolved pathways to degrade structurally diverse compounds, from lignin-derived aromatics to long-chain alkanes and synthetic dyes, often under ambient conditions (temperature 20°C–40°C and atmospheric pressure) (Radhika et al., 2023). A striking example is the use of *Pseudomonas putida* engineered for the bioconversion of lignin into pyruvate, achieving five times higher productivity, up to 1.2 g/L/h (Johnson and Beckham, 2015). Similarly, in wastewater treatment, biofilm-driven biotransformation achieves removal efficiencies exceeding 90% for biodegradable COD and up to 70% for certain ECs (Thuptimdang et al., 2021). Fermentation remains the industrial backbone of large-scale microbial production. Modern fermenters range from laboratory-scale bioreactors (1–10 L) to industrial vessels exceeding $500 \times 10^3$ L, producing over 35 million tons annually of amino acids, organic acids, alcohols, and vitamins (Banks et al., 2022; Blöbaum et al., 2023; Fasim et al., 2021). Key fermentation-derived products include lactic acid (Eş et al., 2018), now produced at over 1.2 million tons per year for biodegradable plastics (Kozlovskiy et al., 2017), and citric acid, exceeding 2 million tons annually for food, pharmaceuticals, and cleaning agents (Mores et al., 2021). Advances in fed-batch, continuous fermentation, and metabolic engineering have pushed product yields close to theoretical maxima; for instance, engineered *Corynebacterium glutamicum* strains achieve glutamic acid yields nearing the stoichiometric limit. Enzymatic catalysis, often described as "cell-free biotransformation," exploits purified or immobilized enzymes to perform highly specific reactions, reducing side products and avoiding toxic solvents. Microbial xylanases, for example, hydrolyze hemicellulose (accounting for ~20%–35% of lignocellulosic biomass) into fermentable xylose and arabinose. Thermostable xylanases, with activities up to 1000–1500 U/mg and operational stability at 80°C (Juturu and Wu, 2014), enable cost-effective bleaching in pulp and paper, improving pulp brightness by 4%–6% ISO without chlorine-based chemicals and reducing adsorbable organic halides (AOX) by 30%–50% (Bajpai, 2018; Juturu and Wu, 2014). Beyond xylanases, microbial proteases, lipases, and cellulases collectively represent a global enzyme market exceeding US$8.42 billion in 2025, growing at ~7.3% Compound Annual Growth Rate (CAGR), reflecting expanding applications in textiles, detergents, food, and biofuel sectors (Industrial Enzymes Market, 2025). From a process integration perspective, these microbial technologies deliver tangible environmental benefits. Compared to purely chemical routes, enzymatic and microbial conversions typically operate at $< 10\%$ of the thermal energy demand and generate $> 50\%$ lower greenhouse gas emissions per functional unit (Prasad et al., 2021). In addition, decentralized deployment of microbial systems, such as modular bioelectrochemical reactors, facilitates treatment and resource recovery directly at industrial sites, reducing transport costs and enabling real-time adaptation to variable waste streams. Critically, the effectiveness of these technologies relies on careful selection and engineering of microbial strains and enzymes, optimization of reactor configurations, and integration with upstream and downstream processes. For instance, coupling MFCs with anaerobic digestion can increase overall energy recovery from wastewater by 10%–20%, while simultaneous nitrification–denitrification biofilms can achieve nitrogen removal efficiencies exceeding 85% (Mahmoudi et al., 2024).

## 2.1 Key microbial industrial sectors

The bioenergy sector uses microbial processes to produce bioethanol, biobutanol, biogas, and emerging platforms such as MFCs and microbial electrolysis cells (MECs). Global bioethanol production was valued at US$83.4 billion last year and is increasing at a CAGR of 6.6% to reach US$114.7 billion by 2028, with fermentative processes using engineered *Saccharomyces cerevisiae* and *Zymomonas mobilis* achieving conversion efficiencies approaching 90%–95% of theoretical yield (~0.51–0.53 g ethanol/g glucose). Beyond conventional first-generation biofuels, microbial consortia also enable anaerobic digestion of lignocellulosic residues to biogas, with methane yields exceeding 300–350 mL $CH_4$/g volatile solids, and power densities in MFCs ranging up to 420–460 MW/m$^2$ (Obileke et al., 2021). In food and feed, microbial enzymes such as amylases, proteases, lipases, and xylanases enhance digestibility, flavor development, and shelf life. The global enzyme market for food applications alone reached over US$3.9 billion in 2024, driven by demand for natural processing aids (Food Enzymes Market Report, 2025). Notably, xylanases improve nutrient availability in animal feed by depolymerizing arabinoxylans, increasing metabolizable energy by 3%–8% and reducing feed costs (Ramatsui et al., 2023).

In the pulp and paper industry, microbial xylanases are central to environmentally friendly bleaching processes. Traditionally, chlorine-based chemicals were used, producing chlorinated organic compounds and AOX. By contrast, enzymatic prebleaching reduces AOX by 30%–50%, lowers COD values by 20%–40%, and cuts chlorine dioxide ($ClO_2$) consumption by up to 25%, with improved pulp brightness (Bajpai, 2018). In textiles, microbial cellulases and laccases enable biopolishing and denim finishing, which traditionally required harsh acids, stones, or permanganate oxidation. Enzymatic treatments reduce energy use by ~30%, processing time by 20%–25%, and water consumption by up to 15%–20%, while improving fabric softness and color uniformity (Cavaco-Paulo and Gübitz, 2003).

The most transformative role of microbial systems is in wastewater treatment. Biological treatment processes, such as activated sludge, trickling filters, and anaerobic digesters, remove >90% of biodegradable COD and >85% of ammonia in municipal and industrial effluents (Kathi and El Din Mahmoud, 2024; Kishor et al., 2024). Emerging systems like MFCs couple treatment with electricity generation: laboratory-scale MFCs treating brewery wastewater achieved COD removal efficiencies of >80% and power densities up to 420–460 $MW/m^2$ (Obileke et al., 2021). Moreover, integration with anaerobic membrane bioreactors (MBRs) or resource recovery units enables nitrogen and phosphorus recovery as struvite, supporting circular nutrient management.

## 2.2 Advantages over conventional technologies

The advantages of microbial technologies have been categorized into sustainability, specificity, lower energy consumption, and high-value product formation. Microbial processes are sustainable as they typically operate under mild pH and temperature conditions, translating to energy savings of >50%–80% compared to thermal or chemical routes that often require temperatures above 150°C or corrosive chemicals (Banks et al., 2022). For example, enzymatic bleaching of pulp reduces AOX discharge and lowers $CO_2$ emissions by an estimated 0.2–0.4 tons per ton of pulp produced (Bajpai, 2018). Because of enzyme specificity, they catalyze highly selective reactions, generating minimal side products. Xylanases, for instance, cleave β-1,4-xylosidic bonds in hemicellulose without attacking cellulose, preserving fiber strength, an outcome challenging to achieve with nonspecific chemical oxidants (Juturu and Wu, 2014). Similarly, in MFCs, exoelectrogenic bacteria such as *Geobacter sulfurreducens* selectively oxidize organic matter and transfer electrons to electrodes, enabling direct electricity generation from waste organics (Bazina et al., 2023). Third, lower energy input: Microbial systems reduce thermal and electrical energy demand compared to traditional processes. In wastewater treatment, aerobic activated sludge requires ~0.3–0.6 $kWh/m^3$, whereas MFCs can produce 0.2–1.5 $kWh/m^3$ of usable electricity, offsetting part of the operational energy needs (Bazina et al., 2023; Sonawane et al., 2022). Finally, microbial technologies generate value-added by-products from industrial side streams. Examples include volatile fatty acids and biogas from anaerobic digestion, lactic acid from carbohydrate-rich

**Table 12.1** The comparative evaluation of microbial technologies and chemical processes.

| Aspect | Microbial technologies | Chemical processes |
| --- | --- | --- |
| Mechanism | Biological agents (e.g., bacteria, fungi, enzymes) | Chemical reactions using inorganic/organic reagents |
| Operating conditions | Mild conditions (ambient T and P, with neutral pH) | Extreme conditions (high T, P, and strong pH) |
| Feedstock | Renewable (e.g., biomass, waste, sugars) | Nonrenewable (e.g., synthetic chemicals) |
| Environmental impact | Low-biodegradable by-products | Highly toxic by-products |
| Selectivity and specificity | High | Lower |
| Process speed | Slower | Faster |
| Scalability | Requires optimization | Well-established |
| Cost efficiency | Potentially lower | High capital and operational costs |

effluents, and bioelectricity from MFCs treating food or brewery wastewater (Bazina et al., 2023; Mahmoudi et al., 2024). These coproducts support the economic viability of treatment processes and advance circular economy goals by transforming waste liabilities into resources. A systematic comparative evaluation between microbial and chemical technologies is provided in Table 12.1, which shows that microbial technologies offer higher selectivity, milder conditions, and renewable feedstocks. In contrast, chemical processes provide faster reactions and easier scalability, but with harsher conditions and greater environmental impact.

## 3 Microbial enzymes

Microbial enzymes form the molecular backbone of modern biotechnological processes, translating microbial metabolic potential into targeted, high-efficiency industrial catalysis. Among these, xylanases, cellulases, proteases, and lipases are especially prominent, driving advances in pulp and paper, food and feed, biofuel production, textiles, and specialty chemicals (Fasim et al., 2021). The growing market demand with the global industrial enzymes market valued at over US\$8.42 billion in 2025 and projected to reach US\$12.01 billion by 2030 (Industrial Enzymes Market, 2025) is underpinned by rapid innovations in enzyme engineering, recombinant production, and process integration. The list of the most common industrial microbial enzymes and their functions is provided in Table 12.2. Xylan, the major hemicellulosic polymer constituting ~20%–35% of lignocellulosic biomass, consists of β-1,4-linked xylose units often substituted with arabinose, glucuronic acid, or acetyl groups. Microbial xylanases, primarily belonging to glycoside hydrolase families GH10 and GH11, hydrolyze these β-1,4 linkages with distinct catalytic profiles:

**Table 12.2** List of common enzymes used in industries. The Enzyme Commission (EC) numbers of the enzymes have been collected from the International Union of Biochemistry and Molecular Biology database (ExplorEnz, 2025).

| Enzyme | EC number | Function/application |
|---|---|---|
| **Food industry** | | |
| Amylase | EC 3.2.1.1 | Starch hydrolysis |
| Glucoamylase | EC 3.2.1.3 | Converts starch to glucose |
| Invertase | EC 3.2.1.26 | Converts sucrose into glucose and fructose |
| Lactase (β-galactosidase) | EC 3.2.1.23 | Hydrolyzes lactose in dairy |
| Protease (e.g., papain) | EC 3.4.22.2 | Meat tenderizer, protein breakdown in brewing |
| Pectinase | EC 3.2.1.15 | Clarification of fruit juices and wines |
| Lipase | EC 3.1.1.3 | Flavor enhancement in cheese and dairy processing |
| **Detergent industry** | | |
| Lipase | EC 3.1.1.3 | Removes fat/oil stains |
| Amylase | EC 3.2.1.1 | Removes starch-based stains |
| Cellulase | EC 3.2.1.4 | Fabric softening |
| **Pharmaceutical industry** | | |
| Penicillin acylase | EC 3.5.1.11 | Synthesis of semisynthetic antibiotics |
| Asparaginase | EC 3.5.1.1 | Anticancer agent (used in leukemia treatment) |
| Streptokinase | EC 3.4.99.22 | Clot buster in myocardial infarction |
| Urokinase | EC 3.4.21.73 | Dissolves blood clots |
| **Agriculture** | | |
| Phytase | EC 3.1.3.8 | Releases phosphate from phytic acid in feed |
| Xylanase | EC 3.2.1.8 | Improves the digestibility of plant fibers |
| Cellulase | EC 3.2.1.4 | Break down cellulose in animal feed |
| β-Glucanase | EC 3.2.1.6 | Degrades glucans in cereals, improves feed uptake |
| **Biofuel and biorefinery** | | |
| Cellulase | EC 3.2.1.4 | Converts cellulose to fermentable sugars |
| Hemicellulase | EC 3.2.1.8 | Breaks down hemicellulose into simple sugars |
| Laccase | EC 1.10.3.2 | Lignin degradation |
| Xylanase | EC 3.2.1.8 | Aids in biomass deconstruction |
| **Textile and leather industry** | | |
| Amylase | EC 3.2.1.1 | Desizing agent in textile processing |
| Cellulase | EC 3.2.1.4 | Biopolishing of fabrics |

*Continued*

**Table 12.2** List of common enzymes used in industries. The Enzyme Commission (EC) numbers of the enzymes have been collected from the International Union of Biochemistry and Molecular Biology database (ExplorEnz, 2025)—cont'd

| Enzyme | EC number | Function/application |
|---|---|---|
| Protease | EC 3.4.21.62 | Dehairing and bating of hides |
| Laccase | EC 1.10.3.2 | Dye decolorization, denim bleaching |
| **Cosmetics and personal care** | | |
| Papain (protease) | EC 3.4.22.2 | Exfoliating agent in skincare |
| Lipase | EC 3.1.1.3 | Sebum regulation in acne products |
| Hyaluronidase | EC 3.2.1.35 | Enhances skin permeability |

GH10 enzymes have broader substrate specificity and tolerate substitutions better, while GH11 enzymes are narrower but often more thermally stable (Juturu and Wu, 2014).

Engineering strategies have been central to improving enzymatic action and function. Site-directed mutagenesis, domain shuffling, and directed evolution have produced xylanases with altered pH optima, improved thermostability, and higher catalytic efficiency (Juturu and Wu, 2012; Mu et al., 2024). For example, substituting specific surface residues to increase hydrogen bonding or ionic interactions can raise the melting temperature ($T_m$) by 5°C–10°C, extending enzyme half-life under process conditions (Li et al., 2022a). Fusion of xylanase catalytic domains with carbohydrate-binding modules further enhances substrate affinity and processivity, which is critical when working with complex lignocellulosic matrices (Li et al., 2022b). Recombinant expression systems have also expanded enzyme production capacity and cost-effectiveness. Bacterial systems like *Escherichia coli* and *Bacillus subtilis* allow high-level expression but often lack proper folding machinery for complex eukaryotic enzymes. Filamentous fungi (*Trichoderma reesei*, *Aspergillus niger*) and yeast (*Pichia pastoris*) remain preferred hosts for secreting glycosylated enzymes, achieving volumetric productivities > 0.5–1 g/L/h and enabling cost-competitive enzyme supply (Duman-Özdamar and Binay, 2021; Wei et al., 2021). Beyond xylanases, microbial cellulases (including endoglucanases, exoglucanases, and β-glucosidases) catalyze the depolymerization of cellulose, the most abundant biopolymer on Earth (~35%–50% of plant biomass). Engineered cellulases have achieved specific activities exceeding 200–400 U/mg, with half-lives of several hours at 60°C. Their integration with pretreatment processes in biofuel production has enabled saccharification yields over 80%–90% of theoretical glucose (Dixit et al., 2023). Other microbial enzymes, like proteases and lipases, also hold substantial industrial importance. Proteases, representing ~60% of the global enzyme market, are used in detergents, leather processing, and food industries, with some alkaline proteases operating efficiently at pH 9–11 and temperatures up to 60°C (Gimenes et al., 2021; Wang et al.,

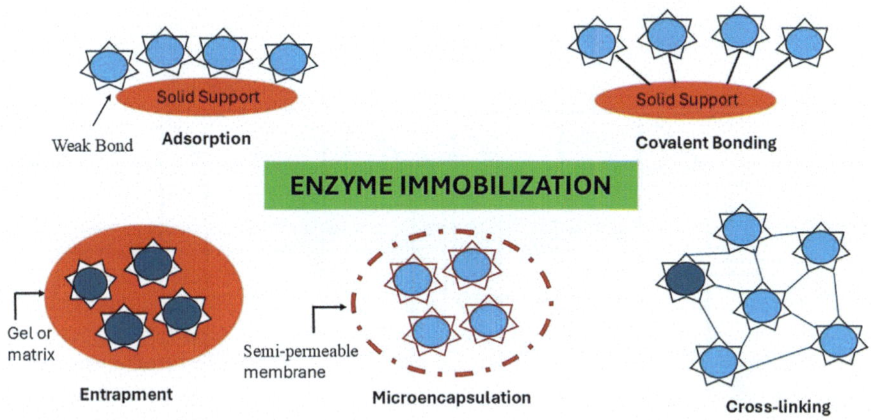


**FIG. 12.2**

Microbial immobilization techniques. Four major types used for representative purposes.

2025). Lipases catalyze ester hydrolysis and synthesis, facilitating biodiesel production (transesterification) with productivities up to 3–5 g/L/h in immobilized systems (Wang et al., 2025).

Immobilization technologies enhance stability and reusability. Enzyme immobilization is the technique of fixing enzymes onto or within solid supports, which further extends enzyme applications by enhancing their stability, reusability, and performance in catalytic processes. Fig. 12.2 represents the different strategies commonly used for enzyme immobilization at an industrial scale. These techniques include adsorption on porous carriers (e.g., silica, resins), covalent binding, and entrapment in polymers (e.g., alginate, polyvinyl alcohol (PVA)). Immobilized xylanases and cellulases have demonstrated operational stability over 10–20 cycles, lowering enzyme costs per unit product by 30%–50% (Bai et al., 2025; Guisan et al., 2022).

## 4 Bioelectrochemical systems

BESs, particularly MFCs, represent a frontier in sustainable industrial biotechnology by directly coupling pollutant removal with energy recovery. These systems exploit the metabolic activity of electroactive bacteria (exoelectrogens) to oxidize organic substrates and transfer electrons to an anode, generating electrical power without external energy input (Bazina et al., 2023). The MFCs consist of two chambers (Fig. 12.3): an anode (anaerobic) and a cathode (aerobic or abiotic), typically separated by an ion exchange membrane. In the anode, microorganisms oxidize organic matter, releasing electrons ($e^-$) and protons. Electrons flow through an external circuit to the cathode, where they reduce a terminal electron acceptor (commonly $O_2$), while

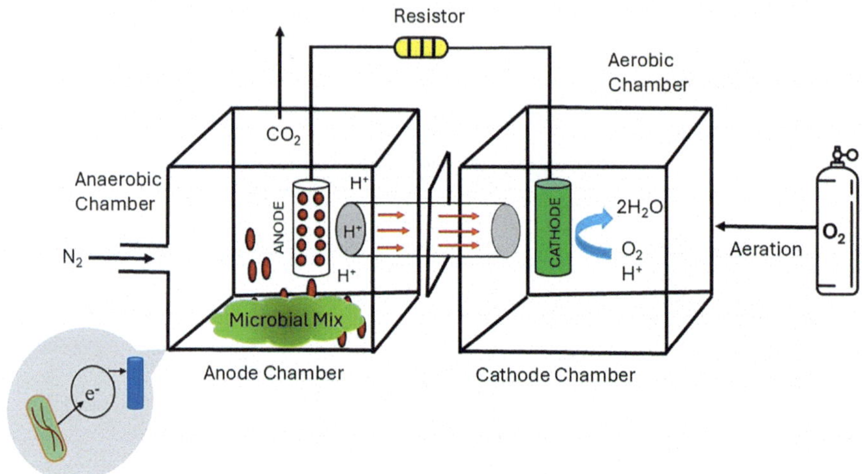


**FIG. 12.3**

A schematic of a dual-chamber MFC shows electron flow and microbial catalysis. *MFC*, microbial fuel cell.

protons migrate through the membrane, sustaining electrical neutrality (Bazina et al., 2023; Obileke et al., 2021).

The key performance metrics for MFCs include:

- Power density, measured in $mW/m^2$ (normalized to anode surface), is now reaching up to 300 $MW/m^2$ with advanced electrode modifications (Bazina et al., 2023)
- CE, representing the fraction of electrons recovered as current from substrate oxidation, often exceeds 80% under controlled conditions (Bazina et al., 2023)
- COD removal efficiency, with laboratory- and pilot-scale systems reporting 50%–80% removal from complex industrial effluents (Kathi and El Din Mahmoud, 2024)

The electrode materials and configurations critically influence the outcomes of MFCs. Carbon-based materials such as graphite felt, carbon cloth, and carbon brushes are widely used due to their high conductivity, chemical stability, and biocompatibility (Agrahari et al., 2022). Modification with nanomaterials, such as nitrogen-doped reduced graphene oxide (rGO), has increased surface area and electron transfer rates, boosting power densities by >20%–50% (Li et al., 2022a). Composite electrodes combining carbon with metal oxides like $TiO_2$ or $MnO_2$ further enhance biocompatibility and redox activity. Oxygen is typically the terminal electron acceptor on the cathode side due to its high redox potential (~0.805 V vs. NHE). Platinum-based catalysts are effective but costly (>US$40/g). Alternatives like activated carbon, manganese oxides, and enzyme-based cathodes reduce costs while maintaining reasonable catalytic activity, though often with lower long-term stability (Baskar et al., 2025). Membranes, such as cation exchange membranes and proton exchange membranes (PEMs), maintain ionic conductivity and prevent oxygen diffusion to the anode. Commercial Nafion membranes offer high proton conductivity but cost approximately US\$1700/m$^2$, representing up to 60% of total reactor cost (Jenani et al., 2024). Research into low-cost alternatives, including chitosan-based and PVA/CS composite membranes, has shown comparable performance at a fraction of the cost (Jenani et al., 2024). MFCs have been applied experimentally in diverse industrial contexts. For instance,

- Brewery wastewater (high COD, low toxicity): power densities ~ 420–460 MW/m$^2$; COD removal >80% (Obileke et al., 2021)
- Textile and dye effluents (containing recalcitrant azo dyes): color removal >70%, COD removal ~ 50%, albeit with lower power densities (~100–200 MW/m$^2$)
- Agroindustrial effluents, including dairy and starch wastewater, have also shown COD reductions of 55%, with the simultaneous generation of bioelectricity or hydrogen (Faria et al., 2017)

Scaling up MFCs presents significant challenges. Laboratory reactors typically operate at 0.1–5 L, while pilot systems have reached >1000 L (Bazina et al., 2023; Faria et al., 2017). Key barriers include internal resistance scaling with reactor size, uneven flow distribution, and biofilm management. Stacked and modular designs partially address these issues, distributing voltage across multiple smaller units to maintain practical power output and system stability. Beyond MFCs, MECs (Arun et al., 2024) and microbial desalination cells (Mishra et al., 2024) extend the concept to hydrogen production and simultaneous desalination, respectively. MECs, for example, apply a small external voltage (<1 V) to drive hydrogen evolution at the cathode, achieving hydrogen production rates of up to 2–3 m$^3$ $H_2$/m$^3$/day under optimized conditions (Arun et al., 2024). From a sustainability perspective, BESs uniquely integrate treatment and energy recovery. Conventional aerobic treatment of wastewater consumes ~0.3–0.6 kWh/m$^3$; MFCs can offset this by producing 0.2–1.5 kWh/m$^3$, depending on substrate strength and system design (Bazina et al., 2023; Obileke et al., 2021; Selvasembian et al., 2022). In addition, direct electron harvesting reduces greenhouse gas emissions by displacing fossil-based electricity and avoids methane losses common in anaerobic digestion.

## 4.1 MFCs scale-up

Despite significant laboratory successes, the commercial realization of microbial technologies such as MFCs and enzyme bioreactors depends on innovations bridging the gap between bench-scale performance and industrial-scale operation. Recent advances span reactor architecture, material science, hybrid process integration, and digital technologies, all aimed at overcoming cost, reliability, and efficiency barriers inherent to scaling. Reactor design advancements have been at the forefront of efforts to increase volumetric power density and treatment capacity while controlling capital and operational costs. Traditional dual-chamber MFCs, while effective in small-scale studies, suffer from high internal resistance and limited scalability (Arun et al., 2024; Selvasembian et al., 2022). In response, stacked configurations electrically connect multiple single MFC units in series or parallel, distributing voltage and maintaining manageable cell dimensions. Pilot-scale stacked systems treating brewery and domestic wastewater have achieved stable operation at scales of >1 $m^3$, with volumetric power densities approaching 5–10 $W/m^3$ (Jadhav et al., 2021). Tubular and upflow designs further optimize hydraulic performance. Tubular MFCs enhance surface-area-to-volume ratios, reducing footprint, while upflow MFCs improve substrate–biofilm contact and mitigate clogging risks, critical for high-strength industrial effluents rich in suspended solids. In addition, photosynthetic MFCs integrate phototrophic organisms in the cathode, supplying oxygen in situ via photosynthesis. Such systems have demonstrated power densities up to 133 $MW/m^2$ under natural light while simultaneously fixing $CO_2$ (Apollon et al., 2021; Sharma et al., 2024). Material innovations address two of the largest cost and performance bottlenecks: electrodes and membranes. Commercial MFCs typically rely on graphite felt, carbon cloth, or platinum catalysts, contributing >50% of total reactor cost (Agrahari et al., 2022). Recent research focuses on low-cost, high-performance alternatives. This includes waste-derived carbon electrodes (e.g., pyrolyzed agricultural residues), which have shown conductivity comparable to commercial carbon (~10–100 S/m) and cost reductions of >70% (Baskar et al., 2025; Jadhav et al., 2021). rGO coatings enhance electron transfer, increasing power density by 20%–50%; nitrogen-doped rGO derived from wastepaper achieved a peak power density of 1468 $MW/m^2$ (Li et al., 2022a). Metal oxide composites, such as $TiO_2$ or $MnO_2$ combined with carbon, improve biocompatibility and catalytic activity while resisting fouling. Membrane costs, notably Nafion (~US\$1700/$m^2$), prompted the development of chitosan-based and PVA/CS composite membranes, offering a 75-fold cost reduction and improved durability under variable pH (Jenani et al., 2024). These material advances collectively reduce capital expenditure and extend system life, directly impacting economic feasibility. Integration with other processes maximizes resource recovery and broadens the application. Hybrid systems coupling MFCs with anaerobic digestion increase overall energy yield: anaerobic digestion produces methane, while MFCs harvest residual biodegradable COD as electricity, boosting net energy recovery by 10%–20% (Selvasembian et al., 2022). Integration with MBRs enables advanced effluent polishing, achieving nutrient and micropollutant removal efficiencies exceeding 85%–90% (Jadhav et al., 2021). Chemical precipitation, e.g., struvite crystallization, recovers nitrogen and phosphorus from MFC effluent, producing marketable fertilizers and reducing downstream treatment costs. Beyond wastewater treatment, enzyme-based bioreactors are also increasingly hybridized with adsorption or membrane separation to improve product recovery and purity, reducing downstream processing costs by 20%–40%. Process intensification and automation are reshaping scalability. Real-time monitoring of pH, redox potential, and biofilm activity enables dynamic control, improving stability and efficiency under variable industrial conditions. Machine learning models and AI optimization predict performance based on historical and sensor data, adjusting operating parameters to maximize power density or COD removal. Digital twins, virtual replicas of physical reactors, allow scenario testing and process optimization without interrupting operation, shortening design cycles and lowering pilot-scale costs by 10%–30% (Baarimah et al., 2024). Collectively, these innovations move microbial technologies closer to commercial deployment, transforming them from laboratory curiosities into robust industrial tools. By redesigning reactors for scale, engineering affordable materials, integrating complementary processes, and leveraging digital intelligence, industries can unlock the full potential of microbial systems while simultaneously achieving pollution control, resource recovery, and energy production in line with circular economy principles.

## 5 Challenges and techno-economic barriers

Despite significant laboratory advances and promising pilot-scale demonstrations, microbial technologies face critical challenges that limit widespread industrial deployment. These challenges span economic, engineering, biological, and market dimensions, highlighting the complexity of translating innovative microbial processes into robust, large-scale systems. One of the most persistent barriers is cost competitiveness. For MFCs, capital expenditure is dominated by electrodes and membranes, which together account for up to 70%–80% of the total reactor cost (Boas et al., 2022). Commercial PEMs, such as Nafion, cost around US\$1700/$m^2$, while platinum catalysts on cathodes cost approximately US\$40/g (Jenani et al., 2024). Although alternative materials like chitosan, PVA membranes, and carbon-based nonprecious electrodes have demonstrated comparable electrochemical performance and reduced costs by >70%, their long-term mechanical stability and fouling resistance under industrial conditions remain underevaluated. Low power density relative to conventional energy technologies further constrains economic viability. Even with recent breakthroughs achieving peak power densities up to 1468 MW/$m^2$ under optimized conditions, volumetric power densities in full-scale systems often drop below 10–20 W/$m^3$, insufficient to offset total operating energy demands in large wastewater treatment plants (Boas et al., 2022; Selvasembian et al., 2022). This drop largely arises from increased internal resistance, uneven substrate distribution, and mass transfer limitations inherent to scale-up. Biofouling and pH gradients also challenge long-term stability. In continuous-flow MFCs treating industrial effluents, biofilm overgrowth can block electrode pores and reduce effective surface area, decreasing power output by >30%–50% over several months (Jadhav et al., 2021). At the same time, proton accumulation in the anode and consumption at the cathode can create pH gradients exceeding 2–3 units, reducing microbial activity and electron transfer rates.

Product inhibition, proteolytic degradation, and limited operational stability under industrial stressors (e.g., high substrate concentrations, solvents, temperature fluctuations) remain major barriers for enzyme-based biocatalytic systems. Although immobilization extends enzyme half-life and reusability, with reported operational stability over 10–20 reaction cycles (Guisan et al., 2022), enzyme cost still constitutes up to 30%–50% of the total cost of some bioprocesses, particularly when using complex lignocellulosic substrates. Feedstock and effluent variability in industrial settings compound these challenges.

Industrial wastewater streams often fluctuate in flow rate, organic load (COD), pH, and the presence of inhibitors, demanding adaptive process control and resilient microbial consortia. For example, dye-laden textile effluents or tannery wastewater contain heavy metals and aromatic compounds that inhibit exoelectrogenic bacteria, reducing CE and COD removal by 20%–40% compared to synthetic substrates (Jadhav et al., 2021).

Techno-economic analyses of full-scale MFC implementation underscore these barriers (Tiwari et al., 2024). Even optimistic scenarios estimate the levelized cost of electricity from MFCs to be above US$0.50–1.00/kWh, significantly higher than solar PV (~US$0.03–0.05/kWh) or wind power (~US$0.04–0.06/kWh). However, when environmental benefits (e.g., COD removal, nutrient recovery, reduced greenhouse gas emissions) are monetized, integrated systems can approach cost parity in niche applications. Regulatory and market uncertainties further complicate commercialization. Industrial stakeholders require proven reliability, predictable return on investment, and compliance with discharge and product quality standards, which demand extensive demonstration data rarely available beyond pilot projects. Lastly, system complexity presents an operational barrier. Successful implementation of microbial technologies requires multidisciplinary expertise in microbiology, electrochemistry, reactor engineering, and process control. This complexity increases training needs and can slow adoption in conservative industrial sectors.

## 6 Future prospects

The technical evolution of microbial enzymes and BESs demonstrates that microbial technologies are transitioning from niche laboratory tools to scalable, industrial solutions. This shift is driven by advances at three levels: molecular engineering, reactor design, and systemic integration into circular economy frameworks. At the molecular scale, synthetic biology and metabolic engineering have produced high-activity enzymes such as xylanases and cellulases with activities >1500 U/mg and extended half-lives through active-site stabilization (Juturu and Wu, 2012). Engineered consortia of exoelectrogenic bacteria (e.g., *G. sulfurreducens* + *Shewanella oneidensis*) show 10%–15% higher CEs and maintain power densities >1000 MW/$m^2$ under variable COD conditions. In reactor engineering, modular and stacked MFC configurations have scaled systems from ~0.1 L to >1 $m^3$ while maintaining volumetric power densities of 10–20 W/$m^3$. The use of waste-derived carbon electrodes and N-doped rGO has improved conductivity and biocompatibility, cutting electrode costs by >70%. Hybrid systems further enhance value creation: coupling MFCs with anaerobic digestion increases net energy recovery by 10%–20%, and integrated MBRs achieve >90% nutrient removal, transforming effluents into sources of clean water and struvite. Key technical barriers remain: scale-dependent power density decline, biofouling, membrane durability, and variability in waste streams limit stable full-scale performance, which remains at ~10%–20% of laboratory-scale peaks. In enzyme applications, product inhibition and high production costs hinder large-scale deployment in biorefineries. Future directions must target:

- AI-guided enzyme and strain design to optimize activity, stability, and yield.
- Dynamic control and digital twins to adapt to fluctuating industrial conditions.
- Coupled systems recovering electricity, hydrogen, organic acids, and nutrients.
- Techno-economic optimization, accounting for lifecycle emissions and cost.

Aligning these innovations with policy incentives, recognizing $CO_2$ avoidance, water reuse, and reduced chemical consumption, can accelerate adoption within industrial circular economy strategies.

## AI disclosure

During the preparation of this work, the author(s) used Chat GPT to correct grammar and improve sentence structure. After using this tool/service, the author(s) reviewed and edited the content as needed and take(s) full responsibility for the content of the publication.

## References

Tracking SDG7: The Energy Progress Report, 2024. Available from: https://www.iea.org/reports/tracking-sdg7-the-energy-progress-report-2024.

Agrahari, R., Bayar, B., Abubackar, H.N., Giri, B.S., Rene, E.R., Rani, R., 2022. Advances in the development of electrode materials for improving the reactor kinetics in microbial fuel cells. Chemosphere. 18791298. 290. Available from: https://www.elsevier.com/locate/chemosphere. doi:10.1016/j.chemosphere.2021.133184.

Ajeje, S.B., Hu, Y., Song, G., Peter, S.B., Afful, R.G., Sun, F., Asadollahi, M.A., Amiri, H., Abdulkhani, A., Sun, H., 2021. Thermostable cellulases/xylanases from thermophilic and hyperthermophilic microorganisms: current perspective. Front. Bioeng. Biotechnol. 22964185. 9. Available from: http://journal.frontiersin.org/journal/bioengineering-and-biotechnology#archive. doi:10.3389/fbioe.2021.794304.

Amdeha, E., 2024. Biochar-based nanocomposites for industrial wastewater treatment via adsorption and photocatalytic degradation and the parameters affecting these processes. Biomass Convers. Biorefin. 2190-6815. 14 (19), 23293–23318. doi:10.1007/s13399-023-04512-2.

ExplorEnz, 2025. Enzyme nomenclature database: International Union of Biochemistry and Molecular Biology https://www.enzyme-database.org/

Food Enzymes Market Report, 2025. SR112025A1624 IMARC food enzymes market report. https://www.imarcgroup.com/food-enzymes-market#:~:text=Market%20Overview%3A,5.81%25%20during%202025%2D2033.

Apollon, W., Luna-Maldonado, A.I., Kamaraj, S.K., Vidales-Contreras, J.A., Rodríguez-Fuentes, H., Gómez-Leyva, J.F., Aranda-Ruíz, J., 2021. Progress and recent trends in photosynthetic assisted microbial fuel cells: a review. Biomass Bioenergy. 18732909. 148. Available from: http://www.journals.elsevier.com/biomass-and-bioenergy/. doi:10.1016/j.biombioe.2021.106028.

Arun, J., SundarRajan, P., Pavithra, K.G., Priyadharsini, P., Shyam, S., Goutham, R., Quynh Hoang, L., Pugazhendhi, A., 2024. New insights into microbial electrolysis cells (MEC) and microbial fuel cells (MFC) for simultaneous wastewater treatment and green fuel (hydrogen) generation. Fuel. 00162361. 355, 129530. doi:10.1016/j.fuel.2023.129530.

Baarimah, A.O., Bazel, M.A., Alaloul, W.S., Alazaiza, M.Y.D., Al-Zghoul, T.M., Almuhaya, B., Khan, A., Mushtaha, A.W., 2024. Artificial intelligence in wastewater treatment: research trends and future perspectives through bibliometric analysis. Case Stud. Chem. Environ. Eng. 26660164. 10. Available from: https://www.sciencedirect.com/science/journal/26660164. doi:10.1016/j.cscee.2024.100926.

Bai, H., Dan, Y., Xiaowei, D., 2025. Review of porous microspheres for enzyme immobilization: strategies, applications, and prospects. Int. J. Biol. Macromol. 01418130. 295, 139627. doi:10.1016/j.ijbiomac.2025.139627.

Bajpai, P., 2018. Biological Treatment of Pulp and Paper Mill Effluents. Springer Science and Business Media LLC, pp. 313–369. doi:10.1007/978-981-10-7853-8_16.

Banks, M., Johnson, R., Giver, L., Bryant, G., Guo, M., 2022. Industrial production of microbial protein products. Curr. Opin. Biotechnol. 09581669. 75 (102707). doi:10.1016/j.copbio.2022.102707.

Baskar, S., Jayaprabakar, J., A, R., Rajan, T.S., Kumar, J.A., Nalini Ramachandran, U., Balaji, E., Sambandam, P., Jisha, P.K., Ruban, M., 2025. Possible bio cathode materials for usage in microbial fuel cells towards energy generation and wastewater treatment to sustain environment: a review. Results Eng. 25901230. 25, 104161. doi:10.1016/j.rineng.2025.104161.

Bazina, N., Ahmed, T.G., Almdaaf, M., Jibia, S., Sarker, M., 2023. Power generation from wastewater using microbial fuel cells: a review. J. Biotechnol. 18734863. 374, 17–30. Available from: https://www.elsevier.com/locate/jbiotec. doi:10.1016/j.jbiotec.2023.07.006.

Blöbaum, L., Haringa, C., Grünberger, A., 2023. Microbial lifelines in bioprocesses: from concept to application. Biotechnol. Adv. 07349750. 62, 108071. doi:10.1016/j.biotechadv.2022.108071.

Boas, J.V., Oliveira, V.B., Simões, M., Pinto, A.M.F.R., 2022. Review on microbial fuel cells applications, developments and costs. J. Environ. Manage. 10958630. 307. Available from: https://www.sciencedirect.com/journal/journal-of-environmental-management. doi:10.1016/j.jenvman.2022.114525.

Branduardi, P., 2021. Closing the loop: the power of microbial biotransformations from traditional bioprocesses to biorefineries, and beyond. Microb. Biotechnol. 1751-7915. 14 (1), 68–73. doi:10.1111/1751-7915.13713.

Cavaco-Paulo, A., Gübitz, G., 2003. Textile Processing with Enzymes Textile Processing with Enzymes. Elsevier Inc, Portugal, pp. 1–228. Available from: http://www.sciencedirect.com/science/book/9781855736108. doi:10.1533/9781855738669.

Cruz-Casas, D.E., Aguilar, C.N., Ascacio-Valdés, J.A., Rodríguez-Herrera, R., Chávez-González, M.L., Flores-Gallegos, A.C., 2021. Enzymatic hydrolysis and microbial fermentation: the most favorable biotechnological methods for the release of bioactive peptides. Food Chem. Mol. Sci. 26665662. 3. Available from: https://www.journals.elsevier.com/food-chemistry-molecular-sciences. doi:10.1016/j.fochms.2021.100047.

Dixit, Y., Yadav, P., Sharma, A.K., Pandey, P., Kuila, A., 2023. Multiplex genome editing to construct cellulase engineered Saccharomyces cerevisiae for ethanol production from cellulosic biomass. Renew. Sustain. Energy Rev. 13640321. 187, 113772. doi:10.1016/j.rser.2023.113772.

Duman-Özdamar, Z.E., Binay, B., 2021. Production of industrial enzymes via pichia pastoris as a cell factory in bioreactor: current status and future aspects. Protein J. 18758355. 40 (3), 367–376. Available from: http://www.kluweronline.com/issn/1572-3887. doi:10.1007/s10930-021-09968-7.

Eş, I., Khaneghah, A.M., Barba, F.J., Saraiva, J.A., Sant'Ana, A.S., Hashemi, S.M.B., 2018. Recent advancements in lactic acid production - a review. Food Res. Int. 18737145. 107, 763–770. Available from: https://www.elsevier.com/inca/publications/store/4/2/2/9/7/0. doi:10.1016/j.foodres.2018.01.001.

Fang, W., Xue, S., Deng, P., Zhang, X., Wang, X., Xiao, Y., Fang, Z., 2019. AmyZ1: a novel α-amylase from marine bacterium Pontibacillus sp. ZY with high activity toward raw starches. Biotechnol. Biofuels. 1754-6834. 12 (1). doi:10.1186/s13068-019-1432-9.

Faria, A., Gonçalves, L., Peixoto, J.M., Peixoto, L., Brito, A.G., Martins, G., 2017. Resources recovery in the dairy industry: bioelectricity production using a continuous microbial fuel cell. J. Clean. Prod. 09596526. 140, 971–976. doi:10.1016/j.jclepro.2016.04.027.

Fasim, A., More, V.S., More, S.S., 2021. Large-scale production of enzymes for biotechnology uses. Curr. Opin. Biotechnol. 18790429. 69, 68–76. Available from: http://www.elsevier.com/locate/copbio. doi:10.1016/j.copbio.2020.12.002.

Garavand, F., Daly, D.F.M., Gómez-Mascaraque, L.G., 2023. The consequence of supplementing with synbiotic systems on free amino acids, free fatty acids, organic acids, and some stability indexes of fermented milk. Int. Dairy J. 09586946. 137. Available from: https://www.elsevier.com/inca/publications/store/4/0/5/8/6/0/index.htt. doi:10.1016/j.idairyj.2022.105477.

Gimenes, N.C., Silveira, E., Tambourgi, E.B., 2021. An overview of proteases: production, downstream processes and industrial applications. Sep. Purif. Rev. 15422127. 50 (3), 223–243. Available from: http://www.tandfonline.com/toc/lspr20/current. doi:10.1080/15422119.2019.1677249.

Guisan, J.M., Fernandez-Lorente, G., Rocha-Martin, J., Moreno-Gamero, D., 2022. Enzyme immobilization strategies for the design of robust and efficient biocatalysts. Curr. Opin. Green Sustain. Chem. 24522236. 35, 100593. doi:10.1016/j.cogsc.2022.100593.

El Hachimi, O., Tiwari, B.R., Drogui, P., Brar, S.K., Blais, J.F., 2025. Advances in membrane bioreactor for landfill leachate treatment: a review of characterization, challenges, and novel configurations. J. Environ. Chem. Eng. 22133437. 13 (3). Available from: https://www.sciencedirect.com/science/journal/22133437. doi:10.1016/j.jece.2025.116653.

Industrial Enzymes Market, 2025. 2025 [FB2277] Market and markets. https://www.marketsandmarkets.com/PressReleases/industrial-enzymes.asp.

Intergovernmental Panel on Climate Change (IPCC), 2023. Energy Systems. Cambridge University Press (CUP), pp. 613–746. doi:10.1017/9781009157926.008.

Jadhav, D.A., Mungray, A.K., Arkatkar, A., Kumar, S.S., 2021. Recent advancement in scaling-up applications of microbial fuel cells: from reality to practicability. Sustain. Energy Technol. Assess. 22131388. 45. Available from: http://www.journals.elsevier.com/sustainable-energy-technologies-and-assessments. doi:10.1016/j.seta.2021.101226.

Jankovic, J., Wilkinson, D.P., 2025. Towards a more sustainable future: transitioning from thermochemical to electrochemical processes in clean energy technologies relevant to hydrogen-containing fuels. Can. J. Chem. Eng. 1939019X. 103 (4), 1602–1622. Available from: http://onlinelibrary.wiley.com/journal/10.1002/(ISSN)1939-019X. doi:10.1002/cjce.25549.

Jenani, R., Karishmaa, S., Ponnusami, A.B., Kumar, P.S., Rangasamy, G., 2024. A recent development of low-cost membranes for microbial fuel cell applications. Desalination Water Treat. 19443986. 320 (100698). doi:10.1016/j.dwt.2024.100698.

Johnson, C.W., Beckham, G.T., 2015. Aromatic catabolic pathway selection for optimal production of pyruvate and lactate from lignin. Metab. Eng. 10967184. 28, 240–247. Available from: http://www.elsevier.com/inca/publications/store/6/2/2/9/1/3/index.htt. doi:10.1016/j.ymben.2015.01.005.

Joshi, D., Prakash, G.V., Ahammad, S.Z., Kar, S., Sreekrishnan, T.R., 2022. Development of low power non-thermal plasma jet and optimization of operational parameters for treating dyes and emerging contaminants. Plasma Sci. Technol. 20586272. 24 (10). Available from: http://www.iop.org/EJ/journal/pst. doi:10.1088/2058-6272/ac6e34.

Juturu, V., Wu, J.C., 2012. Microbial xylanases: engineering, production and industrial applications. Biotechnol. Adv. 07349750. 30 (6), 1219–1227. doi:10.1016/j.biotechadv.2011.11.006.

Juturu, V., Wu, J.C., 2014. Microbial cellulases: engineering, production and applications. Renew. Sustain. Energy Rev. 13640321. 33, 188–203. doi:10.1016/j.rser.2014.01.077.

Kathi, S., Mahmoud, A.E.D., 2024. Trends in effective removal of emerging contaminants from wastewater: a comprehensive review. Desalination Water Treat. 19443986. 317 (100258). doi:10.1016/j.dwt.2024.100258.

Kishor, R., Verma, M., Saratale, G.D., Ferreira, L.F.R., Kharat, A.S., Chandra, R., Raj, A., Bharagava, R.N., 2024. Treatment of industrial wastewaters by algae-bacterial consortium with Bio-H2 production: recent updates, challenges and future prospects. Chemosphere. 18791298. 349. Available from: https://www.elsevier.com/locate/chemosphere. doi:10.1016/j.chemosphere.2023.140742.

Kozlovskiy, R., Shvets, V., Kuznetsov, A., 2017. Technological aspects of the production of biodegradable polymers and other chemicals from renewable sources using lactic acid. J. Clean. Prod. 09596526. 155, 157–163. doi:10.1016/j.jclepro.2016.08.092.

Li, P., Li, X., Huang, J., Qu, W., Pan, X., Chen, Q., Jaromír Klemeš, J., Wang, B., Wang, J., Tao, H., 2022a. Nitrogen-doped graphene oxide with enhanced bioelectricity generation from microbial fuel cells for marine sewage treatment. J. Clean. Prod. 09596526. 376, 134071. doi:10.1016/j.jclepro.2022.134071.

Li, Y., Song, W., Yin, X., Rao, S., Zhang, Q., Zhou, J., Li, J., Du, G., Liu, S., 2022b. Enhanced catalytic performance of thermophilic GH11 xylanase by fusing carbohydrate-binding module 9-2 and linker for better synergistic degradation of wheat bran. Process Biochem. 13595113. 121, 349–359. doi:10.1016/j.procbio.2022.07.021.

Mahmoudi, A., Mousavi, S.A., Darvishi, P., 2024. Performance and recent development in sewage sludge-to-bioenergy using microbial fuel cells: a comprehensive review. Int. J. Hydrogen Energy. 03603199. 50, 1432–1455. Available from: https://www.sciencedirect.com/science/journal/03603199. doi:10.1016/j.ijhydene.2023.10.338.

Mishra, S., Dhanda, A., Dubey, B.K., Ghangrekar, M.M., 2024. Enhancing electrokinetics and desalination efficiency through catalysts and electrode modifications in microbial desalination cells. J. Environ. Manage. 10958630. 366. Available from: https://www.sciencedirect.com/science/journal/03014797. doi:10.1016/j.jenvman.2024.121719.

Mores, S., Vandenberghe, L.P.D.S., Magalhães Júnior, A.I., de Carvalho, J.C., de Mello, A.F.M., Pandey, A., Soccol, C.R., 2021. Citric acid bioproduction and downstream processing: status, opportunities, and challenges. Bioresour. Technol. 18732976. 320. Available from: https://www.elsevier.com/locate/biortech. doi:10.1016/j.biortech.2020.124426.

Mu, D., Li, P., Ma, T., Wei, D., Montalbán-López, M., Ai, Y., Wu, X., Wang, Y., Li, X., Li, X., 2024. Advances in the understanding of the production, modification and applications of xylanases in the food industry. Enzyme Microb. Technol. 01410229. 179, 110473. doi:10.1016/j.enzmictec.2024.110473.

Obileke, K., Onyeaka, H., Meyer, E.L., Nwokolo, N., 2021. Microbial fuel cells, a renewable energy technology for bio-electricity generation: a mini-review. Electrochem. Commun. 13882481. 125, 107003. doi:10.1016/j.elecom.2021.107003.

Prasad, R., Kumar, V., Singh, J., Upadhyaya, C.P., 2021. Recent Developments in Microbial Technologies Environmental and Microbial Biotechnology. Springer Nature Singapore. doi:10.1007/978-981-15-4439-2.

Radhika, N.L., Sachdeva, S., Kumar, M., 2023. Development of potential consortia for biotransformation of lignin and polyaromatic hydrocarbons (PAHs). Biomass Convers. Biorefin. 2190–6815. doi:10.1007/s13399-023-04486-1.

Ramatsui, L., Sithole, T., Mzimkulu-Ncoyi, N.H., Malgas, S., Pletschke, B.I., 2023. The use of xylanases as additives to feeds: a mini-review of their effect on feed digestion and growth performance of monogastric animals. In: Microbial Bioprocesses: Applications and Perspectives. Elsevier, South Africa, pp. 83–105. Available from: https://www.sciencedirect.com/book/9780323953320. doi:10.1016/B978-0-323-95332-0.00003-X.

Saravanan, A., Kumar, P.S., Vo, D.V.N., Jeevanantham, S., Karishma, S., Yaashikaa, P.R., 2021. A review on catalytic-enzyme degradation of toxic environmental pollutants: microbial enzymes. J. Hazard. Mater. 18733336. 419. Available from: https://www.elsevier.com/locate/jhazmat. doi:10.1016/j.jhazmat.2021.126451.

Selvasembian, R., Mal, J., Rani, R., Sinha, R., Agrahari, R., Joshua, I., Santhiagu, A., Pradhan, N., 2022. Recent progress in microbial fuel cells for industrial effluent treatment and energy generation: fundamentals to scale-up application and challenges. Bioresour. Technol. 09608524. 346, 126462. doi:10.1016/j.biortech.2021.126462.

Sharma, A., Chhabra, M., Kumar, S., 2024. Performance evaluation of genetically modified microalgae in photosynthetic microbial fuel cells for carotenoids and power generation. J. Environ. Chem. Eng. 22133437. 12 (3), 112751. doi:10.1016/j.jece.2024.112751.

Sonawane, J.M., Mahadevan, R., Pandey, A., Greener, J., 2022. Recent progress in microbial fuel cells using substrates from diverse sources. Heliyon. 24058440. 8 (12), e12353. doi:10.1016/j.heliyon.2022.e12353.

Thuptimdang, P., Siripattanakul-Ratpukdi, S., Ratpukdi, T., Youngwilai, A., Khan, E., 2021. Biofiltration for treatment of recent emerging contaminants in water: current and future perspectives. Water Environ. Res. 1061-4303. 93 (7), 972–992. doi:10.1002/wer.1493.

Tiwari, B.R., Panda, D., Peta, K.R., Brar, S.K., 2024. Microbial electrochemical technologies: a life cycle and technoeconomic perspective. In: Emerging Trends and Advances in Microbial Electrochemical Technologies: Hypothesis, Design, Operation, and Applications. Elsevier, Canada, pp. 131–153. Available from: https://www.sciencedirect.com/book/9780443155574. doi:10.1016/B978-0-443-15557-4.00001-0.

Wang, G., Abdella, A., Fakhari, M., Dong, J., Yang, K.K., Yang, S.T., 2025. Recent advances in engineering microbial lipases for industrial applications. Biotechnol. Adv. 07349750. 83. Available from: https://www.sciencedirect.com/science/journal/07349750. doi:10.1016/j.biotechadv.2025.108658.

Wei, H., Mengyue, W., Fan, A., Haijia, S., 2021. Recombinant protein production in the filamentous fungus Trichoderma. Chinese J. Chem. Eng. 10049541. 30, 74–81. doi:10.1016/j.cjche.2020.11.006.

Yangyang, L., Cen, L., Huang, H., Rao, S., Zhang, Q., Zhou, J., Li, J., Du, G., Liu, S., 2022. Significantly enhanced thermostability of *Aspergillus Niger* xylanase by modifying its highly flexible regions. J. Agric. Food Chem. 0021-8561. 70 (15), 4620–4630. doi:10.1021/acs.jafc.2c01343.

CHAPTER

# 13 Artificial intelligence integration in microbial waste management

**Vibha Gajbe, Aarav Gupte, and Archisha Patel**
*School of Biotechnology & Bioinformatics, D. Y. Patil Deemed to be University, Navi Mumbai, India*

## 1 Introduction

The surge in waste generation is far outpacing the current waste management infrastructure. The world's mounting garbage problem threatens air and water quality, public health, and even economic stability. Rapid industrialization and urbanization have led to a massive rise in waste generation worldwide. According to waste generation statistics, total global waste generation reached approximately 20 billion tons in 2017, with projections suggesting it could rise to 46 billion tons by 2050. Municipal solid waste, in particular, accounted for about 2.7 billion tons by 2019, with a projected increase of 30%–40% over the past 15 years (Maalouf and Mavropoulos, 2023). A recent study by Singh and Singh (Balasubramanian, 2018 Singh and Singh, 2021) indicates that per capita waste generation in India ranges from 0.12 to 0.60 kg per day, with urban cities contributing significantly to the national total of approximately 105,000 metric tons of solid waste per day (Kumar and Chakrabarti, 2010). According to World Bank statistics from 2018, about 33% of the generated waste is handled improperly, often being disposed of in unmonitored dump sites (Kaza et al., 2018). Improper waste management can lead to air and water pollution, disease transmission, and greenhouse gas emissions (Yeboah Antwi and Ofori-Nyarko, 2023). The literature review indicates that improper waste management strategies and inadequate operations are factors contributing to ineffective waste management (Hannan et al., 2013; Malakahmad and Khalil, 2011). This disparity highlights the need for improved waste management practices to address the growing waste management crisis globally. This can be achieved by incorporating AI technologies, which can transform waste management processes toward sustainability.

The conventional strategies, being linear in their execution, lack the capability to meet the highly complex demands of waste valorization, resource recovery, and environmental protection. Traditional waste management methods rely on intricate mathematical equations and complicated calculations to optimize processes. These methods also involve considerations of costs, societal impacts, regulatory frameworks, and governmental limitations. Previously, multivariate statistical analysis

Microbial Technologies for Sustainable Waste Management. https://doi.org/10.1016/B978-0-443-44093-9.00019-0

techniques were utilized to evaluate the removal of various pollutants and the efficiency of biodegradation. Artificial intelligence (AI) approaches have reduced errors without relying on complex mathematical equations and have provided improved accuracy in calculating the regression coefficient ($R^2$) (Malviya and Jaspal, 2021). Therefore, it is crucial to implement effective waste management strategies through the use of AI.

The integration of AI is essential for improving efficiency and sustainability in addressing the global waste crisis. AI technologies can analyze complex processes, predict waste generation volumes, design synthetic microbial consortia, make informed decisions, optimize real-time collection routes, and enhance waste sorting (Yetilmezsoy et al., 2011). Various AI models, such as artificial neural networks (ANNs), support vector machines (SVMs), random forests (RFs), and decision trees (DTs), are efficient in addressing complex challenges that arise during waste management. Every AI model has its unique benefits and drawbacks, and the choice of model should depend on the type of waste management and treatment system to achieve the best results. For instance, ANNs can be trained for precise prediction and classification. These neural networks are capable of processing large datasets that can be trained to conduct geographical analysis of locations (Thomaz and Mahler, 2024; Yetilmezsoy et al., 2011). Conversely, SVMs are tailored for forecasting waste generation and waste classification. Depending on the characteristics of the produced waste, SVMs can categorize waste into organic, hazardous, and recyclable groups (Pillai et al., 2023). Another AI model, DTs, is capable of managing data that contains missing values and irrelevant features. Studies show that DTs have been employed in forecasting waste generation, waste classification, waste reduction, and predicting waste generation patterns and illegal dumping sites (Abdallah et al., 2020). However, RFs can effectively handle noisy datasets and minimize the likelihood of overfitting (Xia et al., 2022). This chapter provides a comprehensive discussion on the integration of different AI models in microbial waste management. The chapter thoroughly discusses the application of AI tools such as ANN, SVM, RF, and DT. It also highlights the challenges and research opportunities related to sustainable waste management systems.

## 2 AI models in microbial waste management

The advancement of various AI models provides scope for improvement in multiple aspects of microbial waste management techniques. The efficiency, predictability, and overall performance of these technologies significantly improve when integrated with different AI tools (Fig. 13.1). Each tool provides a specific advantage over the others (Table 13.1). AI plays a pivotal role in enhancing microbial waste management by minimizing complexities in microbial waste management processes. The integration of AI technologies facilitates process optimization, forecasts the volume of waste generation, designs synthetic microbial consortia, enables better decision-making, and improves waste treatment processes.

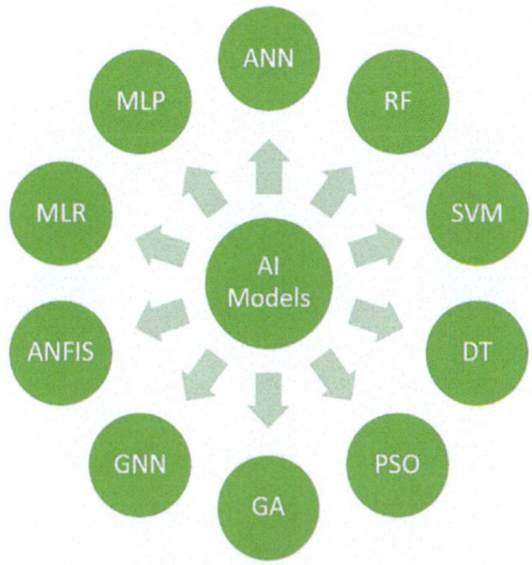


**FIG. 13.1 Different AI models in microbial waste technology.**

AI-driven approaches to microbial waste treatment.

## 2.1 Artificial neural network

Amidst the advancing developments of the modern world, ANNs stand out from conventional rule-based programming. This marks a forward shift in the way we think about solving problems using computers. Just like the functioning of the human brain, these ANN models are built to identify patterns, learn from information, and make decisions using processes analogous to human reasoning. ANNs have been extensively incorporated into waste management systems due to their strong robustness, error tolerance, and ability to accurately depict the complex relationships between multivariate systems. ANN models offer a data-driven approach to understanding complex biological processes that enable efficient waste treatment and energy recovery.

The ANNs mainly consist of an input layer, a hidden layer, and an output layer. ANNs function much like a simplified brain. They take in data through an input layer, where each piece of information is fed into a "neuron," and at each "neuron," the incoming weighted signals are summed up and then passed through an activation function. This function decides if the neuron "fires" and sends a signal to the next layer. This process repeats through hidden layers. Finally, the output layer provides the network's prediction. During training, the ANN compares its prediction to the correct answer, calculates the "error," and uses a process called backpropagation to adjust its connection weights. This adjustment allows the network to learn from its mistakes and improve its accuracy over time, essentially "learning" from data (Maind, 2014). ANNs have been significantly explored to predict the volume of waste generation (Golbaz et al., 2019), waste sorting and waste management (Singh et al., 2017), and the waste-to-energy process (Ezzahra Yatim et al., 2022). ANNs have successfully

**Table 13.1** Artificial neural networks as predictive tools in microbial waste management.

| Waste management process | AI model | Prediction | Result (qualitative) | Result (quantitative) | Reference(s) |
|---|---|---|---|---|---|
| Wastewater treatment (activated sludge, MBRs) | ANN | Effluent quality parameters (BOD, COD, TSS, N, P removal), aeration demand, sludge volume index | Improved removal efficiencies, reduced energy consumption, proactive operational adjustments, enhanced process stability | $R^2$ between observed and predicted SS up to 0.99; $R^2 = 0.96$ for individual models like $PO_4^{-3}$ | (Hamed et al., 2004; Corominas et al., 2018) |
| COD level prediction | ANN | COD effluent concentration for the year 2023 in AIWWTP | ANN, highly successful and a reliable model for forecasting COD levels | $R^2 = 0.9997$ | (Çimen Mesutoğlu and Gök, 2024) |
| BOD level prediction | ANN | Wastewater treatment plant performance | Strong performance with an average flow rate of 1 million $m^3$/day | $R^2 = 0.81$ | (Hamed et al., 2004) |
| Anaerobic digestion | ANN | Biogas production, volatile fatty acid (VFA) concentration, digester stability | Optimized biogas yield, maximized energy recovery, early detection of process upsets, enhanced stability of the digester | $R^2 = 0.998$ | (Yetilmezsoy et al., 2011) |
| Anaerobic digestion | ANN, DFFBP, DCFBP | Biogas yield | Enhanced prediction precision, increased energy recovery | $R^2 = 0.9946$ for DCFBP model | (Li et al., 2024) |
| Solid waste composting | ANN | Temperature profiles, composting duration, maturity indices (C/N ratio, respiration index), and quality of the final compost | Faster composting cycles, higher quality compost, optimized aeration and moisture control, efficient resource management | High accuracy; $R^2 = 0.914$, $R = 0.989$, and low MSE values | (Kulisz and Kujawska, 2020) |

| | | | | | |
|---|---|---|---|---|---|
| Municipal solid waste (MSW) generation | ANN and PSO | Predictive analysis of MSW | Superior performance of ANN–PSO model | $R^2$ = 0.96 (PSO) | |
| $R^2$ = 0.6 (ANN) | (Elshaboury et al., 2021) | | | | |
| Bioremediation (e.g., contaminated soil/water) | ANN | Pollutant degradation rates, optimal conditions for microbial activity, microbial community dynamics, remediation efficiency | Accelerated pollutant removal, design of more effective remediation strategies, better understanding of microbial pathways | Numerical integration of the ANN-embedded rate expression for diesel fuel-spiked soil removal revealed an excellent fit ($R^2 > 0.99$) to the dynamics | (Prasad and Mathur, 2006) |
| Microbial fuel cells (MFCs) | ANN | Power density, voltage generation, current output, COD removal efficiency, long-term performance | Optimized MFC design and operation, maximized bioenergy recovery, improved waste conversion efficiency | Predicted power density output with a minimum error of 5.76 ± 3.16%; R = 0.95 for power production, 99.66% accuracy in data prediction for polarization curves | (Lesnik and Liu, 2017; Zakir Hossain et al., 2023; de Ramón-fernández et al., 2020) |
| Plastic waste management | ANN, SVM, RF | Generation rate of different types of plastic waste in Dhanbad, India | ANN reflect best prediction accuracy | $R^2$ = 0.75; RMSE = 9.53 | (Kumar et al., 2018) |

helped researchers model complex relationships between microbial communities and environmental variables. Based on environmental data, ANNs offer better insights into the dynamics of microbial populations and their role in waste treatment processes (Ţiru et al., 2023). The capability of ANNs to study subtle patterns and correlations makes them apt for process optimization and control. By feeding real-time operational data into trained ANN models, operators gain highly accurate predictions of key performance indicators, such as pollutant removal efficiencies, biomass growth rates, and bioreactor performance. This predictive power allows for proactive adjustments to operational parameters, which help save time and optimize conditions for microbial activity, prevent costly process upsets, and enhance the overall system (Nagpal et al., 2024; Frontistis et al., 2023). There are numerous applications where ANN models have been integrated into waste management techniques, proving to increase efficiency and improve functionality (Table 13.1).

## 2.2 Support vector machine

SVMs are a commonly utilized AI tool in the management of microbial waste. SVMs are specifically designed to forecast waste production and categorize waste as organic, hazardous, recyclable, or nonrecyclable. These algorithms serve as effective tools for data analysis. SVMs are capable of handling both linear and nonlinear classification tasks, even when the sample size is small and limited. In terms of effectiveness, SVMs are well-known for their proficiency in managing high-dimensional data and intricate decision boundaries (Gupta et al., 2024). Moreover, SVMs are nonparametric classifiers that excel in solving classification challenges by maximizing class separation and minimizing errors. SVMs are valuable for predicting the performance of microbial communities to estimate parameters such as biological oxygen demand (BOD), chemical oxygen demand (COD), and nitrogen levels in wastewater treatment processes (Narayanan et al., 2024). In waste generation prediction, SVMs can forecast weekly solid waste generation with accuracy. Preprocessing input variables using wavelet transform enhances the accuracy and robustness of the model (Fagnani and Guimarães, 2017).

## 2.3 Random forest

RF is another frequently used machine learning (ML) algorithm in forecasting waste generation. The technique combines multiple DTs to build an effective predictive tool. RF can handle complex datasets and nonlinear relationships (Breiman, 2001). This quality of RF makes it suitable for applications in microbial waste management and environmental sustainability. By utilizing this ML technique, researchers can uncover complex relationships between microbial features and their interactions, enhancing the understanding of microbial processes involved in waste degradation. This method can improve the efficiency of microbial waste management strategies, promoting environmental sustainability and biotechnological innovation in waste management through microbial communities. RF has a strong capacity to manage missing data, making it a flexible and dependable option for various data analysis tasks. RF has proven effective in a wide range of applications, demonstrating its ability to model and predict in situations involving noisy datasets (Cutler et al., 2012). In the context of organic waste processing, RF has proven effective in enhancing composting processes by accurately forecasting efficiency and decomposition rates, despite fluctuations in parameters such as temperature, moisture content, and organic waste composition (Wang et al., 2024). This attribute explains their extensive use in various applications such as automatic categorization, data forecasting, and supervised learning, making them useful for improved microbial waste management techniques (Hastie et al., 2009).

### 2.4 Decision trees

DTs are commonly employed to forecast waste production, categorize waste, and identify behavior patterns related to waste generation, as well as locate unauthorized dumping sites (Abdallah et al., 2020). Waste categorization helps in deciding the appropriate treatment methods and pathways. The benefits of utilizing DTs include straightforward interpretation of outcomes, minimal computational expenses, and the capacity to handle data that includes missing values and irrelevant characteristics (Kotsiantis, 2013). Nonetheless, this approach can be prone to overfitting the data.

## 3 Application of AI in microbial waste management

AI is increasingly being applied in microbial waste management to enhance efficiency, sustainability, and compliance with environmental standards. AI is used to improve waste prediction, collection, sorting, and recycling by optimizing resource utilization and minimizing ecological impact. Fig. 13.2 depicts the schematic work flow for waste management systems.

### 3.1 Waste generation prediction

ML algorithms have the capability to precisely forecast regions that produce large amounts of waste. These AI models can create predictive analyses to determine waste generation rates by considering factors such as population density, economic development, house size, consumption patterns, and lifestyle choices, while also determining the time intervals for predictions (daily, weekly, monthly, etc.). Both long-term and short-term forecasting can be developed for predicting waste generation (Fagnani and Guimarães, 2017). Waste generation prediction is the most widely investigated application. ANNs and SVMs are the most commonly used ML models in these applications.

Kulisz and Kujawska (2020), in a study, demonstrated an effective application of ANN models for forecasting municipal waste generation in Poland. The study, analyzing socioeconomic data factors from 2003 to 2019, demonstrates highly accurate

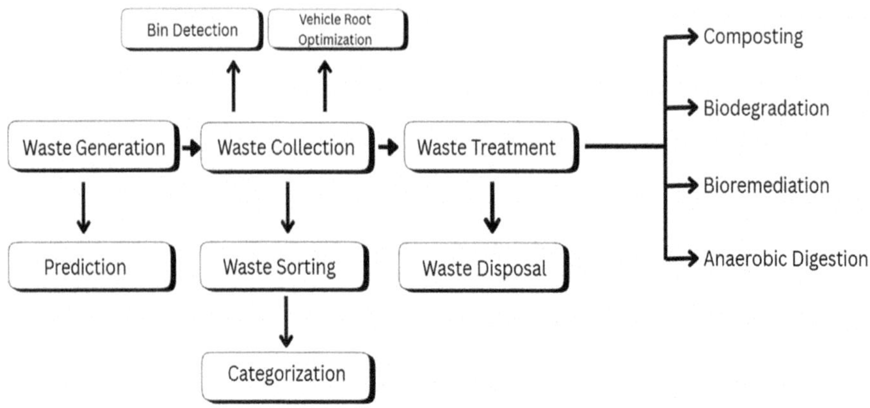


**FIG. 13.2**

Workflow representation of the smart waste management process.

predictions with regression values of 0.914 and 0.989, respectively, and low mean squared error values. The high R values exhibited a strong correlation between predicted and actual waste generation data, indicating that the ANN model is reliable for forecasting. Moreover, the lower MSE values signify that the predictions made by the ANN model are close to the actual values, enhancing the credibility of the forecast. The model forecasts a 2% increase in waste generation by the year 2024. This projection helps in planning and managing waste more effectively in the future (Kulisz and Kujawska, 2020).

## 3.2 Waste collection

The solid waste collection and transport system acts as a central link between resource sources, final disposal sites, and resource recovery management, which is crucial for facilitating the regular movement of urban materials and energy. On the one hand, the swift rise in waste production has resulted in inadequate back-end processing capabilities. Conversely, the operational scope of waste collection and transport systems is rapidly expanding, leading to increased management complexity. As a result, optimizing the routes for waste collection and transportation can significantly lower transportation expenses and enhance operational efficiency.

## 3.3 Waste classification and sorting systems

Numerous research efforts have concentrated on the categorization of waste materials for automated sorting systems that eliminate the need for manual waste separation. The majority of these investigations employed ANNs to identify different waste categories. One particular study utilized hyperspectral imaging combined with multilayered ANNs to detect various types of plastics in electronic waste. The process developed demonstrated an accuracy of 99% in identifying these materials (Tehrani and Karbasi, 2018).

The research conducted by Abbasi et al. (2013) illustrates the application of SVM using the partial least square (PLS) method. The goal was to enhance the model's accuracy through the PLS technique. The integration of PLS and SVM resulted in an $R^2$ value of 0.87, whereas the nonhybrid model registered an $R^2$ of 0.76. These findings indicate an improved predictive capability, as well as effective management of intricate and nonlinear input variables associated with the SVM model. It is important to note that throughout the modeling process, the computation time decreased from 42 s to 29 s. On similar lines, Sakr and his colleagues (2016) proposed automating waste sorting by utilizing ML algorithms to classify waste solely based on images. The researchers evaluated the performance of SVM against convolutional neural network (CNN) to assess the speed of waste classification. Each algorithm generates a distinct classifier that categorizes waste into three primary groups: plastic, paper, and metal, using only $256 \times 256$ colored PNG images of the waste. The results revealed that the SVM model attained a high image classification accuracy of 94.8%, while the CNN showed only 83%. Similarly, Lanorte et al. (2017) investigated the use of SVM for identifying agricultural plastic waste and assisting in the planning of waste facilities and route mapping. SVM was effective in classifying images and differentiating between crops and plastic waste, achieving an accuracy of 94.5%. In the management of agricultural waste, the RF method aids in determining which types of waste are suitable for the production of biomass energy and advancing the valorization process. Pandey and his colleagues worked to integrate IoT technology with RF algorithms to enhance biomass valorization and agricultural waste management. They collected real-time data on the types, patterns, and quantities of waste from agricultural fields. The RF model is capable of classifying whether the agricultural waste (from livestock or crops) can be converted into biomass and is most effective for energy production. This study provided real-time monitoring, efficient conversion of waste to biomass, informed decision-making, and ultimately improved energy production efficiency while reducing waste disposal costs (Pandey et al., 2024).

## 3.4 Bin detection

Waste collection encounters challenges such as elevated transportation costs, with trucks frequently servicing waste bins that are only partially filled, leading to an ineffective use of resources (Ramos et al., 2018). Various factors influence waste accumulation, making it challenging to forecast the fill levels of waste bins. To enhance the efficiency of waste collection, numerous studies have employed ML algorithms for waste bin detection, which can be viewed as a classification issue. The data inputs typically originate from level or image sensors installed within a waste bin, and feature extraction is performed using image processing techniques. Hannan et al. (2013) utilized an ANN alongside a Hough transform model or Gabor wavelet filters to categorize the solid waste levels in the bin, which were classified into five categories:

empty, medium, full, flow, and overflow. According to the information gathered from the sensor located within the container, Rutqvist et al. (2020) approached the waste bin level detection as a classification problem with two categories: emptying and nonemptying. They chose ANN, k-nearest neighbors (KNNs), Logistic regression (LR), SVM, DT, and RF as their classification algorithms, with RF yielding the best performance with an F1 score of 0.938. It is understood that ML techniques enhance the precision of classification, and smart detection of waste bin levels can minimize the driving distance for trucks, leading to cost savings for management agencies and a decrease in carbon dioxide emissions.

### 3.5 Vehicle and route optimization

The success of solid waste management systems greatly depends on the optimization of routes, and ML techniques can significantly improve this aspect. The challenges involved in optimizing routes, which include substantial costs related to labor, resources, and unpredictable operational expenses, have a major impact on waste collection. Waste management firms can optimize garbage collection routes to decrease expenses. By utilizing ML methods like KNNs, RF, and DTs, the effectiveness of routing optimization can be enhanced. KNNs can identify similar regions and predict the anticipated waste production in a particular area by examining historical data on waste generation trends, geographical information, and other important variables. To boost routing efficiency and reduce overall costs, RF considers numerous factors, including distance, traffic conditions, time limitations, and trends in waste generation. Another ML method applicable for enhancing routing optimization is the DT, which shares some characteristics with RF. DT builds a decision framework with possible outcomes that resemble a branching tree structure. To establish the most efficient collection routes, DT can evaluate historical data on waste production, traffic behavior, and geographical information.

Various studies have created optimization models for waste collection and route planning. Vu et al. (2019) introduced ANN and regression models to establish a routine waste collection schedule. Ferreira et al. (2014) integrated ANN and Multiple Linear Regression (MLR) models to estimate the necessary collection frequency across various locations. By including socioeconomic and demographic factors in the model, a reduction of 10% in collection frequency was noted. This approach helped avoid servicing locations with empty bins, leading to decreased environmental impact and lower collection costs.

## 4 ML for process optimization and performance prediction

Modeling and optimization of process variables are crucial in the design and operation of waste-to-energy technologies. The operational parameters can be modeled using various ML algorithms.

## 4.1 Wastewater treatment efficiency

ML algorithms enhance the accuracy of wastewater treatment processes by predicting pollutant concentrations, thus improving treatment efficiency. For instance, ANNs have been successfully incorporated into wastewater treatment processes to assess the performance of parameters like BOD, COD, total suspended solids (TSS), and the removal of nitrogen and phosphorus. In the biological treatment of wastewater, the integration of ANNs demonstrated proficient prediction of parameters such as BOD, COD, suspended solids (SS), and even the removal of nitrogen and phosphorus. Hamed et al. incorporated an ANN to predict wastewater treatment plant performance, focusing on BOD and SS concentrations, utilizing data from a major treatment plant in Cairo, Egypt. The developed ANN models predicted BOD and SS concentrations in wastewater treatment plant effluent. The ANN was also used to optimize operational conditions like aeration and sludge recirculation (Hamed et al., 2004). Mahanna et al. (2024) evaluated four ML techniques—RF, LR, Gradient Boosting (GB), and Support Vector Regression (SVR)—for measuring COD, BOD, and SS at the Al Hayer Wastewater Treatment Plant in Saudi Arabia. Among the models, RF was found to be the best-performing tool for analyzing COD, with an accuracy of 91%–95% for detecting SS performance. Gradient boosting was identified as the most effective model for measuring BOD performance, with a high accuracy of 92%. The RF model outperformed other suggested models in predicting the removal efficiency of COD and SS, according to the data ($R^2 = 0.95$ and 0.91 in the testing stage). The research highlighted the effectiveness of AI tools, particularly RF and gradient boosting, in predicting key performance indicators of wastewater treatment plants.

Studies show that AI models can be trained to improve the performance of waste management systems. In a study, Corominas et al. highlighted the application of computer-based techniques such as ANNs and fuzzy logic to enhance the efficiency of wastewater treatment plants. ANNs learned the intricate nonlinear patterns of the data and historical operational parameters, providing real-time guidance for operators, which improved removal efficiency and reduced energy consumption (Corominas et al., 2018). In addition, the RF technique has proven to be a highly accurate tool for improving water quality by identifying fecal contamination sources in microbial waste management systems. A study conducted by Roguet et al. (2018) illustrated the application of RF analysis using 16S rRNA gene amplicon sequences to detect sources of fecal pollution, focusing on *Clostridiales* and *Bacteroidales*. The researchers compiled a dataset to facilitate fecal source identification through RF classification. The classifiers for *Clostridiales* and *Bacteroidales* primarily consisted of sequences that showed distinct distributions among sources such as sewage, cow, deer, pig, cat, and dog. Each classifier accurately identified human and specific animal sources in about 90% of the fecal and environmental samples analyzed. The RF predictions demonstrated excellent reproducibility, with low average standard deviations of 0.05% for *Clostridiales* and 0.04% for *Bacteroidales* across trials. RF classification with 16S rRNA gene amplicons offered a rapid, sensitive, and accurate solution for identifying host microbial signatures to detect human and animal fecal contamination in environmental samples.

## 4.2 Composting

Composting is a significant process in organic waste treatment. After composting, the waste is converted to odorless and hygienic form. ML algorithms can be trained and modeled to control, optimize, and predict maturity forecasts and the level of nutrients during the composting process (Kujawa et al., 2014). The study performed by Soto-Paz et al. (2020) optimizes biowaste composting mixed with sugarcane filter cake using ANNs and particle swarm optimization (PSO). The integration of ANN and PSO significantly reduced the composting time while maximizing the physicochemical quality of the final product. The study focused on two operational parameters: mixing ratio and turning frequency. Optimal conditions were found at a 76.9:23.1 ratio and two turnings weekly, suggesting an 80-day composting process for final product quality enhancement. Better prediction of anaerobic–aerobic phases for turning, moisture content, and responsive management further improved the composting process and product quality. The integration of ANN and PSO was found to be a valuable tool for defining design criteria and operational conditions during the biowaste composting process. This can lead to more efficient compost production and better management of biowaste, contributing to sustainable agricultural practices.

ML models have also been effectively used in predicting maturity in organic waste composting. These models have received significant attention due to their tremendous potential to enhance compost quality and efficiency. Wang et al. (2024) utilized ML models such as RF, ANN, SVR, and DT to predict the germination index. RF and ANN have been successfully applied to predict the capabilities for the germination index, achieving $R^2$ values greater than 0.90. However, SVR achieved $R^2 < 0.6$, and DT had $R^2 < 0.8$, suggesting lower prediction accuracy compared to RF and ANN. Furthermore, the important factors that influenced compost maturity were composting time, temperature, and pH. The study showed that pH significantly affected the germination index, indicating its crucial role in the model's ability to accurately predict maturity. The RF and ANN models are the most recommended and highly effective models for predicting the germination index in composting, with composting time being the most influential factor. This offers a reliable process for smart composting management (Wang et al., 2024). Combining diverse ML models that can monitor critical parameters can lead to faster maturity times and improved compost quality.

## 4.3 Performance evaluation of microbial fuel cells

Studies have proven that neural network models have been successfully employed in forecasting important performance metrics of microbial fuel cells (MFCs), including power density, voltage, current generation, and COD removal efficiency (Abdollahfard et al., 2023). The integration of ANNs in MFCs provides a robust framework for optimizing and predicting the performance of bioelectrochemical systems. ANNs were utilized to establish a mathematical relationship between wastewater characteristics, biofilm communities, and reactor performance in MFCs. The performance of ANN models in predicting MFC outcomes is influenced by factors such as pH, temperature, ionic strength, and medium nitrogen concentration. ANN models that incorporated biotic interaction predictions predicted reactor performance outcomes more accurately than those that did not. Lesnik and Liu (2017) employed ANNs to evaluate the performance of 33 MFCs under diverse conditions, including eight different substrates and three types of wastewater. The results showed that ANN-based models that included biotic interactions accurately predicted changes more effectively than models that did not incorporate these interactions. This indicates the significance of microbial community dynamics in optimizing MFC performance. Moreover, the average percent error in predicting power density by the ANN was reported to be 16.01 ± in both biofilm community composition and reactor performance. This proves that ANNs are useful and reliable tools for predicting MFC performance under various conditions, which is crucial for advancing the field of bioelectrochemical systems. de Ramón-Fernández et al. (2020) studied ceramic MFCs and found that biofilm growth, influenced by feed flow rate, affects their performance. They used an ANN model to simulate the impact of flow rate on power output from ceramic MFCs using human urine. Three algorithms—quasi-Newton, Levenberg–Marquardt, and conjugate gradient—were compared for prediction accuracy and convergence time. The results showed that all algorithms accurately simulated power production, with Levenberg–Marquardt achieving the highest accuracy of 95% and the fastest convergence at 7.8 s. This highlights the effectiveness of ANNs in predicting energy harvesting from ceramic MFCs under varying flow rates.

## 5 Process outcome prediction

Measuring and quantifying the valuable by-products of waste management, such as biogas and renewable energy, is crucial for maximizing efficiency and sustainability. At the same time, it is equally important to monitor the adverse by-products and fugitive emissions, which contribute to air pollution. By accurately assessing both the beneficial and detrimental outputs, we can achieve optimal waste management practices that protect the environment while harnessing valuable resources.

### 5.1 Bioreactor performance prediction

Several studies have been conducted to predict the volume and composition of different by-products produced during waste management systems. ANNs were successfully used to estimate the volume and quality of biogas produced (such as methane content) from the anaerobic digestion of organic waste like agricultural residues, municipal solid waste, and sewage sludge. Yetilmezsoy et al. (2011) used a two three-layered ANN model to predict the rates of biogas and methane production.

The study highlighted accurate predictions that aid in optimizing the performance of the digester, maximizing energy recovery, and maintaining stable operation. The model's performance suggests that it can serve as a reliable tool for optimizing anaerobic treatment processes and can effectively assist in managing and improving the efficiency of wastewater treatment systems. The proposed ANN-based models produced smaller deviations and exhibited superior predictive accuracy with satisfactory determination coefficients.

In another study, Chen et al. (2022) utilized ANN to model biogas production in integrated anaerobic-aerobic bioreactors. For the anaerobic digestion process, $R^2$ values of 0.998, 0.983, and 0.990 were achieved for COD removal (%), methane purity (%), and methane yield, respectively. In contrast, during the aerobic process, high $R^2$ values of 0.997, 0.981, and 0.995 were recorded for COD removal, BOD removal, and TSS removal. The ANN architecture utilized for the anaerobic process was 6–12–3, while for the aerobic process, it was 7–6–3. The ANN models developed were able to accurately predict the effluent properties for the Integrated Anaerobic-Aerobic Bioreactor (IAAB) plant with minimal errors. The trained ANN model can serve as a decision support system for operators to anticipate the performance of the integrated anaerobic–aerobic bioreactor system. In addition, the application of various ML algorithms has significantly enhanced the prediction of biogas production in industrial-scale anaerobic digestion plants. These algorithms have demonstrated their ability to accurately model complex relationships between input parameters and biogas output. In a recent study, Yildirim and Ozkaya (2023) compared five ML techniques—RF, ANN, KNN, SVR, and XGBoost—to forecast the operational parameters and biogas production quantities. It was reported that the RF method produced the best prediction with the highest accuracy ($R^2 = 0.9242$) out of all five different ML algorithms tested.

## 5.2 Enhancing microbial biodegradation

RF models can predict biodegradation efficiency by measuring the performance of microbial consortia or individual strains in breaking down organic or inorganic waste. A comparable study conducted by Hemalatha et al. (2023) employed RF to predict microbes capable of breaking down plastics, achieving noteworthy accuracy in identifying potential microbes responsible for plastic degradation. They created a dataset featuring microbes such as *Streptococcus*, *Micrococcus*, and *Staphylococcus*, specifically aimed at forecasting plastic degradation. The research implemented various ML algorithms for making predictions, including RF, SVM, KNN, and DT. Of all the models evaluated, RF demonstrated the greatest accuracy at 99.1%.

RF can model relationships between input variables, such as type of pollutant, microbial community, pH, and temperature, and output variables such as the percentage degradation rate of diverse pollutants, including dyes, heavy metals, and pharmaceutical drugs. Moreover, RF is capable of classifying whether a chemical substance is likely to be biodegradable based on its structural and physicochemical characteristics. AI-driven monitoring systems facilitate real-time data collection and analysis, allowing for dynamic adjustments in bioremediation conditions such as pH and nutrient levels. ANNs can predict and model microbial bioremediation processes for the treatment of contaminated soil and water. They link contaminant type and concentration with microbial community structure, nutrient amendments, environmental conditions (e.g., carbon-to-nitrogen ratios and moisture content), and the rate of pollutant degradation. Mahanty et al. (2025) demonstrated hybrid modeling with ANN to forecast insitu bioremediation of diesel fuel from diesel fuel-spiked soil. The experiment was conducted over 138 days with moisture content ranging from 8% to 15% and carbon-to-nitrogen ratios of 120 and 180, respectively. The instantaneous rate of removal was modeled as an ANN function of moisture content, C/N ratio, time, and concentration of diesel fuel. The double hidden layer presented the best fitness with $R^2$ values of 0.996 and 0.995 for the test subset and the entire dataset, respectively. The numerical integration of the ANN-embedded rate expression for diesel fuel-spiked soil removal demonstrated a better fit ($R^2 > 0.99$) to the dynamics. The high correlation coefficient values indicate the reliability of ANN in modeling the bioremediation of diesel-contaminated soil. These AI systems can predict bioremediation outcomes, enabling proactive management of bioremediation strategies, thus improving overall performance. In another study, Prasad and Mathur performed in situ bioremediation using ANN to optimize pumping rates by replacing traditional simulation models and reducing computational efforts. The computational efficiency of the ANN approach was highlighted, as it required only 3% of the simulations needed by two other genetic algorithm variants. This efficiency not only reduced the computational burden but also provided a practical solution for field-scale applications of in-situ bioremediation. The developed ANN technique not only minimized the total costs involved in the remediation process and contaminant concentrations but also prevented the migration of contaminants, making it a viable option for groundwater remediation efforts (Prasad and Mathur, 2006).

## 6 Conclusions

The merging of AI and microbial science is proving to be a crucial moment in the global effort to address growing waste issues. This chapter explores how AI models can enhance waste management practices. Conventional methods fall short in providing effective solutions for planning and managing solid waste. The use of AI models such as ANN, SVM, RF, and DTs has been extensively examined. ANNs are the most commonly used model in waste management due to their ability to effectively handle complex and nonlinear relationships. In contrast, SVMs excel in automated waste categorization and exhibit minimal errors in waste classification. In addition, RF is capable of handling noisy datasets, while DTs are designed to work with data that contains missing values. The choice of AI model can be determined based on the type of waste. AI can improve the identification and classification of waste, as well as optimize the processes of collection, sorting, transportation, disposal, and postdisposal monitoring at designated dumping sites. AI technology can significantly advance the sorting of waste. It can accurately identify and segregate recyclables from mixed waste streams, which increases recycling rates and helps alleviate the pressure on landfills (Capodaglio and Callegari, 2025). The integration of AI technologies enhances waste treatment processes overall. This comprehensive strategy not only tackles the intricacies of waste management but also aligns with the principles of a circular economy.

## 7 Challenges and future scope

AI presents significant advancements in microbial waste management; however, challenges such as data privacy, high implementation costs, and the need for skilled personnel persist. Addressing these issues is crucial for maximizing the potential of AI technologies in sustainable waste management practices. Policymakers should work toward creating a regulatory environment that supports the adoption of smart waste management solutions. This can be achieved by establishing standards for data privacy, ensuring access to technology, and providing funding opportunities for municipalities to implement these innovations (Lakhouit, 2025).

The trajectory of AI in microbial waste management techniques is toward a future of unmatched innovation, automation, and efficiency. In addition to its existing prediction and optimization potential, the future will witness the widespread integration of more advanced AI paradigms. Deep learning, specifically convolutional and recurrent neural networks, will enable the analysis of complex, high-dimensional "omics" data (e.g., metagenomics, metabolomics) and facilitate the deciphering of efficient and targeted waste degradation (Hernández Medina et al., 2022). Reinforcement learning holds great potential for creating genuinely autonomous and self-optimizing waste treatment plants, where AI models will learn optimal control policies in real-time via trial and error, dynamically adapting to changing conditions and disturbances (Nievas et al., 2024). The use of digital twin technology driven by AI will enable virtual simulation and testing of advanced scenarios, refining processes prior to implementation (Shah et al., 2024).

## References


Abbasi, M., Abduli, M.A., Omidvar, B., Baghvand, A., 2013. Forecasting municipal solid waste generation by hybrid support vector machine and partial least square model. Int. J. Environ. Res. 17356865. 7 (1), 27–38. Available from http://ijer.ut.ac.ir/images/Issues/FullText.pdf.

Abdallah, M., Talib, M.A., Feroz, S., Nasir, Q., Abdalla, H., Mahfood, B., 2020. Artificial intelligence applications in solid waste management: a systematic research review. Waste Manag. 18792456. 109, 231–246. Available from: https://www.elsevier.com/locate/wasman. doi:10.1016/j.wasman.2020.04.057.

Abdollahfard, Y., Sedighi, M., Ghasemi, M., 2023. A new approach for improving microbial fuel cell performance using artificial intelligence. Sustainability. 2071-1050. 15 (2), 1312. doi:10.3390/su15021312.

Balasubramanian, M., 2018. Municipal solid waste management in India: status, problems and challenges. IJEWM 21 (4), 253. doi:10.1504/IJEWM.2018.093435.

Breiman, L., 2001. Random Forests. Mach. Learn. 0885-6125. 45 (1), 5–32. doi:10.1023/a:1010933404324.

Capodaglio, A.G., Callegari, A., 2025. Use, potential, needs, and limits of AI in wastewater treatment applications. Water. 2073-4441. 17 (2), 170. doi:10.3390/w17020170.

Chen, W.Y., Chan, Y.J., Lim, J.W., Liew, C.S., Mohamad, M., Ho, C.D., Usman, A., Lisak, G., Hara, H., Tan, W.N., 2022. Artificial neural network (ANN) modelling for biogas production in pre-commercialized integrated anaerobic-aerobic bioreactors (IAAB). Water. 20734441. 14 (9). Available from: https://www.mdpi.com/2073-4441/14/9/1410/pdf. doi:10.3390/w14091410.

Çimen Mesutoğlu, Ö., Gök, O., 2024. Prediction of COD in industrial wastewater treatment plant using an artificial neural network. Sci. Rep. 2045-2322. 14 (1). doi:10.1038/s41598-024-64634-z.

Cutler, A., Cutler, D.R., Stevens, J.R., 2012. Random Forests Ensemble Machine Learning: Methods and Applications. Springer US, United States, pp. 157–175. Available from. doi:10.1007/9781441993267_5. 10.1007/978-1-4419-9326-7.

de Ramón-Fernández, A., Salar-García, M.J., Ruiz Fernández, D., Greenman, J., Ieropoulos, I.A., 2020. Evaluation of artificial neural network algorithms for predicting the effect of the urine flow rate on the power performance of microbial fuel cells. Energy. 03605442. 213, 118806. doi:10.1016/j.energy.2020.118806.

Elshaboury, N., Abdelkader, E.M., Al-Sakkaf, A., Alfalah, G., 2021. Predictive analysis of municipal solid waste generation using an optimized neural network model. Processes. 2227-9717. 9 (11), 2045. doi:10.3390/pr9112045.

Ezzahra Yatim, F., Boumanchar, I., Srhir, B., Chhiti, Y., Jama, C., Ezzahrae M'hamdi Alaoui, F. 2022. Waste-to-energy as a tool of circular economy: prediction of higher heating value of biomass by artificial neural network (ANN) and multivariate linear regression (MLR). Waste Manag. 0956053X. (153), 293–303. doi: 10.1016/j.wasman.2022.09.013.

Fagnani, E., Guimarães, J.R., 2017. Waste management plan for higher education institutions in developing countries: the continuous improvement cycle model. J. Clean. Prod. 09596526. 147, 108–118. doi:10.1016/j.jclepro.2017.01.080.

Ferreira, J.A., Figueiredo, M.C., Oliveira, J.A., 2014. Forecasting household packaging waste generation: a case study. In: Lecture Notes in Computer Science (Including Subseries Lecture Notes in Artificial Intelligence and Lecture Notes in Bioinformatics)Vol. 3. Springer Verlag, Portugal, pp. 523–538. 16113349. http://springerlink.com/content/0302-9743/copyright/2005/8581. doi:10.1007/978-3-319-09150-1_38.

Frontistis, Z., Lykogiannis, G., Sarmpanis, A., 2023. Artificial neural networks in membrane bioreactors: a comprehensive review—overcoming challenges and future perspectives. Science. 2413-4155. 5 (3), 31. doi:10.3390/sci5030031.

Golbaz, S., Nabizadeh, R., Sajadi, H.S., 2019. Comparative study of predicting hospital solid waste generation using multiple linear regression and artificial intelligence. J. Environ. Health Sci. Eng. 2052336X. 17 (1), 41–51. Available from: https://rd.springer.com/journal/40201. doi:10.1007/s40201-018-00324-z.

Gupta, R., Ouderji, Z.H., Uzma, Yu, Z., Sloan, W.T., You, S., 2024. Machine learning for sustainable organic waste treatment: a critical review. Npj Mater. Sustain. 2948-1775. 2 (1). doi:10.1038/s44296-024-00009-9.

Hamed, M.M., Khalafallah, M.G., Hassanien, E.A., 2004. Prediction of wastewater treatment plant performance using artificial neural networks. Environ. Model. Softw. 19 (10), 919–928. doi:10.1016/j.envsoft.2003.10.005.

Hannan, M.A., Arebey, M., Begum, R.A., Mustafa, A., Basri, H., 2013. An automated solid waste bin level detection system using Gabor wavelet filters and multi-layer perception. Resour. Conserv. Recycl. 09213449. 72, 33–42. doi:10.1016/j.resconrec.2012.12.002.

Hastie, T., Tibshirani, R., Friedman, J., 2009, The elements of statistical learning. doi:10.1007/978-0-387-84858-7.

Hemalatha, N., Akhil, W., Raji Vinod, T.A., 2023. Computational Prediction of Plastic Degrading Microbes Using Random Forest. Springer Science and Business Media LLC, pp. 285–293. doi:10.1007/978-981-19-7892-0_22.

Hernández Medina, R., Kutuzova, S., Nielsen, K.N., Johansen, J., Hansen, L.H., Nielsen, M., Rasmussen, S., 2022. Machine learning and deep learning applications in microbiome research. ISME Commun. 2730-6151. 2 (1). doi:10.1038/s43705-022-00182-9.

Kaza, S., Yao, L.C., Bhada-Tata, P., Van Woerden, F., 2018. What A Waste 2.0: A Global Snapshot of Solid Waste Management to 2050. World Bank, Washington, DC. doi:10.1596/978-1-4648-1329-0.

Kotsiantis, S.B., 2013. Decision trees: a recent overview. Artif. Intell. Rev. 0269-2821. 39 (4), 261–283. doi:10.1007/s10462-011-9272-4.

Kujawa, S., Nowakowski, K., Tomczak, R.J., Dach, J., Boniecki, P., Weres, J., Mueller, W., Raba, B., Piechota, T., Rodríguez Carmona, P.C., 2014. Neural image analysis for maturity classification of sewage sludge composted with maize straw. Comput. Electron. Agric. 01681699. 109, 302–310. Available from: https://www.elsevier.com/inca/publications/store/5/0/3/3/0/4. doi:10.1016/j.compag.2014.08.014.

Kulisz, M., Kujawska, J., 2020. Prediction of municipal waste generation in Poland using neural network modeling. Sustainability. 2071-1050. 12 (23), 10088. doi:10.3390/su122310088.

Kumar, A., Samadder, S.R., Kumar, N., Singh, C., 2018. Estimation of the generation rate of different types of plastic wastes and possible revenue recovery from informal recycling. Waste Manag. 0956053X. 79, 781–790. doi:10.1016/j.wasman.2018.08.045.

Kumar, S., Chakrabarti, T., 2010. Effective municipal solid waste management in India. InTech. doi:10.5772/8459.

Lakhouit, A., 2025. Revolutionizing urban solid waste management with AI and IoT: a review of smart solutions for waste collection, sorting, and recycling. Results Eng. 25901230. 25 (104018). doi:10.1016/j.rineng.2025.104018.

Lanorte, A., De Santis, F., Nolè, G., Blanco, I., Loisi, R.V., Schettini, E., Vox, G., 2017. Agricultural plastic waste spatial estimation by Landsat 8 satellite images. Comput. Electron. Agric. 01681699. 141, 35–45. Available from: https://www.elsevier.com/inca/publications/store/5/0/3/3/0/4. doi:10.1016/j.compag.2017.07.003.

Lesnik, K.L., Liu, H., 2017. Predicting microbial fuel cell biofilm communities and bioreactor performance using artificial neural networks. Environ. Sci. Technol. 15205851. 51 (18), 10881–10892. Available from: http://pubs.acs.org/journal/esthag. doi:10.1021/acs.est.7b01413.

Li, Y., Lu, M., Campos, L.C., Hu, Y., 2024. Predicting Biogas Yield after microwave pretreatment using artificial neural network models: performance evaluation and method comparison. ACS ES T Eng. 26900645. 4 (10), 2435–2448. Available from: https://pubs.acs.org/page/aeecco/about.html. doi:10.1021/acsestengg.4c00276.

Ll. Corominas, M.G.-B., Villez, K., Olsson, G., Cortés, U., Poch, M., 2018. Transforming data into knowledge for improved wastewater treatment operation: a critical review of techniques. Environ. Model. Softw. 13648152. 106, 89–103. doi:10.1016/j.envsoft.2017.11.023.

Maalouf, A., Mavropoulos, A., 2023. Re-assessing global municipal solid waste generation. Waste Manag. Res. 0734-242X. 41 (4), 936–947. doi:10.1177/0734242x221074116.

Mahanna, H., El-Rashidy, N., Kaloop, M.R., El-Sapakh, S., Alluqmani, A., Hassan, R., 2024. Prediction of wastewater treatment plant performance through machine learning techniques. Desalination Water Treat. 19443986. 319, 100524. doi:10.1016/j.dwt.2024.100524.

Mahanty, B., Behera, S.K., Godio, A., Chiampo, F., 2025. Hybrid modeling with artificial neural networks for predicting in-situ bioremediation dynamics of diesel fuel-spiked soil. Water, Air, and Soil Pollution. 15732932. 236 (5). Available from: https://www.springer.com/journal/11270. doi:10.1007/s11270-025-07940-0.

Maind, S.B., 2014. Research paper on basic of artificial neural network. Int. J. Recent Innov. Trends Comput. Commun. 2 (1), 96–100.

Malakahmad, A., Khalil, N.D., 2011. Malaysia Solid waste collection system in Ipoh city. In: ICBEIA 2011 - 2011 International Conference on Business, Engineering and Industrial Applications, pp. 174–179. doi:10.1109/ICBEIA.2011.5994236.

Malviya, A., Jaspal, D., 2021. Artificial intelligence as an upcoming technology in wastewater treatment: a comprehensive review. Environ. Technol. Rev. 2162-2515. 10 (1), 177–187. doi:10.1080/21622515.2021.1913242.

Nagpal, M., Siddique, M.A., Sharma, K., Sharma, N., Mittal, A., 2024. Optimizing wastewater treatment through artificial intelligence: recent advances and future prospects. Water Sci. Technol. 0273-1223. 90 (3), 731–757. doi:10.2166/wst.2024.259.

Narayanan, D., Bhat, M., Paul, N.R.S., Khatri, N., Saroliya, A., 2024. Artificial intelligence driven advances in wastewater treatment: evaluating techniques for sustainability and efficacy in global facilities. Desalination Water Treat. 19443986. 320, 100618. doi:10.1016/j.dwt.2024.100618.

Nievas, N., Pagès-Bernaus, A., Bonada, F., Echeverria, L., Domingo, X., 2024. Reinforcement learning for autonomous process control in industry 4.0: advantages and challenges. Appl. Artif. Intell. 0883-9514. 38 (1). doi:10.1080/08839514.2024.2383101.

Pandey, P., Sindhuja, K., Muthukumaran, D., Kanna, C.K., Raj, T.S., Srinivasan, S., 2024. Integrated IoT and random forest algorithm solutions for agricultural waste management and biomass valorization. In: 2024 1st International Conference on Innovations in Communications, Electrical and Computer Engineering, ICICEC 2024. 9798350376517. Institute of Electrical and Electronics Engineers Inc., India. http://ieeexplore.ieee.org/xpl/mostRecentIssue.jsp?punumber=10808176. doi:10.1109/ICICEC62498.2024.10808352.

Pillai, K.S., Sneha, M.L., Aiswarya, S., Anand, A.B., Prasad, G., 2023. EDP sciences India municipal solid waste management: a review of machine learning applications. In: E3S Web of Conferences, p. 455. 22671242. https://www.e3s-conferences.org/. doi:10.1051/e3sconf/202345502018.

Prasad, R.K., Mathur, S., 2006. In-situ bioremediation of contaminated groundwater using artificial neural network. World Environ. Water Resour. Congr. 2006, 1–10. doi:10.1061/40856(200)140.

Ramos, T.R.P., de Morais, C.S., Barbosa-Póvoa, A.P., 2018. The smart waste collection routing problem: alternative operational management approaches. Expert Syst. Appl. 09574174. 103, 146–158. doi:10.1016/j.eswa.2018.03.001.

Roguet, A., Eren, A.M., Newton, R.J., McLellan, S.L., 2018. Fecal source identification using random forest. Microbiome. 20492618. 6 (1). Available from: http://www.microbiomejournal.com/. doi:10.1186/s40168-018-0568-3.

Rutqvist, D., Kleyko, D., Blomstedt, F., 2020. An automated machine learning approach for smart waste management systems. IEEE Trans. Ind. Inform. 1551-3203. 16 (1), 384–392. doi:10.1109/tii.2019.2915572.

Sakr, G.E., Mokbel, M., Darwich, A., Khneisser, M.N., Hadi, A., 2016. Comparing deep learning and support vector machines for autonomous waste sorting. In: IEEE International Multidisciplinary Conference on Engineering Technology, IMCET 2016. 9781509052813, pp. 207–212. Institute of Electrical and Electronics Engineers Inc., Lebanon. 10.1109/IMCET.2016.7777453.

Shah, K.B., Visalakshi, S., Guragain, D.P., Panigrahi, R., 2024. Advancing smart city sustainability with internet of things and artificial intelligence aided low-cost digital twin systems for waste management. Microsyst. Technol. 14321858. Available from: https://www.springer.com/journal/542. doi:10.1007/s00542-024-05827-4.

Singh, S., Mehta, K.S., Bhattacharya, N., Prasad, J., Lakshmi, S.K., Subramaniam, K.V., Sitaram, D., 2017. Identifying uncollected garbage in urban areas using crowdsourcing and machine learning. In: TENSYMP 2017 - IEEE International Symposium on Technologies for Smart Cities. 9781509062553. Institute of Electrical and Electronics Engineers Inc., India. doi:10.1109/TENCONSpring.2017.8070078.

Singh, S. Singh, J., 2021. Solid waste management in India: a review. Int. J. Eng. Sci. Tech. 13 (1), 54–60. doi:10.4314/ijest.v13i1.8S.

Soto-Paz, J., Alfonso-Morales, W., Caicedo-Bravo, E., Oviedo-Ocaña, E.R., Torres-Lozada, P., Manyoma, P.C., Sanchez, A., Komilis, D., 2020. A new approach for the optimization of biowaste composting using artificial neural networks and particle swarm optimization. Waste Biomass Valorization. 1877265X. 11 (8), 3937–3951. Available from: http://www.springer.com/engineering/journal/12649. doi:10.1007/s12649-019-00716-8.

Tehrani, A., Karbasi, H., 2018. 2017 A novel integration of hyper-spectral imaging and neural networks to process waste electrical and electronic plastics 2018. In: 2017 IEEE Conference on Technologies for Sustainability, SusTech, pp. 1–5. 9781538604519. Institute of Electrical and Electronics Engineers Inc., Canada. doi:10.1109/SusTech.2017.8333533.

Thomaz, I.P.L., Mahler, C.F., 2024. Artificial neural network models to support urban waste management: a technological resource that drives the achievement of sustainable development goals. Eng. Sanit. Ambient. 14134152. 29. Available from: https://www.scielo.br/j/esa/a/m9GVkT6VqZs7h9DVJDvTLSy/?lang=en&format=pdf. doi:10.1590/S1413-415220230085.

Ţiru, A.E., Vasiliev, I., Diaconu, L., Vilanova, R., Voipan, D., Ratnaweera, H., 2023. Integration of ANN for accurate estimation and control in wastewater treatment. In: IEEE International Conference on Emerging Technologies and Factory Automation, ETFA. doi:10.1109/ETFA54631.2023.10275569. 19460759. Institute of Electrical and Electronics Engineers Inc, Romania. http://ieeexplore.ieee.org/xpl/conhome.jsp?punumber=1000260.

Vu, H.L., Ng, K.T.W., Bolingbroke, D., 2019. Time-lagged effects of weekly climatic and socio-economic factors on ANN municipal yard waste prediction models. Waste Manag. 18792456. 84, 129–140. Available from: https://www.elsevier.com/locate/wasman. doi:10.1016/j.wasman.2018.11.038.

Wang, N., Yang, W., Wang, B., Bai, X., Wang, X., Xu, Q., 2024. Predicting maturity and identifying key factors in organic waste composting using machine learning models. Bioresour. Technol. 09608524. 400, 130663. doi:10.1016/j.biortech.2024.130663.

Xia, W., Jiang, Y., Chen, X., Zhao, R., 2022. Application of machine learning algorithms in municipal solid waste management: a mini review. Waste Manag. Res. 0734-242X. 40 (6), 609–624. doi:10.1177/0734242x211033716.

Yeboah Antwi, B., Ofori-Nyarko, A., 2023. A Prefatorial View of Solid Waste Management. IntechOpen. doi:10.5772/intechopen.1002660.

Yetilmezsoy, K., Ozkaya, B., Cakmakci, M., 2011. Artificial intelligence-based prediction models for environmental engineering. Neural Netw. World. 12100552. 21 (3), 193–218. doi:10.14311/nnw.2011.21.012.

Yildirim, O., Ozkaya, B., 2023. Prediction of biogas production of industrial scale anaerobic digestion plant by machine learning algorithms. Chemosphere. 00456535. 335, 138976. doi:10.1016/j.chemosphere.2023.138976.

Zakir Hossain, S.M., Sultana, N., Haji, S., Mufeez, S.T., Janahi, S.E., Ahmed, N.A., 2023. Modeling of microbial fuel cell power generation using machine learning-based super learner algorithms. Fuel. 00162361. 349. Available from: http://www.journals.elsevier.com/fuel/. doi:10.1016/j.fuel.2023.128646.

CHAPTER

# 14 Advanced microbial techniques for wastewater treatment

**T.G. Ashwini, Anum Haleema Ali, and Niladri Shekhar Samanta**
*Department of Petroleum Engineering, Presidency University, Bengaluru, India*

## 1 Introduction

Microbial immobilization has become a pivotal method in environmental biotechnology for enhancing the performance and resilience of biological wastewater treatment systems. This approach involves anchoring microbial cells or enzymes onto or within support materials, which allows for higher local concentrations of biomass and extended microbial activity. As a result, it leads to improved pollutant removal, enhanced process stability, and better resistance to environmental stress. The choice of carrier material is essential for maintaining cell viability and ensuring long-term treatment efficiency. Both naturally derived and synthetic carriers, including alginate, chitosan, silica, and various polymers, have been widely explored. Recent innovations have focused on integrating materials like nanocomposites and biochar to improve microbial compatibility and mechanical stability (Ansari and Husain, 2012).

A variety of immobilization methods have been applied, ranging from conventional techniques such as adsorption, encapsulation, and entrapment to more advanced systems like layer-by-layer (LBL) assembly and electrospinning. These methods are designed to protect microbial cells while facilitating efficient mass transfer and maintaining high biological activity. Immobilized systems offer distinct advantages over traditional suspended-growth systems, such as a lower risk of biomass loss, enhanced tolerance to toxic substances, and support for slow-growing but essential microbial communities. This is particularly useful for biological nitrogen removal processes, which rely on sensitive microbial populations such as anammox and aerobic denitrifiers.

Studies have shown that immobilized microbial systems can significantly improve the treatment of complex wastewater, especially those with high nitrogen content. Carriers composed of materials like polyvinyl alcohol (PVA) and sodium alginate (SA) have exhibited excellent performance in supporting microbial activity and maintaining consistent nitrogen removal efficiency over multiple treatment cycles. Moreover, these systems show potential for treating a wide range of wastewater sources, including domestic, agricultural, and industrial effluents. Given their

**Microbial Technologies for Sustainable Waste Management. https://doi.org/10.1016/B978-0-443-44093-9.00012-8**

versatility and efficiency, microbial immobilization technologies are expected to play a key role in the future of sustainable wastewater management (Ahmad et al., 2017).

This chapter mainly deals with the emerging trends in immobilized microbial technology for wastewater treatment, including the selection of immobilization carriers such as natural carrier materials, nanocarrier materials, metal–organic frameworks (MOFs), and biochar material carriers. This chapter also vividly describes immobilized microbial technologies, including the LBL assembly technique and electrostatic spinning techniques. In addition to this, nitrogen removal by anammox and aerobic bacteria is discussed in this chapter. Finally, further exploration of microbial techniques for polluted water treatment and other discussions may be very appropriate for readers of this chapter.

## 2 Emerging trends in immobilized microbial technology for wastewater treatment

Immobilized microbial technology has emerged as an innovative and promising method that has made significant strides in wastewater treatment and other environmental remediation fields. Given its considerable potential for efficient pollutant removal and operational stability, this technology has attracted extensive research attention in recent years. Despite these advancements, several critical challenges persist, necessitating further in-depth investigation and solutions.

A prominent issue that still needs to be addressed is the relatively high cost associated with the development of immobilization carriers. In addition, limitations concerning the mechanical durability and lifespan of these carriers have obstructed their sustained application in large-scale industrial processes. Furthermore, the successful industrial-scale implementation and commercialization of immobilized microbial systems have yet to be fully realized. These obstacles highlight the urgent need for the creation of low-cost, efficient, and durable immobilization carriers that are suitable for practical applications.

It has been observed that both the type of immobilization carrier and the selected immobilization technique directly impact the viability, activity, and functional performance of the immobilized microorganisms. Therefore strategically optimizing the carrier materials and immobilization methodologies is deemed essential for enhancing microbial activity and overall wastewater treatment efficiency.

Recently, substantial academic interest has been dedicated to the synthesis and selection of effective immobilization carriers, as reflected in the significant increase in scholarly publications on this topic. Between 2020 and 2023, a total of 341 research articles focused on immobilization carriers were published. This trend indicates a continuous and growing interest in the field, with both the quantity of annual publications and citation frequency steadily increasing, particularly after 2018. These statistics were obtained from the Web of Science database using the keyword "Immobilized microbial carrier" as of December 31, 2023.

While traditional immobilization methods, such as adsorption, cross-linking, covalent bonding, and embedding, have been extensively investigated and used, the introduction of advanced immobilization techniques has broadened the scope and applicability of this technology. New-generation immobilization methods, including LBL self-assembly and electrostatic spinning, have gained significant traction recently due to their ability to enhance microbial loading capacity and functional stability.

A considerable research focus has also targeted the immobilization of specific bacterial groups, including anaerobic ammonia-oxidizing (anammox) bacteria and aerobic denitrifying (AD) bacteria. These bacteria have been recognized as key contributors to autotrophic nitrogen removal and aerobic denitrification processes, respectively. Consequently, the immobilization of anammox and AD bacteria has become a significant research hotspot, as evidenced by the clustered visualization of keyword cooccurrence generated using VOSviewer. The bibliometric data supporting this visualization were derived from the Web of Science database, focusing on the topics "Immobilized microbial" and "Wastewater Treatment."

Moreover, the utility of immobilized microbial technology has been extended beyond conventional wastewater treatment to include the remediation of complex and refractory wastewater. By immobilizing specialized bacterial strains capable of degrading persistent organic pollutants, highly efficient treatment of industrial and recalcitrant wastewater has become feasible. These findings collectively demonstrate that research in this field is progressively evolving toward the development of multifunctional, durable, and economically viable microbial immobilization systems for diverse environmental applications. The annual publication and citation trends (2010–2023) for immobilized microbial carrier studies are depicted in Fig. 14.1.

Moreover, the application of immobilized microbial technology has expanded beyond conventional wastewater treatment to include the remediation of complex and refractory wastewater. By immobilizing specialized bacterial strains capable of degrading persistent organic pollutants, highly efficient treatment solutions for industrial and recalcitrant wastewater have been developed. These advancements highlight a significant evolution in research within this field, focusing on the creation of multifunctional, durable, and economically viable microbial immobilization systems suitable for a variety of environmental applications.

## 3 Selection of immobilization carriers

The choice of immobilization carrier is crucial to the performance of immobilized microbial technology. This carrier serves as a support matrix for microbial cells and significantly influences various factors, including pollutant removal efficiency, operational costs, microbial stability, and the overall reusability of the treatment system (Blachowicz et al., 2022). The properties of the carrier—such as specific

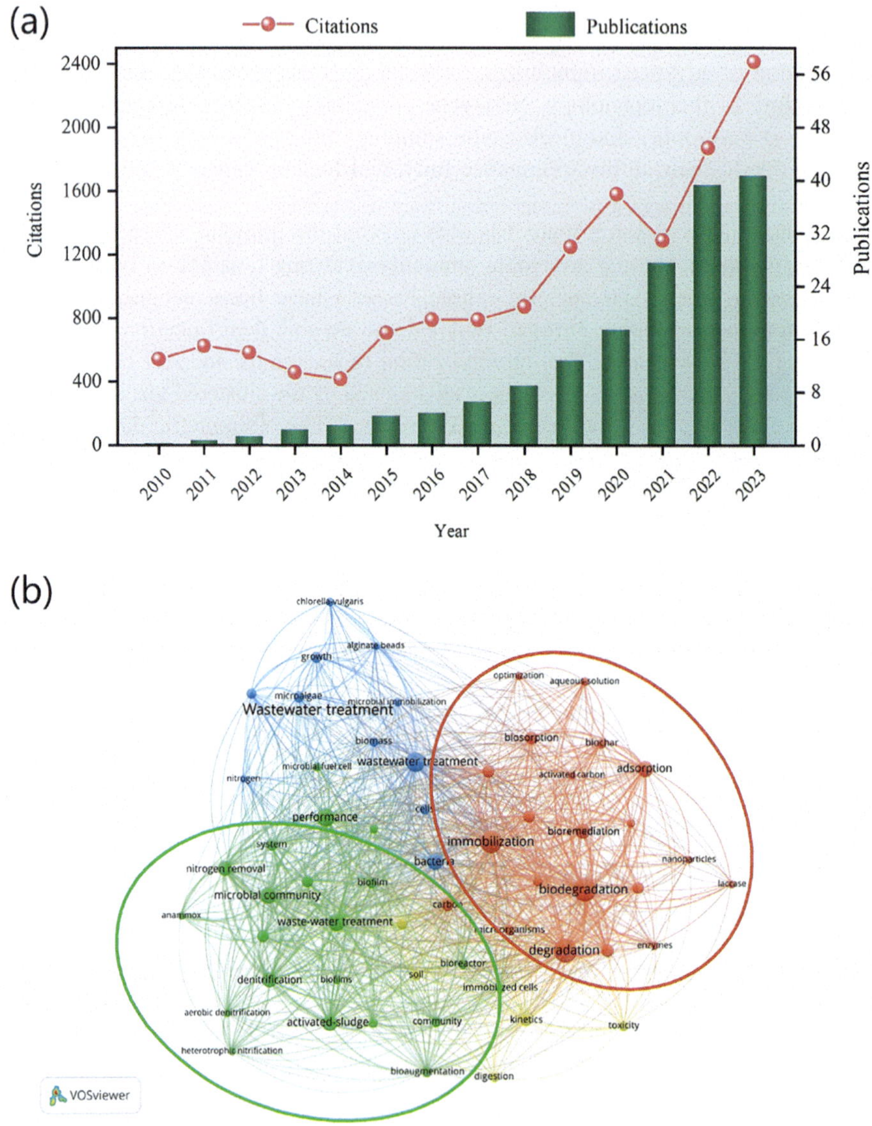


**FIG. 14.1**

Bibliometric analysis of immobilized microbial carrier research. (A) Annual publication and citation trends (2010–2023) for immobilized microbial carrier studies (Web of Science search term: "Immobilized microbial carrier"; data retrieved December 31, 2023). (B) Keyword cooccurrence analysis of immobilized microorganism applications in wastewater treatment (VOSviewer visualization of Web of Science data for "Immobilized microbial" AND "Wastewater Treatment").

*-From Hou, L., Hu, K., Huang, F., Pan, Z., Jia, X., Liu, W., Yao, X., Yang, Z., Tang, P., Jun, L., 2024. Advances in immobilized microbial technology and its application to wastewater treatment: a review. Bioresour. Technol. 09608524. 413, 131518. https://doi.org/10.1016/j.biortech.2024.131518.*

surface area, pore size, structural configuration, chemical stability, and mechanical strength—directly affect microbial activity and the physicochemical behavior of the immobilization system (Boland et al., 2001).

A high specific surface area provides numerous active sites, enhancing microbial adhesion and allowing for greater biomass loading. This results in increased microbial concentration and higher metabolic activity. In addition, a larger surface area facilitates improved mass transfer of nutrients and metabolic by-products, promoting microbial growth and proliferation.

The size of the pores within the carrier material is also vital; a well-calibrated porous structure is essential. While larger pores promote effective nutrient diffusion and microbial transport, excessively small pores can restrict microbial movement and lead to the accumulation of metabolites. Therefore selecting an optimal pore size is crucial for maintaining microbial viability and functional performance.

Surface functional groups, such as hydroxyl, amino, and carboxyl moieties, significantly influence microbial attachment. These functional groups mediate interactions between the surfaces of microbial cells and the carrier material, affecting both immobilization efficiency and the long-term stability of the microorganisms (Bouabidi et al., 2019). Chemical stability is another critical attribute, as immobilization carriers must maintain their structure and function under variable environmental conditions, including extreme pH levels, temperature fluctuations, and the presence of chemical agents. Materials with high chemical resistance ensure sustained activity of immobilized microorganisms over extended periods (Cheng et al., 2024).

Mechanical strength is equally important, as it determines the carrier's ability to withstand shear stress, pressure, and operational wear. Adequate mechanical integrity prevents carrier breakage or deformation during reactor operation, ensuring continuous microbial functionality. Carriers with robust mechanical properties allow for repeated usage and regeneration, contributing to lower operational costs and enhanced process sustainability.

An ideal immobilization carrier should exhibit characteristics such as ease of preparation, cost-effectiveness, biocompatibility, high mechanical and chemical stability, efficient mass transfer, and resistance to microbial degradation (Cho et al., 2018). Traditionally, immobilization carriers have been classified into inorganic and organic types, with organic carriers further subdivided into natural and synthetic polymers (Chowdhury and Nakhla, 2022). While conventional carriers have demonstrated successes, they often fall short in meeting the combined criteria of low cost, environmental friendliness, durability, and consistent performance.

To address these limitations, current research has shifted toward advanced materials, including composite carriers, nanostructured materials, MOFs, and carbon-based materials derived from natural biomass. These novel carriers exhibit superior adsorption capacities, enhanced durability, and better reusability, making them highly promising for sustainable pollutant removal applications (Datta et al., 2013).

## 3.1 Natural carrier materials

Natural carrier materials are widely employed for microbial immobilization due to their eco-friendliness, biocompatibility, renewability, and low cost. Common examples include **alginate**, **chitosan**, **cellulose**, and **biochar**. These materials can retain microbial viability and activity while providing structural support in various wastewater treatment processes.

**Alginate**, a linear copolymer derived from brown algae, is extensively used to immobilize microbial cells due to its gel-forming ability in the presence of divalent cations like $Ca^{2+}$. The calcium alginate matrix enables gentle cell entrapment, maintains high porosity, and allows the diffusion of nutrients and metabolites, making it ideal for both batch and continuous bioreactors (Diogo Januário et al., 2020).

**Chitosan**, derived from chitin, exhibits excellent film-forming, antimicrobial, and chelating properties. It is often used in bead or membrane form for entrapping microbial cells, especially nitrifiers and denitrifiers. Its amino groups allow for chemical modifications, enhancing microbial affinity and structural stability under fluctuating pH and salinity conditions (Eom et al., 2022).

**Cellulose-based carriers**, especially microcrystalline cellulose and cellulose acetate, offer high surface area, water retention capability, and good mechanical properties. These characteristics favor microbial adhesion and biofilm formation (Fu et al., 2022). In addition, cellulose can be functionalized with various chemical groups to improve immobilization efficiency and resistance to biodegradation.

**Biochar**, a porous, carbon-rich material obtained via pyrolysis of agricultural waste, has gained significant attention as a low-cost immobilization substrate. Its high surface area and pore volume support microbial colonization, while its surface functional groups facilitate microbial adhesion. Moreover, biochar can simultaneously adsorb pollutants, providing a dual remediation mechanism (Guo et al., 2022).

Despite their advantages, natural materials often face challenges related to mechanical strength, biodegradability, and performance reproducibility. Researchers are addressing these limitations through chemical modifications and composite blending with synthetic materials to enhance longevity and microbial retention capacity.

## 3.2 Nanocarrier materials

Nanocarrier materials provide an advanced platform for microbial immobilization by offering an extremely high surface area, tunable surface chemistry, and improved interaction with microbial cells. These materials can increase immobilization density, accelerate mass transfer, and improve overall process stability (Hao et al., 2022).

Magnetic nanoparticles, especially $Fe_3O_4$ (magnetite), are among the most widely studied nanocarriers due to their easy separation via magnetic fields and compatibility with microbial systems. When functionalized with amino or carboxyl groups, these nanoparticles show enhanced microbial adhesion and biofilm stability (Hou et al., 2024). Immobilized cells on magnetic nanocarriers also demonstrate increased pollutant degradation rates in fluidized bed reactors and biofilters.

Carbon-based nanomaterials, such as graphene oxide and carbon nanotubes, provide unique properties like high conductivity, antimicrobial resistance, and a large number of oxygen-containing groups. These properties facilitate the electrochemical stimulation of immobilized microbes, especially in microbial fuel cells (MFCs) and advanced oxidation systems (Hsia et al., 2008).

Furthermore, nanosilica and polymeric nanospheres have also been used for encapsulating bacteria, especially for the removal of nitrogenous waste and pharmaceuticals from wastewater (Somu et al., 2022). These nanocarriers can be engineered to release or retain specific substrates, enhancing microbial selectivity and robustness.

However, potential issues such as toxicity, aggregation, difficult recovery, and cost still limit their full-scale application. Eco-friendly synthesis and surface modifications are being explored to overcome these limitations (Hou et al., 2024). The schematic representation of the novel immobilized microbial carrier design is depicted in Fig. 14.2.

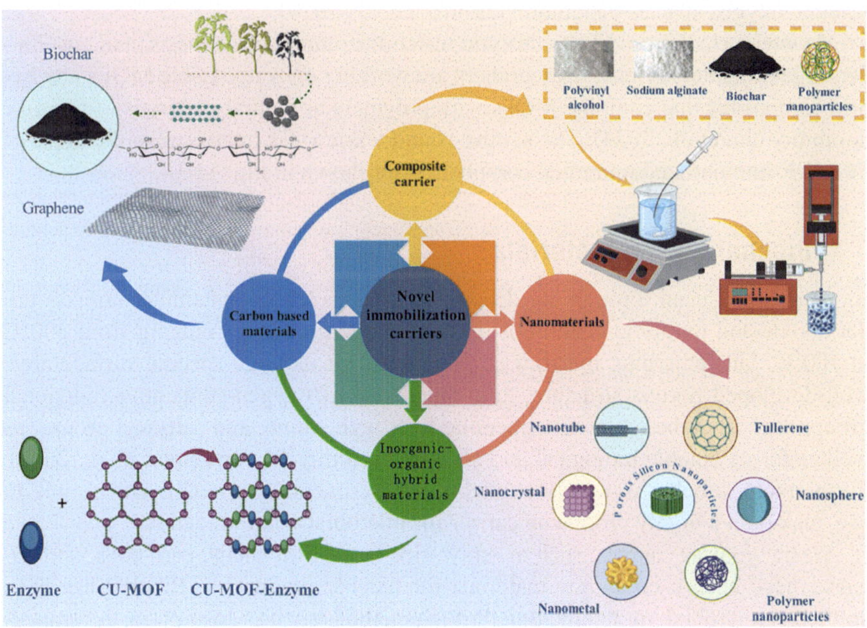


**FIG. 14.2**

Design of immobilized microbial carrier. Schematic representation of a novel immobilized microbial carrier design.

*-From Hou, L., Hu, K., Huang, F., Pan, Z., Jia, X., Liu, W., Yao, X., Yang, Z., Tang, P., Jun, L., 2024. Advances in immobilized microbial technology and its application to wastewater treatment: a review. Bioresour. Technol. 09608524. 413, 131518. https://doi.org/10.1016/j.biortech.2024.131518.*

## 3.3 Metal–organic frameworks

MOFs are crystalline, porous materials made from metal ions coordinated to organic linkers. They possess extremely high surface areas, customizable pore sizes, and unique chemical functionalities, making them ideal candidates for next-generation microbial immobilization.

MOFs can immobilize microorganisms through adsorption, encapsulation, or in situ growth. For example, ZIF-8, a zinc-based MOF, has been successfully used to encapsulate *Escherichia coli* and *Bacillus subtilis*, offering enhanced protection against thermal and chemical stresses while maintaining metabolic activity (Hou et al., 2024). In wastewater treatment, MOFs have been used to immobilize nitrifiers, denitrifiers, and sulfate-reducing bacteria. Their porosity facilitates nutrient diffusion, while the metal ions may serve catalytic or cofactor roles. A study using a Cu-MOF for immobilizing Pseudomonas species showed improved nitrate removal efficiency and better tolerance to salinity (Isaka et al., 2007).

Moreover, MOFs can adsorb copollutants such as heavy metals or dyes, providing a synergistic approach. Some MOFs even exhibit photocatalytic or redox activity, further enhancing pollutant degradation by activating microbial metabolism or reactive oxygen species generation (Jasińska et al., 2019).

However, challenges like hydrolytic instability, complex synthesis, and cost hinder large-scale deployment. Researchers are working on water-stable MOFs and hybrid composites that combine MOFs with polymers or natural materials to improve durability (Jin et al., 2024). The synthesis and wastewater treatment applications of Fe-MOF-immobilized microbial composites are shown in Fig. 14.3.

## 3.4 Biochar carrier materials

Biochar is produced through the decomposition of carbon-containing biomass under anoxic and pyrolytic conditions, typically at temperatures ranging from 300°C to 700°C. This emerging material is recognized for its large specific surface area, well-developed porous structure, and abundance of oxygen-containing functional groups such as carbonyl (C=O), phenolic hydroxyl, amino, and carboxyl groups, as well as for its notable biological and chemical stability (Kampschreur et al., 2009). Furthermore, biochar is widely available, environmentally friendly, and simple to use, making it a highly promising carrier for microbial immobilization.

Waste biomass sources, such as straw, rice husk, fruit shells, corn cobs, and bagasse, have been used as raw materials for biochar production. Biochar has been applied in microbial immobilization through methods such as adsorption, gel embedding, and combined adsorption–embedding techniques (Kou et al., 2021). The immobilization of microorganisms on biochar surfaces occurs through various mechanisms, including physical adsorption, charge interactions, and chemical bonding.

Physical adsorption is facilitated by the graded porosity and extensive surface area of biochar, allowing for the retention of microbial cells on both the external and internal surfaces of the carrier. This adsorption is driven by weak intermolecular

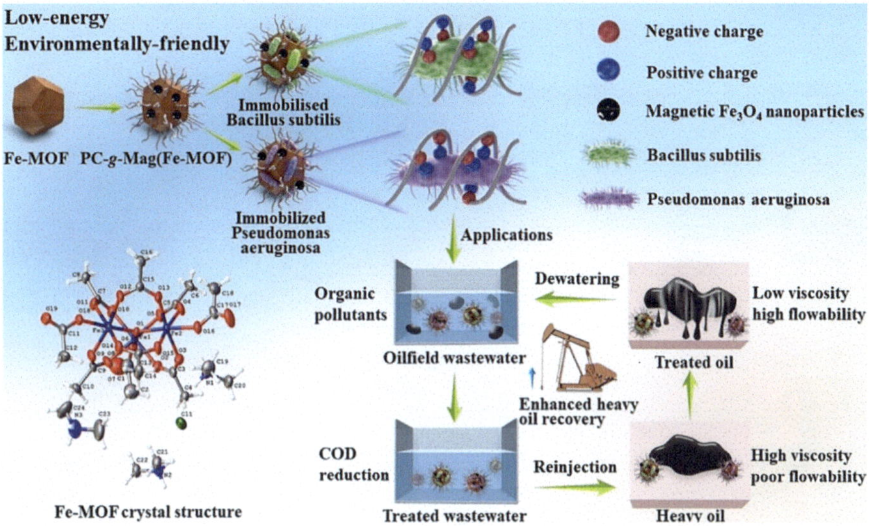


**FIG. 14.3**

Synthesis and wastewater treatment applications of Fe-MOF-immobilized microbial composites.

*-From Huang, L., Ye, J., Xiang, H., Jiang, J., Wang, Y., Li, Y., 2020. Enhanced nitrogen removal from low C/N wastewater using biodegradable and inert carriers: performance and microbial shift. Bioresour. Technol. 300. https://doi.org/10.1016/j.biortech.2019.122658.*

forces, such as van der Waals and electrostatic interactions, which support microbial attachment and promote interactions between the carrier surface and surrounding contaminants. Once immobilized on biochar surfaces, microorganisms tend to form biofilms, significantly enhancing their survivability and pollutant degradation (Li et al., 2022). Fig. 14.4 depicts the biochar synthesis and microbial encapsulation from natural biomass.

A substantial number of aromatic compounds and surface-bound functional groups, including phenolic hydroxyl, amino, and carboxyl moieties, are typically present in biochar. These features enable the simultaneous adsorption of both cationic and anionic species and also facilitate covalent interactions with microbial cell wall components, ensuring stable microbial attachment and retention. The effectiveness of biochar as a microbial immobilization matrix in wastewater treatment processes has been attributed to a dual mechanism involving adsorption phenomena—such as $\pi$–$\pi$ interactions, Lewis acid–base bonding, electrostatic forces, and pore-filling—and microbial degradation activity (Fig. 14.5) (Hou et al., 2024).

For instance, modified walnut shell biochar used to immobilize microbial strain XL-2 achieved an $NH_4^+$ removal rate more than three times greater than the free strain, primarily due to enhanced colonization facilitated by irregular pore structures and polar surface groups. In addition, biochar-based systems have demonstrated

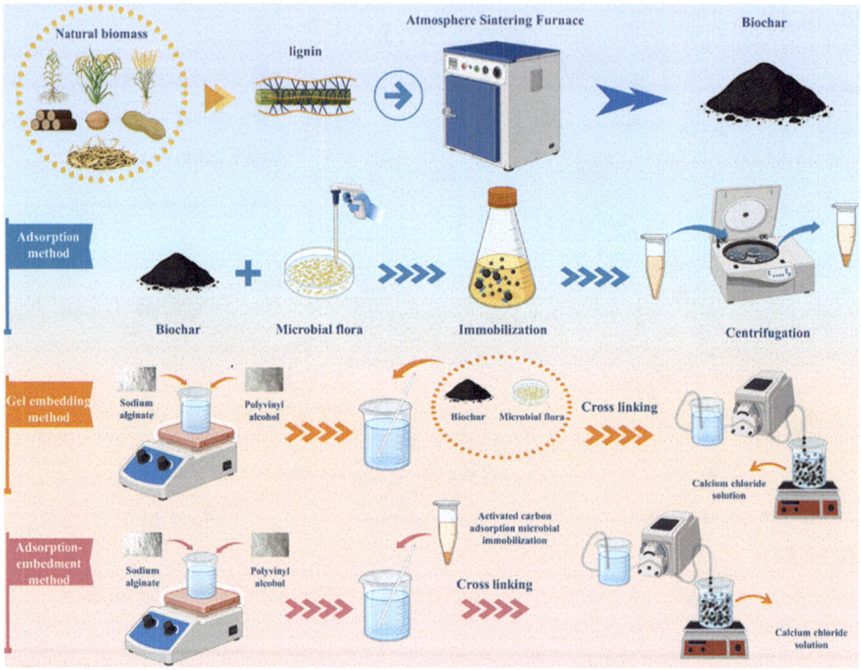


**FIG. 14.4**

Biochar synthesis. Biochar synthesis and microbial encapsulation from natural biomass.

*-From Hou, L., Hu, K., Huang, F., Pan, Z., Jia, X., Liu, W., Yao, X., Yang, Z., Tang, P., Jun, L., 2024. Advances in immobilized microbial technology and its application to wastewater treatment: a review. Bioresour. Technol. 09608524. 413, 131518. https://doi.org/10.1016/j.biortech.2024.131518.*

notable effectiveness in sustaining denitrifying microbial communities under low-temperature conditions. In one such case, using biochar beads as carrier media led to significantly elevated enzymatic activities and electron transport efficiency, along with a marked reduction in nitrous oxide emissions at $6 \pm 2°C$ (Zhou et al., 2022).

Biochar has also proven effective in the bioremediation of complex organics and heavy metals. For example, *Acinetobacter venetianus* immobilized on biochar removed diesel contaminants with up to 94% efficiency, and *Bacillus* SDB4 immobilized on a similar matrix removed over 97% of sulfamethoxazole and 96% of zinc ions. These improvements were largely attributed to synergistic interactions involving biosorption, biotransformation, and ion exchange, all supported by the mineral-rich and functionalized surface of the biochar (Zhou et al., 2010).

Moreover, the type of biochar used can significantly influence treatment performance. Bamboo-based carriers, for example, retained higher nonylphenol removal efficiencies across multiple reuse cycles than their wood-based counterparts or free microbial systems. The presence of biochar was also found to stabilize microbial

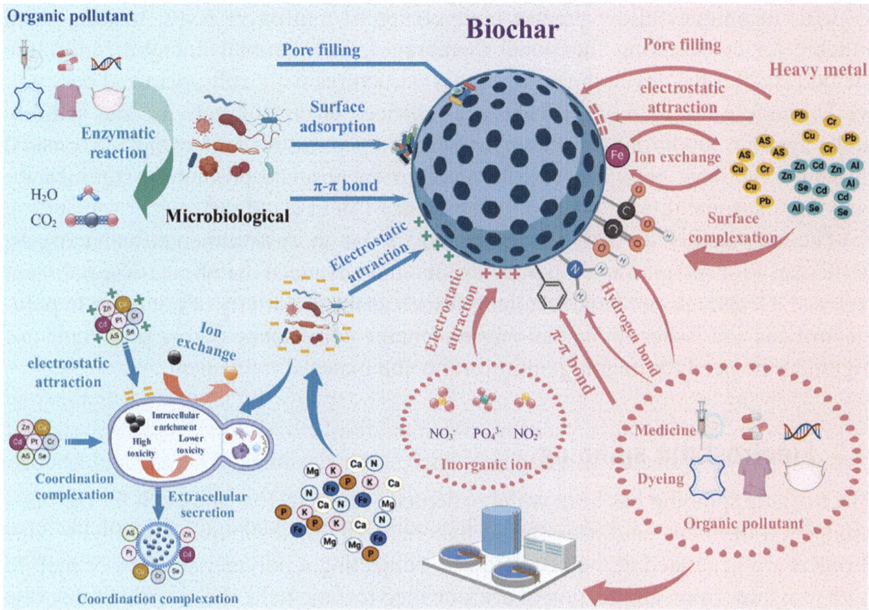


**FIG. 14.5**

Pollutant removal mechanisms of biochar-immobilized microbial systems.

*-From Hou, L., Hu, K., Huang, F., Pan, Z., Jia, X., Liu, W., Yao, X., Yang, Z., Tang, P., Jun, L., 2024. Advances in immobilized microbial technology and its application to wastewater treatment: a review. Bioresour. Technol. 09608524. 413, 131518. https://doi.org/10.1016/j.biortech.2024.131518.*

communities and encourage sustained growth. Its high surface area, strong adsorption capabilities, and ability to support electron transfer processes contribute to enhanced pollutant removal and improved microbial resilience.

Typically produced from natural biomass sources such as wood, straw, or fruit shells, biochar varies in performance based on its feedstock and pyrolysis conditions. Hence, optimizing thermal treatment parameters is critical to tailoring its surface characteristics for high-efficiency microbial immobilization and pollutant removal (Zhuo et al., 2024).

# 4 Novel immobilized microbial technology

## 4.1 Sequential film deposition via LBL for microbial immobilization

The LBL assembly technique involves the alternating deposition of molecular layers onto microbial or material surfaces, governed by forces such as electrostatic attraction, hydrogen bonding, and covalent interactions. This structured build-up leads to the formation of supramolecular assemblies with controlled architecture, functional adaptability, and reliable performance (Yu et al., 2019).

LBL techniques allow precise engineering of multilayer films, making them suitable for constructing functional membranes. In microbial immobilization, the method enables the development of support structures with well-organized pore networks, promoting microbial adherence, enhanced enzymatic activity, and stability under variable conditions. For example, zein and pectin-based microcapsules created through LBL have been successfully used to encapsulate probiotics, significantly boosting their viability and metabolic efficiency (Yu et al., 2020).

The adaptability of LBL has expanded its use in environmental technologies, particularly for integrating biological agents into advanced membranes for pollutant removal. Despite its potential, challenges such as the sensitivity of microbes to polyelectrolytes and issues in maintaining membrane consistency during prolonged use highlight the need for ongoing optimization and material refinement.

### 4.2 Electrostatic spinning

Electrostatic spinning has been widely adopted as an effective approach for the mass production of micro- and nanofibrous materials. In this technique, fine Taylor cone droplets are generated at the needle tip by subjecting a polymer solution or melt to high pressure. Upon application of a strong electrostatic field, the electric force overcomes the surface tension, causing the formation of a fine jet from the droplet. As this jet traverses the field, it undergoes elongation and thinning, eventually depositing onto a grounded collector in the form of a fibrous mat.

The resulting electrospun membranes are characterized by a high specific surface area, adjustable fiber morphology and size, controllable surface wettability, excellent permeability, and ease of production. Due to these features, they have been extensively applied in areas such as biomedical devices, energy storage, environmental engineering, filtration systems, personal protection, and food technology.

In recent years, the immobilization of bioactive compounds through electrospinning has gained momentum. Both natural polymers (e.g., chitosan, gelatin, zein, pectin) and synthetic polymers (e.g., polycaprolactone, PVA, polyethylene oxide, Polylactic Acid (PLA)) have been employed to fabricate suitable fibrous matrices. For example, chitosan/PVA-based nanofibers have been utilized to encapsulate antioxidant peptides, achieving encapsulation efficiencies above 94% while maintaining their functional activity after immobilization (Yu et al., 2023).

Electrospun composites consisting of polycaprolactone and silica nanoparticles have been applied for the immobilization of *Ochrobactrum anthropi*, an AD strain. The embedded bacteria achieved a total nitrogen removal rate of 5.59 mg/L/h, significantly outperforming their suspended counterparts. Enhanced microbial activity and stability over five reuse cycles were also demonstrated.

In addition, polyacrylonitrile-based fibrous carriers have been used to immobilize aerobic denitrifiers for nitrate removal from high-nitrogen wastewater. A carbon-to-nitrogen ratio of seven or higher enabled total nitrogen removal rates exceeding 90%.

High microbial retention and a shift in microbial community structure toward efficient denitrifying taxa were observed.

These performance gains have been attributed to the finely tunable structural attributes of the electrospun fibers, such as diameter, surface roughness, wettability, porosity, and mechanical integrity. Through precise manipulation of spinning parameters and posttreatment methods, microbial adherence and proliferation on the fibrous matrix can be optimized, enabling long-term and stable nitrogen removal performance in wastewater treatment applications.

However, despite their versatility and effectiveness, certain technical limitations restrict the large-scale application of electrostatic spinning. These include low throughput, inconsistent fiber uniformity, and challenges associated with integrating multifunctionality into a single spinning process. Overcoming these barriers remains essential for the broader deployment of electrospun fibrous systems in environmental biotechnologies (Mohamad et al., 2015).

## 5 Biological nitrogen removal from wastewater by immobilization technology

### 5.1 Gel bed embedding of anammox bacteria for nitrogen removal

The anammox process, a pioneering method for autotrophic nitrogen removal, utilizes slow-growing anaerobic bacteria, which rely on $CO_2$ as their carbon source. Despite their promising potential, the slow growth rate of anammox bacteria poses significant challenges for practical applications in wastewater treatment systems. Immobilization technology has emerged as an effective strategy to overcome these limitations. By embedding the bacteria in a solid matrix or gel bed, it becomes possible to increase bacterial retention, enhance their tolerance to environmental stresses (e.g., temperature shifts, oxygen levels, and pH changes), and improve the overall stability of the system.

Several studies have shown that immobilized bacteria exhibit superior solid–liquid separation, stress resistance, and longevity when compared to free-floating cells. In fact, immobilization has been considered a more viable alternative to conventional biofilm and granulation techniques. Notably, Candidatus Brocadia sinica (*Ca. B. sinica*) and Candidatus Jettenia caeni (*Ca. J. caeni*) embedded in PVA/SA gel beads showed impressive growth rates under high nitrogen load conditions, with doubling times of 2.1 and 3.9 days, respectively. This indicates that immobilization can support rapid biomass growth in the presence of high-load nitrogen, making it an effective solution for improving nitrogen removal efficiency.

The choice of carrier materials plays a significant role in optimizing the performance of immobilized anammox bacteria. PVA/SA composites are commonly used, as they offer an ideal balance of porosity, stability, and nutrient diffusion. Formulations with 12% PVA and 2% SA are particularly effective in facilitating bacterial adhesion and enhancing denitrification performance. Studies have shown that the use of these composite materials can lead to a substantial increase in the abundance of *Planctomycetes*, the group of bacteria responsible for the anammox process. This increase in abundance—from 11.98% to 54.77%—directly correlates with improved nitrogen removal efficiency.

Immobilization also significantly improves the efficiency of anammox systems under challenging conditions. For example, activated sludge immobilized in PVA–SA gel beads achieved a remarkable 88.9% total nitrogen removal rate after 6 months. Various types of sludge (including anoxic, municipal, and food wastewater sludge) have been successfully used as carriers for anammox bacteria, achieving reactor startup times ranging from 108 to 157 days, depending on the specific sludge type (Najim et al., 2024).

However, one of the primary challenges facing the widespread implementation of immobilization technology is temperature sensitivity. The performance of immobilized anammox systems is known to decline significantly below temperatures of 10°C. To address this issue, several studies have focused on improving the resilience of these systems at low temperatures. One such study demonstrated that a partially nitrifying–anammox system, where *Nitrospira inopinata* and anammox biomass were coembedded with carbon black, was able to achieve complete nitrogen removal at 10°C. Despite these advances, full-scale studies on low-temperature anammox systems are still limited, and further research is needed to determine the practical feasibility of this approach.

Moreover, there are still challenges related to the shear stress imposed by agitation and the high costs of producing immobilized carriers. Future research should focus on developing cost-effective, durable materials to improve the scalability and sustainability of immobilization technologies for wastewater treatment (Ni et al., 2023).

## 5.2 Immobilization of AD bacteria for nitrogen removal

AD bacteria are another promising class of microorganisms for nitrogen removal in wastewater. Unlike anammox bacteria, AD bacteria can simultaneously perform nitrification and denitrification in the presence of oxygen, offering advantages such as greater pH stability and simplified operational requirements. However, these bacteria are typically present in low abundance in activated sludge systems, which limits their effectiveness. To overcome this limitation, immobilization has been applied to enhance bacterial concentration and improve performance under varied operational conditions.

The immobilization of AD bacteria has been shown to preserve their functionality and increase their resilience against environmental stresses, such as fluctuating oxygen levels. One study demonstrated that *Pseudomonas* sp. HS-N62, immobilized in PVA/SA with activated carbon, achieved 61%–76% nitrate removal across a range of temperatures. In addition, composite mycelial pellets (CMP) have been used as carriers to provide high surface areas for microbial adhesion. Immobilization of

*Pseudomonas* sp. Y1 in CMP led to a significant increase in nitrogen removal, with up to 98.07% removal of ammonium ($NH_4^+$) and 92.57% removal of phenol over a 150-day period.

Recent advancements have also focused on utilizing mixed AD bacterial consortia to improve system performance. These consortia benefit from metabolic synergy, which enhances nitrogen removal efficiency and overall stability. In particular, biochar-immobilized mixed AD bacteria have shown impressive results, with nitrate removal efficiencies of 39%–53% under low-temperature conditions (10°C–25°C). These systems are especially useful in regions with cooler climates, where conventional denitrification processes may not be effective (Patil and Rathod, 2022).

In addition, immobilized AD bacteria have demonstrated reduced nitrous oxide ($N_2O$) emissions, which are a significant concern in traditional denitrification processes due to their potent greenhouse effect. For example, *Pseudomonas stutzeri* PCN-1, when immobilized, reduced $N_2O$ emissions by 36.77%, providing a more sustainable solution for nitrogen removal. However, challenges such as competition from heterotrophic bacteria and operational variations (e.g., dissolved oxygen and pH) must be addressed to further enhance system efficiency. Optimized reactor designs and process control strategies will be essential for maximizing the potential of AD bacteria immobilization in wastewater treatment (Dhara et al., 2024). The characterization of mycelial pellets and their use as carriers for microbial immobilization are represented in Fig. 14.6.

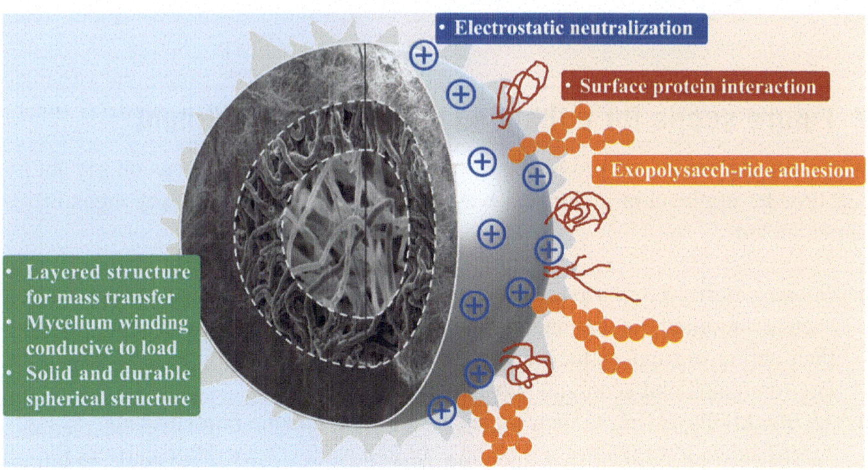


**FIG. 14.6**

Mycelial pellet and its characterization as a carrier. Characterization of mycelial pellet carriers for microbial immobilization.

*-From Hou, L., Hu, K., Huang, F., Pan, Z., Jia, X., Liu, W., Yao, X., Yang, Z., Tang, P., Jun, L., 2024. Advances in immobilized microbial technology and its application to wastewater treatment: a review. Bioresour. Technol. 09608524. 413, 131518. https://doi.org/10.1016/j.biortech.2024.131518.*

## 6 Immobilized microbial technology for enhanced pollutant removal in refractory wastewater

Immobilized microbial technology has emerged as a viable method for treating refractory pollutants such as heavy metals, phenols, and persistent organics. Unlike traditional free-cell systems, immobilized microorganisms demonstrate enhanced pollutant tolerance, higher degradation rates, and better operational stability due to their encapsulation within protective matrices (Sharma et al., 2023).

In recent studies, *Bacillus licheniformis* immobilized on PVA–SA composite beads achieved a $Pb^{2+}$ adsorption capacity of 113.84 mg/g and demonstrated 98% removal efficiency for 200 mg/L lead solutions, retaining over 90% efficiency after five cycles (Yu et al., 2023). In another instance, *Pseudomonas*, *Proteiniphilum*, and *Corynebacterium* immobilized on lignite-activated coke removed 91.7% chemical xygen emand (COD), 88.1% total organic carbon (TOC), and 72.1% $NH_4^+$ from coking wastewater within 12 h, aided by microbial synergism and mass transfer efficiency (Samanta et al., 2023a, 2023b).

Phenol degradation has also seen substantial improvements. When *Brevibacterium* sp. and *Stenotrophomonas acidaminiphila* were immobilized on natural carriers like areca husk and luffa sponge, they achieved 100% phenol removal (1000 mg/L) in under 30 h (Samanta et al., 2024). Furthermore, laccase from *Myrothecium roridum*, immobilized on rapeseed cake, effectively degraded dyes such as Acid Orange 7 and Amaranth, achieving over 70% removal in 24 h and maintaining 50% efficiency after five reuse cycles (Niladri Shekhar Samanta et al., 2023a).

## 7 Future scope for enhanced wastewater treatment

While current results are promising, future advancements must focus on scalability and broader application under real wastewater conditions. A few key areas of research include:

a. *Biomass-derived and waste-based carriers*: Low-cost, sustainable materials, such as modified biochar and agricultural residues, are being evaluated for their ability to support microbial growth and pollutant adsorption, offering both environmental and economic advantages.
b. *MOFs*: MOFs, owing to their large surface area, tunable porosity, and functionalization capacity, are being explored as next-generation carriers for microbial immobilization. Their compatibility with living cells and durability under wastewater conditions remain active research areas (Niladri Shekhar Samanta et al., 2023a; Samanta and Purkait, 2025).
c. *Micro-/nanofabrication techniques.* LBL self-assembly and electrospinning have shown promise in tailoring microenvironments for microbial attachment. LBL enables the sequential deposition of polyelectrolyte layers, enhancing

microbial encapsulation and retention, while electrospinning creates nanofibrous scaffolds that facilitate biofilm formation and nutrient exchange.

**d.** *Synthetic and engineered consortia*: The design of engineered microbial consortia, including aerobic denitrifiers and anammox bacteria, offers a means to target multiple pollutants simultaneously. Omics tools (metagenomics, transcriptomics) are increasingly being used to understand the interactions that drive enhanced degradation.

**e.** *Real wastewater validation and pilot studies*: Most findings are based on synthetic or spiked samples. Therefore adapting these technologies to treat complex, real-world wastewater—often containing salts, inhibitors, or fluctuating loads—is essential. Pilot-scale studies will bridge this gap and guide design criteria for full-scale systems (Das et al., 2024; Samanta et al., 2023c; Samanta et al., 2021).

## 8 Conclusions

Immobilized microbial technology holds great potential for transforming wastewater treatment, especially in handling refractory contaminants. The integration of sustainable carriers, nanofabrication techniques, and microbial consortia enables high efficiency, reusability, and robustness. Case studies have confirmed that such systems can achieve over 90% removal of metals and organics under optimal conditions. Despite these successes, challenges like material degradation, microbial viability loss, and reduced performance in real effluents must be addressed. Advanced materials like MOFs and techniques like electrospinning and LBL are opening new avenues for developing next-generation systems. To fully realize their industrial application, future research must emphasize durability, cost-effectiveness, and validation through pilot-scale operations.

## AI disclosure

During the preparation of this work, the author(s) used ChatGPT to support content organization, phrasing suggestions, and language refinement. To maintain originality, AI assistance was utilized for rephrasing content, reducing textual similarity, polishing the scientific writing style, and suggesting alternative headings and paragraph restructuring. After using this tool/service, the author(s) reviewed and edited the content as needed and take(s) full responsibility for the content of the publication.

## References

Ahmad, M., Liu, S., Mahmood, N., Mahmood, A., Ali, M., Zheng, M., Ni, J., 2017. Synergic adsorption-biodegradation by an advanced carrier for enhanced removal of high-strength nitrogen and refractory organics. ACS Appl. Mater. Interfaces. 19448252. 9 (15), 13188–13200. Available from: http://pubs.acs.org/journal/aamick. doi: 10.1021/acsami.7b01251.

Ansari, S.A., Husain, Q., 2012. Potential applications of enzymes immobilized on/in nano materials: a review. Biotechnol. Adv. 07349750. 30 (3), 512–523. doi:10.1016/j.biotechadv.2011.09.005.

Blachowicz, T., Grzybowski, J., Ehrmann, A., 2022. Influence of agglomerations on magnetic properties of polymer matrices filled with magnetic nanoparticles. Mater. Today Proc. 22147853. 67, 792–796. https://www.sciencedirect.com/science/journal/22147853. doi:10.1016/j.matpr.2022.07.362.

Boland, E., Wnek, G., Simpson, D., Pawlowski, K., Bowlin, G., 2001. Tailoring tissue engineering scaffolds using electrostatic processing techniques: a study of poly(glycolic acid) electrospinning. J. Macromol. Sci. Part A. 1060-1325. 38 (12), 1231–1243. doi:10.1081/MA-100108380.

Bouabidi, Z.B., El-Naas, M.H., Zhang, Z., 2019. Immobilization of microbial cells for the biotreatment of wastewater: a review. Environ. Chem. Lett. 16103661. 17 (1), 241–257. Available from: https://www.springer.com/journal/10311. doi:10.1007/s10311-018-0795-7.

Cheng, B., Sakurai, J., Hata, S., Oka, C., 2024. Insights into thermally-induced disruption of magnetic-nanoparticle agglomerates. Alexandria Eng. J. 11100168. 96, 72–81. Available from: https://www.sciencedirect.com/science/journal/11100168. doi:10.1016/j.aej.2024.03.101.

Cho, K., Choi, M., Lee, S., Bae, H., 2018. Negligible seeding source effect on the final ANAMMOX community under steady and high nitrogen loading rate after enrichment using poly(vinyl alcohol) gel carriers. Chemosphere. 18791298. 208, 21–30. Available from: www.elsevier.com/locate/chemosphere. doi:10.1016/j.chemosphere.2018.05.155.

Chowdhury, M.M.I., Nakhla, G., 2022. Enhanced mainstream nitrogen removal from synthetic wastewater using gel-immobilized anammox in fluidized bed bioreactors: process performance and disintegration mechanisms. Sci. Total Environ. 00489697. 811, 151373. doi:10.1016/j.scitotenv.2021.151373.

Das, P.P., Dhara, S., Samanta, N.S., Purkait, M.K., 2024. Ozonation process for the treatment of wastewater contaminants. ACS Symp. Ser. 1475, 93–109. Available from: https://pubs.acs.org/series/symposium. doi:10.1021/bk-2024-1475.ch005.

Datta, S., Christena, L.R., Rajaram, Y.R.S., 2013. Enzyme immobilization: an overview on techniques and support materials. 3 Biotech. 21905738. 3 (1), 1–9. Available from: https://www.springer.com/journal/13205. doi:10.1007/s13205-012-0071-7.

Dhara, S., Samanta, N.S., Das, P.P., Purkait, M.K., 2024. Biological nitrogen removal and recovery from wastewater. In: Development in Waste Water Treatment Research and Processes: Role of Environmental Microbiology in Industrial Wastewater Research. Elsevier, India, pp. 159–179. Available from: https://www.sciencedirect.com/book/9780443136092. doi:10.1016/B978-0-443-13609-2.00019-7.

Diogo Januário, E.F., de Camargo Lima Beluci, N., Vidovix, T.B., Vieira, M.F., Bergamasco, R., Vieira, A.M.S., 2020. Functionalization of membrane surface by layer-by-layer self-assembly method for dyes removal. Process Saf. Environ. Prot. 09575820. 134, 140–148. Available from: http://www.elsevier.com/wps/find/journaldescription.cws_home/713889/description#description. doi:10.1016/j.psep.2019.11.030.

Eom, J., Kwak, Y., Nam, C., 2022. Electrospinning fabrication of magnetic nanoparticles-embedded polycaprolactone (PCL) sorbent with enhanced sorption capacity and recovery speed for spilled oil removal. Chemosphere. 00456535. 303 (Pt 3), 135063. doi:10.1016/j.chemosphere.2022.135063.

Fu, Y., Liao, Y., Li, P., Li, H., Jiang, S., Huang, H., Sun, W., Tianyan, L., Hai, Y., Kangkang, L., Hao, L., Jia, B., Tianyi, M., 2022. Layer structured materials for ambient nitrogen fixation. Coord. Chem. Rev. 00108545. 460, 214468. doi:10.1016/j.ccr.2022.214468.

Guo, F., Xu, F., Cai, R., Li, D., Xu, Q., Yang, X., Wu, Z., Wang, Y., He, Q., Ao, L., Vymazal, J., Chen, Y., 2022. Enhancement of denitrification in biofilters by immobilized biochar under low-temperature stress. Bioresour. Technol. 09608524. 347, 126664. doi:10.1016/j.biortech.2021.126664.

Hao, Z.L., Ali, A., Ren, Y., Su, J.F., Wang, Z., 2022. A mechanistic review on aerobic denitrification for nitrogen removal in water treatment. Sci. Total Environ. 18791026. 847. Available from: www.elsevier.com/locate/scitotenv. doi:10.1016/j.scitotenv.2022.157452.

Hou, L., Hu, K., Huang, F., Pan, Z., Jia, X., Liu, W., Yao, X., Yang, Z., Tang, P., Jun, L., 2024. Advances in immobilized microbial technology and its application to wastewater treatment: a review. Bioresour. Technol. 09608524. 413, 131518. doi:10.1016/j.biortech.2024.131518.

Hsia, T.H., Feng, Y.J., Ho, C.M., Chou, W.P., Tseng, S.K., 2008. PVA-alginate immobilized cells for anaerobic ammonium oxidation (anammox) process. J. Ind. Microbiol. Biotechnol. 14765535. 35 (7), 721–727. doi:10.1007/s10295-008-0336-7.

Isaka, K., Date, Y., Sumino, T., Tsuneda, S., 2007. Ammonium removal performance of anaerobic ammonium-oxidizing bacteria immobilized in polyethylene glycol gel carrier: anammox bacteria immobilized in gel carrier. Appl. Microbiol. Biotechnol. 01757598. 76 (6), 1457–1465. doi:10.1007/s00253-007-1106-6.

Jasińska, A., Góralczyk-Bińkowska, A., Soboń, A., Długoński, J., 2019. Lignocellulose resources for the Myrothecium roridum laccase production and their integrated application for dyes removal. Int. J. Environ. Sci. Technol. 1735-1472. 16 (8), 4811–4822. doi:10.1007/s13762-019-02290-x.

Jin, Y., Liu, D., Xiong, W., Wu, Z., Xiao, G., Wang, S., Su, H., 2024. Enhancing nitrogen removal performance using immobilized aerobic denitrifying bacteria by modified polyvinyl alcohol/sodium alginate (PVA/SA). Chemosphere. 00456535. 357, 141954. doi:10.1016/j.chemosphere.2024.141954.

Kampschreur, M.J., Temmink, H., Kleerebezem, R., Jetten, M.S.M., van Loosdrecht, M.C.M., 2009. Nitrous oxide emission during wastewater treatment. Water Res. 00431354. 43 (17), 4093–4103. Available from: www.elsevier.com/locate/watres. doi:10.1016/j.watres.2009.03.001.

Kou, L., Huang, T., Zhang, H., Wen, G., Li, N., Wang, C., Lu, L., 2021. Mix-cultured aerobic denitrifying bacterial communities reduce nitrate: novel insights in micro-polluted water treatment at lower temperature. Sci. Total Environ. 00489697. 796, 148910. doi:10.1016/j.scitotenv.2021.148910.

Li, L., Liang, T., Zhao, M., Lv, Y., Song, Z., Sheng, T., Fang, M., 2022. A review on mycelial pellets as biological carriers: wastewater treatment and recovery for resource and energy. Bioresour. Technol. 09608524. 355, 127200. doi:10.1016/j.biortech.2022.127200.

Mohamad, N.R., Marzuki, N.H.C., Buang, N.A., Huyop, F., Wahab, R.A., 2015. An overview of technologies for immobilization of enzymes and surface analysis techniques for immobilized enzymes. Biotechnol. Biotechnol. Equip. 13102818. 29 (2), 205–220. Available from: http://www.tandfonline.com/doi/pdf/10.1080/13102818.2015.1008192. doi:10.1080/13102818.2015.1008192.

Najim, A.A., Radeef, A.Y., al-Doori, I., Jabbar, Z.H., 2024. Immobilization: the promising technique to protect and increase the efficiency of microorganisms to remove contaminants. J. Chem. Technol. Biotechnol. 10974660. 99 (8), 1707–1733. Available from: http://onlinelibrary.wiley.com/journal/10.1002/(ISSN)1097-4660. doi:10.1002/jctb.7638.

Ni, S., Li, C., Zhang, W., Niu, D., Zhi, J., Wang, C., Jiang, X., Ren, J., 2023. Immobilization of purified enzyme EreB in metalorganic framework (MOF) mesopores for erythromycin degradation. Environ. Res. 00139351. 237 (Pt 2), 117023. doi:10.1016/j.envres.2023.117023.

Patil, S.S., Rathod, V.K., 2022. Combined effect of enzyme co-immobilized magnetic nanoparticles (MNPs) and ultrasound for effective extraction and purification of curcuminoids from Curcuma longa. Ind. Crops Prod. 09266690. 177. Available from: www.elsevier.com/inca/publications/store/5/2/2/8/2/5. doi:10.1016/j.indcrop.2021.114385.

Samanta, N.S., Mondal, P., Purkait, M.K., 2023a. Waste-Based Zeolites and Their Advanced Composites for Wastewater and Environmental Remediation Applications. Springer Science and Business Media LLC, pp. 451–490. doi:10.1007/978-3-031-42731-2_14.

Samanta, N.S., Mondal, P., Purkait, M.K., 2023b. Nanofiltration technique for the treatment of industrial wastewater. In: Advanced Application of Nanotechnology to Industrial Wastewater. Springer Nature, India, pp. 165–190. Available from: https://link.springer.com/book/10.1007/978-981-99-3292-4. doi:10.1007/978-981-99-3292-4_9.

Samanta, N.S., Mondal, P., Dhara, S., Bora, U., Purkait, M.K., 2023c. Fabrication of LD-slag derived zeolite Y coated polysulfone (PSf) membrane for decontamination of groundwater. Chem. Eng. J. 478. doi:10.1016/j.cej.2023.147330

Samanta, N.S., Das, P.P., Purkait, M.K., 2024. Performance of anammox process for wastewater treatment. In: Anammox Process: Technological Advancement and Application in Industrial Wastewater Treatment Plant. Elsevier, India, pp. 25–44. Available from: https://www.sciencedirect.com/book/9780443192098. doi:10.1016/B978-0-443-19209-8.00008-9.

Samanta, N.S., Banerjee, S., Mondal, P., Anweshan, Bora, U., Purkait, M.K., 2021. Preparation and characterization of zeolite from waste Linz-Donawitz (LD) process slag of steel industry for removal of Fe3+ from drinking water. Adv. Powder Technol. 32 (9), 3372–3387. doi:10.1016/j.apt.2021.07.023.

Samanta, N.S., Purkait, M.K., 2025. Preparation of zeolite by green synthesis from steel industry by-product LD-slag and its application for photodegradation. Mater. Res. Bull. 190. doi:10.1016/j.materresbull.2025.113492

Sharma, M., Samanta, N.S., Chakraborty, A., Purkait, M.K., 2023. Simultaneous treatment of industrial wastewater and resource recovery using microbial fuel cell. In: Resource Recovery in Industrial Waste Waters. Elsevier, India, pp. 621–637. Available from: https://www.sciencedirect.com/book/9780323953276. doi:10.1016/B978-0-323-95327-6.00002-6.

Somu, P., Narayanasamy, S., Gomez, L.A., Rajendran, S., Lee, Y.R., Balakrishnan, D., 2022. Immobilization of enzymes for bioremediation: a future remedial and mitigating strategy. Environ. Res. 10960953. 212. Available from: http://www.elsevier.com/inca/publications/store/6/2/2/8/2/1/index.htt. doi:10.1016/j.envres.2022.113411.

Yu, X., Shi, J., Khan, A., Yun, H., Zhang, P., Zhang, P., Kakade, A., Tian, Y., Pei, Y., Jiang, Y., Huang, H., Kejia, W., Xiangkai, L., 2020. Immobilized-microbial bioaugmentation protects aerobic denitrification from heavy metal shock in an activated-sludge reactor. Bioresour. Technol. 09608524. 307, 123185. doi:10.1016/j.biortech.2020.123185.

Yu, X., Mao, C., Wang, W., Kulshrestha, S., Zhang, P., Usman, M., Zong, S., Hilal, M.G., Fang, Y., Han, H., Xiangkai, L., 2023. Reduction of metronidazole in municipal wastewater and protection of activated sludge system using a novel immobilized Aspergillus tabacinus LZ-M. Bioresour. Technol. 09608524. 369, 128509. doi:10.1016/j.biortech.2022.128509.

Yu, Y., An, Q., Zhou, Y., Deng, S., Miao, Y., Zhao, B., Yang, L., 2019. Highly synergistic effects on ammonium removal by the co-system of Pseudomonas stutzeri XL-2 and modified walnut shell biochar. Bioresour. Technol. 18732976. 280, 239–246. Available from: www.elsevier.com/locate/biortech. doi:10.1016/j.biortech.2019.02.037.

Zhou, L., Li, G., An, T., Li, Y., 2010. Synthesis and characterization of novel magnetic Fe3O4/ polyurethane foam composite applied to the carrier of immobilized microorganisms for wastewater treatment. Res. Chem. Intermed. 15685675. 36 (3), 277–288. doi:10.1007/ s11164-010-0134-5.

Zhou, Z., Chao, H., He, W., Su, P., Song, J., Yang, Y., 2022. Boosting the activity of enzymes in metal-organic frameworks by a one-stone-two-bird enzymatic surface functionalization strategy. Appl. Surf. Sci. 01694332. 586. Available from: http://www.journals.elsevier.com/applied-surface-science/. doi:10.1016/j.apsusc.2022.152815.

Zhuo, S.N., Zhang, W., Ren, H.Y., Liu, B.F., 2024. Highly efficient degradation of acetaminophen via nano zero-valent iron biochar with periodate system at low temperature. Bioresour. Technol. 18732976. 395. Available from: https://www.sciencedirect.com/science/journal/09608524. doi:10.1016/j.biortech.2024.130349.

CHAPTER

# Microbial innovations in waste management

15

**Aparna Pandey, Debarati Paul, Samarth Gohel, and Sourish Bhattacharya**

*Process Design and Engineering Division, CSIR-Central Salt and Marine Chemicals Research Institute, Bhavnagar, India*

## 1 Introduction

The accelerating expansion of urban areas, intensifying industrial operations, and soaring global population growth have collectively fueled an unprecedented surge in waste production. Waste is defined as "any material that is discarded, unwanted, or no longer useful" and is classified into municipal solid waste (MSW) (household/commercial), industrial waste (manufacturing by-products), hazardous waste (ignitable, corrosive, reactive, toxic), organic waste (biodegradable, such as food/yard trimmings), and specialized streams such as biomedical waste, e-waste, and construction debris. Municipal solid waste, industrial effluents, agricultural remnants, and hazardous materials are accumulating at alarming rates, overwhelming conventional waste management systems (Leow et al., 2018). Globally, MSW reached ~2.3 billion tons in 2023, projected to rise to ~3.8 billion tons by 2050 under current trends (Wikipedia Waste Management). In India alone, ~62 million tons of solid waste is generated annually, making it the world's third-largest waste producer.

For decades, landfilling and incineration have served as the primary strategies for waste disposal; however, their environmental footprints are becoming increasingly evident. Sanitary landfilling, although widely adopted for its low technical demands and capacity to handle large volumes, generates persistent leachate rich in heavy metals and organics that can contaminate groundwater long after site closure (Yaashikaa et al., 2022). Anaerobic degradation within landfills produces methane, a greenhouse gas with ~30-fold higher warming potential than $CO_2$, and $CO_2$, both of which contribute significantly to climate change (Yaashikaa et al., 2022). Landfill liners, though designed to contain leachate for centuries, eventually degrade, posing risks of soil and aquifer pollution. Labeled as simple and cost-effective, landfills still require vast land, road infrastructure, long-term monitoring, and present challenges such as disease vectors, odor, noise, and property devaluation.

Incineration reduces waste volume by over 85% but emits fine particulates, heavy metals, acid gases, $NO_x$, dioxins, and furans, even with advanced flue-gas controls (Cole-Hunter et al., 2020; N.R.C. Incineration, 2000). Waste-to-energy (WtE)

**Microbial Technologies for Sustainable Waste Management. https://doi.org/10.1016/B978-0-443-44093-9.00007-4**

incinerators incur high capital and operating costs, extended payback periods, and foster infrastructure lock-in for ≥25 years. Hazardous waste incineration consumes significant energy, yet residual ash and fly ash retain toxic metals and acid gas risks, requiring hazardous landfill disposal (N.R.C. Incineration, 2000). Chemical treatment methods for industrial effluents often involve intensive reagent use, generate secondary brine/sludge, and incur high operational costs. Many chemical recycling and treatment strategies are energy-intensive and require complex logistics, with limited scalability for mixed waste at volume (Cai et al., 2023). Life-cycle analyses consistently show that incineration's energy recovery does not always offset its atmospheric emissions and waste ash toxicity, and landfilling defers but does not eliminate environmental burden (Assamoi and Lawryshyn, 2012; Cole-Hunter et al., 2020). Beyond environmental consequences, these methods are economically burdensome, demanding substantial land, infrastructure, and long-term monitoring. With escalating regulatory pressures and the growing emphasis on sustainability, a paradigm shift toward more eco-conscious, cost-effective, and integrative waste management solutions has become imperative, a turn that foregrounds the capacity of microbial technologies to transform waste into resources while ensuring environmental integrity (Leow et al., 2018).

Microorganisms, namely, bacteria, fungi, algae, and archaea, boast a remarkable metabolic repertoire shaped by millennia of evolutionary adaptation. They possess notable metabolic versatility, enabling them to degrade or transform organic waste into nontoxic end-products such as $CO_2$, water, and biomass, while also oxidizing or reducing inorganic contaminants like heavy metals. Bioremediation is broadly defined as any process wherein a biological system, typically bacteria, microalgae, fungi, or archaea, is used to remove environmental pollutants from air, water, soil, or industrial effluents in natural or engineered settings (Ayilara and Babalola, 2023). Fungi (mycoremediation), bacteria, microalgae, and cyanobacteria have been shown to secrete extracellular enzymes such as laccases, peroxidases, hydrolases, and mono-/dioxygenases that break down a wide range of pollutants, including polycyclic aromatic hydrocarbons (PAHs), pesticides, dyes, plastics, and petroleum hydrocarbons. Archaea contribute to chemical cycling processes such as ammonia oxidation, denitrification, sulfur oxidation, and methanogenesis, thus playing essential roles in pollutant transformation in soils, sediments, and wastewater systems. Microbial bioremediation is recognized as an eco-friendly, cost-effective, and sustainable alternative to conventional physicochemical treatments, with advantages, including on-site application, minimal secondary pollution, and lower infrastructure costs (P. Singh et al., 2020). Advances in microbial biotechnology, such as enzyme prediction tools, metagenomics, and microbial consortia engineering, have enhanced efficiency, scalability, and applicability across municipal, industrial, agricultural, and hazardous waste streams.

Microbial decomposition of organic waste via composting and anaerobic digestion (AD) enables the conversion of food scraps, agricultural residues, and sewage into stabilized biomass, $CO_2$, and methane-rich biogas (Meilander and Caporaso, 2024). Bacteria and fungi facilitate heavy metal detoxification through biosorption, bioaccumulation, enzymatic reduction, precipitation, and microbially induced carbonate precipitation, effectively immobilizing cadmium, lead, chromium, and arsenic in contaminated soils and waters (Joshi et al., 2023). WtE conversion using microbial systems, such as AD for biogas and microbial fermentation of biomass into bioethanol, offers renewable energy production from carbon-rich wastes. Microbes capable of plastic degradation produce specialized enzymes like PETases and laccases that cleave polymers into monomers for further assimilation. In industrial and municipal effluent treatment, microbes such as *Shewanella oneidensis* and sulfate-reducing consortia remove organic contaminants, heavy metals, and chlorinated compounds via reduction, biosorption, and enzymatic pathways (Z. Liu et al., 2018; Yin et al., 2022). Certain bacteria, notably marine hydrocarbonoclastic species like *Alcanivorax borkumensis*, produce extracellular biosurfactants that emulsify oil, enhancing biodegradation in spill-affected environments (Deng et al., 2024).

Engineered microbes created through synthetic biology are being tailored to perform environmental tasks such as pollutant degradation and biosensing by redesigning metabolic pathways using modular genetic circuits (Han et al., 2022). CRISPR–Cas genetic engineering enables precision editing of microbial genomes to enhance the degradation of complex pollutants or improve metal-detoxification capabilities in bacteria like *Pseudomonas putida* (Zhang et al., 2022). High-throughput omics techniques—genomics, proteomics, and metabolomics—now allow comprehensive characterization and optimization of microbial communities and enzyme systems used in bioremediation and wastewater treatment. Bioelectrochemical systems (BESs), including microbial fuel cells (MFCs) and microbial electrolysis cells (MECs), harness electroactive bacteria (EAB) to simultaneously degrade organic waste and generate electricity or valuable chemicals (Chandrasekhar et al., 2021).

Despite advances, large-scale microbial remediation faces challenges such as fluctuating environmental conditions (pH, temperature, oxygen), competition from native microbes, and waste stream variability. High toxin levels can inhibit microbial growth, and public concerns over Genetically Modified Organisms (GMOs) add regulatory complexity. Emerging solutions include encapsulating engineered microbes, biosensor-based monitoring, hybrid bioaugmentation, biostimulation strategies, and risk-assessment frameworks. As global momentum builds toward achieving sustainable development goals, microbial innovations in waste management are evolving from experimental interventions to mainstream technologies. Their capacity to integrate pollutant detoxification with resource recovery, the circular bioeconomy, and climate resilience positions them as critical in the next generation of environmental engineering.

## 2 Recycling and circular economy

Global waste volumes have escalated under a traditional linear economy model—"take, make, dispose"—which relies heavily on extraction, manufacturing, and disposal, often neglecting environmental impacts. To address this unsustainable trajectory, the circular economy (CE) paradigm proposes a systemic shift, emphasizing the continual circulation of materials through reuse, remanufacturing, and recycling, thereby "eliminating waste and pollution" (Adami and Schiavon, 2021). In CE, waste is transformed from an endpoint into a resource for new production cycles, supporting economic, social, and ecological objectives. Recycling, the cornerstone of this model, extends material life, reduces dependence on virgin resources, and fosters job creation and innovation in waste management (Barbhuiya et al., 2024). In the context of sustainable waste management, recycling is no longer a singular act but a gateway to broader circular strategies that incorporate environmental, economic, and social dimensions. To understand the transformative potential of this shift, it is essential to explore how recycling integrates into the broader circular framework and how various sectors, ranging from MSW to industrial by-products, can align with CE principles to enable systemic change.

Solid wastes—including food, plastics, metals, textiles, wood, and glass—are expected to increase by ~59% to 3.4 billion tons by 2050 if current trends continue under a linear "take–make–dispose" model. The CE framework advocates slowing, closing, and narrowing resource loops by recovering materials and energy from waste, extending product life, and emphasizing the 10 Rs: refuse, rethink, reduce, reuse, repair, refurbish, remanufacture, repurpose, recycle, and recover. Under CE, waste is viewed not as a burden but as a resource, enabling environmental, economic, and social benefits via resource efficiency, emissions reduction, green jobs, and cost savings (Akomea-Frimpong et al., 2024) (Fig. 15.1).

A biobased CE treats organic solid waste as a feedstock for aerobic and AD, producing chemicals, nutrients, compost, and biogas, thus advancing sustainable biorefinery models. Resource recovery from organic waste supports nutrient recycling (e.g., nitrogen), renewable energy bioproducts (e.g., biogas, volatile fatty acids), and integrated processes to improve system economics. Beyond composting and

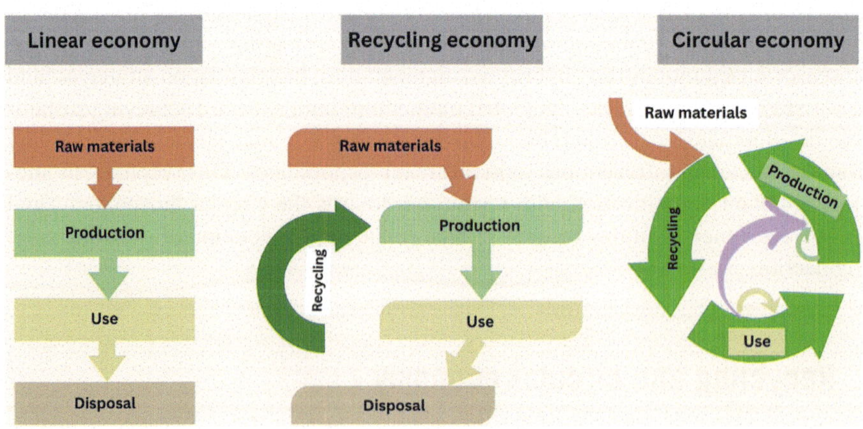


**FIG. 15.1**

Comparison of linear, recycling, and circular economy models.

digestion, CE strategies seek to valorize organic residues across municipal, agricultural, and industrial streams, aligning with Sustainable Development Goals (SDG) goals and biocircular visions (Wainaina et al., 2020).

Recycling metals such as aluminum and glass can save 95% of the energy and reduce 95% of air pollution relative to producing virgin materials, while creating 6–13 jobs per thousand tons compared to <1 job in landfill operations (Wikipedia Recycling). However, closed-loop recycling remains limited: only ~9% of global inflows are recycled, which, when accounting for nonrecoverable flows, equates to just 27% of value recirculated and represents only ~1.4% of global GDP (Rashid, 2025). Deployment of a CE at scale also faces challenges: fragmented policy frameworks, lack of producer responsibility, behavioral inertia, and logistics complexity, although instruments like Extended Producer Responsibility (EPR), green taxes, and digital material flow monitoring can catalyze systemic shifts (Rashid, 2025).

## 3 WtE technologies

WtE technologies convert waste materials into usable energy forms (electricity, heat, or biofuels), thereby reducing waste volumes and supporting CE goals. These systems encompass a range of processes—biological, thermal, and chemical—that valorize both organic and inorganic waste, integrating seamlessly into municipal and industrial resource management strategies (Fig. 15.2). WtE supports a CE by transforming emissions-intensive disposal systems into energy-generating assets,

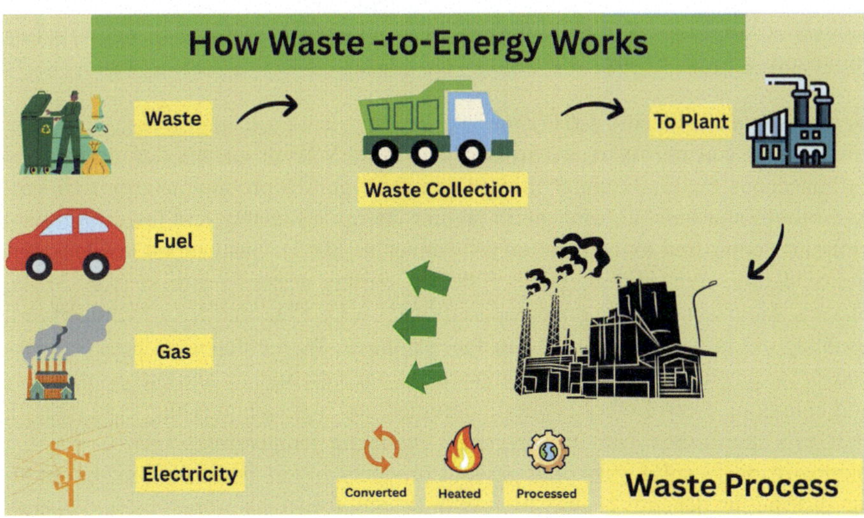


**FIG. 15.2**

Flowchart of the waste-to-energy conversion process.

aligning with sustainable development targets and climate mitigation goals. To address the mounting pressure of waste accumulation and resource scarcity, a range of WtE technologies have emerged to convert organic and inorganic waste into usable energy forms.

## 3.1 Thermal conversion technologies

Thermal conversion technologies use heat to transform waste materials into energy, fuels, or usable by-products. Common methods include incineration, pyrolysis, and gasification, each operating at different temperatures and oxygen levels. These processes help reduce waste volume and recover energy, making them vital in sustainable WtE systems.

### *3.1.1 Incineration with energy recovery (WtE)*

Incineration with energy recovery combusts MSW at high temperatures (>850°C), converting it into heat to produce steam that drives turbines for electricity or combined heat and power systems (Bertone et al., 2024). Modern plants reduce waste volume by ~90%–95% and weight by ~75%, significantly diverting refuse from landfills (Mata-Lima et al., 2021). Advanced flue-gas treatment with activated carbon injection, baghouse filters, Selective Catalytic Reduction (SCR)/Selective Non-Catalytic Reduction (SNCR) for $NO_x$, and scrubbers captures >99% of particulates, heavy metals, and dioxins (Bertone et al., 2024). This process offsets fossil energy use, mitigates methane emissions from landfills, and lowers net $CO_2$ per kWh compared to coal, though it is still higher than renewables or natural gas (Mata-Lima et al., 2021). Overall, incineration with energy recovery offers a reliable WtE solution that significantly reduces landfill volume, recovers usable energy, and minimizes environmental impact when supported by robust pollution controls.

### *3.1.2 Gasification and pyrolysis*

Gasification and pyrolysis are advanced thermal WtE processes that decompose carbonaceous materials under limited or no oxygen to produce energy-rich syngas, biooil, and biochar, often with greater energy efficiency and lower pollutant emissions compared to incineration (Abedin et al., 2025). Gasification involves partial oxidation—typically at 700°C–1200°C—turning waste feedstocks (e.g., MSW, plastics, biomass) into syngas composed of $H_2$, CO, $CH_4$, and light hydrocarbons, enabling use in turbines, engines, or fuel synthesis. With efficiencies between 45% and 65%, gasification generally outperforms incineration in net energy recovery and emissions control when properly designed and maintained (Dong et al., 2019). Pyrolysis, performed without oxygen at moderate temperatures (300°C–700°C), thermally cracks solid waste into gaseous products, biooil, and solid biochar; product distribution depends heavily on reactor design, feedstock properties, and operating parameters. Biooil from plastic waste pyrolysis boasts high calorific value and potential as a fuel or feedstock for chemicals, while biochar offers opportunities in soil amendment and carbon sequestration (Dai et al., 2022). Comprehensive system optimization covering reactor design, catalyst use, emissions handling, and techno-economic integration is essential to realize the full potential of gasification and pyrolysis in sustainable WtE frameworks.

## 3.2 Biological conversion technologies

Biological conversion technologies use microorganisms to break down organic waste into useful products like biogas, bioethanol, or compost. These processes occur under controlled aerobic or anaerobic conditions, depending on the method used. Common technologies include AD (biogas), waste-to-bioethanol, and other biofuels. They are sustainable and environmentally friendly, especially suited for biodegradable waste.

### 3.2.1 Anaerobic digestion (biogas)

AD is a biological process that enables the breakdown of organic waste, including food scraps, agricultural residues, manures, and sewage sludge, under oxygen-free conditions to produce biogas and digestate products (Richard et al., 2019). The AD process unfolds through four key biochemical stages: hydrolysis, acidogenesis, acetogenesis, and methanogenesis, conducted by diverse microbial consortia, culminating in a methane-rich biogas containing 50%–75% $CH_4$ alongside $CO_2$ (Figueredo et al., 2023). Biogas serves as a renewable energy source, suitable for the generation of heat, electricity, and vehicle fuel following treatment and grid integration.

### 3.2.2 Waste-to-bioethanol and other biofuels

Waste-derived feedstocks, such as MSW, paper sludge, food scraps, garden clippings, and agricultural residues, offer abundant, nonfood biomass for bioethanol and other biofuels (Awogbemi and Kallon, 2022). Through pretreatment methods such as steam explosion, dilute acid or alkali treatments, and hot water treatments, the tough lignocellulosic structure is broken down, releasing fermentable sugars. For example, steam-exploded paper sludge yielded ~0.39 L ethanol/kg with minimal chemical inputs. Simultaneous saccharification and fermentation using yeasts such as *Saccharomyces cerevisiae* can achieve high conversion efficiencies (> 80%) and ethanol concentrations near 40 g/L from food waste and paper waste streams (Prasetyo et al., 2011). At the pilot scale, high-solid fermentation of virgin pulp paper sludge reached ethanol titers of up to 86.2 g/L (≈80% of the theoretical maximum) under optimized conditions (van Dyk et al., 2025). Aside from ethanol, wastes also produce biogas and biobutanol. For example, acid–alkaline pretreated green waste yielded 33.7% ethanol and significant biogas via AD. Industrial-scale butanol production has even emerged from whiskey by-products using ABE fermentation (Sofokleous et al., 2022).

## 3.3 Bioelectrochemical and electroactive systems

Bioelectrochemical and electroactive systems harness the metabolic activity of microorganisms to generate electrical current or drive chemical reactions. These systems, such as MFCs, use EAB to transfer electrons to electrodes, converting organic waste into electricity or valuable products.

### 3.3.1 Microbial fuel cells and bioelectrochemical systems

MFCs are specialized BESs that utilize EAB to oxidize organic substrates such as acetate, glucose, domestic wastewater, food scraps, or lignocellulosic residues and convert chemical energy directly into electricity via electron transfer to the anode (Apollon et al., 2025). Structurally, MFCs typically consist of an anode chamber and a cathode chamber separated by a proton exchange membrane; microbial metabolism in the anodic biofilm produces electrons and protons, which flow through an external circuit and the membrane, respectively, generating current, while oxygen at the cathode reduces protons to water (Wang et al., 2022). Power densities reported in advanced MFC setups reach 4–8 $W/m^2$ (or up to ~25 $W/m^2$ on nonconventional substrates like vegetable and distillery waste) (Apollon et al., 2025). A dual benefit arises from simultaneous wastewater treatment via high COD removal and nutrient recovery and sustainable energy generation, reducing the energy needs of conventional treatment systems (Palanisamy et al., 2019).

Expanding BES variations, such as MECs, microbial desalination cells, and microbial electrosynthesis, supports the production of hydrogen, methane, volatile fatty acids, struvite, and valuable biosynthesis products (N. K. Singh et al., 2024). Despite progress, challenges remain in scaling up, developing cost-effective electrodes, addressing limited power output, managing feedstock variability, and ensuring operational robustness. Nonetheless, ongoing research that merges electrode innovation, microbial engineering, and system integration is bringing MFCs and BESs closer to commercial viability for decentralized energy production, wastewater treatment, and resource recovery applications.

## 3.4 Hybrid/integrated systems

Hybrid or integrated systems combine BESs—such as MFCs or MECs—with complementary technologies like AD, constructed wetlands, algal photobioreactors, and nutrient recovery units to enhance energy recovery, pollutant removal, and overall process efficiency. For example, coupling AD and MEC enables simultaneous enhancement of biogas production and in situ upgrading, addressing issues like volatile fatty acid toxicity and impurity-laden biogas (Ao et al., 2024). Integrating MFCs with constructed wetlands (CW–MFC) has also proven effective for decentralized wastewater treatment: pilot systems combining anaerobic filters, floating and vertical wetlands reduced COD by ~65%, $BOD_5$ by 82%, $N\text{–}NH_3$ by 93%, TP by 87%, turbidity by 94%, and coliforms by three logs, while generating voltages up to 557 mV (open circuit) and power densities up to ~0.325 $MW/m^2$ (Colares et al., 2022).

Innovations beyond physical integration include algae-MFC hybrids, nutrient-recovery systems, and cascade designs. Photo-BESs utilizing algae-supported cathodes avoid expensive catalysts, increase nitrogen/ammonia removal (by ~59% compared to standalone MFCs), and enable simultaneous bioelectricity and biomass production, although challenges like membrane fouling and light-cycle dependencies remain (Capodaglio et al., 2021). The synergistic potential of hybrid BES integrations spanning AD, wetlands, algae, nutrient recovery, and cascade designs offers a promising route toward self-sustaining, decentralized platforms capable of simultaneous waste treatment, energy recovery, and resource valorization.

## 4 An insight into organic municipal waste and its prospects

Organic municipal solid waste (OMSW), often referred to as the organic fraction of municipal solid waste, includes biodegradable materials such as kitchen scraps, vegetable and fruit peels, food leftovers, garden waste, and biosolids. In low- and middle-income countries, this organic fraction typically constitutes 50%–70% of the total MSW stream (Z. Yang et al., 2024). Globally, over 2 billion tons of municipal waste are generated annually, with nearly 45% of this inadequately managed (Olhoff et al., 2023). Organic waste in unmanaged dumpsites undergoes anaerobic decomposition, emitting methane, a greenhouse gas ~25× more potent than $CO_2$ (IPCC AR6, 2021). Moreover, this decomposition produces foul odors, attracts pests, and generates leachate that pollutes groundwater, threatening public health and ecosystems. With projections indicating a 66% increase in global waste volumes by 2050 and a financial burden that could reach USD 600 billion annually due to environmental and health impacts, OMSW requires urgent intervention through sustainable and circular management approaches.

Traditional OMSW treatment methods, such as composting and AD, have long been adopted due to their simplicity and ecological benefits. Composting, an aerobic biodegradation process, transforms organic residues into stable humus-like material that enhances soil fertility and structure (Terashima et al., 2009). In-vessel composting systems improve process control, while vermicomposting enhances nutrient quality through worm activity. Similarly, AD stabilizes OMSW under oxygen-free conditions, generating biogas composed primarily of methane and carbon dioxide, which can be used as a renewable energy source (Zhao et al., 2011). The residual digestate from AD is rich in nitrogen, phosphorus, and organic matter and has shown potential as an organic fertilizer capable of partially replacing synthetic inputs.

Emerging valorization strategies are now focusing on integrated biorefinery approaches to convert OMSW into high-value products such as enzymes, organic acids, biosurfactants, and biofuels. Bioconversion using enzymes like cellulases, pectinases, and proteases enables the breakdown of lignocellulosic waste into fermentable sugars, which are subsequently transformed into ethanol, butanol, or other biobased chemicals via microbial fermentation. Pilot-scale experiments and operational case studies have validated the potential of OMSW-based biorefineries. Key challenges like feedstock inconsistency, high enzyme costs, and limited infrastructure hinder OMSW valorization, especially in developing regions. Addressing these requires collaborative action, supportive policies, and investment in decentralized technologies. Integrating advanced bioprocesses with digital management can turn OMSW from a waste issue into a pillar of the CE, offering a vital route toward sustainable, climate-resilient urban systems.

## 5 Different pollutants and their toxicity to living organisms

Environmental pollutants span air, water, soil, and radioactive agents, each posing serious risks to living organisms through distinct exposure routes and mechanisms. Understanding how these pollutants impair health and ecosystem function is essential for crafting effective mitigation and remediation strategies.

### 5.1 Air pollutants

Air pollutants such as fine particulate matter ($PM_{2.5}$), $PM_{10}$, ozone ($O_3$), nitrogen dioxide ($NO_2$), sulfur dioxide ($SO_2$), carbon monoxide (CO), volatile organic compounds (VOCs), and heavy metals pose significant health risks to humans and ecosystems (Romieu et al., 2007). $PM_{2.5}$ triggers oxidative stress and systemic inflammation by generating reactive oxygen species and activating inflammatory pathways, contributing to cardiovascular diseases, including hypertension, atherosclerosis, arrhythmia, and heart failure (Pryor et al., 2022). These particles penetrate deep into the lungs, impair lung function, exacerbate Chronic Obstructive Pulmonary Disease (COPD) and asthma, and increase mortality. Gaseous pollutants such as $NO_2$ and $O_3$ irritate airways through oxidative mechanisms, reduce antioxidant defenses, and impair respiratory and cardiovascular health (Romieu et al., 2007). CO binds to hemoglobin, causing systemic hypoxia that affects heart and brain functions. VOCs like benzene and formaldehyde are mutagenic and carcinogenic, causing DNA damage and affecting the liver, kidneys, and nervous systems. Heavy metals and PAHs bioaccumulate, impair neurological development, weaken immunity, and increase cancer risk (Reimann and Lewis, 2007). Overall, air pollution's toxicity stems from oxidative stress, inflammation, genetic damage, and disrupted physiological processes, leading to widespread, multiorgan harm and serious environmental consequences (Miller, 2020).

### 5.2 Water pollutants

Water bodies worldwide are contaminated by a complex mix of heavy metals (such as mercury, lead, cadmium, arsenic, and chromium), pesticides, petroleum hydrocarbons, excess nutrients, salinity, and pharmaceuticals, posing severe threats to aquatic life and human health. Mercury, especially as methylmercury, bioaccumulates in fish and marine mammals, causing neurological damage, reproductive failure, and even death in both wildlife and humans through seafood consumption. Lead and cadmium exposure disrupts fish growth, impairs gill and kidney function, disturbs enzyme systems, and reduces reproductive capacity. Arsenic exposure results in anemia, tissue damage, immune dysfunction, and carcinogenic effects in aquatic organisms and humans (Hama Aziz et al., 2023; Shou et al., 2022). Chromium, nickel, and copper interfere with fish respiration and metabolic processes. Pesticide runoff causes endocrine disruption, behavioral changes, and mass mortalities in fish, amphibians, and aquatic invertebrates (Ray and Shaju, 2023). Nutrient overenrichment of nitrogen and phosphorus leads to eutrophication, oxygen depletion, algal blooms, fish kills, and toxin release harmful to all aquatic life. Petroleum hydrocarbons (especially PAHs) bioaccumulate in shellfish, disrupt reproduction, and exert carcinogenic and endocrine-disrupting effects (Clean Coastal Waters, 2000). Elevated salinity disrupts osmotic balance and stunts growth in freshwater species (Sarma et al., 2020). Collectively, these pollutants induce oxidative stress, enzyme inhibition, hormonal dysfunction, neurotoxicity, and ecosystem collapse, threatening biodiversity, food sources, and human health through water and food supply contamination.

## 5.3 Soil pollutants

Soils are contaminated by toxic heavy metals such as lead (Pb), cadmium (Cd), arsenic (As), mercury (Hg), chromium (Cr), nickel (Ni), and copper (Cu), which inhibit microbial activity, disrupt nutrient cycling, and pose serious risks to crop productivity and food safety (Almotairy, 2024). High levels of Cd, Pb, and As damage plant physiological processes like photosynthesis, respiration, root development, and seed germination, causing chlorosis, stunted growth, and reduced yields (Ding et al., 2022). Heavy metal toxicity also alters soil pH, disrupts microbial communities, reduces enzymatic activity essential for organic matter decomposition, and affects nitrogen fixation and denitrification processes. Soil microorganisms' reproductive and membrane functions are impaired by Cd, while fungal spore formation and earthworm health are also negatively impacted (Angon et al., 2024; Ding et al., 2022). Synthetic pesticide residues such as neonicotinoids, organophosphates, herbicides, and fungicides accumulate in soil, causing DNA damage, enzyme inhibition, disrupted growth and reproduction in soil fauna (e.g., earthworms, springtails), and reduced microbial diversity (Beaumelle et al., 2023). Persistent pesticide accumulation can last for years or decades, altering soil community structure and reducing its fertility and resilience. Together, heavy metals and persistent agrochemicals undermine soil ecosystem health, impeding plant growth, degrading biodiversity, "poisoning" the base of food chains, and threatening human and environmental wellbeing (Chibuike and Obiora, 2014).

### 5.4 Radioactive pollutants

Radioactive contaminants, including radionuclides like cesium-137, strontium-90, iodine-131, plutonium isotopes, and radon, pose significant risks to ecosystems and human health due to their persistence and ionizing radiation emissions. In plants, chronic exposure to radionuclides can inhibit growth, disrupt leaf and stem anatomy, and reduce reproduction even at dose rates of 40–760 μGy/day, with higher rates causing necrosis and impaired photosynthesis (Yankauskas et al., 2024). Aquatic and marine organisms accumulate radioactive elements, leading to oxidative damage, mutations, and impaired immune responses, as evidenced in wetlands around Chernobyl (Videvall et al., 2023). Wildlife exposed to radioisotopes in contaminated zones exhibit reduced population density, impaired reproduction (e.g., aspermy in rodents), smaller brain size, and increased incidence of genetic abnormalities (Kesäniemi et al., 2019). In humans, acute exposures (0.5–5 Sv) can cause radiation sickness, leukopenia, organ failure, and death, while long-term exposures, even at lower doses, elevate cancer risk, cause DNA strand breaks, chromosomal aberrations, and thyroid disorders from iodine-131 uptake. Radioactive isotopes bioaccumulate and biomagnify up food chains, increasing exposure risk for predators, including humans consuming contaminated seafood. Collectively, these effects spanning morphological, genetic, reproductive, immunological, and ecological domains underscore the potent toxicity of radioactive pollutants across all levels of life (Videvall et al., 2023).

## 6 Types of remediation

Environmental remediation refers to the removal, reduction, or neutralization of pollutants from soil, water, or air to restore ecological balance and safeguard human health. Remediation technologies are selected based on the pollutant type, the extent of contamination, site characteristics, and regulatory requirements. The main types of remediation can be categorized in the following.

### 6.1 Physical remediation

Physical remediation utilizes mechanical and thermal approaches, including soil replacement, soil washing, thermal desorption, soil vapor extraction (SVE), electrical resistance heating (ERH), and magnetic and hydrodynamic separation, to remove or isolate contaminants from soil systems (Raffa et al., 2021). Soil replacement (excavation and back-fill) and deep spading dilute contaminant concentrations by adding clean soil, effectively reducing exposure in localized contamination zones. Soil washing, which uses physical separation techniques (screening, gravity, and magnetic methods) and aqueous extraction with surfactants or chelating agents, removes heavy metals and organics from fine particles, after which coarse clean soil can be returned (Priya et al., 2023). Thermal desorption applies heat (90°C–900°C) via blankets or in situ heaters to volatilize organic contaminants and some metals, which are captured in vapor treatment systems. SVE pulls volatile pollutants from unsaturated zones using vacuum wells and treats the off-gas to prevent air release. ERH works in situ by heating the subsurface to vaporize contaminants, enabling recovery through steam and vapor capture systems. Hydrodynamic and magnetic separation methods segregate contaminated fine particles based on size, density, and magnetic properties, concentrating pollutants and reducing the contaminated volume. While these physical methods often deliver rapid, high-efficiency contaminant removal, they tend to be costly and energy-intensive and may disrupt soil structure, making them most suitable for limited-area or heavily polluted sites (L. Xu et al., 2024).

## 6.2 Chemical remediation

Chemical remediation involves using chemical agents such as acids, chelators, oxidants/reductants, and stabilization compounds to transform, mobilize, or immobilize contaminants within soils (Sánchez-Castro et al., 2023). Chelating agents like Ethylenediaminetetraacetic acid (EDTA), Ethylenediamine-N,N'-disuccinic acid (EDDS), L-Glutamic acid N,N-diacetic acid (GLDA), and citric acid form soluble complexes with heavy metals (e.g., Pb, Cd, Ni, Cu), enhancing extraction efficacy. EDTA can remove up to 92% of Cd, while more biodegradable agents like GLDA and EDDS offer reduced toxicity and soil impact (Zheng et al., 2022). Chemical soil washing leverages surfactants, acids, or chelates to rinse contaminants from soils, followed by sedimentation or filtration, effectively targeting metals and organics, though it may degrade soil structure and pose secondary risks (Raffa et al., 2021). Oxidative treatments, including Fenton's reagent ($H_2O_2/Fe^{2+}$), persulfate, and permanganate, break down organic pollutants like petroleum hydrocarbons and solvents; integrating chelators (e.g., citrate) can boost degradation from ~51% to ~93%. In situ chemical reduction using zero-valent iron or zinc stabilizes and precipitates metals by reducing them to inert forms via permeable reactive barriers. Stabilization/solidification methods add binders such as cement, biochar, zeolite, or activated carbon to immobilize pollutants, reducing mobility and ecological risk while preserving soil structure (Raffa et al., 2021; Zheng et al., 2022). Although generally faster and effective for high-contamination scenarios, chemical remediation must carefully balance treatment efficiency, soil health, secondary impacts, and costs for sustainability and long-term viability (Ifon et al., 2019).

## 6.3 Biological remediation (bioremediation)

Biological remediation, or bioremediation, employs microorganisms, including bacteria, fungi, and archaea, as well as plants, to detoxify or immobilize soil pollutants, notably heavy metals, petroleum hydrocarbons, and other organic toxins (Arantza et al., 2022). Endophytic bacteria and plant-growth-promoting rhizobacteria enhance phytoremediation by increasing heavy metal uptake, tolerance, and growth through mechanisms like biosorption, bioaccumulation, and the secretion of siderophores and organic acids (Thirumalaivasan et al., 2024). Strains such as *Azotobacter chroococcum*, *Bacillus*, and *Pseudomonas* have demonstrated significant removal or stabilization of lead, cadmium, copper, and zinc in contaminated soils (Adil et al., 2024). Mycorrhizal fungi form symbiotic relationships with plant roots, binding heavy metals in fungal biomass and enhancing phytostabilization by lowering metal bioavailability. Bioaugmentation, which involves adding specialized microbial strains combined with phytoremediation, has shown improved removal of both heavy metals and petroleum hydrocarbons; for example, inoculation with *Pseudomonas aeruginosa* and alfalfa removed 68% of petroleum hydrocarbons from cocontaminated soil in controlled trials (Agnello et al., 2016). Genetically engineered microbes (GEMs), such as modified *Bacillus subtilis,* have achieved >99% removal efficiencies for metals such as lead, copper, and mercury in laboratory conditions. In addition, combining biochar with microbial or plant-based remediation enhances pollutant adsorption, reduces toxicity, and supports microbial degradation in soil ecosystems (Chen et al., 2024).

## 6.4 Thermal remediation

Thermal remediation employs heat to volatilize, mobilize, or destroy soil contaminants through methods like thermal desorption, in situ thermal conductive heating (TCH), and ERH (O'Brien et al., 2018). Ex situ thermal desorption, using temperatures from approximately 90°C to 900°C, effectively removes a wide range of volatile and semivolatile organic pollutants, including petroleum hydrocarbons and PAHs, within relatively short treatment durations. In situ ERH applies alternating current via electrodes to heat the subsurface to water's boiling point (>100°C), enabling volatile contaminant capture through vapor extraction wells, achieving rapid soil and groundwater cleanup in under 40 days. In situ TCH uses heater wells to gently transfer heat into soil, vaporize contaminants at depth, and drive them toward extraction points, with efficacy influenced by soil properties and well configuration of sustainable and effective bioremediation technologies.

## 6.5 Enzymatic oxidation

Microorganisms exploit oxidative enzymes such as laccases, peroxidases (e.g., lignin peroxidase, manganese peroxidase), and catalase-peroxidases to degrade complex organic pollutants like PAHs, dyes, phenols, antibiotics, and chlorophenols via electron transfer reactions (Sun et al., 2024). However, thermal methods can significantly alter soil organic matter, texture, pH, nutrient levels, and microbial diversity, especially at higher temperatures and extended durations (O'Brien et al., 2018). Field studies show that while microbial alpha diversity may recover post-treatment, key functional groups may shift, impacting ecological restoration. Despite high contaminant removal rates, including 99.99% of Total Petroleum Hydrocarbons

(TPH) and heavy metals in silty sand at ~700°C within 48 h, thermal techniques are energy-intensive, costly, and best suited for heavily contaminated or site-specific remediation (Kim et al., 2022). Therefore careful assessment that balances cleanup goals with soil function preservation is essential when planning thermal remediation projects.

## 7 Different microbes used as bioremediation agents

Microorganisms play a central role in the natural detoxification of polluted environments through a process known as bioremediation. Their diverse metabolic capabilities allow them to degrade, transform, immobilize, or accumulate a wide variety of pollutants, including heavy metals, petroleum hydrocarbons, pesticides, and radionuclides. Bacteria, fungi, and even algae have evolved unique biochemical mechanisms, such as biosorption, enzymatic degradation, redox transformation, and bioaccumulation, that enable them to survive in and remediate contaminated habitats.

## 8 Mechanisms of microbial bioremediation

Microbial bioremediation operates through a variety of well-defined mechanisms that enable the breakdown, transformation, or immobilization of pollutants in contaminated environments. These mechanisms include enzymatic oxidation and reduction, which chemically modify toxic compounds into less harmful forms, as well as strategies like bioaugmentation and biostimulation that enhance microbial activity in situ. Other processes, such as bioleaching, biosorption, and bioaccumulation, enable microbes to mobilize, bind, or internalize heavy metals and organics, while microbial-induced precipitation facilitates the conversion of soluble contaminants into stable, insoluble forms. Together, these mechanisms form the foundation of sustainable and effective bioremediation technologies.

### 8.1 Microbial bioremediation using enzymes

Microorganisms exploit oxidative enzymes such as laccases, peroxidases (e.g., lignin peroxidase, manganese peroxidase), and catalase-peroxidases to degrade complex organic pollutants like PAHs, dyes, phenols, antibiotics, and chlorophenols via electron transfer reactions that generate radical intermediates and reduce molecular oxygen to water (C. Di Dong et al., 2023; Unuofin et al., 2019). These multicopper oxidases and heme-containing peroxidases, produced primarily by white-rot fungi—and also found in bacteria—cleave aromatic rings and lignin-like structures without requiring external cofactors such as $H_2O_2$ or $Mn^{2+}$ in the case of laccases, thereby enhancing substrate specificity and catalytic resilience across diverse pH and temperature ranges (Unuofin et al., 2019).

## 8.2 Enzymatic reduction

Microbial enzymatic reduction uses specialized enzymes, such as reductive dehalogenases and chromate reductases, to detoxify pollutants by transferring electrons under anaerobic or facultative conditions. Organohalide-respiring bacteria, such as *Dehalococcoides* spp., employ membrane-bound reductive dehalogenases (e.g., TceA, PceA) to sequentially dechlorinate chlorinated solvents (PCE → TCE → DCE → ethene) with hydrogen as the electron donor (Nijenhuis and Zinder, 2005). Fungi, such as *Purpureocillium lilacinum,* produce NADH-dependent reductive enzymes that reduce toxic Cr(VI) to the less harmful Cr(III), with NADH acting as an electron donor (González et al., 2021). These enzymatic pathways convert highly toxic halogenated chemicals and metals into benign forms, enabling effective in situ remediation of soil and groundwater contaminants.

## 8.3 Bioaugmentation

Bioaugmentation involves introducing specialized microbial strains or consortia into contaminated environments to enhance pollutant degradation beyond native microbial capacities. For instance, inoculating petroleum-contaminated soil with alkane-degrading strains (e.g., *Arthrobacter* and *Pseudomonas*) significantly improved diesel degradation and plant colonization in maize rhizospheres, achieving up to 80% hydrocarbon removal compared to controls (Ummara et al., 2021). Bacterial consortia containing *Aeromonas* and *Shewanella* demonstrated superior resistance to copper, zinc, and nickel, with reductions of ~50%–62% of heavy metals in cocultures versus monocultures (Khidr et al., 2025). Filamentous fungal consortia have also been successfully used to bioaugment heavy metal-polluted soils, improving PAH removal (up to 36%) and mercury uptake (Dusengemungu et al., 2022). Bioaugmentation thus offers targeted, high-efficiency remediation tailored to specific pollutants and site conditions.

## 8.4 Biostimulation

Biostimulation enhances existing soil microbial communities by adding nutrients (e.g., nitrogen, phosphorus), electron donors, or slow-release organic amendments such as biochar, biosolids, compost, or agricultural residues to accelerate the degradation of hydrocarbons, chlorinated compounds, and other pollutants without introducing external microbes. Controlled nutrient amendments (e.g., C/N/P ratios of ~100:10:1) have effectively increased total petroleum hydrocarbon removal to >75% within weeks by stimulating native Proteobacteria, Actinobacteria, and Firmicutes populations (Mekonnen et al., 2024). Innovations like chitosan-encapsulated slow-release nutrient capsules have doubled hydrocarbon degradation (43% → 80% in 7 days) in groundwater systems (Wang et al., 2023), demonstrating the power of tailored biostimulation strategies for effective in situ bioremediation (Fig. 15.3).

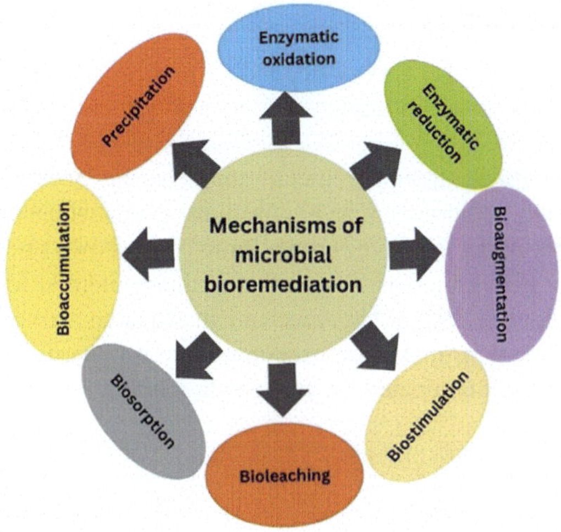


**FIG. 15.3**

Key mechanisms in microbial bioremediation processes.

## 8.5 Bioleaching

Bioleaching employs acidophilic microbes such as *Acidithiobacillus ferrooxidans*, *Acidithiobacillus thiooxidans*, and *Leptospirillum ferriphilum* to oxidize sulfide minerals and solubilize metals like copper, nickel, zinc, and gold from ores or contaminated soils under acidic conditions (Fu et al., 2008). These bacteria generate ferric iron ($Fe^{3+}$) and sulfuric acid through iron and sulfur oxidation, which chemically leach metals into solution. In shake-flask and pilot studies, mixed cultures of *A. ferrooxidans* and *L. ferriphilum* have increased copper extraction from chalcopyrite by ~30%–46% compared to pure cultures (Zhang et al., 2008). This eco-friendly, low-energy process offers a sustainable alternative to traditional smelting for metal recovery and soil detoxification.

## 8.6 Biosorption

Biosorption is a passive, physicochemical process where microbial biomass, whether live or dead, adsorbs heavy metals (e.g., $Pb^{2+}$, $Cd^{2+}$, $Cu^{2+}$, $Zn^{2+}$) via reactive cell-wall functional groups like carboxyls and amines. Both Gram-positive and Gram-negative bacteria show high sorption capacities (e.g., *Pseudomonas pseudoalcaligenes* binds 272 mg Pb/g dry biomass at pH 5) (Leung et al., 2001), and extracellular polymeric substances (EPS) from *Pseudomonas fluorescens* adsorb metals even more efficiently (e.g., $Zn^{2+}$, $Pb^{2+}$). Fungal and algal biomasses, such as *Aspergillus* spp. and *Spirulina,* also exhibit rapid biosorption (e.g., 90% Pb within 1 h) (Ayele et al., 2021). This low-cost, reversible, and high-capacity mechanism offers a sustainable approach to removing toxic metals from wastewater and soil.

### 8.7 Bioaccumulation

Bioaccumulation is an active, energy-dependent process where living microorganisms transport and sequester pollutants such as heavy metals and radionuclides inside their cells, often using proteins like metallothioneins for intracellular binding. Engineered *E. coli* expressing human metallothioneins can accumulate up to 6.4 mg $Cd^{2+}$/g cell dry weight, demonstrating high-affinity metal sequestration (Thai et al., 2023). Naturally occurring *Bacillus cereus* strains have shown up to 79% lead accumulation (~194 mg $Pb^{2+}$/g biomass), underlining their potential for in situ remediation (Sharma and Shukla, 2021). This mechanism relies on active transport systems and intracellular chelation, making bioaccumulation a powerful tool for removing pollutant loads from contaminated environments without external chemical inputs (Adil et al., 2024).

### 8.8 Precipitation

Microbial-induced precipitation transforms dissolved contaminants into stable, solid forms such as sulfides, phosphates, and carbonates through metabolic alterations in pH, redox conditions, or enzyme secretion. Ureolytic bacteria (e.g., *Sporosarcina pasteurii*) hydrolyze urea via urease, increasing local pH and promoting calcium carbonate formation that coprecipitates heavy metals like Cd, Pb, and Zn (Joshi et al., 2023). Sulfate-reducing bacteria (e.g., *Desulfovibrio* spp.) generate hydrogen sulfide under anaerobic conditions, precipitating metals such as $Cd^{2+}$, $Pb^{2+}$, and $Hg^{2+}$ as insoluble sulfides. Iron-oxidizing bacteria (e.g., *Shewanella*) induce hydrous ferric oxide precipitation, immobilizing contaminants, including $UO_2^{2+}$, $Sr^{2+}$, and $AsO_4^{3}$ (Ferris et al., 2021). Such in situ precipitation mechanisms immobilize metals efficiently and establish long-term stability within the soil matrix.

## 9 Factors affecting microbial bioremediation

Microbial bioremediation is a complex and dynamic process influenced by multiple environmental, chemical, and biological factors. The efficiency of pollutant degradation depends not only on the presence of competent microbial communities but also on optimal conditions such as pH, temperature, moisture, nutrient availability, and oxygen levels.

### 9.1 pH

pH significantly influences microbial activity and enzymatic efficiency in bioremediation. Most bioremediating bacteria and fungi function optimally at near-neutral pH (6.5–7.5); extremes can denature enzymes or disrupt membrane transport processes. The degradation of petroleum hydrocarbons by *P. putida* significantly decreases below pH 5.5 and above pH 9.0, highlighting the need for near-neutral conditions (Fatima Kareem Shandookh et al., 2024).

## 9.2 Temperature

Temperature affects microbial metabolism and enzymatic turnover. Hydrocarbon-degrading bacteria display optimal activity between 25°C and 35°C, with significantly reduced degradation at 10°C. Thermophilic microbes, such as *Geobacillus* spp., maintain high bioremediation rates at elevated temperatures, which are useful for composting-based systems (Pandey et al., 2016).

## 9.3 Moisture content

Moisture regulates microbial mobility, nutrient solubility, and substrate diffusion in soils. An optimal moisture range (40%–70% water-holding capacity) enhances biodegradation, while extremes (dry or saturated) suppress oxygen diffusion and microbial activity.

## 9.4 Nutrient availability

Microbial growth depends on the balance of carbon, nitrogen, and phosphorus. Hydrocarbon degradation improves significantly when the C/N/P ratio is maintained at ~100:10:1.

## 9.5 Oxygen availability

Oxygen is a critical electron acceptor for aerobic biodegradation enzymes (e.g., mono- and dioxygenases). Soil bioremediation rates plummet under low $O_2$ conditions; techniques like bioventing or biosparging are often used to ensure adequate oxygenation (Das and Chandran, 2011).

# 10 Microbial enzymes used in bioremediation

Microbial enzymes are crucial catalysts in bioremediation, enabling the breakdown of toxic pollutants into less harmful forms. These enzymes, produced by bacteria and fungi, target contaminants like hydrocarbons, dyes, and pesticides with high specificity and efficiency. Table 15.1 summarizes key enzymes, their microbial sources, and modes of action.

# 11 Molecular approaches for validating microbial remediation

Molecular approaches provide precise tools to monitor and validate microbial remediation by detecting specific genes, enzymes, and active microbial populations. Techniques such as quantitative polymerase chain reaction (qPCR), metagenomics,

**Table 15.1** Microbial enzymes involved in bioremediation: Sources and mechanisms.

| Enzyme | Function in bioremediation | Microbial source | References |
|---|---|---|---|
| Laccase | Oxidizes phenols, dyes, and PAHs through multicopper oxidase activity | *Pleurotus ostreatus*, *Rhodococcus erythropolis* | (J. Yang et al., 2017) |
| Lignin peroxidase (LiP) and manganese peroxidase (MnP) | Breaks down lignin-like structures and high-molecular-weight pollutants through $H_2O_2$ activation | White-rot fungi (*Trametes versicolor*, *Phanerochaete chrysosporium*) | (Xu et al., 2013) |
| Reductive dehalogenase | Removes halogens from chlorinated compounds, detoxifying them | *Dehalococcoides* spp. | (Ruttimann-Johnson et al., 1994) |
| Chromate reductase | Reduces Cr(VI) to less toxic Cr(III) | *Shewanella oneidensis* | (Xu et al., 2013) |
| Nitroreductase | Reduces nitroaromatic compounds using NAD(P)H | *Escherichia coli*, *Enterobacter cloacae* | (Baborová et al., 2006) |
| Hydrolases (esterase, lipase) | Hydrolyzes ester bonds in pesticides and plastics | *Pseudomonas putida*, *Bacillus subtilis* | (Das et al., 2016; Kumar and Arora, 2022) |

and stable isotope probing (SIP) offer real-time insights into biodegradation processes, enabling accurate assessment and optimization of environmental clean-up strategies.

## 11.1 Polymerase chain reaction and qPCR for functional gene tracking

Polymerase chain reaction (PCR) and qPCR are essential tools for detecting and monitoring specific catabolic genes in bioremediation sites. For example, PCR with degenerate primers has identified reductive dehalogenase genes from *Dehalobacter* and *Desulfitobacterium* in 1,2-DCA-contaminated aquifers (Marzorati et al., 2010). Degenerate primer PCR screens also revealed a deep diversity of dehalogenase genes in contaminated and pristine samples, highlighting novel biodegradation potential (Hug and Edwards, 2013).

## 11.2 Community profiling (Terminal-Restriction Fragment Length Polymorphism (T-RFLP), Automated Ribosomal Intergenic Spacer Analysis (ARISA))

T-RFLP is a reliable molecular fingerprinting technique for monitoring microbial community dynamics during bioremediation (De Vrieze et al., 2018). Although not directly linked to functional gene activity, community fingerprinting via techniques like T-RFLP and ARISA provides a high-throughput snapshot of microbial succession during remediation. Shifts in operational taxonomic unit patterns can validate community responses to pollutant degradation.

## 11.3 SIP and metagenomics

SIP and metagenomics are advanced molecular techniques used to identify and characterize active pollutant-degrading microbes by linking metabolic function to microbial identity in complex environmental samples. SIP, using $^{13}C$-labeled compounds, enables the linkage of active degraders to catabolic pathways. In PCB-polluted sediments, SIP identified *Polaromonas* and *Pseudonocardia* as key PCB degraders, with metagenomics confirming their dioxygenase genes (Dang et al., 2023).

## 11.4 Metatranscriptomics, proteomics, and metabolomics

Metatranscriptomics, proteomics, and metabolomics are omics-based molecular approaches that provide real-time insights into microbial function during bioremediation. Metatranscriptomics analyzes gene expression (mRNA), proteomics identifies active proteins and enzymes, and metabolomics detects the intermediate and final products of pollutant degradation, together offering a comprehensive view of microbial activity and pathway functionality in contaminated environments (Dang et al., 2023).

# 12 Other bioremediation metabolites produced by microbes

During bioremediation, microbes not only degrade pollutants but also release a variety of secondary metabolites that enhance the effectiveness of the process. These metabolites play roles in pollutant solubilization, metal chelation, bioavailability enhancement, and ecological stabilization. Compounds such as biosurfactants, organic acids, siderophores, and exopolysaccharides act synergistically with microbial enzymatic processes to improve pollutant removal, especially in complex or low-bioavailability environments.

# 13 Recent advancements in microbial bioremediation

Recent advancements in microbial bioremediation have significantly expanded the scope and efficiency of pollution cleanup using microorganisms. Modern tools in genetic engineering, synthetic biology, omics technologies, and computational modeling are enabling the design of highly specialized and adaptable microbial systems. These innovations help tackle complex contaminants, including heavy metals, persistent organic pollutants, and petroleum derivatives, which traditional methods often struggle to remediate.

One of the most impactful developments is the creation of GEMs. Researchers have modified bacteria like *P. putida* and *Escherichia coli* to express additional degradative enzymes or pathways, allowing them to target specific pollutants with improved efficiency. These engineered strains demonstrate enhanced resistance to toxicity and increased metabolic versatility (Samin et al., 2014). Furthermore, CRISPR–Cas gene editing tools have revolutionized how specific biodegradation pathways are activated or suppressed within microbial genomes. This precise editing allows researchers to optimize catabolic genes responsible for breaking down pollutants, creating tailor-made microbes for specific site conditions. Another promising advancement is the use of synthetic microbial consortia, where scientists rationally assemble communities of different microbial species that work synergistically. Each microbe within the consortium performs a unique function, collectively leading to more complete and robust pollutant degradation. Such synthetic ecosystems have proven effective in degrading complex organic mixtures and heavy metals (Wu et al., 2024).

The integration of metagenomics and metatranscriptomics enables researchers to uncover and monitor the roles of uncultured microbes and their active genes during remediation. These omics approaches provide insights into microbial diversity, community structure, and real-time functional gene expression in polluted environments (Behera et al., 2020). Moreover, biosensor-enabled microbial strains have been engineered to emit detectable signals—such as fluorescence or luminescence—when exposed to specific pollutants. These biosensors allow on-site, real-time monitoring of pollutant presence and microbial activity, offering valuable feedback for process control (X. Liu et al., 2010). Innovations in immobilized microbial systems and 3D biofilm bioreactors have further enhanced remediation efficiency. By attaching microbes to carriers or encapsulating them in biopolymer matrices, researchers have improved microbial stability and pollutant contact in dynamic treatment systems. Finally, the integration of artificial intelligence (AI) and machine learning models into microbial bioremediation research is gaining momentum. These tools help predict microbial behavior, degradation pathways, and pollutant interactions under varying environmental conditions, significantly improving remediation planning and system optimization. Together, these recent advancements are reshaping the future of environmental clean-up, making microbial bioremediation more precise, scalable, and effective for both industrial and ecological applications.

## 14 Future perspectives and conclusions

The field of microbial bioremediation is poised for a transformative leap, owing to technological convergence and growing environmental urgency. Advanced synthetic biology tools, such as CRISPR–Cas systems and modular genetic circuits, now allow precise reprogramming of microbial genomes for enhanced pollutant degradation, metal detoxification, and biosensing. Simultaneously, the integration of high-throughput omics—genomics, proteomics, metabolomics, and transcriptomics—is providing deep insights into microbial community structure, function, and adaptability under various environmental conditions. These technologies are paving the way for engineered microbial consortia that can adapt to fluctuating pH, temperature, salinity, and toxin concentrations, improving the robustness of bioremediation systems.

Moreover, the future of microbial remediation lies in smart, real-time responsive systems. The use of bioelectrochemical platforms, such as MFCs and MECs, presents dual benefits: effective waste degradation and concurrent generation of energy or value-added products. Nanoengineering advances, including enzyme immobilization and polymer-coated microbial cells, are enhancing catalytic efficiency and environmental resilience. Meanwhile, encapsulation technologies protect GEMs from native competition and environmental stress, enabling their survival and functionality in situ. The integration of microbial processes with CE models will further enable the transformation of waste into useful resources, such as bioplastics, fertilizers, and green fuels, fostering resource efficiency and climate resilience.

Looking ahead, addressing regulatory concerns and biosafety risks associated with GEMs will be critical to ensuring public trust and global applicability. Policy frameworks must evolve to accommodate risk-based approvals, region-specific protocols, and transparent monitoring systems. In parallel, low-cost, modular bioreactors suited to decentralized and rural applications will be essential, especially in low- and middle-income countries where conventional infrastructure is lacking. In conclusion, microbial innovations offer a holistic, adaptable, and sustainable path forward, redefining how society perceives waste, pollution, and environmental recovery in the 21st century.

## 15 Recommendations

To accelerate the transition from laboratory-scale innovation to field-level implementation, a multidimensional strategy is essential. Governments and industry stakeholders must prioritize the scale-up of microbial remediation technologies through pilot and demonstration projects in real waste treatment contexts, particularly targeting MSW, landfill leachate, agroindustrial effluents, and e-waste dumps. Investment in modular and decentralized treatment units, such as anaerobic digesters coupled with MFCs or biofilters, can provide localized solutions with minimal maintenance, making them ideal for rural and periurban communities. In addition, there is a pressing need to integrate digital biosensing tools and AI-driven microbial monitoring systems into treatment workflows to enable real-time analysis and predictive maintenance, thereby improving system reliability and transparency.

Policy and regulation must be realigned to encourage the safe deployment of GEMs under well-defined risk assessment frameworks. This includes adopting international biosafety protocols, providing training to environmental agencies, and creating clear guidelines for postrelease monitoring and emergency response. Public awareness and stakeholder engagement should form the backbone of any rollout strategy; by demystifying microbial processes and highlighting their benefits, community resistance can be transformed into participation. Academic and industrial research partnerships should be incentivized to develop location-specific microbial consortia and biocatalysts tailored to local waste streams. Financial mechanisms such as green bonds, tax incentives, and innovation grants must support startups and industries adopting these technologies. Finally, fostering interdisciplinary education and capacity building in environmental microbiology, biochemical engineering, and ecological policy will ensure a skilled workforce ready to drive this microbial revolution forward.

## References

Abedin, T., Pasupuleti, J., Paw, J.K.S., Tak, Y.C., Islam, M.R., Basher, M.K., Nur-E-Alam, M., 2025. From waste to worth: advances in energy recovery technologies for solid waste management. Clean Technol. Environ. Policy. 16189558. Available from: https://www.springer.com/journal/10098. doi:10.1007/s10098-025-03204-x.

Adami, L., Schiavon, M., 2021. From circular economy to circular ecology: a review on the solution of environmental problems through circular waste management approaches. Sustainability. 2071-1050. 13 (2), 925. doi:10.3390/su13020925.

Agnello, A.C., Bagard, M., van Hullebusch, E.D., Esposito, G., Huguenot, D., 2016. Comparative bioremediation of heavy metals and petroleum hydrocarbons co-contaminated soil by natural attenuation, phytoremediation, bioaugmentation and bioaugmentation-assisted phytoremediation. Sci. Total Environ. 00489697. 563–564, 693–703. doi:10.1016/j.scitotenv.2015.10.061.

Akomea-Frimpong, I., Tetteh, P.A., Ofori, J.N.A., Tumpa, R.J., Pariafsai, F., Tenakwah, E.S., Asogwa, I.E., Vanapalli, K.R., Adu-Gyamfi, G., Kukah, A.S., Tenakwah, E.J., 2024. A bibliometric review of barriers to circular economy implementation in solid waste management. Discov. Environ. 27319431. 2 (1). Available from: https://link.springer.com/journal/44274. doi:10.1007/s44274-024-00050-4.

Almotairy, H., 2024. Heavy Metal Contamination in Soil: Implications for Crop Resilience and Abiotic Stress Management. IntechOpen. doi:10.5772/intechopen.114978.

Angon, P.B., Islam, M.S., KC, S., Das, A., Anjum, N., Poudel, A., Suchi, S.A., 2024. Sources, effects and present perspectives of heavy metals contamination: soil, plants and human food chain. Heliyon. 24058440. 10 (7). Available from: https://www.sciencedirect.com/science/journal/24058440. doi:10.1016/j.heliyon.2024.e28357.

Ao, T.J., Liu, C.G., Sun, Z.Y., Zhao, X.Q., Tang, Y.Q., Bai, F.W., 2024. Anaerobic digestion integrated with microbial electrolysis cell to enhance biogas production and upgrading in situ. Biotechnol. Adv. 07349750. 73. Available from: https://www.sciencedirect.com/science/journal/07349750. doi:10.1016/j.biotechadv.2024.108372.

Apollon, W., Rusyn, I., Paucar, N.E., Hibbert, M., Kamaraj, S.K., Sato, C., 2025. Energy recovery from organic wastes using microbial fuel cells: traditional and nonconventional organic substrates. Resources. 20799276. 14 (3). Available from: www.mdpi.com/journal/resources. doi:10.3390/resources14030047.

Arantza, S.J., Hiram, M.R., Erika, K., Chávez-Avilés, M.N., Valiente-Banuet, J.I., Fierros-Romero, G., 2022. Bio- and phytoremediation: plants and microbes to the rescue of heavy metal polluted soils. SN Appl. Sci. 25233971. 4 (2). Available from: https://springer.com/snas. doi:10.1007/s42452-021-04911-y.

Assamoi, B., Lawryshyn, Y., 2012. The environmental comparison of landfilling vs. incineration of MSW accounting for waste diversion. Waste Manag. 0956053X. 32 (5), 1019–1030. doi:10.1016/j.wasman.2011.10.023.

Awogbemi, O., Kallon, D.V.V., 2022. Valorization of agricultural wastes for biofuel applications. Heliyon. 24058440. 8 (10), e11117. doi:10.1016/j.heliyon.2022.e11117.

Ayele, A., Haile, S., Alemu, D., Kamaraj, M., Cameselle, C., 2021. Comparative utilization of dead and live fungal biomass for the removal of heavy metal: a concise review. Sci. World J. 1537-744X. 2021, 1–10. doi:10.1155/2021/5588111.

Ayilara, M.S., Babalola, O.O., 2023. Bioremediation of environmental wastes: the role of microorganisms. Front. Agron. 26733218. 5. Available from: www.frontiersin.org/journals/agronomy#. doi:10.3389/fagro.2023.1183691.

Baborová, P., Möder, M., Baldrian, P., Cajthamlová, K., Cajthaml, T., 2006. Purification of a new manganese peroxidase of the white-rot fungus Irpex lacteus, and degradation of polycyclic aromatic hydrocarbons by the enzyme. Res. Microbiol. 09232508. 157 (3), 248–253. doi:10.1016/j.resmic.2005.09.001.

Balíková, K., Vojtková, H., Duborská, E., Kim, H., Matúš, P., Urík, M., 2022. Role of exopolysaccharides of Pseudomonas in heavy metal removal and other remediation strategies. Polymers. 2073-4360. 14 (20), 4253. doi:10.3390/polym14204253.

Barbhuiya, S., Das, B.B., Adak, D., 2024. A comprehensive review on integrating sustainable practices and circular economy principles in concrete industry. J. Environ. Manag. 03014797. 370, 122702. doi:10.1016/j.jenvman.2024.122702.

Beaumelle, L., Tison, L., Eisenhauer, N., Hines, J., Malladi, S., Pelosi, C., Thouvenot, L., Phillips, H.R.P., 2023. Pesticide effects on soil fauna communities—A meta-analysis. J. Appl. Ecol. 13652664. 60 (7), 1239–1253. Available from: http://onlinelibrary.wiley.com/journal/10.1111/(ISSN)1365-2664. doi:10.1111/1365-2664.14437.

Behera, B.K., Chakraborty, H.J., Patra, B., Rout, A.K., Dehury, B., Das, B.K., Sarkar, D.J., Parida, P.K., Raman, R.K., Rao, A.R., Rai, A., Mohapatra, T., 2020. Metagenomic analysis reveals bacterial and fungal diversity and their bioremediation potential from sediments of River Ganga and Yamuna in India. Front. Microbiol. 1664302X. 11. Available from: https://www.frontiersin.org/journals/microbiology#. doi:10.3389/fmicb.2020.556136.

Bertone, M., Stabile, L., Buonanno, G., 2024. An overview of waste-to-energy incineration integrated with carbon capture utilization or storage retrofit application. Sustainability. 2071-1050. 16 (10), 4117. doi:10.3390/su16104117.

Cai, P.-T., Chen, T., Chen, B., Wang, Y.-C., Ma, Z.-Y., Yan, J.-H., 2023. The impact of pollutant emissions from co-incineration of industrial waste in municipal solid waste incinerators. Fuel. 00162361. 352, 129027. doi:10.1016/j.fuel.2023.129027.

Capodaglio, A.G., Bolognesi, S., Cecconet, D., 2021. Sustainable, decentralized sanitation and reuse with hybrid nature-based systems. Water. 2073-4441. 13 (11), 1583. doi:10.3390/w13111583.

Chandrasekhar, K., Kumar, A.N., Raj, T., Kumar, G., Kim, S.H., 2021. Bioelectrochemical system-mediated waste valorization. Syst. Microbiol. Biomanuf. 26627663. 1 (4), 432–443. Available from: https://www.springer.com/journal/43393. doi:10.1007/s43393-021-00039-7.

Chen, Y., He, X., Gao, J., Wang, F., Hou, Y., Cai, Q., Liu, Q., 2024. Biochar assisted bioremediation of soils with combined contamination of petroleum hydrocarbons and heavy metals: a review. Appl. Soil Ecol. 09291393. 204, 105720. doi:10.1016/j.apsoil.2024.105720.

Chibuike, G.U., Obiora, S.C., 2014. Heavy metal polluted soils: effect on plants and bioremediation methods. Appl. Environ. Soil Sci. 1687-7667. 2014, 1–12. doi:10.1155/2014/752708.

Clean Coastal Waters, 2000. Available from: https://doi.org/10.17226/9812.

Climate Change 2022: mitigation of climate change. Available from: https://www.ipcc.ch/report/ar6/wg3/.

Colares, G.S., Dell'Osbel, N., Paranhos, G., Cerentini, P., Oliveira, G.A., Silveira, E., Rodrigues, L.R., Soares, J., Lutterbeck, C.A., Rodriguez, A.L., Vymazal, J., Machado, Ê.L., 2022. Hybrid constructed wetlands integrated with microbial fuel cells and reactive bed filter for wastewater treatment and bioelectricity generation. Environ. Sci. Pollut. Res. 16147499. 29 (15), 22223–22236. Available from: https://link.springer.com/journal/11356. doi:10.1007/s11356-021-17395-5.

Cole-Hunter, T., Johnston, F.H., Marks, G.B., Morawska, L., Morgan, G.G., Overs, M., Porta-Cubas, A., Cowie, C.T., 2020. The health impacts of waste-to-energy emissions: a systematic review of the literature. Environ. Res. Lett. 17489326. 15 (12). Available from: https://iopscience.iop.org/article/10.1088/1748-9326/abae9f. doi:10.1088/1748-9326/abae9f.

Dai, L., Zhou, N., Yuancai, L., Cheng, Y., Wang, Y., Liu, Y., Cobb, K., Chen, P., Lei, H., Ruan, R., 2022. Pyrolysis technology for plastic waste recycling: a state-of-the-art review. Prog. Energy Combust. Sci. 03601285. 93, 101021. doi:10.1016/j.pecs.2022.101021.

Dang, H., Ewald, J.M., Mattes, T.E., 2023. Genome-resolved metagenomics and metatranscriptomics reveal insights into the ecology and metabolism of anaerobic microbial communities in PCB-contaminated sediments. Environ. Sci. Technol. 15205851. 57 (43), 16386–16398. Available from: http://pubs.acs.org/journal/esthag. doi:10.1021/acs.est.3c05439.

Das, A., Bhattacharya, S., Gangotri Panchanan, B.S.N., Nambiar, P., 2016. Production, characterization and Congo red dye decolourizing efficiency of a laccase from Pleurotus ostreatus MTCC 142 cultivated on co-substrates of paddy straw and corn husk. J. Genet. Eng. Biotechnol. 1687157X. 14 (2), 281–288. doi:10.1016/j.jgeb.2016.09.007.

Das, N., Chandran, P., 2011. Microbial degradation of petroleum hydrocarbon contaminants: an overview. Biotechnol. Res. Int. 2090-3138. 2011, 1–13. doi:10.4061/2011/941810.

De Vrieze, J., Ijaz, U.Z., Saunders, A.M., Theuerl, S., 2018. Terminal restriction fragment length polymorphism is an "old school" reliable technique for swift microbial community screening in anaerobic digestion. Sci. Rep. 20452322. 8 (1). Available from: www.nature.com/srep/index.html. doi:10.1038/s41598-018-34921-7.

Deng, S., Wang, B., Zhang, H., Ruixue, Q., Sun, S., You, Q., She, Y., Zhang, F., 2024. Degradation and enhanced oil recovery potential of Alcanivorax borkumensis through production of bio-enzyme and bio-surfactant. Bioresour. Technol. 09608524. 400, 130690. doi:10.1016/j.biortech.2024.130690.

Ding, C., Chen, J., Zhu, F., Chai, L., Lin, Z., Zhang, K., Shi, Y., 2022. Biological toxicity of heavy metal(loid)s in natural environments: from microbes to humans. Front. Environ. Sci. 2296-665X. 10. doi:10.3389/fenvs.2022.920957.

Dong, C.D., Tiwari, A., Anisha, G.S., Chen, C.W., Singh, A., Haldar, D., Patel, A.K., Singhania, R.R., 2023. Laccase: a potential biocatalyst for pollutant degradation. Environ. Pollut. 18736424. 319. Available from: https://www.journals.elsevier.com/environmental-pollution. doi:10.1016/j.envpol.2023.120999.

Dong, J., Tang, Y., Nzihou, A., Chi, Y., 2019. Key factors influencing the environmental performance of pyrolysis, gasification and incineration Waste-to-Energy technologies. Energy Convers. Manag. 01968904. 196, 497–512. doi:10.1016/j.enconman.2019.06.016.

Dusengemungu, L., Gwanama, C., Simuchimba, G., Mubemba, B., 2022. Potential of bioaugmentation of heavy metal contaminated soils in the Zambian Copperbelt using autochthonous filamentous fungi. Front. Microbiol. 1664-302X. 13. doi:10.3389/fmicb.2022.1045671.

Ferris, F.G., Szponar, N., Edwards, B.A., 2021. 6.2 Applications of Microbially Induced Mineral Precipitation, Groundwater Microbiology.

Figueredo, M.B., Do Nascimento Silva, A., de Araújo Fontoura, J.R., Monteiro, R.L.S., Antônio Felipe Santos, M., 2023. Anaerobic Digestion of Solid Waste A Review of the State of the Art.

Fu, B., Zhou, H., Zhang, R., Qiu, G., 2008. Bioleaching of chalcopyrite by pure and mixed cultures of Acidithiobacillus spp. and Leptospirillum ferriphilum. Int. Biodeterior. Biodegrad. 09648305. 62 (2), 109–115. doi:10.1016/j.ibiod.2007.06.018.

González, J.F.C., Rodríguez, I.A., Figueroa, Y.T., Oliveras, P.L., Flores, R.M., Pérez, A.S.R., 2021. Biotransformation of chromium (Vi) via a reductant activity from the fungal strain purpureocillium lilacinum. J. Fungi. 2309608X. 7 (12). Available from: https://www.mdpi.com/2309-608X/7/12/1022/pdf. doi:10.3390/jof7121022.

Gul, I., Adil, M., Lv, F., Li, T., Chen, Y., Lu, H., Ahamad, M.I., Lu, S., Feng, W., 2024. Microbial strategies for lead remediation in agricultural soils and wastewater: mechanisms, applications, and future directions. Front. Microbiol. 1664302X. 15. Available from: https://www.frontiersin.org/journals/microbiology#. doi:10.3389/fmicb.2024.1434921.

Hama Aziz, K.H., Mustafa, F.S., Omer, K.M., Hama, S., Hamarawf, R.F., Rahman, K.O., 2023. Heavy metal pollution in the aquatic environment: efficient and low-cost removal approaches to eliminate their toxicity: a review. RSC Adv. 20462069. 13 (26), 17595–17610. Available from: http://pubs.rsc.org/en/journals/journal/ra. doi:10.1039/d3ra00723e.

Han, P., Teo, W.Z., Yew, W.S., 2022. Biologically engineered microbes for bioremediation of electronic waste: wayposts, challenges and future directions. Eng. Biol. 2398-6182. 6 (1), 23–34. doi:10.1049/enb2.12020.

Hug, L.A., Edwards, E.A., 2013. Diversity of reductive dehalogenase genes from environmental samples and enrichment cultures identified with degenerate primer PCR screens. Front. Microbiol. 1664302X. 4. Available from: http://www.frontiersin.org/Journal/FullText.aspx?ART_DOI=10.3389/fmicb.2013.00341&x=y#h9. doi:10.3389/fmicb.2013.00341.

Ifon, B.E., Togbé, A.C.F., Tometin, L.A.S., Suanon, F., Yessoufou, A., Ifon, B.E., Togbé, A.C.F., Tometin, L.A.S., Suanon, F., Yessoufou, A., 2019. Metal-contaminated soil remediation: phytoremediation, chemical leaching and electrochemical remediation. Metals Soil—Contam. Remediat. doi:10.5772/INTECHOPEN.81223.

Joshi, S., Gangola, S., Bhandari, G., Bhandari, N.S., Nainwal, D., Rani, A., Malik, S., Slama, P., 2023. Rhizospheric bacteria: the key to sustainable heavy metal detoxification strategies. Front. Microbiol. 1664302X. 14. Available from: https://www.frontiersin.org/journals/microbiology#. doi:10.3389/fmicb.2023.1229828.

Kesäniemi, J., Jernfors, T., Lavrinienko, A., Kivisaari, K., Kiljunen, M., Mappes, T., Watts, P.C., 2019. Exposure to environmental radionuclides is associated with altered metabolic and immunity pathways in a wild rodent. Mol. Ecol. 1365294X. 28 (20), 4620–4635. Available from: http://onlinelibrary.wiley.com/journal/10.1111/(ISSN)1365-294X. doi:10.1111/mec.15241.

Khidr, R., Qurbani, K., Muhammed, V., Salim, S., Abdulla, S., Wsw, H., 2025. Synergistic effects of indigenous bacterial consortia on heavy metal tolerance and reduction. Environ. Geochem. Health. 0269-4042. 47 (3). doi:10.1007/s10653-025-02392-1.

Kim, C., Koh, T., Lee, D., Park, D., 2022. Assessment of heavy metal and oil-contaminated silty sand treatment by electrical resistance heating method. Appl. Sci. 2076-3417. 12 (9), 4630. doi:10.3390/app12094630.

Kumar, A., Arora, P.K., 2022. Biotechnological applications of manganese peroxidases for sustainable management. Front. Environ. Sci. 2296665X. 10. Available from: https://journal.frontiersin.org/journal/environmental-science. doi:10.3389/fenvs.2022.875157.

Leow, C.W., Van Fan, Y., Chua, L.S., Muhamad, I.I., Klemes, J.J., Lee, C.T., 2018. A review on application of microorganisms for organic waste management. Chem. Eng. Trans. 22839216. 63, 85–90. Available from: http://www.aidic.it/cet/18/63/015.pdf. doi:10.3303/CET1863015.

Leung, W.C., Chua, H., Lo, W., 2001. Biosorption of heavy metals by bacteria isolated from activated sludge. In: Applied Biochemistry and Biotechnology—Part A Enzyme Engineering and Biotechnology. Humana Press, China, pp. 171–184. 02732289 http://www.humanapress.com/JournalTOC.pasp?issn=0273-228991-93. doi:10.1385/ABAB:91-93:1-9:171.

Liu, X., Germaine, K.J., Ryan, D., Dowling, D.N., 2010. Whole-cell fluorescent biosensors for bioavailability and biodegradation of polychlorinated biphenyls. Sensors. 14248220. 10 (2), 1377–1398. Available from: http://www.mdpi.com/1424-8220/10/2/1377/pdf. doi:10.3390/s100201377Ireland.

Liu, Z., Li, L., Li, Z., Tian, X., 2018. Removal of sulfate and heavy metals by sulfate-reducing bacteria in an expanded granular sludge bed reactor. Environ. Technol. 1479487X. 39 (14), 1814–1822. Available from: http://www.tandf.co.uk/journals/titles/09593330.asp. doi:10.1080/09593330.2017.1340347.

Marzorati, M., Balloi, A., De Ferra, F., Daffonchio, D., 2010. Identification of molecular markers to follow up the bioremediation of sites contaminated with chlorinated compounds. Methods Mol. Biol. (Clifton, N.J.). 19406029. 668, 219–234. doi:10.1007/978-1-60761-823-2_15.

Mata-Lima, H., Silva, D.W., Nardi, D.C., Klering, S.A., de Oliveira, T.C.F., Morgado-Dias, F., 2021. Waste-to-energy: an opportunity to increase renewable energy share and reduce ecological footprint in small island developing states (SIDS). Energies. 19961073. 14 (22). Available from: https://www.mdpi.com/1996-1073/14/22/7586/pdf. doi:10.3390/en14227586.

Meilander, J., Caporaso, J.G., 2024. The microbiome science of composting and human excrement composting: a review. arXiv. 23318422. Available from: https://arxiv.org. doi:10.48550/arXiv.2409.07376.

Mekonnen, B.A., Aragaw, T.A., Genet, M.B., 2024. Bioremediation of petroleum hydrocarbon contaminated soil: a review on principles, degradation mechanisms, and advancements. Front. Environ. Sci. 2296665X. 12. Available from: https://journal.frontiersin.org/journal/environmental-science. doi:10.3389/fenvs.2024.1354422.

Miller, M.R., 2020. Oxidative stress and the cardiovascular effects of air pollution. Free Radic. Biol. Med. 18734596. 151, 69–87. Available from: www.elsevier.com/locate/freeradbiomed. doi:10.1016/j.freeradbiomed.2020.01.004.

N.R.C. Incineration, 2000. Waste Incineration Overview. Available from: https://www.ncbi.nlm.nih.gov/books/NBK233614/.

Nijenhuis, I., Zinder, S.H., 2005. Characterization of hydrogenase and reductive dehalogenase activities of Dehalococcoides ethenogenes strain 195. Appl. Environ. Microbiol. 71 (3), 1664–1667. doi:10.1128/AEM.71.3.1664-1667.2005.

O'Brien, P.L., DeSutter, T.M., Casey, F.X.M., Khan, E., Wick, A.F., 2018. Thermal remediation alters soil properties – a review. J. Environ. Manag. 10958630. 206, 826–835. Available from: https://www.sciencedirect.com/journal/journal-of-environmental-management. doi:10.1016/j.jenvman.2017.11.052.

Olhoff, A., Christensen, J., Lamb, W.F., Pathak, M., Kuramochi, T., den Elzen, M., Fransen, T., Rogelj, J., Portugal-Pereira, J., Burton, J., Bustamante, M., 2023. Emissions Gap Report 2023: Broken Record–Temperatures hit new highs, yet world fails to cut emissions (again). United Nations Environment Programme, Nairobi.

Palanisamy, G., Jung, H.Y., Sadhasivam, T., Kurkuri, M.D., Kim, S.C., Roh, S.H., 2019. A comprehensive review on microbial fuel cell technologies: processes, utilization, and advanced developments in electrodes and membranes. J. Clean. Prod. 09596526. 221, 598–621. Available from: https://www.journals.elsevier.com/journal-of-cleaner-production. doi:10.1016/j.jclepro.2019.02.172.

Pandey, P., Pathak, H., Dave, S., 2016. Microbial ecology of hydrocarbon degradation in the soil: a review. Res. J. Environ. Toxicol. 18193420. 10 (1), 1–15. doi:10.3923/rjet.2016.1.15.

Prasetyo, J., Naruse, K., Kato, T., Boonchird, C., Harashima, S., Park, E.Y., 2011. Bioconversion of paper sludge to biofuel by simultaneous saccharification and fermentation using a cellulase of paper sludge origin and thermotolerant Saccharomyces cerevisiae TJ14. Biotechnol. Biofuels. 17546834. 4. doi:10.1186/1754-6834-4-35.

Priya, A.K., Muruganandam, M., Ali, S.S., Kornaros, M., 2023. Clean-up of heavy metals from contaminated soil by phytoremediation: a multidisciplinary and eco-friendly approach. Toxics. 23056304. 11 (5). Available from: http://www.mdpi.com/journal/toxics. doi:10.3390/toxics11050422.

Pryor, J.T., Cowley, L.O., Simonds, S.E., 2022. The physiological effects of air pollution: particulate matter, physiology and disease. Front. Public Health. 22962565. 10. Available from: http://journal.frontiersin.org/journal/public-health/section/public-health-education-and-promotion#archive. doi:10.3389/fpubh.2022.882569.

Raffa, C.M., Chiampo, F., Shanthakumar, S., 2021. Remediation of metal/metalloid-polluted soils: a short review. Appl. Sci. 2076-3417. 11 (9), 4134. doi:10.3390/app11094134.

Rashid, A., 2025. How circular is the linear economy? Analysing circularity, resource flows and their relation to GDP. arXiv. 23318422. Available from: https://arxiv.org. doi:10.48550/arXiv.2505.13048.

Ray, S., Shaju, S.T., 2023. Bioaccumulation of pesticides in fish resulting toxicities in humans through food chain and forensic aspects. Environ. Anal. Health Toxicol. 2671-9525. 38 (3), e2023017. doi:10.5620/eaht.2023017.

Reimann, S., Lewis, A.C., 2007. Anthropogenic VOCs. Volatile organic compounds in the atmosphere, 33–81.

Richard, E.N., Hilonga, A., Machunda, R.L., Njau, K.N., 2019. A review on strategies to optimize metabolic stages of anaerobic digestion of municipal solid wastes towards enhanced resources recovery. Sustain. Environ. Res. 24682039. 1 (1). Available from: https://link.springer.com/journal/42834. doi:10.1186/s42834-019-0037-0.

Romieu, I., Castro-Giner, F., Kunzli, N., Sunyer, J., 2007. Air pollution, oxidative stress and dietary supplementation: a review. Eur. Respir. J. 0903-1936. 31 (1), 179–197. doi:10.1183/09031936.00128106.

Ruttimann-Johnson, C., Cullen, D., Lamar, R.T., 1994. Manganese peroxidases of the white rot fungus Phanerochaete sordida. Appl. Environ. Microbiol. 00992240. 60 (2), 599–605. Available from: http://aem.asm.org/. doi:10.1128/aem.60.2.599-605.1994.

Samin, G., Pavlova, M., Arif, M.I., Postema, C.P., Damborsky, J., Janssen, D.B., 2014. A Pseudomonas putida strain genetically engineered for 1,2,3-trichloropropane bioremediation. Appl. Environ. Microbiol. 10985336. 80 (17), 5467–5476. Available from: http://aem.asm.org/content/80/17/5467.full.pdf. doi:10.1128/AEM.01620-14.

Sánchez-Castro, I., Molina, L., Prieto-Fernández, M.-Á., Segura, A., 2023. Past, present and future trends in the remediation of heavy-metal contaminated soil—remediation techniques applied in real soil-contamination events. Heliyon. 24058440. 9 (6), e16692. doi:10.1016/j.heliyon.2023.e16692.

Sarma, K., Dey, A., Kumar, S., Chaudhary, B.K., Mohanty, S., Kumar, T., Prasad, S.S., Bhatt, B.P., 2020. Effect of salinity on growth, survival and biochemical alterations in the freshwater fish labeo rohita (Hamilton 1822). Indian J. Fish. 09706011. 67 (2), 41–47. Available from: http://epubs.icar.org.in/ejournal/index.php/IJF/article/download/86894/39930. doi:10.21077/ijf.2019.67.2.86894-06.

Shandookh, F.K., Mohammed, M.K., Jabbar, A.D., 2024. Bioremediation of hydrocarbon pollutants by Pseudomonas putida under optimal conditions. Adv. Life Sci. 2310-5380. 11 (4), 761. doi:10.62940/als.v11i4.1963.

Sharma, B., Shukla, P., 2021. Lead bioaccumulation mediated by Bacillus cereus BPS-9 from an industrial waste contaminated site encoding heavy metal resistant genes and their transporters. J. Hazard. Mater. 03043894. 401, 123285. doi:10.1016/j.jhazmat.2020.123285.

Shou, Y., Zhao, J., Zhu, Y., Qiao, J., Shen, Z., Zhang, W., Han, N., Núñez-Delgado, A., 2022. Heavy metals pollution characteristics and risk assessment in sediments and waters: the case of Tianjin, China. Environ. Res. 00139351. 212 (Pt A), 113162. doi:10.1016/j.envres.2022.113162.

Singh, N.K., Mathuriya, A.S., Mehrotra, S., Pandit, S., Singh, A., Jadhav, D., 2024. Advances in bioelectrochemical systems for bio-products recovery. Environ. Technol. 1479487X. 45 (19), 3853–3876. Available from: http://www.tandf.co.uk/journals/titles/09593330.asp. doi:10.1080/09593330.2023.2234676.

Singh, P., Singh, V.K., Singh, R., Borthakur, A., Madhav, S., Ahamad, A., Kumar, A., Pal, D.B., Tiwary, D., Mishra, P.K., 2019. Bioremediation: a sustainable approach for management of environmental contaminants. In: Abatement of Environmental Pollutants: Trends and Strategies. Elsevier, India, pp. 1–23. Available from: http://www.sciencedirect.com/science/book/9780128180952. doi:10.1016/B978-0-12-818095-2.00001-1.

Sofokleous, M., Christofi, A., Malamis, D., Mai, S., Barampouti, E.M., 2022. Bioethanol and biogas production: an alternative valorisation pathway for green waste. Chemosphere. 00456535. 296, 133970. doi:10.1016/j.chemosphere.2022.133970.

Sun, X., Zhao, L., Huang, M., Hai, J., Liang, X., Chen, D., Liu, J., 2024. In-situ thermal conductive heating (TCH) for soil remediation: a review. J. Environ. Manag. 03014797. 351, 119602. doi:10.1016/j.jenvman.2023.119602.

Terashima, M., Goel, R., Komatsu, K., Yasui, H., Takahashi, H., Li, Y., Noike, T., 2009. CFD simulation of mixing in anaerobic digesters. Bioresour. Technol. 09608524. 100 (7), 2228–2233. doi:10.1016/j.biortech.2008.07.069.

Thai, T.D., Lim, W., Na, D., 2023. Synthetic bacteria for the detection and bioremediation of heavy metals. Front. Bioeng. Biotechnol. 22964185. 11. Available from: http://journal.frontiersin.org/journal/bioengineering-and-biotechnology#archive. doi:10.3389/fbioe.2023.1178680.

Thirumalaivasan, N., Gnanasekaran, L., Kumar, S., Durvasulu, R., Sundaram, T., Rajendran, S., Nangan, S., Kanagaraj, K., 2024. Utilization of fungal and bacterial bioremediation techniques for the treatment of toxic waste and biowaste. Front. Mater. 2296-8016. 11. doi:10.3389/fmats.2024.1416445.

Ummara, U., Noreen, S., Afzal, M., Ahmad, P., 2021. Bacterial bioaugmentation enhances hydrocarbon degradation, plant colonization and gene expression in diesel-contaminated soil. Physiol. Plant. 13993054. 173 (1), 58–66. Available from: http://onlinelibrary.wiley.com/journal/10.1111/(ISSN)1399-3054. doi:10.1111/ppl.13171.

Unuofin, J.O., Okoh, A.I., Nwodo, U.U., 2019. Aptitude of oxidative enzymes for treatment of wastewater pollutants: a laccase perspective. Molecules. 14203049. 24 (11). Available from: https://www.mdpi.com/1420-3049/24/11/2064/pdf. doi:10.3390/molecules24112064.

van Dyk, J., Görgens, J.F., van Rensburg, E., 2025. Valorisation of industrial waste for second-generation ethanol production: towards the scale-up of high-solids paper sludge fermentation. Waste Biomass Valoriz. 1877265X. 16 (8), 4449–4461. Available from: https://www.springer.com/journal/12649. doi:10.1007/s12649-025-02962-5.

Videvall, E., Burraco, P., Orizaola, G., 2023. Impact of ionizing radiation on the environmental microbiomes of Chornobyl wetlands. Environ. Pollut. 02697491. 330, 121774. doi:10.1016/j.envpol.2023.121774.

Wainaina, S., Awasthi, M.K., Sarsaiya, S., Chen, H., Singh, E., Kumar, A., Ravindran, B., Awasthi, S.K., Liu, T., Duan, Y., Kumar, S., Zhang, Z., Taherzadeh, M.J., 2020. Resource recovery and circular economy from organic solid waste using aerobic and anaerobic digestion technologies. Bioresour. Technol. 18732976. 301. Available from: www.elsevier.com/locate/biortech. doi:10.1016/j.biortech.2020.122778.

Wang, J., Ren, K., Zhu, Y., Huang, J., Liu, S., 2022. A review of recent advances in microbial fuel cells: preparation, operation, and application. BioTech. 2673-6284. 11 (4), 44. doi:10.3390/biotech11040044.

Wang, J., Zhang, Y., Liu, T., Shi, Y., Ding, Y., Zhang, Y., Weiqing, X., Zhang, X., Wang, Y., Dong, L., 2023. A biodegradable chitosan-based polymer for sustained nutrient release to stimulate groundwater hydrocarbon-degrading microflora. Chemosphere. 00456535. 344, 140346. doi:10.1016/j.chemosphere.2023.140346.

Wu, Z.H., Li, F., Wang, F., Jin, R., Li, Y., Li, S., Zhou, Z., Jia, P., Li, J.T., 2024. A synthetic bacterial consortium improved the phytoremediation efficiency of ryegrass on polymetallic contaminated soil. Ecotoxicol. Environ. Saf. 10902414. 282. Available from: https://www.sciencedirect.com/science/journal/01476513. doi:10.1016/j.ecoenv.2024.116691.

Xu, J.Z., Zhang, J.L., Hu, K.H., Zhang, W.G., 2013. The relationship between lignin peroxidase and manganese peroxidase production capacities and cultivation periods of mushrooms. Microb. Biotechnol. 17517915. 6 (3), 241–247. doi:10.1111/j.1751-7915.2012.00365.x.

Xu, L., Zhao, F., Xing, X., Peng, J., Wang, J., Mingfei Ji, B.L.L., 2024. A review on remediation technology and the remediation evaluation of heavy metal-contaminated soils. Toxics. 2305-6304. 12 (12), 897. doi:10.3390/toxics12120897.

Yaashikaa, P.R., Kumar, P.S., Nhung, T.C., Hemavathy, R.V., Jawahar, M.J., Neshaanthini, J.P., Rangasamy, G., 2022. A review on landfill system for municipal solid wastes: insight into leachate, gas emissions, environmental and economic analysis. Chemosphere. 18791298. 309. Available from: www.elsevier.com/locate/chemosphere. doi:10.1016/j.chemosphere.2022.136627.

Yang, J., Li, W., Ng, T.B., Deng, X., Lin, J., Ye, X., 2017. Laccases: production, expression regulation, and applications in pharmaceutical biodegradation. Front. Microbiol. 1664302X. 8 (MAY). Available from: http://journal.frontiersin.org/article/10.3389/fmicb.2017.00832/full. doi:10.3389/fmicb.2017.00832.

Yang, Z., Larsen, O.C., Muhayodin, F., Jiawei, H., Xue, B., Rotter, V.S., 2024. Review of anaerobic digestion models for organic solid waste treatment with a focus on the fates of C, N, and P. Energy Ecol. Environ. 2363-7692. doi:10.1007/s40974-024-00343-7.

Yankauskas, A., Larionova, N., Shatrov, A., Toporova, A., 2024. The effect of radionuclide and chemical contamination on morphological and anatomical parameters of plants. Plants. 2223-7747. 13 (20), 2860. doi:10.3390/plants13202860.

Yin, Y., Liu, C., Zhao, G., Chen, Y., 2022. Versatile mechanisms and enhanced strategies of pollutants removal mediated by Shewanella oneidensis: a review. J. Hazard. Mater. 03043894. 440, 129703. doi:10.1016/j.jhazmat.2022.129703.

Zhang, N., He, J., Muhammad, A., Shao, Y., 2022. CRISPR/Cas9–mediated genome editing for Pseudomonas fulva, a Novel Pseudomonas species with clinical, animal, and plant–associated isolates. Int. J. Mol. Sci. 1422-0067. 23 (10), 5443. doi:10.3390/ijms23105443.

Zhang, Y.S., Qin, W.Q., Wang, J., Zhen, S.J., Yang, C.R., Zhang, J.W., Nai, S.S., Qiu, G.Z., 2008. Bioleaching of chalcopyrite by pure and mixed culture. Trans. Nonferrous Met. Soc. China. 10036326. 18 (6), 1491–1496. doi:10.1016/S1003-6326(09)60031-5.

Zhao, W., Huppes, G., van der Voet, E., 2011. Eco-efficiency for greenhouse gas emissions mitigation of municipal solid waste management: a case study of Tianjin, China. Waste Manag. 0956053X. 31 (6), 1407–1415. doi:10.1016/j.wasman.2011.01.013.

Zheng, X.J., Li, Q., Peng, H., Zhang, J.X., Chen, W.J., Zhou, B.C., Chen, M., 2022. Remediation of heavy metal-contaminated soils with soil washing: a review. Sustainability. 20711050. 14 (20). Available from: http://www.mdpi.com/journal/sustainability/. doi:10.3390/su142013058.

CHAPTER 16

# Sustainable waste management: Challenges, opportunities, and future roadmap

**Barun Kanoo[a] and Shruti Sharma[b]**

[a]*Department of Civil Engineering, National Institute of Technology Silchar, Silchar, Assam, India*

[b]*Dillon Consulting Limited, Vancouver, BC, Canada*

## 1 Introduction

Due to population growth, rapid industrialization, and urbanization, the amount of solid waste generated in India has increased manifold over the past few decades. The increase in the quantity of solid waste poses a significant environmental challenge in terms of treatment and management, as well as for human well-being. Hence, to address its growing concerns and promote a healthy environment, sustainable waste management is the need of the hour.

### 1.1 Overview of India's waste generation trends and its impact

The total quantity of solid waste generated in India is 1,70,339 TPD, as reported by the Central Pollution Control Board (CPCB) in 2021. However, the amount of solid waste treated or processed remains relatively low, i.e., 54%. The waste hierarchy demonstrates that landfilling is the least preferred method for waste management. In addition, the landfilling of solid waste (CPCB, 2022) also poses risks of water pollution, greenhouse gas (GHG) emissions, bad odor, fire hazards, and rodent proliferation. However, landfill disposal accounts for almost 50% of the total waste treated or processed (CPCB, 2022). A critical factor in the Indian context of waste management is the rising population and the increase in per capita waste generation, with India already surpassing China, reaching 1.4 billion in 2023 (Ministry of Jal Shakti, 2022; CPCB, 2022). According to CPCB, per capita waste generation has shown an increasing trend, rising from 98.8 g/capita/day to 123.5 g/capita/day over the last 5 years (2017–2022) (CPCB, 2022). Hence, the application of sustainable waste management practices at the source of generation, coupled with stringent policies, is the way ahead in the near future.

Microbial Technologies for Sustainable Waste Management. https://doi.org/10.1016/B978-0-443-44093-9.00016-5

## 1.2 Relevance of sustainable waste management in the Indian context and its impact on attaining Sustainable Development Goals

The strategy for implementing nationwide policies for effective waste management in India should be based on sustainable development, with an emphasis on equity in resource consumption among present and future generations. Hence, proper planning is required to identify a viable waste management option that integrates economic, technical, social, and environmental factors. To achieve the Sustainable Development Goals (SDGs) (SDG-11 and SDG-12) at the national level, quantifying waste or specific waste streams is crucial for assessing environmental pressures and addressing them. Regular data from waste audits help assess progress toward the SDGs. In the national indicator framework of the SDGs, indicators such as the generation and treatment of waste include 8.4.1, 11.6.1, and 12.4.2, which represent the proportion of waste recycled vs. waste generated, the percentage of waste processed, and hazardous waste generated per capita and the proportion of hazardous waste treated, respectively (Ministry of Statistics and Programme Implementation, n.d.). The list of global SDG indicators related to waste management is provided in Table 16.1.

**Table 16.1** List of global SDG indicators related to the national framework for waste management (CPCB, 2022).

| Target | Global SDG indicators |
|---|---|
| Target 11.6: Reduce the environmental impact of cities on a per capita basis, with a focus on air quality and municipal and other waste management by the year 2030 | Indicator 11.6.1: Fraction of municipal solid waste (MSW) collected in cities and managed in facilities out of the total MSW generated |
| Target 12.4: Effective management of chemicals and all wastes throughout their life cycle as per international frameworks, and reduction of releases into air, water, and soil to minimize adverse impacts on human health and the environment by the year 2020 | Indicator 12.4.1: Number of parties to international multilateral environmental agreements on hazardous waste and other chemicals that meet their commitments and obligations in transmitting information as required by each relevant agreementIndicator 12.4.2: (a) Hazardous waste generated on a per capita basis and (b) fraction of hazardous waste treated by type of treatment |
| Target 12.5: Significant reduction in waste generation through prevention, reduction, recycling, and reuse by the year 2030 | Indicator 12.5.1: National recycling rate and tons of materials recycled |

*Source: Central Pollution Control Board, 2016. Annual report 2021–2022 on implementation of Solid Waste Management Rules.*

### 1.3 Scope of the study

The chapter reviews the current status of municipal solid waste (MSW) management in India, with an emphasis on policy frameworks, national missions, and technological interventions. It examines systemic gaps, implementation challenges, and the role of governance, community participation, and the informal sector in achieving sustainable outcomes. Furthermore, the study draws lessons from global and Indian success stories to identify scalable strategies and future pathways aligned with the SDGs.

## 2 Current status of MSW management in India

India is a developing country with a growing economy and population. Hence, waste generation is inevitable during developmental and manufacturing activities. According to the Ministry of Housing and Urban Affairs, it is estimated that the solid waste generated in Indian cities will be approximately 435 million tons by 2050 (Ministry of Housing and Urban Affairs, 2021a). The concerning factor is that around 46% of the collected waste is disposed of in landfill sites or leaks into the waste supply chain (CPCB, 2022). Therefore several initiatives have been launched in India to address the growing concern of solid waste management (SWM), such as the Swachh Bharat Mission (SBM) in 2014. Under SBM Urban 2.0 (2021–2026), the target of achieving garbage-free Indian cities by 2026 has been set. Hence, to ensure compliance with existing laws and regulations, it is necessary to understand the challenges specific to cities and the gaps at each stage of the waste treatment plan.

### 2.1 Composition of MSW in India

Solid waste generation in India ranges from 0.1 to 0.5 kg per capita per day, depending on the size of the city (Meena et al., 2023). The major fraction of MSW generated in India is the organic fraction, which accounts for 48%–50% (Kharvel, 2012; Meena et al., 2022). The remaining fraction primarily consists of recyclables (17.5%) and inerts (31%) (Kharvel, 2012). These mass percentages can vary spatially depending on the involvement of the informal sector in that region.

### 2.2 Current practices in India

Urban cities in India are experiencing rapid growth, including development, population, and income. Consequently, the rate of resource consumption has increased significantly, leading to a rise in waste generation in recent times. Recent policies in the context of SWM in India include source segregation and collection. Hence, per capita generation of solid waste is higher in urban areas than in rural areas. SBM 2.0 emphasizes creating garbage-free cities, scientific landfill management, and the bioremediation of legacy waste in dumpsites. However, due to a lack of infrastructure, such as processing facilities in many Indian cities, many municipalities still rely on open dumping. The situation worsens due to a lack of public awareness, which impedes proper segregation and, thereby, the further processing of waste. The dependence of waste generation on urbanization and rising incomes underscores the need to consider socioeconomic factors in formulating a strategy for waste management in India, with a focus on urban areas (Kapoor and Chakma, 2024).

### 2.3 Environmental, economic, and social implications of the current practice of MSW management

Critical challenges with respect to SWM in India include a lack of teeth in the policies. One important aspect of waste collection is the lack of infrastructure, which leads to mixed waste streams. This creates a significant challenge for implementing efficient recycling or disposal methods. The environmental consequences of unscientific waste dumping include GHG emissions, water and soil contamination, gaseous emissions from open burning, and biodiversity loss (Meena et al., 2022; Cooper, 2024; Mahajan, 2023).

Mismanagement of waste can have severe economic implications as well. A significant amount of precious metals and other resources is lost to landfills if they are not recovered or recycled, or if circular-economy approaches are not adopted. The surrounding area near dumpsites is also affected. Property rates are lower around the dumpsites. In addition, there are always hidden external costs associated with environmental damage, pollution-related healthcare costs, and environmental restoration (Berger, 2025; Abubakar et al., 2022).

MSW management in India can also give rise to social inequalities. The classic example is the "Not in My Backyard" movement, where nearby residents object to the establishment of waste management facilities in their area. Generally, waste management facilities are located near low-income and marginalized communities, which exposes them to odor, making them prone to diseases and pollution. Apart from that, the role of the informal sector in waste management in India cannot be ignored. However, while the informal sector contributes to resource recovery from waste, it also exposes workers to several health risks, adverse neonatal outcomes, exploitation, and social stigma (Berger, 2025; Abubakar et al., 2022; Fernando and Zutshi, 2023).

## 3 Policy frameworks and national initiatives

### 3.1 Policy frameworks

#### *3.1.1 Solid Waste Management Rules, 2016*

In accordance with the Indian Constitution, SWM falls under state jurisdiction and is incorporated into the 12th Schedule *via* the Constitution (74th Amendment) Act, 1992. In the year 2000, the Ministry of Environment and Forests (now the Ministry of Environment, Forest and Climate Change, MoEF&CC) introduced the MSW (Management and Handling) Rules, 2000, mandating municipal authorities to manage solid waste generated in cities and towns (Ministry of Urban Development, 2016). These rules outlined responsibilities for collection, segregation, storage, transportation, processing, and disposal, holding local authorities accountable for efficient waste management. However, the 2000 rules faced implementation challenges, including inadequate public awareness, technical limitations, infrastructural deficits, and weak enforcement, thus limiting their effectiveness. To address these shortcomings, the MoEF&CC revised them with the SWM Rules, 2016, which expanded the scope and introduced robust mechanisms. A few salient features of the SWM Rules, 2016, are listed as follows (CPCB, 2016):

1. Extended waste management practices beyond municipal boundaries to include urban outgrowths, pilgrimage sites, and historical places.
2. Defined mandatory duties for stakeholders, including waste generators, central ministries, pollution control boards, local authorities, and manufacturers.
3. Established timelines for infrastructure development and conducted periodic reviews to ensure effective implementation.
4. Prioritized resource recovery through recycling, composting, and waste-to-energy over traditional disposal methods.

### 3.1.2 Plastic Waste Management Rules and amendments

The Plastic Waste Management (PWM) Rules, introduced in 2016, are significant in addressing the country's escalating plastic pollution. The PWM Rules, 2016, revised the Plastic Waste (Management and Handling) Rules, 2011 (Ministry of Environment, Forest and Climate Change, 2016c). The PWM Rules, 2016, aim to minimize plastic waste generation, promote sustainable alternatives, and ensure environmentally friendly disposal methods through measures such as extended producer responsibility (EPR), under which producers, importers, and brand owners (PIBOs) are held accountable for managing plastic waste. The 2016 Rules emphasized waste segregation at the source, collection, reuse, and recycling, and a ban on plastic carry bags below 50 μm in thickness (Ministry of Environment, Forest and Climate Change, 2016c). Annual reporting to State Pollution Control Boards (SPCBs) was also mandated for the periodic review of the on-ground implementation (Singh and Biswas, 2023). Subsequently, amendments have been made to the PWM Rules, 2016, to address the gaps and emerging challenges, such as single-use plastics and microplastics.

The first amendment was introduced in 2018 and further detailed EPR compliance, including the establishment of waste collection systems, buy-back schemes for multilayered plastic pouches, and the phasing out of nonrecyclable/nonenergy-recoverable multilayered plastics within 2 years (Ministry of Environment, Forest and Climate Change, 2018c). It also strengthened the responsibilities of urban local bodies (ULBs) in developing and implementing waste management infrastructure. The second amendment came into effect in 2021, with a phase-wise ban on single-use plastics, including plastic cutlery and polystyrene, effective from July 1, 2022 (Ministry of Environment, Forest and Climate Change, 2021a). Under this amendment, the minimum thickness of plastic carry bags was increased to 75 μm by September 30, 2021, and then further to 120 μm by December 31, 2022. In addition, rural areas under Gram Panchayats were also included in the PWM Rules, 2016, further broadening their scope.

Subsequently, the amendment made in 2022 further refined EPR guidelines and introduced a centralized online portal for PIBOs, set recycling and reuse targets, and labeling standards for compostable plastics (IS/ISO 17088:2021) and recycled plastics (IS 14534:1998) (Ministry of Environment, Forest and Climate Change, 2022a). An important feature of this amendment is the "Polluter Pays Principle" announcement, which imposes environmental compensation on PIBOs for failing to meet EPR targets to safeguard the environment (Ministry of Environment, Forest and Climate Change, 2022a). This principle holds polluters liable for compensating for ecological damage, regardless of intent, thus fostering accountability.

The 2024 amendment, notified on October 30, 2024, defined biodegradable plastics as those that can be degraded biologically without leaving microplastics behind (1–1000 μm) and banned nonrecyclable multilayered plastic sachets/pouches (Ministry of Environment, Forest and Climate Change, 2024b). It also mandated the labeling of compostable plastics as "compostable only under industrial composting." As part of efforts to manage plastic waste, the Food Safety and Standards Authority of India mandated the inclusion of 30% recycled plastic in polyethylene terephthalate packaging for the beverage industry by the financial year 2026. However, challenges persist along the pathway of PWM Rules, including low public awareness, inadequate infrastructure for recycling, and reliance on informal recyclers (Hossain et al., 2022). Details on PWM Rules and their amendments are listed in Table 16.2.

**Table 16.2** Salient features of Plastic Waste Management Rules, 2016, and respective amendments.

| Rules | Key features | Timeline | References |
| --- | --- | --- | --- |
| Plastic Waste Management Rules, 2016 | • Revised version of the 2011 Rules<br>• Banned plastic carry bags with <50 μm thickness<br>• Mandated waste segregation and collection for urban areas<br>• Launched EPR for PBIOs<br>• Set requirements for PIBO registration and annual SPCB reporting | March 18, 2016 | (Ministry of Environment, Forest and Climate Change, 2016c) |

| Rules | Key features | Timeline | References |
|---|---|---|---|
| Plastic Waste Management (Amendment) Rules, 2018 | • Detailed EPR compliance<br>• Introduced buy-back schemes for multilayered plastics<br>• Improved enforcement mechanisms<br>• Strengthened ULBs' responsibilities for waste infrastructural development | March 27, 2018 | (Ministry of Environment, Forest and Climate Change, 2018c) |
| Plastic Waste Management (Amendment) Rules, 2021 | • Phase-wise ban on single-use plastics<br>• Promoted use of compostable and biodegradable plastics<br>• Increased thickness of plastic carry bags to 75 and 120 μm<br>• Broadened scope of 2016 Rules, included rural areas under Gram Panchayats | August 12, 2021 | (Ministry of Environment, Forest and Climate Change, 2021a) |
| Plastic Waste Management (Second Amendment) Rules, 2022 | • Announced labeling standards for recycling and compostable plastics<br>• Launched the polluter pays principle for EPR noncompliance<br>• Introduced a centralized EPR interface<br>• Initiated recycling and reuse targets | February 16, 2022 | (Ministry of Environment, Forest and Climate Change, 2022a) |
| Plastic Waste Management (Amendment) Rules, 2024 | • Laid down the definition of biodegradable plastics<br>• Banned nonrecyclable multilayered plastics<br>• Strengthened EPR compliance | October 30, 2024 | (Ministry of Environment, Forest and Climate Change, 2024b) |

### *3.1.3 Hazardous and Other Wastes (Management and Transboundary Movement) Rules, 2016*

Under the Environment Protection Act, 1986, the Hazardous and Other Wastes (Management and Transboundary Movement) Rules were notified on April 4, 2016, to ensure the safe handling, generation, processing, storage, transportation, treatment, and disposal of hazardous wastes in India (Ministry of Environment, Forest and Climate Change, 2016b). As per the Gazette of India, hazardous waste refers to any waste that, by virtue of its physicochemical properties, reactivity, toxicity, flammability, or corrosiveness, may pose a threat to the environment, whether on its own or in combination with other wastes (Ministry of Environment, Forest and Climate Change, 2016b). This revised version of the 2008 Rules introduced stricter regulations to prevent illegal transboundary movement, promoted safe and environmentally sound management, and encouraged the recycling and recovery of hazardous and other wastes. For effective implementation, the responsibilities of various stakeholders involved in regulating the management and disposal of hazardous wastes, including waste generators, SPCBs, CPCB, MoEF&CC, and other local authorities, were also detailed in the 2016 Rules. A total of 10 amendments have been made to this rule to improve compliance and address specific waste streams such as waste tires and used oil. The key features of the Hazardous and Other Wastes (Management and Transboundary Movement) Rules, 2016, and their subsequent amendments are listed in Table 16.3.

**Table 16.3** Key features of the Hazardous and Other Wastes (Management and Transboundary Movement) Rules, 2016, and subsequent amendments.

| Rules | Key features | Timeline | References |
|---|---|---|---|
| Hazardous and Other Wastes (Management and Transboundary Movement) Rules, 2016 | • Superseded 2008 Rules under the Environment Protection Act 1989<br>• Defined "hazardous wastes"<br>• Distributed responsibilities to various stakeholders<br>• Prohibited import of specific hazardous wastes<br>• Reexport of illegal hazardous waste is at their cost within 90 days of the waste's arrival in the country | April 4, 2016 | (Ministry of Environment, Forest and Climate Change, 2016b) |
| First Amendment | • Permitted industries in special economic zones (SEZs) to import solid plastic wastes | July 6, 2016 | (Ministry of Environment, Forest and Climate Change, 2016a) |
| Second Amendment | • Explained the types of units that are allowed to import a specific type of hazardous waste, in particular, the import of "Solid plastic waste polymethyl methacrylate" | February 28, 2017 | (Ministry of Environment, Forest and Climate Change, 2017) |

| Rules | Key features | Timeline | References |
|---|---|---|---|
| Third Amendment | • Permitted import of used electrical and electronic assemblies for research, testing, and project work by Department of Scientific and Industrial Research (DSIR)-approved research units<br>• Allowed importing used plant and machinery with at least 5 years of residual life for manufacturing electronic items<br>• Mandated documentation like Form 6, Chartered Engineer Certificate, and details of past imports for tracking and compliance | June 11, 2018 | (Ministry of Environment, Forest and Climate Change, 2018a) |
| Fourth Amendment | • Defined the term "waste collector" and exempted occupiers from obtaining separate waste authorization, who do not need consents under the Water or Air Acts, provided they hand over waste only to authorized handlers per CPCB guidelines<br>• Allowed original equipment manufacturers to reimport defective or nonfunctional electronic items exported from India within a year<br>• Excused silk waste exports from requiring MOEF&CC permission | March 1 2019 | (Ministry of Environment, Forest and Climate Change, 2019) |
| Fifth Amendment | • Mandated recognition, registration, and industrial skill development activities for workers involved in hazardous waste management<br>• Directed annual monitoring of health to ensure the health and safety of the workers | October 9, 2020 | (Ministry of Environment, Forest and Climate Change, 2020) |

Continued

**Table 16.3** Key features of the Hazardous and Other Wastes (Management and Transboundary Movement) Rules, 2016, and subsequent amendments—cont'd

| Rules | Key features | Timeline | References |
| --- | --- | --- | --- |
| Seventh Amendment | • Focused on formalizing the utilization and management of waste tires<br>• Introduced EPR for waste tires and mandated registration of the involved entities on the portal developed by CPCB | July 21, 2022 | (Ministry of Environment, Forest and Climate Change, 2022b) |
| Eight Amendment | • Announced EPR for the sound management of used oil and mandated the registration of involved entities, including producers, collecting agents, recyclers, and used oil importers, on the CPCB's portal<br>• Detailed responsibilities of various stakeholders for the effective implementation of the launched EPR | September 18, 2023 | (Ministry of Environment, Forest and Climate Change, 2023) |
| 10th Amendment | • Focused on strengthening the enforcement by explicitly specifying the penalties for noncompliance under Section 15 of the Environment Protection Act<br>• In particular, Rule 40 and Paragraph 11 of Schedule IX were substituted to include the penalty for noncompliance | November 12, 2024 | (Ministry of Environment, Forest and Climate Change, 2024a) |

### 3.1.4 National Green Tribunal directives

The National Green Tribunal (NGT), established in 2010 under the NGT Act, is a quasijudicial body responsible for handling matters related to environmental protection and the conservation of natural resources, including forests (Rengarajan et al., 2018). The tribunal consists of subject matter experts across law, science, and environmental policy. The NGT is uniquely positioned to resolve complex environmental disputes efficiently and with scientific depth.

In the context of SWM, the NGT plays a pivotal role in ensuring compliance with the SWM Rules, 2016, and subsequent amendments. It actively monitors the implementation of SWM strategies across states, issuing binding directives to municipal bodies, state authorities, and local governance institutions to address gaps and improve waste-handling practices (Ahluwalia and Patel, 2018). The tribunal also functions as an oversight mechanism, evaluates the progress of waste management programs, and provides detailed compliance reports to relevant regulatory agencies (Gill, 2019).

A landmark moment in the evolution of India's SWM legal framework was the NGT judgment in Almitra H. Patel vs. Union of India, which set a precedent for waste governance (Das and Gargate, 2023). The tribunal underscored the constitutional right to a clean and healthy environment under Article 21 and issued strict timelines for implementing the SWM Rules (Das and Gargate, 2023). It mandated source segregation of waste, promoted waste-to-energy initiatives, and banned the open burning of waste, including at landfill sites. The directives to the Delhi Municipal Corporation and other urban bodies following this judgment have had far-reaching implications for urban waste governance nationwide.

## 3.2 National initiatives

### 3.2.1 SBM: Urban and rural

The narrative of SWM in India remains incomplete without the transformative impact of the SBM, a flagship initiative launched by the Government of India on October 2, 2014. With the intention of enhancing sanitation and waste management nationwide while aligning with the SDGs, SBM has redefined cleanliness as a national priority (Ministry of Minority Affairs, Ministry of Minority Affairs, Ministry of Minority Affairs, n.d.). The mission has driven systemic change across urban and rural landscapes through multistakeholder collaboration involving local authorities, state governments, private entities, and communities.

The program was structured into two sub-missions: (1) SBM-Gramin and (2) SBM-Urban. The initiative addresses the distinct needs of rural and urban communities, focusing on eradicating open defecation, improving sanitation infrastructure, and promoting sustainable waste management practices through door-to-door waste collection, source segregation, scientific waste processing, and public awareness campaigns to foster behavioral change and education on cleanliness (Ghosh, 2016). The first phase of SBM, spanning from 2014 to 2019, aimed to achieve a "Clean India" by October 2, 2019. This ambitious 5-year plan yielded remarkable outcomes, establishing SBM as a pioneering pan-India initiative.

By 2019, MSW treatment capacity surged from 18% (2600 tons per day of waste generation) in 2014 to approximately 70%, with urban areas achieving 97% door-to-door waste collection and 85% source segregation at the ward level, a significant leap from negligible coverage in 2014 (Ministry of Housing and Urban Affairs, 2021c). Approximately 90,000 informal waste workers were also integrated into the formal waste management value chain, enhancing livelihoods and system efficiency. In terms of sanitation, all urban areas across 35 states and Union Territories, except one ULB in West Bengal, were declared Open Defecation Free (ODF) by 2019.

Meanwhile, in rural India, over 100 million individual household toilets were constructed, and nearly 600,000 villages achieved ODF status (Ministry of Housing and Urban Affairs, 2021c; Ministry of Jal Shakti, 2022). These achievements reflected SBM's success in mobilizing resources, fostering community participation, and leveraging a "Jan Andolan" to institutionalize cleanliness. However, to sustain and expand these gains, the Government of India launched SBM 2.0 on October 1, 2021, with a vision to achieve a "Garbage-Free" nation by October 1, 2026 (Ministry of Housing and Urban Affairs, 2021c). The second phase emphasizes comprehensive waste management and environmental sustainability. It focuses on 100% source segregation, collection, transportation, and scientific processing of all waste types, including construction and demolition waste and plastic waste, alongside remediation of legacy dumpsites to convert them into green zones (Singh, 2022). In addition, SBM 2.0 prioritizes the safe containment, transportation, processing, and disposal of wastewater and fecal sludge, particularly in smaller cities, to prevent pollution of groundwater and water bodies. Furthermore, the program aims to maintain and strengthen ODF achievements across urban and rural areas through robust infrastructure and behavioral reinforcement.

### 3.2.2 Galvanizing Organic Bio-Agro Resources Dhan: Organic waste-to-wealth

The Government of India has launched the Galvanizing Organic Bio-Agro Resources Dhan (GOBARdhan) initiative to leverage the vast availability, market potential, and entrepreneurial opportunities of agricultural waste and cattle dung, particularly in rural ecosystems (Ministry of Jal Shakti, Ministry of Jal Shakti, Government of India GOBARdhan, n.d.). This program aims to convert organic waste into valuable products such as biogas, compressed biogas, and biocompressed natural gas to maximize resource recovery while minimizing the environmental impact of existing waste management practices. To promote participation from households, cooperatives, and entrepreneurs, the initiative includes several schemes, such as the Sustainable Alternative Towards Affordable Transportation by the Ministry of Petroleum and Natural Gas, the Agri Infrastructure Fund by the Department of Agriculture Cooperation and Farmers Welfare, and the Waste to Energy (WtE) Scheme by the Ministry of New and Renewable Energy. A successful implementation of GOBARdhan is expected to significantly contribute to India's climate change mitigation goals, notably the transition to net-zero by 2070 (Dhiman and Mukherjee, 2025). Experts and policymakers have also highlighted the long-term benefits of redirecting organic waste, including enhanced energy security, access to clean energy, entrepreneurial opportunities, and improved waste management and sanitation in rural economies.

### 3.2.3 Atal Mission for Rejuvenation and Urban Transformation (AMRUT) and smart cities mission

Launched in the year 2015, the Atal Mission for Rejuvenation and Urban Transformation (AMRUT) enhanced urban infrastructure across 500 cities and towns in India, targeting sectors such as SWM, stormwater drainage, green spaces, and nonmotorized urban transport (Mishra, 2025). The AMRUT scheme can be seen as complementary to SBM, as this program emphasizes infrastructure development while SBM focuses on cleanliness and behavioral change.

Following the success of AMRUT 1.0, AMRUT 2.0 was launched in 2021 to further improve urban service delivery. Similarly, the Smart Cities Mission, also launched in 2015, enhanced the quality of life in 100 selected cities through sustainable urban development and smart technologies (Kumar et al., 2020). For SWM, it promoted advanced technologies such as artificial intelligence (AI), the Internet of Things (IoT), and data analytics, alongside strategies such as retrofitting, redevelopment, and pan-city initiatives (Singh et al., 2024). Over 50 cities adopted these technologies, improving waste collection efficiency, route optimization, and digitization through tools like Automatic Vehicle Location systems.

### 3.2.4 National Mission on Sustainable Habitat

The National Mission on Sustainable Habitat (NMSH) is one of eight sub-missions under the National Action Plan on Climate Change, launched in 2010 (Ministry of Environment, Forest and Climate Change, 2018b). NMSH promotes sustainable and climate-resilient urban development. NMSH 2.0 defines a sustainable habitat as an ecosystem balancing adequate shelter, basic services, infrastructure, and livelihood opportunities with environmental, socioeconomic, and disaster-resilient features, including equality and inclusiveness (Ministry of Housing and Urban Affairs, 2021b). To achieve these goals, NMSH identifies key areas such as waste management, green cover, biodiversity, energy efficiency, green buildings, and water management.

For SWM, NMSH recommended strategies to city authorities, including waste segregation, adherence to the 4R principles (reduce, reuse, recycle, and recover), and the use of recycled materials over virgin materials (Ministry of Housing and Urban Affairs, 2021b). It advised state governments to promote the use of refuse-derived fuels (RDF) in cement factories to reduce coal dependency and GHG emissions. The cities with populations exceeding 0.5 million were directed to develop roadmaps for collecting, processing, and reusing construction and demolition waste (Ministry of Housing and Urban Affairs, 2021b). In addition, NMSH emphasized remediating legacy dumpsites to mitigate emissions, operating landfills scientifically, and incentivizing waste reduction at the source through on-site wet waste processing and dry waste segregation. It is also recommended that material recovery facilities (MRFs) be established in cities with over 0.5 million residents to sort recyclables and segregated combustible fractions/RDF for enhanced resource efficiency (Ministry of Housing and Urban Affairs, 2021b).

### 3.2.5 National Action for Mechanized Sanitation Ecosystem

The National Action for Mechanized Sanitation Ecosystem (NAMASTE) eradicated environmentally unsafe sewers, assessed septic tank cleaning practices, and promoted the safety and dignity of sanitation workers (Ministry of Social

Justice and Empowerment, 2025). It was launched in 2023 as a Central Sector Scheme by the Ministry of Social Justice and Empowerment and the Ministry of Housing and Urban Affairs. NAMASTE aimed for zero fatalities in sanitation work, ensured no direct contact with human fecal matter, and empowered workers through skilled training and Self-Help Groups for sanitation enterprises (Hindustan Times, 2025). It provided health insurance under Ayushman Bharat—Pradhan Mantri Jan Arogya Yojana, distributed personal protective equipment kits, and supplied safety devices to Emergency Response Sanitation Units. The scheme identified sewer and septic tank workers, provided occupational training, and offered capital subsidies for mechanized equipment, aiming to achieve 100% mechanical desludging by 2025. In 2023–2024 NAMASTE included components of the Self-Employment Scheme for Rehabilitation of Manual Scavengers and in 2024–2025 extended them to waste pickers, thereby enhancing its scope for SWM (Department of Social Justice and Empowerment, Department of Social Justice and Empowerment, Department of Social Justice and Empowerment, n.d.).

# 4 Technological interventions and innovations

## 4.1 Composting, vermicomposting, and anaerobic digestion

Composting is the aerobic biodegradation of organic waste by bacteria, worms, fungi, and other organisms. It significantly reduces the mass and volume of organic waste. This process requires favorable conditions, such as pH, temperature, particle size, moisture content, and C/N ratio. The process involves the recycling of organic waste by converting it into a valuable product called compost, also known as humus, which serves as a soil conditioner. Composting has the primary benefit of not generating any secondary pollutants (Haggar, 2007).

Organic waste composting is conducted at both large and decentralized scales. Large-scale composting methods include windrow composting and aerated static pile composting, whereas decentralized composting methods include in-vessel composting, drum composting, and rotary drum composting (Tchobanoglous et al., 2014). Large-scale composting facilities can process the MSW generated by an entire city. Nevertheless, in recent times, municipalities are mostly inclined toward decentralized composting systems due to the subsequent drawbacks of large-scale composting facilities, such as: (1) the large amount of MSW generated necessitates sorting because the waste is mixed during collection and transportation, (2) there is an additional cost involved in the collection and transportation of waste to sites that are typically located outside the cities, (3) challenges faced during the mixing of MSW with additives and bulking agents on a larger scale, and (4) the initial investment, operational cost, and expenses associated with hiring technical personnel are higher (Rao and Parsai, 2023). In-vessel composting systems are typically used to process household- or community-level HWBW at the source of generation. In-vessel composting offers superior process and odor management, increased efficiency, and is economically advantageous, requiring less space (Enebe and Erasmus, 2023).

Vermicomposting converts organic waste into a nutrient-rich soil conditioner, much like conventional composting. However, in vermicomposting, earthworms are used to augment the microbial community and produce nutrient-rich mucus, unlike in conventional composting. Its benefits include faster processing, lower cost, and the production of a superior-quality soil conditioner (Enebe and Erasmus, 2023; Tan et al., 2025). The major drawback of vermicomposting is the inability to attain the thermophilic temperature range; hence, pathogen reduction in the vermicompost is uncertain. In addition, it requires more maintenance than composting. Therefore an integrated system combining vermicomposting followed by conventional composting has also been observed to achieve faster decomposition and produce high-quality compost (Ndegwa, 2001; Kaur et al., 2023).

Anaerobic digestion involves the valorization of organic waste in the absence of oxygen to produce biogas, which mainly includes methane (60%–70%), carbon dioxide (30%–40%), and small traces of hydrogen and hydrogen sulfide. Biogas finds its application in heating and cooking. It is observed that anaerobic digestion can valorize various types of organic waste more effectively than other biomass conversion methods, such as composting (Milano et al., 2016; Xu et al., 2018). The process is divided into four stages: hydrolysis, fermentation, acetogenesis, and methanogenesis (Hu et al., 2020; Mihi et al., 2024). Hydrolysis involves the conversion of proteins, lipids, and carbohydrates into simple monomers by extracellular enzymes. The products formed after hydrolysis are further degraded into volatile fatty acids during the acidogenesis or fermentation stage. This process also produces hydrogen, alcohol, and carbon dioxide. In the third stage, i.e., acetogenesis, volatile fatty acids and alcohols undergo oxidation to form dihydrogen and carbon dioxide. In the last stage, acetate is converted into methane and carbon dioxide (Malinauskaite et al., 2017).

## 4.2 Waste-to-energy technologies

WtE technologies utilize waste in the form of heat, electricity, or fuel through thermal processes such as combustion, pyrolysis, gasification, and plasma arc gasification. Incineration is a commonly used method in developed nations for managing MSW. It is characterized as a controlled process in which waste is combusted in excess air at temperatures above 900°C, producing heat that drives steam turbines to generate electricity (Malinauskaite et al., 2017). Pyrolysis is the thermal decomposition of the feed at a temperature of 500°C–800°C, in which volatile gases and solid char are produced in an oxygen-free environment (Gupta et al., 2015). Gasification is the thermal conversion of solid waste into "producer gas" or "syngas" in the presence of partial oxygen, which is basically lower than the stoichiometric requirement. Generally, syngas contains carbon dioxide, hydrogen, and traces of methane, which is used as an alternative fuel due to its high calorific value (Lombardi et al., 2015).

### 4.3 Thermochemical conversion of waste

Hydrothermal treatment is a novel technique that involves heating a closed reaction vessel containing an aqueous solution or moist waste to create a high-temperature and high-pressure environment. The high-pressure condition may be achieved by either applying external pressure or utilizing the vapor pressure generated within the system (Yang and Park, 2019). The categorization of hydrothermal processes based on temperature and different products is as follows: hydrothermal carbonization (temperature = 180°C–250°C, product: hydrochar), hydrothermal liquefaction (temperature = 250°C–373°C, product: liquid slurry), and hydrothermal gasification or supercritical water gasification (temperature > 373.95°C, pressure > 22.06 MPa, product: combustible gas) (Basso et al., 2016). Hydrothermal treatment is primarily used for resource recovery from food waste and sewage sludge. Hydrothermal treatment does not require prior drying of the waste and relies on the material's inherent moisture for optimal effectiveness. Furthermore, the process does not require any chemicals (Tekin et al., 2014).

### 4.4 Digital technologies and IoT for waste management

Rapid urbanization and industrialization have led to increased environmental pollution in urban areas recently. Traditional environmental engineering approaches need to evolve by integrating AI, including machine learning, deep learning, and the IoT, to develop innovative solutions, as traditional approaches for waste management are unable to cope with the ever-increasing volume of MSW generated recently. AI-driven techniques optimize resource management, control environmental pollution, enhance environmental sustainability, and provide data-driven insights for policy-making. AI enables end users to optimize collection routes, automated sorting, and recycling of MSW, reducing fuel costs, emissions, and manpower (Shennib et al., 2024; Muhammed Mahshook Ali et al., 2024; Iyiola et al., 2024). IoT also improves MSW management by providing real-time data on the status of bin capacity, waste fill levels, location coordinates, etc. (Lakhouit, 2025; Fatorachian et al., 2025). It ensures data transfer to centralized monitoring units, thereby optimizing the whole process. It helps predict waste generation trends, forecast waste volume, etc. However, introducing these digital technologies faces challenges such as high initial costs and maintenance, privacy and security concerns, and inaccuracies in predictive behavior due to a lack of accurate data (Lakhouit, 2025).

## 5 Systemic gaps and challenges

The sustainable management of solid waste has been a pressing challenge for the Indian economy amid rapid development across urban, periurban, and rural areas

(Chaudhary et al., 2021). The central and state governments, municipalities, and local bodies have taken various initiatives comprising policy frameworks, technical interventions, and legal structures to ensure the environmentally sound management of generated solid waste and to provide a pollution-free environment to the public. However, the impact of these initiatives has been limited due to numerous roadblocks, including a lack of infrastructure, workforce, and financial constraints; limited public participation; insufficient awareness; and ineffective policy enforcement (Kapoor and Chakma, 2024).

Though it is difficult to say which roadblocks contribute most, inadequate infrastructure is one of the primary challenges in the cascade of SWM's sustainability goals. Several ULBs have limited infrastructural services for managing waste generation, collection, and segregation (Kumar et al., 2024). In addition, there is a shortage of waste-collection vehicles and workforce to handle waste. As a result, the visuals of dumpsters overflowing with vast piles of garbage, breeding mosquitoes and flies, are common in urban and periurban areas. However, the consequences of inadequate SWM infrastructure extend beyond visible issues such as overflowing dumpsters and roadside litter, as they also influence public attitudes and behaviors, often leading to a decline in civic responsibility and environmental consciousness (Venugopal, 2024). Dumping waste in open landfills and water bodies is one such outcome that reflects both infrastructural gaps and public disengagement. Further, poor source segregation of waste significantly hampers the recovery of recyclable materials and the recycling of organic fractions (Menon and Dubey, 2024). As a result, large volumes of mixed waste end up in landfills, causing unwanted piling up that demands costly technical interventions and land reclamation efforts. Moreover, this mismanagement poses serious health risks to nearby communities due to exposure to harmful pollutants and disease vectors. The lack of sufficient funding also hinders the infrastructure required for the sustainable management of solid waste. For instance, in developing nations, approximately 80%–95% of their finances reportedly go only toward waste collection and transportation. Further, in many low- and middle-income nations, garbage collection is limited to as low as 10% in suburban regions, which poses environmental and public health risks, with the primary vulnerable group being children. These systemic gaps spanning infrastructure, finances, governance, and public participation highlight why sustainable waste management in India remains a formidable challenge. Addressing these barriers requires not only improved investment and institutional capacity but also innovative, community-driven approaches. The following section presents global success stories that demonstrate how such challenges can be overcome, offering lessons relevant to the Indian context.

# 6 Lessons learned: Success stories around the globe

## 6.1 Case Study 1: Teocelo, Veracruz, Mexico

While SWM is often a critical issue in developing countries due to its socioeconomic and environmental impacts, several success stories demonstrate how collaboration between communities and government officials can transform SWM practices and thus enhance the quality of life. One such example is the semiurban community of Teocelo in the state of Veracruz, Mexico. Teocelo is a semiurban municipality with a population of 16,327 and has a per capita MSW generation of 0.24 kg/resident/day (De Medina-Salas et al., 2020). Since 2001, the Teocelo municipality has developed a simple and effective waste management program that has earned national recognition as a model for developing countries. Earlier, the city faced typical challenges of developing regions, such as limited financial resources and dominance of organic waste. However, Teocelo succeeded through strong community participation and effective public policies. Here, the residents separate waste into organic and inorganic fractions, with only the inorganic portion entering the collection system. A significant portion of organic waste is composted at home, while the remainder is processed at a municipal vermicomposting plant that produces marketable fertilizer used in local agriculture, school gardens, and municipal green spaces. The program's success is said to be driven by high citizen engagement, with 90% of residents expressing pride in waste separation, driven by social collaboration (32.60%), environmental care (27.90%), and moral obligation (20.93%) (De Medina-Salas et al., 2020). The key motivations include environmental improvement (40.75%) and fertilizer production (35.19%). The vermicomposting plant also serves as an educational hub and has offered workshops since 2010 to reinforce community awareness. Various awards have been given to Teocelo, including the "2016 Ecological Merit Award for Environmental Education," "2002 National Government and Management Award," and "2009 Let's Clean Our Mexico." Through the integration of waste separation, composting, and public awareness, the Teocelo municipality demonstrated how community-driven initiatives and policies can enhance environmental sustainability and quality of life, offering a replicable model for other developing regions.

## 6.2 Case Study 2: San Fernando, Pampanga, Philippines

In the bustling city of San Fernando, Pampanga, a first-class city in the Philippines with over 300,000 residents, a silent revolution in waste management unfolded at the grassroots level (Pilipinas Dito sa, 2025). Once struggling with overflowing landfills and inefficient systems, the city transformed into a global model for zero-waste practices. San Fernando diverted 80% of its waste from landfills in 2018, up from just 12% in 2012 (Zero Waste City, 2020). This remarkable journey, rooted in the barangays (villages) that form the city's heart, demonstrated a story of community determination, innovative policies, and partnerships that reshaped waste management in a developing nation. The foundation for change was the *Philippines' Republic Act 9003* or *the Ecological Solid Waste Management Act of 2000*, which mandated decentralized waste management but was poorly implemented in many regions (Ancheta et al., 2020). In 2011 San Fernando's local government, determined to address its growing waste crisis, partnered with the Mother Earth Foundation (MEF), a nonprofit focused on sustainable waste solutions (Zero Waste City, 2020). Together, they empowered each barangay in the city to manage waste locally. The residents were trained to segregate waste at the source into biodegradable, recyclable, and residual categories. Meanwhile, supported by city ordinances, the barangay leaders enforced compliance through regular inspections and penalties and fostered a culture of accountability (Ancheta et al., 2020). A central part of this effort was the establishment of MRFs. Each barangay received a subsidy of PHP 150,000 (≈USD 2650) to develop MRFs and was equipped with tools such as tri-bikes for collection (Global Alliance for Incinerator Alternatives, 2018a). Biodegradable wastes were composted for local farms, while recyclables were sold to junk shops and industries.

The outcome of these efforts was encouraging. By 2019, residual waste dropped from 85% to 20%, diverting nearly 60,000 tons of waste from landfilling annually. The city also banned plastic bags in 2015, further reducing nonrecyclable waste streams (Global Alliance for Incinerator Alternatives, 2021). In terms of finances, the city reduced its waste management budget from PHP 70 million (~USD 1.23 million) to PHP 34.6 million (~USD 610,000) by 2018, nearly a 50% reduction (Zero Waste City, 2020). In addition to other benefits, new jobs were created (Global Alliance for Incinerator Alternatives, 2017). It is important to highlight that the cultural transformation played a vital role in sustaining the system. MEF's outreach efforts, such as TV shows that rewarded proper segregation and barangay-level contests, turned waste management into a source of local pride (Ancheta et al., 2020).

San Fernando's barangay-level zero-waste system demonstrated that decentralized, community-led models can achieve diversion rates comparable to those of advanced economies without relying on expensive technologies. Its success stemmed from political will, NGO–government collaboration, financial incentives, and active citizen engagement. In the case of India, the parallels are clear. Like San Fernando, Indian cities struggle with dependence on landfills, weak segregation compliance, and high transportation costs. By adopting ward-level decentralization, strengthening partnerships with civil society, and investing in localized composting and recycling infrastructure, Indian municipalities can replicate these results. San Fernando proved that even in resource-constrained conditions, waste can be turned into opportunity when communities take ownership of the system.

## 6.3 Case Study 3: Kamikatsu, Japan

Kamikatsu is a small mountainous town with approximately 1500 residents, located in central Tokushima on Shikoku Island (Nakam, Nakam, Nakam, n.d.). Though it is a small town, it has become a global symbol of sustainable waste management in the past few years (Shenyoputro, 2023). In 2003 the rural community of Kamikatsu set an ambitious goal to become Japan's first "zero-waste" town by 2020, eliminating the need for incineration or landfills (Nakam, Nakam, Nakam, n.d.). Although complete zero-waste could not be achieved, this small town achieved a recycling rate of 81% by 2020, far surpassing Japan's national average of 20% (Shenyoputro, 2023). Kamikatsu's journey in waste management demonstrated how extreme source separation, policy innovation, and strong community engagement can overcome geographic and demographic constraints.

The catalyst for change was the closure of Kamikatsu's outdated incinerator in 2001 due to toxic emissions (Klean Industries, 2025). With limited financial resources and no room for landfills in its forested landscape, the town adopted a community-driven recycling strategy. Drawing inspiration from a lecture by American chemist Dr. Paul Connett, Kamikatsu launched its Zero Waste Declaration, supported by the nonprofit Zero Waste Academy. The residents embraced a meticulous system of sorting waste into 45 categories, ranging from aluminum cans and glass bottles to paper flyers, plastic caps, and thermometers (Karunanidhi, 2025). Unlike most Japanese towns reliant on incineration, Kamikatsu has no garbage trucks. The inhabitants personally deliver their sorted waste to the Zero Waste Center, a facility built from reclaimed materials that functions as both a processing hub and an educational space. Organic waste is composted at home using subsidized electric composters. At the same time, reusable items are deposited at the "Kuru-kuru" shop, enabling items such as clothing and tableware to circulate within the community (Shenyoputro, 2023). Local businesses also contribute to recycling by making zero-waste products and repurposing organic by-products.

By 2020, total waste generation was reduced by 65%, from 150 tons in 2000 to 54 tons, and disposal costs were significantly lowered through recycling revenues and a points-based incentive system (Shenyoputro, 2023). Even the aging population did not hinder participation; residents embraced the system as a civic responsibility. The Zero Waste Center also attracts visitors, spreading knowledge of sustainable practices beyond the town. In summary, the residents of Kamikatsu demonstrated that even small, resource-limited communities can achieve world-class waste diversion through rigorous source separation, active citizen participation, and innovative reuse systems.

## 6.4 Case Study 4: Pune, Maharashtra, India

Pune is a vibrant city in the state of Maharashtra, home to over three million people and generating nearly 2258 tons of waste per day (Saifi and Jha, 2024). It is one of India's fastest-growing urban centers and is well-known for its transformative approach to waste management. In the absence of an efficient waste management system, it had been grappling with overflowing dumpsites and streets littered with garbage (Centre for Public Impact, 2021). However, in recent years, through a pioneering partnership between the Pune Municipal Corporation (PMC) and the Solid Waste Collection and Handling (SWaCH) cooperative, the city has earned several accolades and national recognition for decentralized waste segregation and composting (The Circulate Initiative, 2024). An estimated 60 MT of waste is diverted from landfills daily, with 80%–85% of it recycled and processed (Centre for Public Impact, 2021).

The evolution of the SWaCH cooperative dates back to 1993, when informal waste pickers, often marginalized women, united to form the Kagad Kach Patra Kashtakari Panchayat, a trade union that advocates for their rights and recognition (Saifi and Jha, 2024). Later, in 2005 their efforts led to the formation of the operational cooperative

"SWaCH," India's first wholly owned waste picker cooperative. The PMC signed an agreement with SWaCH in 2008 and set an example of public–private partnership for SWM. Under this initiative, >3000 waste pickers, primarily women, were empowered to provide door-to-door waste collection services to >600,000 households in the city (United Nations Environment Programme, 2018). The inhabitants were also made aware of waste segregation and encouraged to segregate waste at the source into wet and dry fractions. The cooperative workers' responsibilities included sorting recyclables like paper, plastics, and metals for sale to scrap dealers, while channeling wet waste to composting and biomethanation facilities.

The success of the city lies in its decentralized approach to SWM. Reportedly, the city has established 165 decentralized composting plants and 25 biogas plants to process wet waste into biogas and compost, which are used to power street lights and support local agricultural activities (First Green, 2023). A few processing plants include Hanjer Biotech (1000 TPD), Ajinkya Biofert (200 TPD), and Disha Waste Management (100 TPD) (SWaCH Cooperative, 2025). The Red Dot campaign, initiated by SWaCH, further improved sanitary waste disposal by educating residents to mark hazardous waste, thereby protecting waste pickers' health (Saifi and Jha, 2024). However, the impact of this partnership goes beyond the environmental gains. SWaCH's model has formalized waste pickers' roles, providing them with identification cards, safety gear, and household user fees, thus enhancing their dignity and income. In addition, community engagement through awareness campaigns and initiatives such as the Zero Garbage Ward Contest has fostered a culture of environmental stewardship. The city's efforts have saved millions in waste management costs and reduced reliance on landfills, with only 20% of waste requiring disposal. Despite challenges like resistance in slum areas, Pune's model demonstrates how inclusive and decentralized systems can address urban waste challenges in developing nations. It could be said that its success offers a blueprint for other Indian cities, demonstrating that collaboration among communities, informal workers, and local governments can transform waste into a resource while improving quality of life.

## 7 Conclusions

India's SWM landscape reflects the complex interplay between rapid urbanization, changing consumption patterns, institutional capacity, and evolving governance frameworks. As discussed across the legislative review, scheme analysis, and case-based illustrations presented in this chapter, the magnitude of waste generation in the country is not merely an environmental concern but a multidimensional development challenge with direct implications for public health, resource efficiency, social equity, and climate resilience. The analysis across sections highlights that while India has made significant normative and policy-level progress, outcomes on the ground remain uneven and deeply contingent on governance effectiveness. The evolution of India's waste management regime, from the Municipal Solid Waste Rules to the Solid Waste Management Rules, Plastic Waste Management Rules, and their subsequent amendments, signals an apparent policy shift toward sustainability, source segregation, decentralization, and EPR. National missions and schemes have further strengthened the institutional mandate for improved waste governance. However, as discussed in the chapter, regulatory ambition has frequently outpaced implementation capacity. Inadequate infrastructure, fragmented institutional responsibilities, financial constraints, and weak monitoring mechanisms continue to undermine the transformative potential of these frameworks, particularly at the ULB level.

A central insight emerging from the chapter is the indispensable role of the informal waste sector in India's waste economy. Waste pickers and community-based cooperatives contribute substantially to recycling, material recovery, and cost reduction, yet remain structurally marginalized within formal systems. The case-based evidence and programmatic initiatives discussed herein demonstrate that inclusive models, where informal workers are formally recognized, protected, and integrated, yield both environmental and socioeconomic benefits. Their exclusion, conversely, represents a missed opportunity for advancing circular economy objectives and social justice simultaneously.

The chapter also underscores the growing relevance of circular economy principles in India's waste discourse. Initiatives promoting segregation at source, organic waste processing, recycling markets, and producer accountability indicate a gradual transition away from linear, disposal-centric approaches. Nevertheless, this transition remains partial and uneven. Behavioral change, citizen participation, and market development for secondary resources emerge as critical enablers, without which technical and regulatory interventions risk remaining symbolic. From a policy perspective, the findings point to the need for a more integrated and outcome-oriented approach to SWM. Strengthening interinstitutional coordination, enhancing municipal capacity, ensuring sustained financing, and embedding social inclusion within policy design are essential for bridging the persistent gap between policy intent and implementation reality. Aligning waste management strategies with broader climate action, urban sustainability, and livelihood frameworks can further enhance coherence and long-term impact.

In conclusion, sustainable SWM in India cannot be achieved through isolated regulations or projects. It requires a holistic governance framework that combines robust policy instruments, institutional accountability, technological appropriateness, and inclusive participation. If effectively operationalized, India's waste management system holds the potential to transform waste from an environmental burden into a resource that supports sustainable urban development, economic resilience, and environmental stewardship. The challenge ahead lies not in the absence of policy vision but in translating that vision into consistent, inclusive, and scalable practice.

## References

Abubakar, I.R., Maniruzzaman, K.M., Dano, U.L., AlShihri, F.S., AlShammari, M.S., Ahmed, S.M.S., Al-Gehlani, W.A.G., Alrawaf, T.I., 2022. Environmental sustainability impacts of solid waste management practices in the global south. Int. J. Environ. Res. Public Health. 16604601. 19 (19). Available from: http://www.mdpi.com/journal/ijerph. doi:10.3390/ijerph191912717.

Ahluwalia, I.J., Patel, U., 2018. Solid Waste Management in India: an Assessment of Resource Recovery and Environmental Impact (No. 356). Working Paper. Indian Council for Research on International Economic Relations. Available from: https://icrier.org/pdf/Working_Paper_356.pdf.

Ancheta, A., de Villa de Castro, B., Castillo, R.M., Garcia, M.R.V.C., Garcia, M.N.Z., 2020. Streamlining waste handling approaches to zero waste: community-based experiences in the city of San Fernando, Pampanga, Philippines. J. Nat. Stud. 19 (1), 49–65.

Basso, D., Patuzzi, F., Castello, D., Baratieri, M., Rada, E.C., Weiss-Hortala, E., Fiori, L., 2016. Agro-industrial waste to solid biofuel through hydrothermal carbonization. Waste Manag. 47, 114–121. doi:10.1016/j.wasman.2015.05.013.

Berger, M., 2025. Waste Not! How Nanotechnologies Can Increase Efficiencies Throughout Society Waste Not! How Nanotechnologies Can Increase Efficiencies Throughout Society. Bloomsbury Publishing Plc, United States, pp. 1–517. Available from https://books.rsc.org/books/monograph/2314/Waste-Not-How-Nanotechnologies-Can-Increase. doi:10.1039/9781837677290.

Centre for Public Impact, 2021. Waste Management Cooperative: pune, India. https://centreforpublicimpact.org/public-impact-fundamentals/waste-management-cooperative-pune-india/.

Chaudhary, P., Garg, S., George, T., Shabin, M., Saha, S., Subodh, S., Sinha, B., 2021. Underreporting and open burning—the two largest challenges for sustainable waste management in India. Resour. Conserv. Recycl. 175. doi:10.1016/j.resconrec.2021.105865.

Cooper, J., 2024. Waste management showdown: environmental implications of landfilling vs. incineration. J. Environ. Hazards 08, 248.

CPCB, 2016. Salient features of Solid Waste Management Rules. https://cpcb.nic.in/salient-features-of-msw-rules/.

CPCB, 2022. Annual Report 2021–2022 on Implementation of Solid Waste Management Rules. https://cpcb.nic.in/uploads/MSW/MSW_AnnualReport_2021-22.pdf.

Das, P., Gargate, G., 2023. Legal jurisprudence of solid waste management in India: development through the decades. Int. J. Glob. Environ. Issues. 14666650. 22 (2-3), 268–294. Available from: http://www.inderscience.com/ijgenvi. doi:10.1504/IJGENVI.2023.134083.

De Medina-Salas, L., Castillo-González, E., Giraldi-Díaz, M.R., Fernández-Rosales, V., Rodríguez, C.M., 2020. A successful case in waste management in developing countries. J. Pollut. Eff. Control 8 (2). doi:10.35248/2375-4397.20.8.242.

Department of Social Justice and Empowerment, n.d. Scheme guidelines of waste scheme. https://www.nskfdc.nic.in/en/system/files/Scheme%20guidelines%20of%20Waste%20Scheme..._compressed.pdf.

Dhiman, S., Mukherjee, G., 2025. Waste to Wealth: Emerging Technologies for Sustainable Development Waste to Wealth: Emerging Technologies for Sustainable Development. CRC Press, pp. 1–346. doi:10.1201/9781003449102.

Enebe, M.C., Erasmus, M., 2023. Vermicomposting technology—a perspective on vermicompost production technologies, limitations and prospects. J. Environ. Manag. 345. doi:10.1016/j.jenvman.2023.118585.

Fatorachian, H., Kazemi, H., Pawar, K., 2025. Digital technologies in food supply chain waste management: a case study on sustainable practices in smart cities. Sustainability (Switz.) 17 (5). doi:10.3390/su17051996.

Fernando, S.J., Zutshi, A., 2023. Municipal solid waste management in developing economies: a way forward. Cleaner Waste Syst. 27729125. 5. doi:10.1016/j.clwas.2023.100103.

First Green, 2023. Solid Waste Management in Pune: a Model of Excellence. https://firstgreen.co/solid-waste-management-in-pune-a-model-of-excellence/.

Ghosh, S.K., 2016. Swachhaa Bharat Mission (SBM)—a paradigm shift in waste management and cleanliness in India. Procedia Environ. Sci. 18780296. 35, 15–27. doi:10.1016/j.proenv.2016.07.002.

Gill, G.N., 2019. The National Green Tribunal: evolving adjudicatory dimensions. Environ. Policy Law. 0378777X. 49 (2-3), 153–162. doi:10.3233/EPL-190151.

Global Alliance for Incinerator Alternatives, 2017. Filipino Barangays—Leading the way in zero waste models. https://www.no-burn.org/filipino-barangays-leading-the-way-in-zero-waste-models/.

Global Alliance for Incinerator Alternatives, 2018a. https://www.no-burn.org/implementing-zero-waste-challenging-but-doable/. Implementing Zero Waste: Challenging but Doable.

Global Alliance for Incinerator Alternatives, 2021. Zero Waste Systems: Small Investment, Big Payoff. https://www.no-burn.org/wp-content/uploads/Zero-Waste-Cost-Effectiveness-Fact-SheetENGLISH-1-1.pdf.

Gupta, N., Yadav, K.K., Kumar, V., 2015. A review on current status of municipal solid waste management in India. J. Environ. Sci. (China) 37, 206–217. doi:10.1016/j.jes.2015.01.034.

Haggar, S.E., 2007. Sustainable Industrial Design and Waste Management—Cradle-to-Cradle for Sustainable Development. Elsevier Academic Press, London UK.

Hindustan Times, 2025. Hindustan Times Centre's NAMASTE scheme nears two years; 91,000 sanitation workers identified. https://www.hindustantimes.com/india-news/centres-namaste-scheme-nears-2-years-91-000-sanitation-workers-identified-101748670083848.html

Hossain, R., Islam, M.T., Shanker, R., Khan, D., Locock, K.E.S., Ghose, A., Schandl, H., Dhodapkar, R., Sahajwalla, V., 2022. Plastic waste management in India: challenges, opportunities, and roadmap for circular economy. Sustainability. 2071-1050. 14 (8), 4425. doi:10.3390/su14084425.

Hu, Y., Bassi, A., Xu, C.C., 2020. Energy from biomass. In: Letcher, T.M. (Ed.), Future Energy: Improved, Sustainable and Clean Options for Our Planet, Elsevier. pp. 447–471. doi:10.1016/B978-0-08-102886-5.00021-9.

Iyiola, C.O., Shakantu, W., Daniel, E.I., 2024. Digital technologies for promoting construction and demolition waste management: a systematic review. Buildings 14 (10). doi:10.3390/buildings14103234.

Kapoor, A., Chakma, N., 2024. Challenges of Solid Waste Management in Urban India. EAC-PM Working Paper Series. Available from: https://eacpm.gov.in/wp-content/uploads/2024/05/Solid_Waste_management_Updated.pdf.

Karunanidhi, R., 2025. Citizen Consumer and Civic Action Group, Small town, Big Impact: japan's Kamikatsu zero-waste journey. https://www.cag.org.in/blogs/small-town-big-impact-japans-kamikatsu-zero-waste-journey.

Kaur, H., Hussain, S.J., Mir, R.A., Verma, V.C., Naik, B., Kumar, P., Dubey., 2023. Nanofertilizers—emerging smart fertilizers for modern and sustainable agriculture. Biocatal. Agric. Biotechnol. 54. doi:10.1016/j.bcab.2023.102921.

Kharvel, A.R., 2012. Sustainable Solid Waste Management in India (MS thesis). Department of Earth and Environmental Engineering, Foundation School of Engineering & Applied Science, Columbia University, New York.

Klean Industries, 2025. Kamikatsu zero waste initiatives.

Kumar, G., Vyas, S., Sharma, S.N., Dehalwar, K., 2024. Challenges of environmental health in waste management for peri-urban areas. Sustain. Dev. Goals Ser. Part F3313, 149–168. doi:10.1007/978-3-031-60684-7_9.

Kumar, H., Singh, M.K., Gupta, J.M., 2020. Moving towards smart cities: solutions that lead to the Smart City Transformation Framework. Technol. Forecast. Soc. Change 153. doi:10.1016/j.techfore.2018.04.024.

Lakhouit, A., 2025. Revolutionizing urban solid waste management with AI and IoT: a review of smart solutions for waste collection, sorting, and recycling. Results Eng. 25. doi:10.1016/j.rineng.2025.104018.

Lombardi, L., Carnevale, E., Corti, A., 2015. A review of technologies and performances of thermal treatment systems for energy recovery from waste. Waste Manag. 37, 26–44. doi:10.1016/j.wasman.2014.11.010.

Mahajan, R., 2023. Environment and health impact of solid waste management in developing countries: a review. Curr. World Environ. 23208031. 18 (1), 18–29. doi:10.12944/cwe.18.1.3.

Malinauskaite, J., Jouhara, H., Czajczyńska, D., Stanchev, P., Katsou, E., Rostkowski, P., Thorne, R.J., Colón, J., Ponsá, S., Al-Mansour, F., Anguilano, L., Krzyżyńska, R., López, I.C., Vlasopoulos, A., Spencer, N., 2017. Municipal solid waste management and waste-to-energy in the context of a circular economy and energy recycling in Europe. Energy 141, 2013–2044. doi:10.1016/j.energy.2017.11.128.

Meena, M.D., Dotaniya, M.L., Meena, B.L., Rai, P.K., Antil, R.S., Meena, H.S., Meena, L.K., Dotaniya, C.K., Meena, V.S., Ghosh, A., Meena, K.N., Singh, A.K., Meena, V.D., Moharana, P.C., Meena, S.K., Srinivasarao, C., Meena, A.L., Chatterjee, S., Meena, D.K., Prajapat, M., Meena, R.B., 2023. Municipal solid waste: opportunities, challenges and management policies in India: a review. Waste Manag. Bull. 29497507. 1 (1), 4–18. doi:10.1016/j.wmb.2023.04.001.

Meena, M.D., Narjary, B., Sheoran, P., Jat, H.S., Joshi, P.K., Chinchmalatpure, A.R., 2022. Changes in physical and chemical properties of saline soil amended with municipal solid waste compost and chemical fertilizers in a mustard–pearl millet cropping system. Land Degrad. Dev. 1099145X. 33 (10), 1677–1688. doi:10.1002/ldr.4256.

Menon, U., Dubey, B.K., 2024. Opportunities and challenges in resource recovery from waste. In: Yadav, V., Shrotriya, S. (Eds.), Waste-to-Wealth: Resource Recovery and Value-Added Products for Sustainable Development, CRC Press. pp. 1–21. doi:10.1201/9781003327646-1.

Mihi, M., Ouhammou, B., Aggour, M., Daouchi, B., Naaim, S., Mers, E.M.E., Kousksou, T., 2024. Modeling and forecasting biogas production from anaerobic digestion process for sustainable resource energy recovery. Heliyon 10 (19). doi:10.1016/j.heliyon.2024.e38472.

Milano, J., Ong, H.C., Masjuki, C., Lam, M.K., Loh, P.K., Vellayan, V., 2016. Microalgae biofuels as an alternative to fossil fuel for power generation. Renew. Sustain. Energy Rev. 58, 180–197. doi:10.1016/j.rser.2015.12.150.

Ministry of Environment, Forest and Climate Change, 2016a. Hazardous and Other Wastes (Management and Handling) Rules – july Amendment.

Ministry of Environment, Forest and Climate Change, 2016b. Hazardous and Other Wastes (Management and Transboundary Movement) Rules.

Ministry of Environment, Forest and Climate Change, 2016c. Plastic Waste Management Rules.

Ministry of Environment, Forest and Climate Change, 2017. Hazardous and Other Wastes (Management and Handling) Rules—February Amendment.

Ministry of Environment, Forest and Climate Change, 2018a. Hazardous and Other Wastes (Management and Handling) Rules—June Amendment.

Ministry of Environment, Forest and Climate Change, 2018b. National plan on climate change.

Ministry of Environment, Forest and Climate Change, 2018c. Plastic Waste Management (Amendment) Rules.

Ministry of Environment, Forest and Climate Change, 2019. Hazardous and Other Wastes (Management and Handling) Rules—March Amendment.

Ministry of Environment, Forest and Climate Change, 2020. Hazardous and Other Wastes (Management and Handling) Rules—Fifth Amendment Rules.
Ministry of Environment, Forest and Climate Change, 2021a. Plastic Waste Management (Amendment) Rules
Ministry of Environment, Forest and Climate Change, 2022a. Draft Plastic Waste Management Rules, 2022 (Draft).
Ministry of Environment, Forest and Climate Change, 2022b. Hazardous and Other Wastes (Management and Handling) Rules—Sixth Amendment Rules.
Ministry of Environment, Forest and Climate Change, 2023. New Delhi Hazardous and Other Wastes (Management and Handling) Rules—Second Amendment Rules.
Ministry of Environment, Forest and Climate Change, 2024a. Hazardous and Other Wastes (Management and Handling) Rules—Tenth Amendment Rules.
Ministry of Environment, Forest and Climate Change, 2024b. Plastic Waste Management (Amendment) Rules.
Ministry of Housing and Urban Affairs, 2021a. National Mission for Sustainable Habitat.
Ministry of Housing and Urban Affairs, 2021b. Circular economy in municipal solid and liquid waste.
Ministry of Housing and Urban Affairs, 2021c. Swachh Bharat Mission 2.0. Government of India.
Ministry of Jal Shakti, 2022. Press release. Government of India.
Ministry of Jal Shakti, n.d. Galvanizing Organic Bio-Agro Resources Dhan. https://www.myscheme.gov.in/schemes/gobardhan.
Ministry of Minority Affairs, n.d. Swachh Bharat. Government of India. https://www.minorityaffairs.gov.in/WriteReadData/RTF1984/1521446346.pdf.
Ministry of Social Justice and Empowerment, 2025. NAMASTE scheme.
Ministry of Statistics and Programme Implementation, n.d. Residual accounts and its linkages with Sustainable Development Goals. https://mospi.gov.in/sites/default/files/main_menu/Seminar/Concept%20Note%20of%20Swachhata%20Pakhwada%20Seminar.pdf.
Ministry of Urban Development, 2016. Municipal solid waste management manual.
Mishra, A., 2025. Smart Cities and the Poor—Towards an Agenda for Inclusive Urbanization in India. Routledge India, pp. 1.
Muhammed Mahshook Ali, P., Shijina, B., Sooraj, P., Babu, S., Nainachan, A., 2024. Towards a greener future: IoT-enabled waste management systems. Authorea Preprints. doi:10.22541/au.172971573.30184146/v1.
Nakam, K., n.d. Zero Waste Initiative. https://www.nakam.info/en/works/kamikatsu0/.
Ndegwa, T., 2001. Integrating composting and vermicomposting in the treatment and bioconversion of biosolids. Bioresour. Technol. 76 (2), 107–112. doi:10.1016/S0960-8524(00)00104-8.
Pilipinas Dito sa, 2025. San Fernando, Pampanga leading the way to zero waste https://ditosapilipinas.com/national/news/article/01/31/2025/san-fernando-pampanga-leading-the-way-to-zero-waste/1308.
Rao, J.N., Parsai, T., 2023. A comprehensive review on the decentralized composting systems for household biodegradable waste management. J. Environ. Manag. 345. doi:10.1016/j.jenvman.2023.118824.
Rengarajan, S., Palaniyappan, D., Ramachandran, P., Ramachandran, R., 2018. National Green Tribunal of India—an observation from environmental judgements. Environ. Sci. Pollut. Res. 0944-1344. 25 (12), 11313–11318. doi:10.1007/s11356-018-1763-2.
Saifi, N., Jha, B., 2024. An overview of solid waste management practices in Pune, Maharashtra, India. Nat. Environ. Pollut. Technol. 23 (2), 923–934. doi:10.46488/NEPT.2024.v23i02.027.

Shennib, F., Eicker, U., Schmitt, K., 2024. OpenWasteAI—open data, IoT, and AI for circular economy and waste tracking in resource-constrained communities. IEEE Technol. Soc. Mag. 43 (1), 39–53. doi:10.1109/MTS.2024.3372610.

Shenyoputro, T.E.J., 2023. Reflections on a two-decade journey toward zero waste: a case study of Kamikatsu town, Japan. Front. Environ. Sci. 11. doi:10.3389/fenvs.2023.1171379.

Singh, M., Singh, M., Singh, S.K., 2024. Tackling municipal solid waste crisis in India: insights into cutting-edge technologies and risk assessment. Sci. Total Environ. 917. doi:10.1016/j.scitotenv.2024.170453.

Singh, R. , 2022. Centre for Science and Environment, New Delhi. Toolkit: legacy Waste Management and Dumpsite Remediation to Support Swachh Bharat Mission 2.0, 1–72. https://sbmurban.org/storage/app/media/pdf/sbm_knowledge_center/sbm-20-toolkit-legacy-waste-management-and-dumpsite.pdf.

Singh, S., Biswas, M.K., 2023. Management strategies for single-use plastics: lessons to learn from Indian approach of minimizing microplastic waste. Environ. Sci. Adv. 27547000. 2 (12), 1680–1695. doi:10.1039/d3va00222e.

SWaCH Cooperative, 2025. Compendium of Good Practices in Urban Solid Waste Management: Involving Waste-Pickers to Improve Door-to-Door Collection. https://swachcoop.com/pdf/wastepickerstoimprovedoor-to-doorcollection.pdf.

Tan, Z.S., Manogaran, M.D., Shamsuddin, R., Hakimi, M., Looi, L.W., Woo, K.T., Liew, C.S., Qomariyah, L., 2025. Comparative study on composting and vermicomposting to improve physicochemical properties of digestate with the addition of kitchen waste. Food Technol. Biotechnol. 63 (2), 149–158. doi:10.17113/ftb.63.02.25.8904.

Tchobanoglous, G., Theisen, H., Vigil, S., 2014. Integrated Solid Waste Management—engineering Principles and Management Issues. McGrawHill Education Private Limited, India.

Tekin, K., Karagöz, S., Bektaş, S., 2014. A review of hydrothermal biomass processing. Renew. Sustain. Energy Rev. 40, 673–687. doi:10.1016/j.rser.2014.07.216.

The Circulate Initiative, 2024. City Waste Management Profile: Pune, India. https://www.thecirculateinitiative.org/wp-content/uploads/City-Waste-Management-Profile-Pune-India.pdf.

United Nations Environment Programme, 2018. Empowering waste workers for a cleaner, safer city. https://www.unep.org/news-and-stories/story/empowering-waste-workers-cleaner-safer-city.

Venugopal, K., 2024. Insights into waste management in semi-urban areas: understanding community perspectives. In: Mehra, R., Kumar, P., Hung, B. (Eds.), Municipal Solid Waste Management and Recycling Technologies, IGI Global. pp. 301–325. doi:10.4018/979-8-3693-4054-7.ch011.

Xu, F., Yangyang, L., Xumeng, G., Yang, L., Yebo, L., 2018. Anaerobic digestion of food waste—Challenges and opportunities. Bioresour. Technol. 247, 1047–1058. doi:10.1016/j.biortech.2017.09.020.

Yang, G., Park, S.-J., 2019. Conventional and microwave hydrothermal synthesis and application of functional materials: a review. Materials 12 (7). doi:10.3390/ma12071177.

Zero Waste City, 2020. How a city increased its recycling rate from 12% to 80% in six years and reduced its waste management. https://zerowastecity.com/how-a-city-increased-its-recycling-rate-from-12-to-80-in-6-years-and-reduced-its-waste-management/.

CHAPTER

# 17 Biotechnological approaches for zero waste agriculture

**Tapas Ranjan Das[a], Sundram Singh[b], Abhishek Singh[c], and Pooran Chand[c]**

[a]*Division of Genetics, ICAR-Indian Agricultural Research Institute, New Delhi, India*

[b]*Animal Biochemistry Division, National Dairy Research Institute, Karnal, Haryana, India*

[c]*Department of Genetics and Plant Breeding, Sardar Vallabhbhai Patel University of Agriculture and Technology, Meerut, Uttar Pradesh, India*

## 1 Introduction

Agriculture contributes a significant portion to the GDP of most developing countries. Major crops grown include cereals, pulses, oilseeds, vegetables, etc., which produce significant quantities of waste and residues. The generation of waste is not a problem, but its inefficient management is, which includes the burning of residues and the discarding of vegetable residues in landfills. An increase in air pollution and diseases spread by flies and insects are some of the primary repercussions of this. Managing crop residues and agricultural wastes is crucial due to the increase in intensive farming practices that generate significant amounts of biomass. These practices can lead to an increase in environmental pollution. With proper handling, we can create innovative ways to achieve both sustainable agriculture practices and energy production.

Further, proper management practices can be useful in nutrient recycling, thereby promoting a circular economy in agriculture. Traditional waste management practices often focus on disposal rather than resource recovery. However, zero-waste agriculture aims to create closed-loop systems where all materials are reused or recycled, thus minimizing waste and enhancing sustainability. An estimated 1.3 billion tons of solid waste are produced annually by world cities alone, according to a World Health Organization report. It takes more than simply handling and treating the massive amount of waste produced by our everyday activities to keep it from harming the environment or being unsightly to the general public. Our world's resources are depleting since the majority of our garbage is landfilled without taking the materials' life cycle into account. It has to do with the preservation and use of resources. Waste is no longer waste because it symbolizes our future riches. It will be difficult for scientists to create novel technologies that will turn garbage into useful bioproducts.

The term "biotechnology" encompasses a wide range of disciplines. Biotechnology, according to the European Federation of Biotechnology, is "the integration

Microbial Technologies for Sustainable Waste Management. https://doi.org/10.1016/B978-0-443-44093-9.00018-9

of natural sciences and engineering to achieve the application of organisms, cells, parts thereof, and molecular analogues for products and services." Biobased waste conversion methods have been used in the garbage business since 1995. The technologies can be broadly divided into two groups: one that produces a value-added product while simultaneously reducing the need for subsequent disposal and another that breaks down specific compounds that are highly toxic to prevent or lessen their impact on the environment. Air emissions are also treated using biological techniques.

Due mostly to the rise in energy demand, biotechnology's role in bioenergy sectors has advanced significantly. For example, the International Energy Agency estimates that 104,426 TWh of energy will be consumed worldwide, with fossil fuels meeting 65% of this demand. Due to peak utilization rates in the 2010s, these reserves are almost at the end of their life. Even while new techniques for locating and extracting fossil fuels are continuously being developed, this cannot continue indefinitely because it takes decades for these fuels to be created and they cannot satisfy human hunger. The development of alternative solutions is therefore crucial, and the process of turning solid waste into biofuel is accelerating. Energy crops were formerly thought to be a viable solution, but the fight between fuel and food crops for our limited amount of cultivable land is still up for debate. As a result, organic wastes are currently affordable and sustainable choices or substrates that have spurred significant and important research in several waste management fields since 1995. Wastewater treatment has historically used anaerobic digestion, a promising method that converts waste to energy. However, with numerous industrial-scale anaerobic digestion plants operating successfully and producing energy, the utilization of food and organic wastes with larger solid substrates is currently growing. Similar to "betting on the winning horse," new technologies with a long history of operation are frequently given preference, while developing and emerging techniques with higher productivity are still struggling for adoption. Consequently, it is necessary to take into account a critical analysis that prioritizes innovation and demonstration in this field of study. Although the reactors' scale-up is the subject of intense discussion and investigation, microbial fuel cell-based integration has garnered attention more recently as a means of improving energy recovery and accomplishing pollutant degradation. In addition, digestate (solid waste from anaerobic digesters) needs to be disposed of properly, which can be recycled as a soil additive. Therefore implementing integrated biorefinery technology to create a sustainable organic waste management strategy is multifaceted and the "key" to offering comprehensive solutions.

Although composting has been practiced for generations, commercial composting requires strict supervision to minimize odor and nutrient loss. For composting to become the preferred approach for treating organic waste, particularly in urban settings, odor emissions must be prevented. The long-term viability of a composting plant can also be affected by nutrient loss (nitrogen), which can influence the quality of the compost product. Numerous methods have been studied in this area and show promise for better utility. Given the high-end vermicast produced, vermicomposting remains a desirable composting method. It is interesting to note that antibiotics used in the livestock industry have been shown to break down when composted and end up in animal dung. Composting can be a useful method of treating animal manure because the thermophilic phase, in particular, is associated with a notable decrease in antimicrobials.

Plastics play a significant role in our contemporary way of life. Approximately 300 million tons of plastic are manufactured worldwide, with 95% of that amount derived from fossil fuels, primarily petroleum. Despite their slow rate of degradation in the natural environment, plastics are often regarded as nonbiodegradable. Bioplastics have emerged as a response to the numerous problems this persistence has caused in soil and water matrices. Due to the enduring nature of plastic bags, many countries have imposed restrictions on their use in recent years. Research into using microbes to accelerate the rate of plastic decomposition is gaining popularity as a method to clean up polluted areas. Conversely, the production of bioplastics based on microbially driven polyhydroxybutyrate and lactic acid, which are easily biodegradable, has already advanced significantly; however, the high cost of production must be addressed to ensure their widespread applicability.

## 2 Importance of zero-waste agriculture

Zero-waste agriculture is a sustainable farming system that focuses on minimizing waste while optimizing resource utilization to improve soil health and long-term productivity. It advocates for reducing excessive use of fertilizers and pesticides while promoting the efficient management of agricultural waste, thus transforming organic waste into nutrient-rich by-products. An example of a zero-waste rice-based farming system is one where harvested rice by-products, such as rice straw, offer multiple uses: as a fertilizer, for energy production, and as a substrate in mushroom production.

## 3 Bioplastics from agricultural and dairy waste

Agricultural waste, which consists of complex macromolecules—carbohydrates and proteins—can be converted into simpler forms. Microbial treatment includes the use of fungi or bacteria to degrade lignin and hemicellulose through the action of enzymes like lignin peroxidase, manganese peroxidase, and laccase, converting them into polyhydroxyalkanoates (PHAs), which are a class of biodegradable plastics produced by microbes such as *Cupriavidus necator*, *Pseudomonas oleovorans*, *Alcaligenes latus*, *Halomonas campaniensis,* and *Bacillus megaterium*. These microbes typically produce PHAs as a means of storing carbon and energy under nutrient-limiting conditions by utilizing a wide range of substrates, including agricultural and food wastes. Milk is a rich source of lactose sugar. Two microorganisms, namely, Lava DSM1034 and *Methylobacterium* species ZP24, can be used for the conversion of lactose to PHAs.

## 4 Bioenergy from agricultural and dairy waste

Agricultural biomass contains complex lignocellulosic compounds, which should be broken down into simpler sugars that can later be fermented by microorganisms to produce biofuels such as ethanol, biogas, and biodiesel. Various pretreatment methods include physical methods such as mechanical comminution and steam explosion. Chemical methods include the use of acid and alkali for hydrolysis. Other chemical methods involve the use of oxidizing agents—hydrogen peroxide and ozone—for the breakdown of lignin and hemicellulose. Physicochemical pretreatment methods involve the ammonia fiber explosion method, where biomass is treated with liquid ammonia at high pressure and temperature, followed by rapid pressure release, breaking down lignocellulosic compounds. Other methods include carbon dioxide explosion, which is similar to steam explosion but uses supercritical carbon dioxide to disrupt the biomass structure. Biological pretreatment includes the use of fungi or bacteria to degrade lignin and hemicellulose through the action of enzymes like lignin peroxidase, manganese peroxidase, and laccase.

The anaerobic digestion of a mixture of *mozzarella* cheese whey and buttermilk, supplemented with 5% industrial animal manure pellets, resulted in significant production of hydrogen and methane gases and highlighted a dynamic shift in the microbial populations, with a decrease in lactic acid bacteria and an increase in hydrogen-producing *clostridia* and anaerobic spore-forming bacteria. This research underscores the importance of microbial community dynamics in optimizing biofuel production from dairy waste.

## 5 Genome editing to reduce plant waste

With the exponential increase in the human population, coupled with environmental pollution, energy crises, and climate change, global food production is under severe pressure and needs a shift toward a bioeconomy that ensures food security, minimizes agrifood waste and loss, and reduces dependence on fossil-based resources. To maintain food security, crop yields must nearly triple and remain consistently high. Achieving this objective will require the development of highly resilient crop varieties capable of withstanding multiple stresses. Looking forward, one of the most formidable challenges of the 21st century is to meet the growing food and nutrition demand of the global population while simultaneously minimizing food loss and wastage, protecting the environment, and combating climate change (Tilman et al., 2002; Rockström et al., 2009). Crop loss after harvest and improper distribution exert adverse effects on the environment, contributing to pollution, natural resource degradation, ecological imbalances, and global climate change. Valorizing by-products from on-farm and off-farm lateral streams to crop residues generated during harvesting and agroindustrial processing is vital in enhancing resource utility, reducing the carbon footprint, and bolstering economic sustainability. Recent endeavors to convert food waste into bioplastics through valorization (Kourmentza et al., 2018) could alleviate pressure on land resources and the environment, providing sufficient space for diverse life forms and healthy diets while mitigating biodiversity loss (FAO et al., 2021).

Genome editing tools aim to revolutionize existing plant breeding by protecting the global food supply through the adoption of sustainable production systems and the reduction of environmental pollution (Zhang et al., 2020a, 2020b). Recent advancements in biotechnology have provided us with detailed insights into biochemical and molecular mechanisms capable of DNA alteration, consequently affecting downstream signaling pathways. These genome-editing methods, like CRISPR, offer the innovative ability to induce genetic modifications at designated target sites. This technique offers a win–win solution, meaning it not only minimizes agricultural waste through valorization methods but also aids in producing biofuel from waste.

Lignin present in feedstocks impairs the efficiency of sugar production to a large extent. The high cost of cellulose limits the production and availability of bioethanol. Work toward reducing lignin content and modification of its composition can serve toward make lignocellulosic biomass more amenable to bioenergy conversion (Halpin-Healy et al., 2020; Sticklen, 2008). Latest research has shown that CRISPR can precisely modify genetic materials in a variety of plants, suggesting its potential to simplify the complex processes involved in lignin modification. For example, by expressing cellulase enzymes in *Saccharomyces cerevisiae*, it can be used to break down lignocellulosic material from crop waste (Kroukamp et al., 2018), along with antibiotic selection, which presents an economical, safe, and practical method for generating bioethanol (Lee et al., 2021).

Recent studies have increasingly focused toward the isolation of microbial species that can naturally break down lignin polymers, while simultaneously employing bioengineering methods to enhance the production of enzymes essential for lignin degradation (Bugg et al., 2020; Pollegioni et al., 2015). Furthermore, this chapter provides an extensive examination of CRISPR technology, exploring its utility in reducing plant-derived waste during the preliminary stages of distribution channels, developing microbial strategies for environmental conservation, and minimizing overall ecological contamination. The loss or wastage of agricultural produce results in the underutilization of all valuable resources required for cultivating them. The present yield potential of horticultural food crops is insufficient to meet dietary requirements, a situation further exacerbated by postharvest losses. The causes of food loss and waste include numerous factors, ranging from production and distribution to storage. To maintain the quality of agricultural produce from farm to table, an efficient production and supply chain must be established.

## 6 Biofuel generations and evolution

Conventional biofuels, also known as first-generation biofuels, are made from sugar, starch, or vegetable oil by microbes or their enzymes. The most commonly utilized organism is *S. cerevisiae* (Arous et al., 2017). Some contend that the production of bioethanol or butanol from sugar (sugarcane, sugar beetroot, etc.) or starch (wheat, corn, barley, and potatoes) raises the cost of food worldwide (Al-Azzawi and Jassem, 2016). There could be serious food shortages as a result of the rising demand on the land utilized for food crop production, especially in developing nations where over 800 million people are already undernourished or hungry (FAO, 2017). Furthermore, there will be detrimental effects on the ecology from increased land usage, including excessive water use and fertilizer and pesticide applications (Schenk et al., 2008). Lignocellulosic biomass is used to make second-generation biofuels (Al-Azzawi and Jassem, 2016; Zhu et al., 2020; Carmona-Cabello et al., 2021). Bioethanol and biodiesel produced by Khandaker et al. (2020) include conventional methods with special starch, oil, and sugar crops. Lignocellulose, the primary structural component in crops, including nonfood crops, is notably recalcitrant to decomposition (Schubert, 2006; Cabrol et al., 2017). One major issue with second-generation biofuels (Trabelsi et al., 2018) is that they are too costly to produce in large quantities. In contrast, third-generation biofuels made from algae offer a promising alternative with the potential for more sustainable and cost-effective energy production. Algae-based biofuels include hydrogen produced from green microalgae and microorganisms, bioethanol produced from microalgae and macroalgae (seaweeds), and biodiesel produced from microalgae (Dragone et al., 2010). Fourth-generation biofuels are produced using petroleum-like hydro-refining, sophisticated biochemical processes, or innovative methods such as Joule's unique solar-to-fuel technology, which defies conventional biofuel categories. However, some of these methods are either theoretical or only applicable in laboratory settings (Demirbas, 2011; Milledge and Heaven, 2013). Fig. 17.1 shows a comparative overview of the evolution and technological progression of biofuel production strategies.

## 6.1 Sustainable biochemical conversion of agricultural waste

Compared to thermochemical conversion and direct combustion, biochemical conversion has slower reaction rates and takes place at lower temperatures. Using microbes, fermentation occurs naturally and turns sugars and starches from crops into ethanol (Deeba et al., 2017). To extract more sugars from the lignocellulosic biomass, the process is commonly followed by hydrolyzing the cellulosic feedstock using acid or an enzyme (Dhyani and Bhaskar, 2018). Biodiesel is primarily produced through the chemical transesterification of lipids, which can be of plant or animal origin. Since waste cooking oils and fats are key pollutants in many regions, repurposing them as biodiesel feedstock offers an eco-friendly solution. Biodiesel production relies mainly on four major methods, which are transesterification, thermal cracking (pyrolysis), microemulsions, and direct use and mixing (Shalaby, 2011; Bridgwater, 2012). Fig. 17.2 shows the processes used to produce biodiesel from glycerol. Since it would enable the use of a greater variety of raw materials, the development of microbial fermentation techniques for biodiesel synthesis is becoming more and more popular. Examples include biomass, corn, and sugarcane.

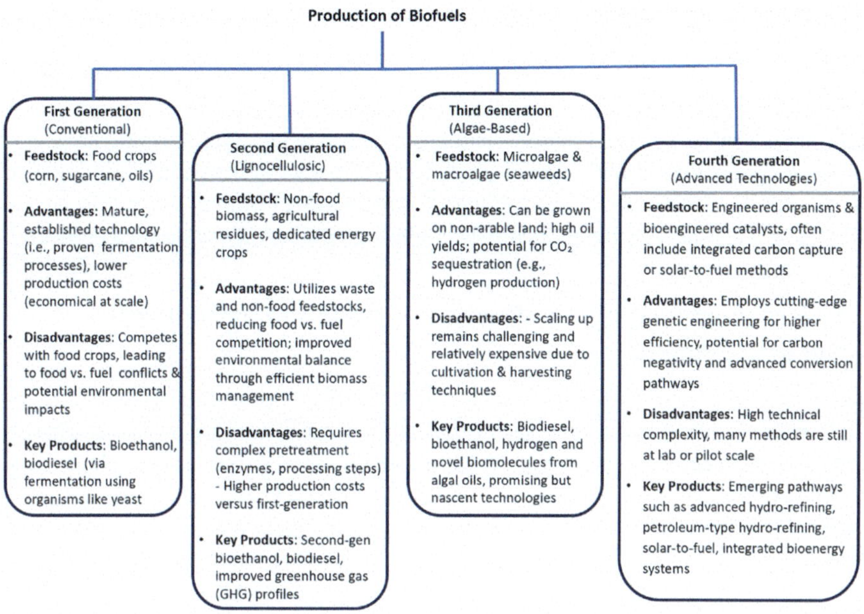


**FIG. 17.1**

Evolution and technological progression of biofuel production strategies.

## 6.2 Advancements in microbial biofuels

Early in the third millennium, surging fossil fuel prices, heightened environmental awareness, and a push for self-sufficiency spurred interest in biofuels (Hwang et al., 2016). Government support has given some nations' companies the economic stability they need to make investments in different types of biofuels. For example, South Africa's 2011 law mandates a 2% bioethanol blend in petrol and 5% biodiesel in diesel, further spurring regional investment and research (OECD-FAO, 2011). Global use of bioethanol was 100 billion liters in 2010, while biodiesel was 20 billion liters (OECD-FAO, 2011). Over the next 10 years, both of these volumes are expected to increase steadily. Although less common, $H_2$ petrol is a renewable biofuel boasting high energy density and burns without emitting any particles or greenhouse gases. Notably, fermentative biological methods are particularly attractive compared to conventional methods due to their reduced energy demands and the ability to convert waste or agricultural by-products into valuable feedstock. In these systems, the hydrogenase enzyme facilitates hydrogen evolution (OECD-FAO, 2011) by catalyzing the reversible oxidation of hydrogen gas (Lynd et al., 2009; Das and Veziroğlu, 2001). Hydrogenases and/or organisms containing hydrogenases have the potential to provide inexpensive and renewable hydrogen for biofuel and fuel cells to produce electricity if properly utilized. Low yields and the expense of photochemical synthesis reactors restrict the current output (Collet et al., 2014). Lactose fermentation by

*Clostridium thermolacticum* can also be used as a cost-effective substitute for waste-derived hydrogen production.

The genetically modified microbes yield increased product production, while specificity emerges as the central focus of biofuel science. As the battle between fuel and food intensifies, higher-level biofuels will become increasingly significant. The degradation of lignocellulose will be prioritized, with forthcoming innovations set to boost the productivity, affordability, and efficiency of microbial biofuel (Ventorino et al., 2015). Fig. 17.2 shows different methods of biofuel production and derived products.

## 6.3 Vermicomposting: An effective approach to agricultural waste management

Growing amounts of trash have led to a number of environmental issues, such as biodiversity loss and soil, water, and air pollution. Global warming is also caused by a variety of greenhouse gases released from solid waste disposal sites. Nowadays, the most popular ways to dispose of solid waste are landfilling and open dumping. Additional techniques for recovering resources from solid wastes include pyrolysis, gasification, anaerobic digestion, composting, and biomethanation. Although various wastes have been utilized in some papers to create nanomaterials, real-world applications for environmental advantages are still being developed (Griffin et al., 2018).

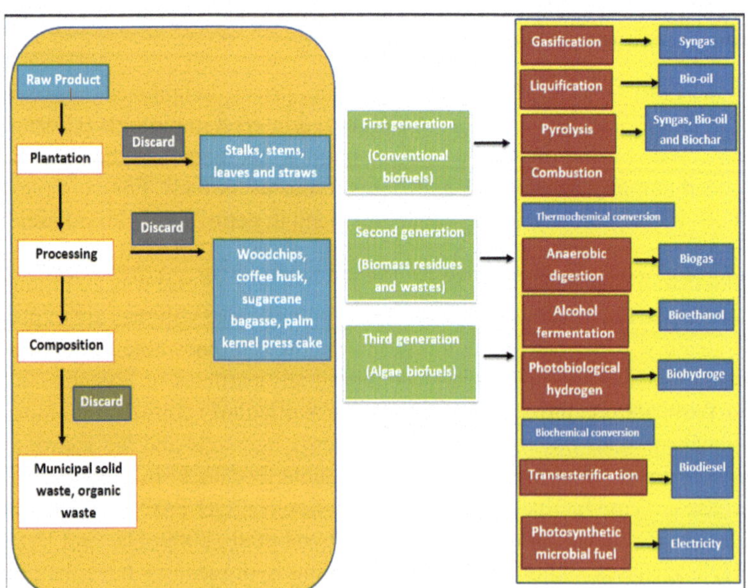


**FIG. 17.2**

Methods of biofuel production and derived products.

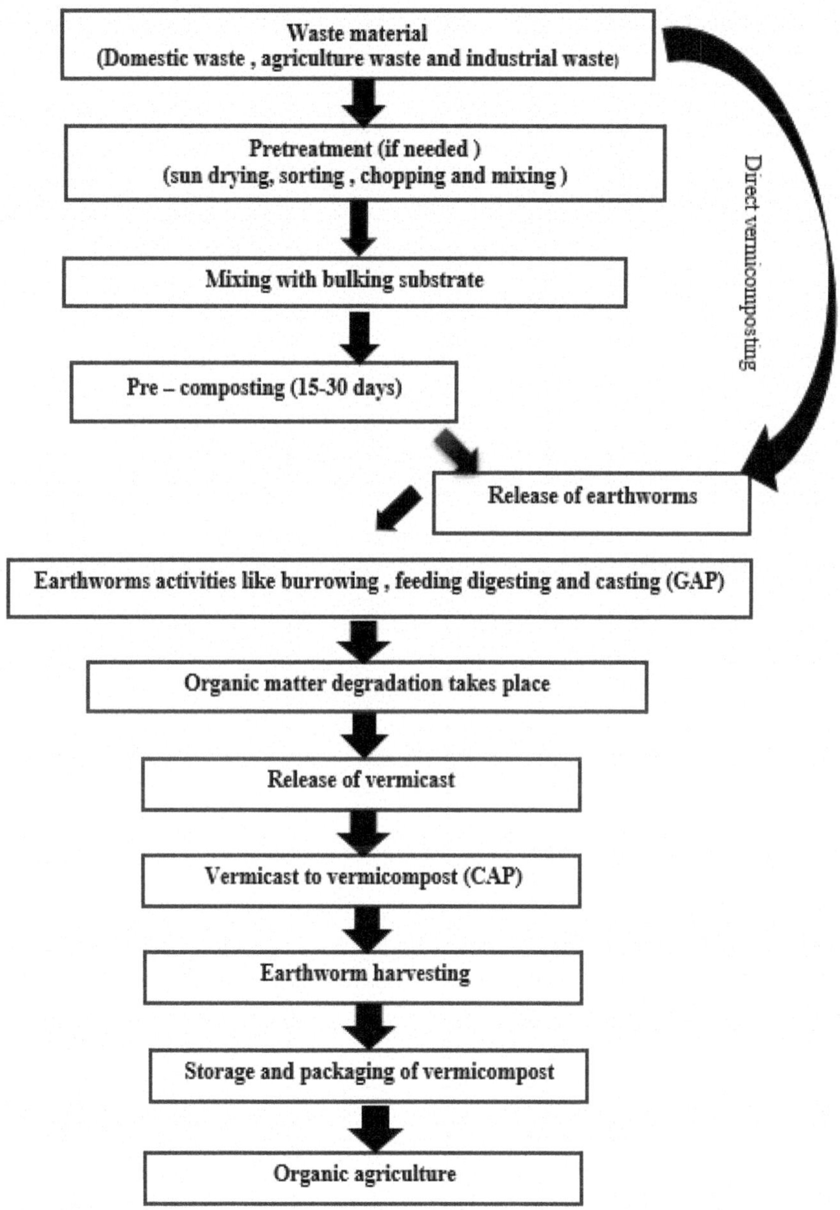


**FIG. 17.3**

Vermicompost pathway.

To manage solid waste without endangering the environment, scientists are always looking for sustainable options. Biodegradable garbage is present in significant amounts in waste generated from several industries. This biodegradable waste can be utilized in vermicomposting, composting, and biomethanation processes, and it may serve as a raw material for a number of biological processes. Owing to its high efficiency and simplicity, vermicomposting has drawn interest from all around the world in recent times (Hussain et al., 2018). Numerous waste materials can be used as source materials for vermicomposting, including animal waste, agricultural waste (Sharma and Garg, 2017, 2018), industrial waste (Lee et al., 2017), and municipal garbage (Soobhany et al., 2017). Fig. 17.3 represents a schematic diagram of the vermicompost pathway using earthworms.

## 7 Using agricultural waste for ethanol production

In industrial fermentation, a range of sugars can serve as a feedstock for bioethanol production. Due to their lignocellulosic composition, starchy materials rich in glucose polymers cannot be fermented using conventional methods (Kumar and Kumar, 2017). Starch (15%–20%), mixed with water, boiled, and treated with enzymes, is broken down into glucose chains, with amylase converting the starch into simple glucose chains during liquefaction. To guarantee fermentation, microbes are added once the mixture has cooled to about 30°C (Konur, 2021). Ethanol is created by hydrolyzing several carbohydrates, including wheat and maize, with enzymes. The maize ethanol industry uses both dry and wet milling techniques because ethanol dry mills are smaller and mainly intended for the production of ethanol and animal feeds (Minteer, 2016; Konur, 2021).

### 7.1 Fermentation-based valorization

The metabolic route of organic substances by bacteria, known as fermentation, is characterized by chemical changes brought on by enzyme activity. Fermentation processing can be classified as either aerobic or anaerobic based on oxygen concentration (Khandaker et al., 2020). Several naturally occurring microorganisms can adapt to use different carbohydrate polymers to produce ethanol. Generally, during the fermentation process, microbes break down various carbohydrates in both the presence and absence of oxygen to generate ethanol (Khandaker et al., 2020). About 40%–48% of the glucose used in fermentation is converted into ethanol with a 46% efficiency, yielding roughly 583 L of pure ethanol from 1000 kg of sugar (Siqueira et al., 2020).

### 7.2 Anaerobic digestion process for biofertilizers and renewable energy

Anaerobic digestion of biological waste is considered a potential process for bioenergy generation (Sikora et al., 2017) and has attracted significant interest as a possible substitute method for utilizing organic farming to produce biofertilizer and fuel.

The same procedures used for biogas production are applied here, where organic substrates are broken down to produce 60%–70% $CH_4$ and 30% $CO_2$ (Sikora et al., 2017). As an initial phase, methane can be produced from municipal solid waste using anaerobic digestion. These molecules can mix with atmospheric nitrogen, ambient $CO_2$, and organic substrate residues, commonly referred to as landfill gas. However, because $CO_2$ and volatile organic pollutants must be eliminated to produce a high-quality commercial product, methane is of lower quality compared to natural gas (Siqueira et al., 2020). To separate natural gas from landfill gases, highly effective separation equipment is required. Among the advantages of producing methanol in liquid form are its low sulfur content, low ash content, and commercial viability. Compared to gaseous fuel, liquid fuel is easier to transport and store. Due to the anaerobic fermentation process, lignocellulosic biomass is utilized in agricultural biofertilizers and liquid fuels (Siqueira et al., 2020).

### 7.3 Production of biohydrogen

Microbial hydrolysis of the complex, microcrystalline lignocellulosic matrix and the conversion of cellulose-like compounds into soluble sugars limit biohydrogen production from crop waste biomass. To promote hydrolysis, the raw material frequently needs to undergo proper pretreatment procedures. Mechanical, physical, chemical, and biological methods form the basis of the primary pretreatments. Before fermentation, a mechanical shredding phase is necessary to increase the organic waste's surface area and decrease its particle size. Consequently, the acidogenic fermentation process favors both solubility and fermentation efficiency. Given the current state of knowledge, more research is necessary to fully comprehend how the compositions and properties of organic substrates affect the performance of biohydrogen synthesis. Crop residue pretreatment procedures also need careful research because the best pretreatment depends on the nature and sources of the organic substrates.

## 8 CRISPR–Cas genome engineering in advancing biofuel production

The escalating levels of $CO_2$ emissions from sources such as automobiles, industrial processes, and power plants have become a critical global issue, significantly contributing to climate change. Despite this, human activities still mainly rely on fossil fuels as their energy source. Biofuels, offering an environmentally friendly and renewable energy solution, have emerged as a promising alternative that can combat climate change while enhancing energy security. Enhancements in microbial biofuel production can be achieved through strategies such as strain optimization, rerouting carbon pathways, and eliminating competing metabolic routes. However, traditional genetic engineering tools often render these methods labor-intensive and slow.

The advent of CRISPR and Cas proteins has revolutionized genome editing by streamlining these processes. This mechanism, originally serving as adaptive immune defenses in bacteria, induces double-stranded breaks (DSBs) at precise genome locations. Yet, early use of heterogeneous CRISPR–Cas systems highlighted issues like DSB-associated lethality and premature Cas9 expression in certain bacterial strains. These hurdles were mitigated by using native CRISPR–Cas systems (Zhang et al., 2019), inducible promoters for Cas protein expression, and modified systems employing alternative Cas variants.

Approaches like CRISPR interference, which inhibit gene function without introducing DSBs, allow for precise regulation of gene expression, while CRISPR activation recruits activators to fine-tune transcription. Agricultural biomass, which is produced in tons of quantity, could be harnessed by key fermentative organisms, such as yeast, for biofuel production. Moreover, to fully support the metabolic needs of these biofuel-producing microbes, the lignocellulose must first be converted to fermentable sugars. This conversion process is challenged by several factors, such as lignin's resistance to microbial degradation, microbes' inability to effectively consume both hexose and pentose sugars, the presence of inhibitory compounds released during lignocellulose hydrolysis, and inefficiencies in simultaneous saccharification.

CRISPR–Cas offers multiple strategies to overcome these hurdles, either by genetically optimizing the potent strains or by modifying the feedstock itself. For example, in feedstock improvement, researchers have utilized this technology to alter the genetic makeup of barley, thereby modifying its lignin structure. Enhancing the release of fermentable sugars is a key goal during lignocellulose pretreatment. One approach to fulfill this is by reducing the lignin content in plants through CRISPR–Cas9 by creating mutations in the caffeic acid O-methyltransferase 1 gene (HvCOMT1) (Lee et al., 2021). Disruption of this gene locus led to a 12% increase in ethanol production compared to the wild-type. By fine-tuning the cell wall composition in key plants such as barley and rice while preserving overall biomass yield, saccharification efficiency is enhanced, resulting in increased biofuel output (Yu et al., 2021). Notably, pretreatment methods that separate cellulose from other plant-derived components along with impurities often rely on expensive chemical, mechanical, or biological methods. Consequently, reforming the lignin biosynthetic pathway becomes critical in overcoming the high pretreatment costs that ultimately affect the economic feasibility of lignocellulosic fuel production. In addition, CRISPR–Cas9 has been instrumental in modifying plant lignin structures by targeting pivotal genes in the lignin pathway (Yu et al., 2021).

Beyond feedstock optimization, CRISPR–Cas has been leveraged to expand the capabilities of microbes through the introduction of novel metabolic pathways that are absent in wild-type phenotypes. For instance, wild-type *Zymomonas mobilis* is unable to utilize pentose sugars, which are a prime component of lignocellulose. To overcome this limitation, researchers reconfigured the cellular metabolism by incorporating a heterologous xylose assimilation pathway from *E. coli*. The incomplete utilization of xylose in such hydrolyzates further limits the broader application of genetically modified *Z. mobilis* in the biofuel sector. In one study, fine-tuning key regulatory components of the xylose pathway with the assistance of the CRISPR–Cas system resulted in a threefold enhancement in xylose utilization in *E. coli* (Zhu et al., 2017).

Another method involves transforming xylose into valuable compounds, such as xylonic acid, through co-fermentation using *Candida glycerinogenes*, which has shown potential for producing lignocellulosic ethanol (Zhu et al., 2019).

The biofuel industry aspires to develop genetically engineered microbes capable of metabolizing both hexose and pentose sugars found in hydrolyzed lignocellulosic feedstock. This can be partly realized through the introduction of heterologous pathways into existing strains or by utilizing co-fermenting microbial strains. Yet, many microbes, including yeast, exhibit a preference for one carbon source when presented with a mixture, a phenomenon attributed to the carbon catabolite repression (CCR) regulatory pathway. The CCR limits the utilization of lignocellulosic biomass by prioritizing one carbon source while repressing others, posing a challenge to fermentation efficiency and productivity. To alleviate CCR, CRISPRi tools have been applied. For example, SpCRISPR-dCas9 was used, enabling two strains of *Clostridium* to co-utilize xylose and glucose from lignocellulosic biomass (Bruder et al., 2016).

Tolerance to ethanol and inhibitors, such as acetic acid and furfural, also plays a vital role in improving biofuel production. Techniques, such as modulating regulatory gene expression, metabolite recovery during fermentation, and disrupting competing pathways, are being explored to enhance stress tolerance in engineered strains, for example, the CHAnGE method, which employs CRISPR–Cas9 with homology-directed repair (Bao et al., 2018).

## 8.1 CRISPR for reducing yield and postharvest losses

Crop yield is influenced by a complex interaction of multiple genes and environmental factors. Key determinants of grain yield include panicle count, grains per panicle, and grain weight. Enhancing hybrid vigor, or heterosis, remains a proven method to boost yields, as hybrid varieties often outperform their parental inbred lines. By harnessing CRISPR technology to alter yield-reducing functional genes in superior rice varieties, researchers have developed more productive rice strains. Gene targeting with sequence-specific nucleases has enabled precise genome modifications for crop improvement. In earlier studies, the DEP1 gene was edited in indica rice to produce novel dep1 mutants, significantly improving yields (Huang et al., 2009).

## 8.2 Food waste valorization by CRISPR

The current growth of agrofood industries has led to an increase in by-product generation. Often left unmanaged, food waste decomposes rapidly, polluting the environment. A promising solution lies in adopting biorefinery models for citrus waste processing, which offers a promising strategy for the elimination of fruit waste (Satari and Karimi, 2018). In southern European countries, where the climate favors citrus cultivation, there is notable interest in converting citrus by-products into value-added products, and established bioeconomy models are already generating bioenergy from these feedstocks (Bedoi´c et al., 2019; Valenti et al., 2020). Ultimately, citrus residues demonstrate considerable potential for bioethanol synthesis due to their rich fermentable sugar content, offering a promising feedstock for sustainable renewable energy.

Lignocellulosic biomass represents a prominent type of biowaste and serves as a sustainable, carbon-based energy resource. However, the biorefinery potential of such biomass is hindered by its resistance to processing, primarily due to the lignin content in plant cell walls. Research shows that agricultural residues from the processing of plant-derived materials after harvest can be transformed into biofuel with the help of CRISPR technology. This can be achieved either by modifying the genome of the plants or by enhancing the capabilities of the microorganisms important for industrial applications (Lee et al., 2021).

## 9 Ethanol pathway optimization by CRISPR–Cas system

CRISPR–Cas technology has also facilitated multiplex targeted genetic modification for ethanol pathway optimization. For example, CRISPR–Cas9 permitted the concurrent removal of three specific genes regulating carbon flux, yielding about 1.4 times higher ethanol than that of the wild strain (Liu et al., 2019). In addition, CRISPRi is being used to redirect metabolism toward biofuel production by suppressing repressor genes or competing pathways (Gutmann et al., 2021). Reprogramming microbial metabolic pathways for ethanol production enhances fermentation yield and titer, making production more economically viable. Enhanced ethanol resistance is another goal that can be achieved by targeting yeast regulatory networks with CRISPR-based genome editing (Cai et al., 2019). Such advancements provide key insights into the regulation of ethanol production and help achieve near-theoretical yields.

Integrating lignocellulose processing via CRISPR–Cas9 has emerged as a cost-efficient biofuel production technology. This consolidated bioprocessing framework integrates multiple steps, such as enzyme synthesis, biomass breakdown, and fermentation within a microbe. For instance, the insertion of a cellulase gene from *Ampullaria gigaspora* into the genome of yeast enabled direct ethanol production (7.53 g/L) from orange peel substrate within 48 h, achieving a 37.7-fold improvement compared to a wild-type strain (Yang et al., 2018). Successful cellulose degradation requires the coordinated secretion of cellulolytic enzymes like endoglucanase, cellobiohydrolase, and glucosidase. CRISPR–Cas9 has also been applied to enhance protein production outside the cell in *Trichoderma reesei* (Fonseca et al., 2020). These implementations highlight the potential of CRISPR–Cas systems toward the advancement of biofuel production from lignocellulosic materials while reducing dependence on fossil fuels.

## 10 Different strategies for agricultural waste management

The waste management system for fruit waste encompasses a range of approaches for managing, discarding, or recycling.

## 10.1 Traditional approaches

Traditional approaches to managing fruit waste employ both biological and chemical processes to break down its organic components into simpler substances such as gases (CO, $CO_2$, $CH_4$, $H_2$, $H_2O$, $H_2S$, and $NH_3$), inert solids (compost, sludge), and liquid, predominantly as leachates (Esparza et al., 2020). Traditional methods rank low in waste management and require less rigorous standards for quality, sanitation, sorting, or storage. This makes them cost-effective and versatile enough to handle a broad spectrum of food waste (Eriksson, 2015). Although such methods yield value-added products like biohydrogen, biogas, compost, electricity, and heat that can substitute fossil fuel-based energy and products, they fall short in preserving the original complex molecules found in agricultural waste, as the material is either consumed, degraded, or lost in the process. Moreover, while these methods aim to reduce the risk to health and the environment from poorly managed fruit waste, they can themselves introduce significant environmental burdens through air pollution, greenhouse gas emissions, and the production of both liquid and solid wastes (Esparza et al., 2020; Leong and Chang, 2022).

### 10.1.1 *Landfilling*

Landfilling is one of the most commonly used traditional practices for food waste disposal, accounting for >95% of cases due to its minimal technical expertise and labor requirements, cost-effectiveness, and applicability to all waste types (Melikoglu et al., 2013; Nanda and Berruti, 2021). In this process, waste materials are buried in designated terrestrial sites, where bacterial decomposition of organic matter produces a mixture of gases, commonly referred to as "landfill gas" or biogas (Esparza et al., 2020). For example, one ton of food waste from landfills can generate roughly 1,25,000 L of biogas (Melikoglu et al., 2013). Some environmental concerns regarding biogas production include the presence of a high amount of methane (60%) along with 40%–50% carbon dioxide by volume. Another significant environmental concern linked to landfilling is the formation of leachates, which are toxic compounds. As precipitation and inherent moisture infiltrate the waste, these leachates (a concentrated mixture of organic matter, in organic salts, ammonia, xenobiotic compounds, and heavy metals) can gather and form deposits at the base of landfill sites. Without adequate engineering controls to prevent leakage, these leachates risk contaminating the surrounding soil and groundwater.

### 10.1.2 *Composting*

Composting is defined as the aerobic breakdown of organic waste by microbes, which leads to the formation of carbon dioxide, water, and a nutrient-rich compost. This compost is stable, abundant in carbon and nitrogen, boosts soil microbial diversity, improves aggregate stability, and augments water retention (Ganesh et al., 2022). In anaerobic zones within the compost heap, methane can also be generated. Producing compost can minimize reliance on chemical inputs, thereby lessening adverse environmental impacts such as soil degradation, increased greenhouse gas emissions, and harm to nontarget organisms. In addition, when applied to soil, compost promotes carbon sequestration and accelerates the growth of plant biomass, which further enhances the plant's ability to absorb and store carbon (Ganesh et al., 2022). However, composting is not considered to be very efficient; it is still superior to incineration, as it involves lower energy consumption and a decrease in greenhouse gas emissions.

### 10.1.3 *Animal feed*

Around the globe, agricultural by-products are repurposed as feed for a range of livestock, including ruminants, pigs, poultry, and fish, which helps offset the resource and energy demands of growing conventional feed crops and mitigates related environmental emissions (Wadhwa et al., 2015; Esparza et al., 2020). This approach effectively transforms substandard agricultural waste into high-value feed products that can ultimately reenter the human food supply chain, earning it a favorable ranking within the waste management hierarchy (Kasapidou et al., 2015; Wadhwa et al., 2015). With global population growth, the demand for animal products is projected to increase by 70%, spurred by rising incomes and urbanization in developing nations. This escalation will likely aggravate existing scarcities of livestock feed in numerous regions (Wadhwa et al., 2015; Wadhwa and Bakshi, 2013). Fruit, vegetables, and other waste rich in antioxidants, vitamins, antimicrobials, dietary fat, fiber, phytochemicals, and other nutrients represent a promising supplementary feed source that can reduce feeding costs and potentially increase farmers' profits (Wadhwa and Bakshi, 2013).

### 10.1.4 *Anaerobic digestion*

Comparable to landfilling, anaerobic digestion generates biogas by breaking down organic waste through anaerobic bacteria. However, rather than occurring in unmanageable landfills, this process is executed in purpose-built bioreactors known as "digestors." These digestors allow precise control of essential parameters such as temperature, pH, C/N ratio, moisture content, the profile of organic acids, and the strategies for nutrient addition. These factors influence the composition and metabolic function of the microbes and, consequently, the makeup of the produced biogas (Esparza et al., 2020; Ji et al., 2017). By fine-tuning these variables, methane yields can be maximized, comparable to the biogas generated in landfills. In addition, the process yields a solid digestate that can be repurposed as animal bedding, employed as fertilizer, or incorporated as an additive in soil enhancements for subsequent treatment. Fruit waste is exceptionally well-suited for anaerobic digestion due to its high moisture content (Eriksson and Spångberg, 2017). Various fruit wastes display different potentials for methane production. Compared to other methods, anaerobic digestion is widely recognized as the top choice for managing agricultural waste. This is largely because it not only produces biogas but also has one of the lowest carbon footprints when processing different food wastes, in contrast with other traditional methods. Moreover, utilizing solid digestate for soil enrichment can lessen reliance on chemical alternatives and their environmental impacts.

## 10.2 Emerging valorization methods

New approaches in agricultural waste valorization focus on retaining the nutrients and bioactive compounds present in the waste, enabling their recovery to use them directly or convert them into more complex, value-added products. This involves isolating these beneficial elements and reincorporating them across diverse distribution networks, including food, cosmetics, pharmaceuticals, and manufacturing industries, to generate better economic value than the original waste while simultaneously minimizing the volume of generated waste (Eriksson, 2015; Esparza et al., 2020).

Consequently, these valorization methods hold a prime position in waste management ranking (Eriksson and Spångberg, 2017). Nevertheless, their widespread adoption is hindered by transportation costs, as they require processing facilities, prevention of decomposition, and scaling-up techniques (Esparza et al., 2020). In addition, the inconsistent supply and varying composition of agricultural waste, due to different geographical conditions, seasons, and processing methods, complicate the conversion processes (Kasapidou et al., 2015; Leong and Chang, 2022). Table 17.1 provides a structured, side-by-side comparison of how different sectors

**Table 17.1** Various industries with their primary waste sources, derived products, and their associated impacts through waste valorization.

| Industry | Waste streams/ source | Examples | Sustainability impact/ benefits |
|---|---|---|---|
| Food industry | Residues produced directly during crop cultivation and postharvest processing | Crop residues, fruit peels, pulp, seeds, and bran | These diverse agricultural wastes can be redirected to minimize waste, enhance resource efficiency, and reduce environmental impact (e.g., lower emissions and waste disposal) |
| Cosmetics industry | Agricultural residues rich in bioactive compounds | Plant extracts, fruit seed oils, peel extracts, and other bioactive components extracted from crops | Recycled ingredients support "green" and sustainable cosmetic formulations by reducing waste and appealing to eco-conscious consumers |
| Pharmaceutical industry | Agricultural by-products that contain a rich profile of bioactive compounds | Extracts rich in antioxidants, nutraceutical compounds, and other plant-derived bioactives | These materials are utilized in developing innovative nutraceuticals and drug formulations, promoting circular economy practices, and reducing reliance on conventional raw materials |
| Manufacturing industry | Versatile agricultural waste streams | Biochar, biosorbents, nanocellulose, bioplastics, biodegradable coatings, and metallic nanoparticles | Agricultural residues offer sustainable raw materials for producing eco-friendly products, while reducing dependence on nonrenewable resources and fossil fuels |

utilize agricultural waste, along with the examples and sustainability benefits associated with each industry.

### 10.2.1 *Integrated biorefinery and advanced extraction techniques*

Agricultural wastes, which are abundant in cellulose, sugars, phenolic substances, and starch, can be converted into economical products for various applications (Ding et al., 2023; Leong and Chang, 2022). Biorefineries integrate physical, biological, and chemical techniques to transform biomass into diverse high-value products (Moreno et al., 2020). This method is quite crucial for achieving zero-waste agriculture, as it ensures every fraction of agricultural products is put to productive use (Slegers et al., 2020; Leong and Chang, 2022). The traditional high-yield extraction techniques involve the use of organic solvents, which can generate toxic waste. However, novel extraction methods, such as microwave-assisted extraction and enzyme-assisted extraction, eliminate the need for organic solvents but, on the other hand, require significant initial investments and industrial-scale tuning and validation (Kandemir et al., 2022).

### 10.2.2 *Substrates for microbial fermentation*

Agricultural wastes, particularly those of fruit origin, serve as an excellent growth medium for microbes as they are rich in sugar and various nutrients, supporting biomass production and valuable metabolites (Leong and Chang, 2022). Moreover, the emergence of targeted fermentation, which harnesses engineered microbial hosts, can result in the production of compounds whose traditional production imposes severe environmental burdens (Chai et al., 2022). Furthermore, this microbe-based method can be merged with the biorefinery system, utilizing microbial biomass as an additional raw material (Slegers et al., 2020). Collectively, these valorization techniques, ranging from integrated biorefineries and novel extraction methods to microbial fermentation, illustrate the wide range of strategies available for reusing agricultural waste. By combining these processes, the agricultural sector can advance toward a zero-waste agriculture paradigm, maximizing the utility of every component of agricultural waste while significantly reducing resource inputs and greenhouse gas emissions.

## 11 Transforming agricultural residues using microalgae biotechnology

Worldwide waste production and rampant natural resource consumption have surged to unsustainable levels. As a result, without a shift to sustainable production and consumption patterns, the associated environmental harms, such as the exhaustion of essential resources and pollution, are expected to intensify as the world's population grows. Microalgae are particularly promising as renewable sources of numerous commercially valuable compounds. They are capable of taking up nutrients from wastewater, ponds, and photobioreactors. This feature helps prevent competition over cultivable land and additional resources for food crops (Cheng et al., 2022; Chew et al., 2017). To overcome the problem of high manufacturing expenses in achieving the economic feasibility of microalgal bioproduction, the overarching concept of a zero-waste microalgal biorefinery has been proposed. This system facilitates the valorization of waste streams by processing and transforming various components of microalgal biomass into multiple products without generating any waste (Cheirsilp and Maneechote, 2022). Furthermore, reducing food waste can have wide-reaching positive effects on natural resources, climate change, and an increase in biodiversity. Food waste alone is estimated to carry a carbon footprint of 3.3 gigatons of $CO_2$ (Socas-Rodríguez et al., 2021). Thus repurposing food waste to create a variety of renewable products that can substitute traditional goods positions microalgal biorefineries as a promising strategy toward improved resource management and waste reduction (Esparza et al., 2020).

## 12 Conclusions

The transition to zero-waste agriculture is a vital step toward addressing the dual challenges of environmental sustainability and resource efficiency. This chapter emphasized biotechnology's critical role in converting agricultural waste into valuable bioproducts such as biofuels, bioplastics, fertilizers, and other industrially relevant materials. Composting, anaerobic digestion, microbial biotransformation, and CRISPR-based genetic engineering have all demonstrated enormous potential for transforming waste streams into economic and environmental advantages. Furthermore, integrating microalgal biorefineries with circular bioeconomy principles helps combat climate change, reduce dependency on fossil fuels, and encourage sustainable agricultural practices. Researchers have made tremendous progress in increasing feedstock usefulness, improving microbial tolerance, and optimizing metabolic pathways for efficient waste-to-value conversion by using cutting-edge microbial engineering and genome editing techniques, particularly CRISPR–Cas systems. Such biotechnological developments not only provide commercially viable alternatives to traditional waste management but also support global aspirations for food security, environmental conservation, and energy resilience. Looking ahead, biotechnology has the potential to advance zero-waste agriculture in several intriguing directions:

1. Integration of AI and precision agriculture: Internet of things (IoT)-enabled precision agriculture, which combines biotechnology and AI, can maximize the use of waste valorization processes at the farm level and expedite waste monitoring.
2. Synthetic biology for custom biocatalysts: To improve yield and shorten processing times, future studies will increasingly concentrate on creating synthetic microbial consortia and designer enzymes for the targeted breakdown and conversion of complex agricultural wastes.

3. CRISPR 2.0 and Multiplex Editing: Base editing and prime editing are two examples of next-generation CRISPR technologies that provide more accurate and multiplexed capabilities, allowing for significant advances in microbial strains and crops for effective biomass utilization.
4. Growth of microalgal biorefineries: As the cost of cultivation and harvesting technologies declines, it is anticipated that microalgal systems will expand, becoming a key component in the production of biofuel, bioplastics, and nutraceuticals while providing integrated wastewater treatment.
5. Policy and market development: To fully achieve zero-waste agriculture, financial incentives, public-private partnerships, and supportive policy frameworks must be reinforced to promote the use of biotechnology solutions.
6. Global collaboration and capacity building: To guarantee that biotechnological advances for waste valorization are affordable and scalable, particularly in developing nations, cross-border research collaborations, knowledge-sharing platforms, and capacity-building initiatives will be essential.

## References

Al-Azzawi, A.M., Jassem, E.K., 2016. Synthesis and characterization of several new succinimides linked to phenyl azo benzothiazole or thiazole moieties with expected biological activity. Iraq. J. Sci. 57, 534–544.

Arous, F., Azabou, S., Triantaphyllidou, I.-E., Aggelis, G., Jaouani, A., Nasri, M., Mechichi, T., 2017. Newly isolated yeasts from *Tunisian microhabitats*: lipid accumulation and fatty acid composition. Eng. Life Sci. 17, 226–236. doi:10.1002/elsc.201500156.

Bao, Z., HamediRad, M., Xue, P., Xiao, H., Tasan, I., Chao, R., Liang, J., Zhao, H., 2018. Genome-scale engineering of Saccharomyces cerevisiae with single-nucleotide precision. Nat. Biotechnol. 36 (6), 505–508.

Bedoi´c, R., Cosi´c, B., Dui´c, N., 2019. Technical potential and geographic distribution of agricultural residues, co-products and by-products in the European Union. Sci. Total Environ. 686, 568–579.

Bridgwater, A.V., 2012. Review of fast pyrolysis of biomass and product upgrading. Biomass Bioenergy 38, 68–94.

Bruder, M.R., Pyne, M.E., Moo-Young, M., Chung, D.A., Chou, C.P., 2016. Extending CRISPR-cas9 technology from genome editing to transcriptional engineering genus Clostridium. Appl. Environ. Microbiol. 82 (20), 6109–6119.

Bugg, T.D.H., Williamson, J.J., Rashid, G.M.M., 2020. Bacterial enzymes for lignin depolymerisation: new biocatalysts for generation of renewable chemicals from biomass. Curr. Opin. Chem. Biol. 55, 26–33. doi:10.1016/j.cbpa.2019.11.007.

Cabrol, L., Marone, A., Tapia-Venegas, E., Steyer, J.P., Ruiz-Filippi, G., Trably, E., 2017. Microbial ecology of fermentative hydrogen producing bioprocesses: useful insights for driving the ecosystem function. FEMS Microbiol. Rev. 41, 158–181.

Cai, P., Gao, J., Zhou, Y., 2019. CRISPR-mediated genome editing in non-conventional yeasts for biotechnological applications. Microb. Cell Fact. 18, 63.

Carmona-Cabello, M., García, I.L., Papadaki, A., Tsouko, E., Koutinas, A., Dorado, M.P., 2021. Biodiesel production using microbial lipids derived from food waste discarded by catering services. Bioresour. Technol. 323, 124597. doi:10.1016/j.biortech.2020.124597.

Chai, K.F., Ng, K.R., Samarasiri, M., Chen, W.N., 2022. Precision fermentation to advance fungal food fermentations. Curr. Opin. Food Sci. 47, 100881. doi:10.1016/j.cofs.2022.100881.

Cheirsilp, B., Maneechote, W., 2022. Insight on zero waste approach for sustainable microalgae biorefinery: sequential fractionation, conversion and applications for high-to-low value-added products. Bioresour. Technol. Rep. 18, 101003. doi:10.1016/j.biteb.2022.101003.

Cheng, C.L., Lo, Y.C., Huang, K.L., Nagarajan, D., Chen, C.Y., Lee, D.J., Chang, J.S., 2022. Effect of pH on biomass production and carbohydrate accumulation of *Chlorella vulgaris* JSC-6 under autotrophic, mixotrophic, and photoheterotrophic cultivation. Bioresour. Technol. (351), 127021. doi:10.1016/j.biortech.2022.127021.

Chew, K.W., Yap, J.Y., Show, P.L., Suan, N.H., Juan, J.C., Ling, T.C., Lee, D.J., Chang, J.S., 2017. Microalgae biorefinery: high value products perspectives. Bioresour. Technol. 229, 53–62.

Collet, P., Lardon, L., Hélias, A., Bricout, S., Lombaert-Valot, I., Perrier, B., Lepine, O., Steyer, J.P., Bernard, O., 2014. Biodiesel from microalgae–life cycle assessment and recommendations for potential improvements. Renew. Energy 71, 525533.

Das, D., Veziroğlu, T.N., 2001. Hydrogen production by biological processes: a survey of literature. Int. J. Hydrog. Energy 26, 13–28. doi:10.1016/s03603199(00)00058-6.

Deeba, F., Raza, I., Muhammad, N., Rahman, H., Ur Rehman, Z., Azizullah, A., Khattak, B., Ullah, F., Daud, M.K., 2017. Chlorpyrifos and lambda cyhalothrin-induced oxidative stress in human erythrocytes: in vitro studies. Toxicol. Ind. Health 33, 297–307.

Demirbas, M.F., 2011. Biofuels from algae for sustainable development. Appl. Energy 88, 3473–3480.

Dhyani, V., Bhaskar, T., 2018. A comprehensive review on the pyrolysis of lignocellulosic biomass. Renew. Energy 129, 695–716. doi:10.1016/j.renene.2017.04.035.

Ding, Z., Ge, Y., Sar, T., Kumar, V., Harirchi, S., Binod, P., Sirohi, R., Sindhu, R., Wu, P., Lin, F., Zhang, Z., Taherzadeh, M.J., Awasthi, M.K., 2023. Valorization of tropical fruits waste for production of commercial biorefinery products—a review. Bioresour. Technol. 374. 421, 132159. doi:10.1016/j.biortech.2025.132159.

Dragone, G., Fernandes, B.D., Vicente, A.A., Teixeira, J.A., 2010. Third generation biofuels from microalgae. In: Méndez-Vilas, A. (Ed.), Current Research. Technology and Education Topics in Applied Microbiology and Microbial Biotechnology. Spain Formatex Research Center, pp. 1355–1366.

Eriksson, M., 2015. Supermarket Food Waste: Prevention and Management with the Focus on Reduced Waste for Reduced Carbon Footprint. Swedish University of Agricultural Sciences, Department of Energy and Technology, pp. 1–97.

Eriksson, M., Spångberg, J., 2017. Carbon footprint and energy use of food waste management options for fresh fruit and vegetables from supermarkets. Waste Manag. 60, 786–799.

Esparza, I., Jim´enez-Moreno, N., Bimbela, F., Ancín-Azpilicueta, C., Gandía, L.M., 2020. Fruit and vegetable waste management: conventional and emerging approaches. J. Environ. Manage. 265, 110510. doi:10.1016/j.jenvman.2020.110510.

FAO, WFP, UNECE, UNICEF, WHO, and WMO, 2021. Regional Overview of Food Security and Nutrition in Europe and Central Asia 2020: Affordable Healthy Diets to Address All Forms of Malnutrition for Better Health. FAO, Rome. doi:10.4060/cb3849en.

FAO, 2017. The future of food and agriculture–trends and challenges. Annual Report. http://www.fao.org/3/i6583e/i6583e.

Fonseca, L.M., Parreiras, L.S., Murakami, M.T., 2020. Rational engineering of the *Trichoderma reesei* RUT-C30 strain into an industrially relevant platform for cellulase production. Biotechnol. Biofuels 13, 93.

Ganesh, K.S., Sridhar, A., Vishali, S., 2022. Utilization of fruit and vegetable waste to produce value-added products: conventional utilization and emerging opportunities—a review. Chemosphere 287 (3), 132221.

Griffin, S., Sarfraz, M., Farida, V., Jawad, M., 2018. No time to waste organic waste: nanosizing converts remains of food pro cessing into refined materials. J. Environ. Manage. 210, 114e21.

Gutmann, F., Jann, C., Pereira, F., Johansson, A., Steinmetz, L.M., Patil, K.R., 2021. CRISPRi screens reveal genes modulating yeast growth in lignocellulose hydrolysate. Biotechnol. Biofuels 14 (1), 1–14.

Halpin-Healy, T.S., Klompe, S.E., Sternberg, S.H., Fernandez, I.S., 2020. Structural basis of DNA targeting by a transposon-encoded CRISPR-Cas system. Nature 577, 271–274.

Huang, X., Qian, Q., Liu, Z., Sun, H., He, S., Luo, D., Xia, G., Chu, C., Li, J., Fu, X., 2009. Natural variation at the DEP1 locus enhances grain yield in rice. Nat. Genet. 41 (4), 494–497.

Hussain, N., Das, S., Goswami, L., Das, P., Sahariah, B., Bhattacharya, S.S., 2018. Intensification of vermitechnology for kitchen vegetable waste and paddy straw employing earthworm consortium: assessment of maturity time, microbial community structure, & and economic benefit. J. Clean. Prod. 182, 414–426.

Hwang, J.-H., Church, J., Lee, S.-J., Park, J., Lee, W.H., 2016. Use of microalgae for advanced wastewater treatment and sustainable bioenergy generation. Environ. Eng. Sci. 33, 882–897.

Ji, C., Kong, C.X., Mei, Z.L., Li, J., 2017. A review of the anaerobic digestion of fruit and vegetable waste. Appl. Biochem. Biotechnol. 183 (3), 906–922.

Kandemir, K., Piskin, E., Xiao, J., Tomas, M., Capanoglu, E., 2022. Fruit juice industry wastes as a source of bioactives. J. Agric. Food Chem. 70 (23), 6805–6832.

Kasapidou, E., Sossidou, E., Mitlianga, P., 2015. Fruit and vegetable co-products as functional feed ingredients in farm animal nutrition for improved product quality. Agriculture (Switzerland) 5 (4), 1020–1034.

Khandaker, M.M., Abdullahi, U.A., Abdulrahman, M.D., Badaluddin, N.A., Mohd, K.S., 2020. Bioethanol production from fruit and vegetable waste by using Saccharomyces cerevisiae. In: Inambao, F. (Ed.), Bioethanol Technologies, IntechOpen Publishing, pp. 1–17. doi: 10.5772/intechopen.94358.

Konur, O., 2021. Biodiesel Fuels: Science, Technology, Health, and Environment. CRC Press, Boca Raton, FL, pp. 448.

Kourmentza, C., Economou, C.N., Tsafrakidou, P., Kornaros, M., 2018. Spent coffee grounds make much more than waste: exploring recent advances and future exploitation strategies for the valorization of an emerging food waste stream. J. Clean. Prod. 172, 980–992. doi:10.1016/j.jclepro.2017.10.088.

Kroukamp, H., Haan, R., Zyl, J., Zyl, W., 2018. Rational strain engineering interventions to enhance cellulase secretion by *Saccharomyces cerevisiae*. Biofuels Bioprod. Biorefin. 12, 108–124.

Kumar, R., Kumar, P., 2017. Future microbial applications for bioenergy production: a perspective. Front. Microbiol. 8, 450. doi:10.3389/fmicb.2017.00450.

Lee, J.H., Won, H.J., Hoang Nguyen Tran, P., Lee, S.-M., Kim, H.-Y., Jung, J.H., 2021. Improving lignocellulosic biofuel production by CRISPR/Cas9-mediated lignin modification in barley. GCB Bioenergy 13, 742–752.

Lee, L.H., YeongWu, T., Shak, K.P.Y., Lim, S.L., Ng, K.Y., Nguyen, M.N., Teoh, W.H., 2017. Sustainable approach to bio transform industrial sludge into organic fertilizer via ver micomposting: a mini-review. J. Chem. Technol. Biotechnol. 7 (93), 925–935.

Leong, Y.K., Chang, J.S., 2022. Valorization of fruit wastes for circular bioeconomy: current advances, challenges, and opportunities. Bioresour. Technol. 359, 127459. doi:10.1016/j.biortech.2022.127459.

Liu, R., Liang, L., Choudhury, A., Garst, A.D., Eckert, C.A., Oh, E.J., Winkler, J., Gill, R.T., 2019. Multiplex navigation of global regulatory networks (MINR) in yeast for improved ethanol tolerance and production. Metab. Eng. 51, 50–58.

Lynd, L.R., Larson, E., Greene, N., Laser, M., Sheehan, J., Dale, B.E., McLaughlin, S., Wang, M., 2009. The role of biomass in America's energy future: framing the analysis. Biofuels Bioprod. Biorefin. 3, 113–123. doi:10.1002/bbb.134.

Melikoglu, M., Lin, C.S.K., Webb, C., 2013. Analysing global food waste problem: pinpointing the facts and estimating the energy content. Cent. Eur. J. Eng. 3 (2), 157–164.

Milledge, J.J., Heaven, S., 2013. A review of the harvesting of microalgae for biofuel production. Rev. Environ. Sci. Biotechnol. 12, 165–178. doi:10.1007/s11157-012-9301-z.

Minteer, S., 2016. Alcoholic Fuels. CRC Press, Boca Raton, FL, pp. 296.

Moreno, A.D., Ballesteros, M., Negro, M.J., 2020. Biorefineries for the valorization of food processing waste. In: Galankis, C. (Ed.), The Interaction of Food Industry and Environment. Academic Press (Elsevier), pp. 155–190.

Nanda, S., Berruti, F., 2021. Municipal solid waste management and landfilling technologies: a review. Environ. Chem. Lett. 19 (2), 1433–1456.

OECD-FAO, 2011. OECD-FAO Agricultural Outlook 2011 (Summary in German). OECD Publishing, Paris. https://www.oecd-ilibrary.org/docserver/agr_outlook2011-sum-de.pdf?expires=1643258219&id=id&accname=guest&checksum=5AE95056CE101D4218EC32D23C222629.

Pollegioni, L., Tonin, F., Rosini, E., 2015. Lignin-degrading enzymes. FEBS J 282, 1190–1213.

Rockström, J., Steffen, W., Noone, K., et al., 2009. A safe operating space for humanity. Nature 461, 472–475.

Satari, B., Karimi, K., 2018. Citrus processing wastes: environmental impacts, recent advances, and future perspectives in total valorization. Resour. Conserv. Recycl 129, 153–167.

Schenk, P.M., Thomas-Hall, S.R., Stephens, E., Marx, U.C., Mussgnug, J.H., Posten, C., Kruse, O., Hankamer, B., 2008. Second generation biofuels: high-efficiency microalgae for biodiesel production. Bioenerg. Res. 1, 20–43. doi:10.1007/s12155-008-9008-8.

Schubert, C., 2006. Can biofuels finally take center stage? Nat. Biotechnol. 24, 777–784.

Shalaby, E.A., 2011. Algal biomass and biodiesel production. In: Stoytcheva, M. (Ed.), Biodiesel-Feedstock and Processing Technologies. IntechOpen Publishing, pp. 111–132.

Sharma, K., Garg, V.K., 2017. Vermi-modificaion of ruminant excreta using *Eisenia fetida*. Environ. Sci. Pollut. Res. 24(2), 19938e45.

Sharma, K., Garg, V.K., 2018. Comparative analysis of vermicom post quality produced from rice straw and paper waste employing earthworm *Eisenia fetida* (Sav.). Bioresour. Technol. 24(8), 7829e36.

Sikora, A., Detman, A., Chojnacka, A., Błaszczyk, M., 2017. Anaerobic digestion: i. A common process ensuring energy flow and the circulation of matter in ecosystems. II. A tool for the production of gaseous biofuels. In: Jozala, A., Ed., Fermentation Processes. IntechOpen Publishing, pp. 271–301. doi:10.5772/64645.

Siqueira, J.G.W., Rodrigues, C., Vandenberghe, L.P.D.S., Woiciechowski, A.L., Soccol, C.R., 2020. Current advances in on-site cellulase production and application on lignocellulosic biomass conversion to biofuels: a review. Biomass Bioenergy 132, 105419.

Slegers, P.M., Olivieri, G., Breitmayer, E., Sijtsma, L., Eppink, M.H.M., Wijffels, R.H., Reith, J.H., 2020. Design of value chains for microalgal biorefinery at industrial scale: process integration and techno-economic analysis. Front. Bioeng. Biotechnol. 8, 1–17.

Socas-Rodríguez, B., Alvarez-Rivera, G., Valdés, A., Ibáñez, E., Cifuentes, A., 2021. Food by-products and food wastes: are they safe enough for their valorization? Trends Food Sci. Technol. 114, 133–147.

Soobhany, N., Gunasee, S., Rago, Y.P., Joyram, H., Raghoo, P., Mohee, R., Garg, V.K., 2017. Spectroscopic, & thermogravimetric and structural characterization analyses for comparing Municipal Solid Waste composts and vermicomposts stability and maturity. Bioresour. Technol. 263, 11e9.

Sticklen, M.B., 2008. Plant genetic engineering for biofuel production: towards affordable cellulosic ethanol. Nat. Rev. Genet. 9 (6), 433–443.

Tilman, D., Cassman, K., Matson, P., et al, 2002. Agricultural sustainability and intensive production practices. Nature 418, 671–677.

Trabelsi, A.B.H., Zaafouri, K., Baghdadi, W., Naoui, S., Ouerghi, A., 2018. Second generation biofuels production from waste cooking oil via pyrolysis process. Renew. Energy 126, 888–896.

Valenti, F., Porto, S.M., Selvaggi, R., Pecorino, B., 2020. Co-digestion of by-products and agricultural residues: a bioeconomy perspective for a Mediterranean feedstock mixture. Sci. Total Environ. 700, 134440. doi:10.1016/j.scitotenv.2019.134440.

Ventorino, V., Aliberti, A., Faraco, V., Robertiello, A., Giacobbe, S., Ercolini, D., Amore, A., Fagnano, M., Pepe, O., 2015. Exploring the microbiota dynamics related to vegetable biomasses degradation and study of lignocellulose-degrading bacteria for industrial biotechnological application. Sci. Rep. 5, 8161. doi:10.1038/srep08161.

Wadhwa, M., Bakshi, M.P.S., 2013. Utilization of Fruit and Vegetable Wastes as Livestockfeed and as Substrates for Generation of Other Value-added Products. FAO, pp. 1–56.

Wadhwa, M., Bakshi, M.P.S., Makkar, H.P.S., 2015. Waste to worth: fruit wastes and by-products as animal feed. In: CAB Reviews: Perspectives in Agriculture, Veterinary Science, Nutrition and Natural Resources Vol. 10. CABI International, pp. 1–26.

Yang, P., Wu, Y., Zheng, Z., Cao, L., Zhu, X., Mu, D., Jiang, S., 2018. CRISPR-Cas9 approach constructing cellulase sestc-engineered Saccharomyces cerevisiae for the production of Orange peel ethanol. Front. Microbiol. 9, 2436.

Yu, H., Liu, C., Dixon, R.A., 2021. A gene-editing/complementation strategy for tissue-specific lignin reduction while preserving biomass yield. Biotechnol. Biofuels 14 (1), 175.

Zhang, Y., Malzahn, A.A., Sretenovic, S., Qi, Y., 2019. The emerging and uncultivated potential of CRISPR technology in plant science. Nat. Plants 5 (8), 778–794. doi:10.1038/s41477-019-0461-5.

Zhang, Y., Pribil, M., Palmgren, M., Gao, C., 2020b. A CRISPR way for accelerating improvement of food crops. Nat. Food 1 (4), 200–205. doi:10.1038/s43016-020-0051-8.

Zhang, Y., Zhang, H., Wang, Z., Wu, Z., Wang, Y., Tang, N., Xu, X., Zhao, S., Chen, W., Ji, Q., 2020a. Programmable adenine deamination in bacteria using a Cas9-adeninedeaminase fusion. Chem. Sci. 11, 1657–1664.

Zhu, M., Sun, L., Lu, X., Zong, H., Zhuge, B., 2019. Establishment of a transient CRISPRCas9 genome editing system in *Candida glycerinogenes* for co-production of ethanol and xylonic acid. J. Biosci. Bioeng. 128 (3), 283–289.

Zhu, P., Abdelaziz, O.Y., Hulteberg, C.P., Riisager, A., 2020. New synthetic approaches to biofuels from lignocellulosic biomass. Curr. Opin. Green Sustain. Chem. 21, 16–21.

Zhu, X., Zhao, D., Qiu, H., Fan, F., Man, S., Bi, C., Zhang, X., 2017. The CRISPR/Cas9-facilitated multiplex pathway optimization (CFPO) technique and its application to improve the *Escherichia coli* xylose utilization pathway. Metab. Eng. 43 (Pt A), 37–45.

# Index

Note: Page numbers followed by "*f*" indicate figures, "*t*" indicate tables.


## A

Absorption 266
Acetogenesis 254*t*
Acetophilic methanogens 259–260
Acetyl xylanesterase 209–210
Acidogenesis 254*t*, 257
Actinobacteria 59–60
AD bacteria. *see* Aerobic denitrifying (AD) bacteria
AD inhibitors 265–266
  ammonia $NH_3$ 265–266
  organic compounds 266
  oxygen 265
  sulfide 265
AD. *see* Anaerobic digestion (AD)
Advanced oxidation processes (AOPs) 90–93, 91*f*
Aerobic bacterial strains 20–21
Aerobic cometabolism 20
Aerobic denitrifying (AD) bacteria 365
  immobilization 376–377, 377*f*
Agriculture 183, 445
  biomass 448, 456
  smart 190–191
  use of nanomaterials 186–187
  zero-waste 447, 463
Agriculture waste 447, 462
  for ethanol production 454–455
  management 452–454, 458–462
  sustainable biochemical conversion 450
Agrifood industry 282–283
Agroindustrial biowastes 279
AI models. *see* Artificial intelligence (AI) models
Air pollutants 394
*Alcanivorax borkumensis* 386–387
Algae-based biofuels 449–450
Algal bioremediation 59
Alginate 113, 115, 118, 368
Ammonia ($NH_3$) 265–266
AMRUT scheme 428–429
Anaerobic digestion (AD) 239–240, 240*f*, 281–282, 284, 299–301, 323–324, 331–333, 391, 393, 448, 454–455, 460
  biogas 266–267
  BMP 239–244
  C/N ratios 247*t*
  design considerations 261–262
  digestate 267–268
  environmental impacts of 268–269
  exoenzymes & endoenzymes 254*t*
  feeding mode 261
  feedstocks for 244–246
  inhibitors in 265–266
  methane production, kinetics 242–243
  methane yield curve 243
  operational parameters 261–262
  pH 262–265
  phase separation 261
  process description/microbiology 252–266, 253*f*
  products of 266–268
  single-stage digester 262*f*
  syntrophy in 260
  temperature 261–262
  two-stage digester 262*f*
Anaerobic filter 264–265
Anaerobic PAH degradation 15
Anammox process 375–376
Animal feed 460
Animal wastes 305–306
ANNs. *see* Artificial neural networks (ANNs)
Anodic reactions 81–82
AOPs. *see* Advanced oxidation processes (AOPs)
Aquatic biomass 203–204
Aquatic plants 203–204
Arabinases 209
Arabinofuranosidase 209
Archaea 386
Arctic microbes 148
Arsenic 394–395
Artificial intelligence (AI) models 341–347, 343*f*, 355–356
Artificial neural networks (ANNs) 342–352, 354–355
  bioreactor performance 353–354
  microbial fuel cells 352–353
  microbial waste management 344*t*
  particle swarm optimization 352
Ascomycetes 214

## B

BA. *see* Bioaugmentation (BA)
*Bacillus cereus* strains 402

Bacteria/bacterial 59
exoelectrogenic 325–326
extremophiles 15
and fungi 386–387
BAC-UF system 89
Basel Convention 1
BDI. *see* Biodegradability index (BDI)
BESs. *see* Bioelectrochemical systems (BESs)
BGL. *see* β-D glucosidase (BGL)
Bioaccumulation 105, 166–168
Bioadsorption systems 85–88
limitations 87–88
waste management 86–87
Bioaugmentation (BA) 57
actinobacterial consortia 65–70
and biostimulation 57–58
Biochar carrier materials 368, 370–373, 372*f*, 373*f*
Biochemical methane potentia (BMP)
chemical oxygen demand 240
elemental composition 240
organic fraction composition 241
theoretical calculation of 241
Bioconversion processes 280–281
Biodegradability index (BDI) 242
Biodegradation 105, 125–126
Biodiesel 449–450
Bioelectrochemical systems (BESs) 170, 283–286, 289, 299, 322–323, 329–335, 387
Bioenergy 448
Bioethanol 449–450
Biofertilizers 285–286
Biofuel 449–454, 451*f*, 452*f*
algae-based 449–450
conventional 449–450
industry 457
production 455–458
Biogas 266–267
absorption 266
cryogenic separation 267
membrane separation 267
Biohydrogen production 290, 455
Bioleaching 9, 163*f*, 163–168, 165*t*, 170–171
direct 163–164
indirect 163–164
Biological membrane filtration 88–90, 89*f*
limitations 89–90
waste management 88–89
Biological oxidation 90–93
limitations 92–93
waste management 90–92
Biomass 201–202
animal & human waste 204–205
categories 202*t*, 202–205
chemical composition 205
classification 202–205
for enzyme production 219–220
mixtures 205
Biopesticides 52
Biophotocatalysis 83–85
mechanism of 83–84
waste management 84–85
Biophotoelectrochemical cell (BPEC) 313–314
Bioplastics 286–287, 447
from agricultural & dairy waste 447
Biorefinery 462
Bioremediation 55, 121, 127, 386, 397–398
biobased PVC 152–153
factors 124–125
microbial consortia/strategies 55–58, 56*f*, 62–64
phylum *Actinomycetota* 59–61
Biosensor-enabled microbial strains 406
Biosorption 105, 166–168
Biostimulation (BS) 56–57
and bioaugmentation 57–58
Biotechnology 322*f*, 445–446
applications 128–129
cryogels 117
nanocell encapsulation 126–127
Biotransformations 323–324
Biotrickling filters (BTFs) 21*f*, 117
BMP. *see* Biochemical methane potentia (BMP)
BPEC. *see* Biophotoelectrochemical cell (BPEC)
*Bradyrhizobium* sp. strain R5 15–16
Brevibacillus parabrevis strain BCP-09 15–16
Brown-rot fungi 214–215
BS. *see* Biostimulation (BS)
BTFs. *see* Biotrickling filters (BTFs)


## C


Cadmium 7, 22
*Candida albicans* 19–20
*Candida tropicalis* 58
Capacitor-based power management 309–311, 310*f*
Carbon catabolite repression (CCR) 457
Carbon NPs 187, 191
Carbon-based materials 330–331
Carbon-based nanomaterials 369
Cardiovascular diseases 394
Cathodic reactions 81–82
Cation exchange membrane (CEM) 22, 24
CBHs. *see* Cellobiohydrolases (CBHs)
CCR. *see* Carbon catabolite repression (CCR)
CE. *see* Circular economy (CE)
Cellobiase 206–207
Cellobiohydrolases (CBHs) 206
Cellobiose dehydrogenase 207

Cellulases 205–206, 214, 222–223
Cellulolytic enzymes 205–208
Cellulose-based carriers 368
CEM. *see* Cation exchange membrane (CEM)
Central Pollution Control Board (CPCB) 417
Chemical composition classification
 inorganic NPs 185
 organic NPs 185
 physical characteristics 186
 production method 186
 structural properties 185–186
 usage 186
Chemical oxygen demand (COD) 86–87, 304
Chemical treatment methods 385–386
Chitosan 368
Chlordane (CLD) 66–67
*Chlorella pyrenoidosa* 92
Chlorinated volatile organic compounds (Cl-VOCs) 20–22
Chlorpyrifos (CP) 68–70
Circular economy (CE) 387–389
 model 171–172, 287–288
CLD. *see* Chlordane (CLD)
CLEAs. *see* Cross-linked enzyme aggregates (CLEAs)
*Clostridiales* & *Bacteroidales* 351–352
Cl-VOCs. *see* Chlorinated volatile organic compounds (Cl-VOCs)
CMP. *see* Composite mycelial pellets (CMP)
COD. *see* Chemical oxygen demand (COD)
Composite mycelial pellets (CMP) 376–377
Composting 459–460
Continuous stirred tank reactors (CSTRs) 263–264
Conventional biofuels 449–450
Conventional remediation techniques 2, 14
Core–shell NPs 185
CP. *see* Chlorpyrifos (CP)
CPCB. *see* Central Pollution Control Board (CPCB)
CRISPR–Cas
 ethanol pathway optimization 458
 genome engineering 455–458
 heterogeneous 455–456
CRISPR–Cas9 technology 456, 458
Crop yield 457
Cross-linked enzyme aggregates (CLEAs) 109–110
Cryogel 117–118
CSTRs. *see* Continuous stirred tank reactors (CSTRs)

## D

DC-MFC. *see* Double-chamber microbial fuel cells (DC-MFC)
Decision trees (DTs) 347
*Dehalococcoides mccartyi* bacteria 8, 20–21
*Dehalococcoides* spp. 400
Deuteromycetes 214
Digestate 267–268
 liquid fraction processing 267
 posttreatment & valorization 267–268
 solid fraction processing 268
Digital technologies 432
Domestic wastewater 302–303, 303*f*
Double-chamber microbial fuel cells (DC-MFC) 22
DTs. *see* Decision trees (DTs)
Dye-decolorizing peroxidase (DyPs) 212

## E

Electrical resistance heating (ERH) 396–397
Electroactive microorganisms 81
Electrochemical capacitors 309
Electrocoagulation 94–95
Electronic waste (e-waste) 159
 composition & classification 160
 conventional approaches limitations 162
 degradation 167
 environmental & economic concerns 160–161
 management techniques 161–162
 physical & chemical methods 161–162
 pyrometallurgical methods 161–162
 recycling 168–171
Electrostatic spinning 374–375
Encapsulation technique
 mass transport 114–115
 material stability 113–114
 sustainability 114
Endo-1,4-β-D-glucanase 206
Endoglucanases 206
Endophytic microorganisms & mechanisms 16*f*, 16–17
Energy harvesting, MFC 302–308
 complex industrial wastes 304–305
 domestic wastewater 302–303, 303*f*
 food wastes 302–304
 landfill leachate 304
 sewage sludge 305
Engineered microbes 387
Environmental impacts, AD
 assessment 268–269
 LCA 269
Environmental pollutants 394

Environmental Protection Agency (EPA) regulations 3
Environmental remediation
biological 397–398
chemical 397
enzymatic oxidation 398–399
physical 396–397
thermal 398
Enzymes 286–287, 327*t*
biomass conversion 205–213
biphenyl bond cleavage 213
catalysis 323–324
degradation 224–226
immobilization 289, 329
strain improvement 219–220
technology advancement 220–221
waste valorization 226–227
EPA regulations. *see* Environmental Protection Agency (EPA) regulations
EPR. *see* Extended producer responsibility (EPR)
EPS. *see* Extracellular polymeric substances (EPS)
ERH. *see* Electrical resistance heating (ERH)
*Escherichia coli* 20–22
Esterase 209–210
Ethanol 454
pathway optimization 458
β-Etherase 212–213
Ex situ bioremediation 55
Ex situ thermal desorption 398
Exoelectrogens 299–301
Exoglucanase 206
Extended producer responsibility (EPR) 421–422
Extracellular polymeric substances (EPS) 113–114

## F

FAO. *see* Food and Agriculture Organization (FAO)
FBR. *see* Fixed-bed reactor (FBR)
Feedstock pretreatment
for AD 244–246
biological 251–252
characterization 246–247
chemical 249–250, 250*f*
need for 247–248, 248*f*
physical 248–249
physicochemical 249–251
Fermentation 323–324, 454
Fermentation-based valorization 454
Fixed-bed reactor (FBR) 264
Fluorescence infrared (FTIR) spectroscopy 152
Food and Agriculture Organization (FAO) 51–52
Food production systems 51
Food waste 462–463
valorization by CRISPR 457–458
Freshwater resources 79–80
Fruit waste 460
FTIR spectroscopy. *see* Fluorescence infrared (FTIR) spectroscopy
Functional group analysis 152
Fungi 58–59

## G

GA. *see* Glutaraldehyde (GA)
GAC-SBR system 86–87
Gaseous pollutants 394
Gasification 268, 431–432
and pyrolysis 390–391
GEMs. *see* Genetically engineered microbes (GEMs); Genome-scale metabolic models (GEMs)
Genetic engineering 52
Genetically engineered microbes (GEMs) 397–398, 406, 408
Genome editing 448–449
Genome-scale metabolic models (GEMs) 168–169
GH. *see* Glycoside hydrolase (GH)
GHG. *see* Greenhouse gas (GHG)
Gibbs free energy 260
Global warming 452–454
β-D glucosidase (BGL) 206–207
α-Glucuronidase 210
Glutaraldehyde (GA) 106–107, 111
Glycoside hydrolase (GH) 206–209
GOBARdhan 428
Gold NPs 186
Greenhouse gas (GHG) 267

## H

Hazardous & toxic wastes 1–2
classification & sources 3–6
environmental & public health impacts 6–7
Hazardous & Other Wastes Rules, 2016 423–424
amendments 424*t*
HCs. *see* Hydrocarbons (HCs)
Heavy metals (HMs) 1–2, 4, 6–7, 9–12
and metalloids 9–12
toxicity 395
Hemicellulases 208–210
Herbaceous biomass 203
HRT. *see* Hydraulic retention time (HRT)
HTP. *see* Hydrothermal pretreatment (HTP)
Hydraulic retention time (HRT) 87, 92

Hydrocarbons (HCs) 1–2, 6, 8, 13–14
  *hexa*-decane 13–14
  remediation 12–15
Hydrodynamic & magnetic separation 396–397
Hydrogenase 451–452
Hydrolysis 254*t*
Hydrothermal pretreatment (HTP) 249
Hydrothermal treatment 432

## I

Immobilized microbial technology 363–366, 366*f*
  biological nitrogen removal 375–378
  carriers selection 365–373
  LBL assembly 373–374
  refractory pollutants treatment 378
  wastewater treatment 364
In situ bioremediation 55
In situ ERH/TCH 398
Incineration 385–386
  energy recovery 390
Industrial
  and municipal effluent treatment 386–387
  waste zones 148–150
  wastewater streams 334
Inorganic NPs 184–185
Insect gut microbiota 145–146
Internet of Things (IoT) 432
Intrinsic bioremediation 55–56
In-vessel composting systems 393
IoT. *see* Internet of Things (IoT)

## K

Kamikatsu 435–436
*Klebsiella pneumoniae* GM-04 19–21
K-nearest neighbors (KNNs) 350

## L

Laccase 211
Landfill 417, 459
  leachate 304
  sites 148–150
Layer-by-layer (LBL) 363
LCA. *see* Life-cycle assessment (LCA)
LCB. *see* Lignocellulosic biomass (LCB)
Lead & cadmium 394–395
LHW. *see* Liquid hot water (LHW)
Life-cycle assessment (LCA) 226–227, 269
Lignin 449
  peroxidase 212
Ligninolytic enzymes 210–213
Lignocellulolytic enzymes
  bacterial 215–216
  biofuels 221–222
  bioplastics & biocomposites 224–225, 225*f*
  food & feed 223
  fungi 213–215
  industrial applications 221*f*, 221–224
  microbial production 213–220
  production techniques 217–219
  pulp & paper industry 223
  recombinant microorganisms 216–217
  textile industry 222–223
Lignocellulose 449–450, 452
Lignocellulosic biomass (LCB) 205*f*, 205, 210–211, 214–217, 244–245, 245*f*, 306, 449–450, 454–455, 458
  biofuels 221–222, 222*f*
  bioplastics 224–225, 225*f*
  challenges 224–226
Lindane (LIN) 57, 65–69
Lipases 286–287
Liquid hot water (LHW) 250–251
Lytic polysaccharide monooxygenases (LPMOs) 207–208

## M

Machine learning (ML) 346–347, 349–353
  biowaste composting 352
  microbial fuel cells 351–352
  wastewater treatment 351–352
Magnetic nanoparticles (MNPs) 121, 123–124, 126–127, 368
Malachite green (MG) 18–20
Manganese peroxidase 211–212
Marshes & swamps 204
Material recovery facilities (MRFs) 429
MBR. *see* Membrane bioreactor (MBR)
MC. *see* Microbial consortia (MC); Moisture content (MC)
MDC. *see* Microbial desalination cell (MDC)
MECs. *see* Microbial electrolysis cells (MECs)
Mediated electron transfer (MET) 81
MEF. *see* Mother Earth Foundation (MEF)
Membrane bioreactor (MBR) 88–89, 95–97
Membrane-less single-chamber MFCs (SC-MFCs) 22–24
Mercury 394–395
MESs. *see* Microbial electrochemical systems (MESs)
MET. *see* Mediated electron transfer (MET)
Metabolomics 405
Metagenomics 405

Metal–organic frameworks (MOFs) 370–371, 371*f*, 378
Metatranscriptomics 405
Methane production, kinetics
  first-order kinetic model 242
  france model 243
  Zwietering–Gompertz model 243
Methane yield curve 243–244, 244*f*
Methanogenesis 254*t*, 259
Methanogens 261–262
Methoxychlor (MTX) 66–67
MFCs. *see* Microbial fuel cells (MFCs)
MG. *see* Malachite green (MG)
Microalgae 204
  bacterial consortia 282–283
  biorefineries 462–464
  biotechnology 462–463
Microbes/microbial 8–13, 16, 25, 55, 57, 61–62, 143–144, 146, 251, 281–283, 287–288, 386, 398–399, 405
  adaptation to polymeric waste 148–150
  approaches 162–167
  biodegradation 143–144
  biofuels 451–452
  biotechnology 279–280, 290–291, 386
  cellulases 325, 328–329
  communities 148
  conversion processes 282–284, 283*f*, 288–289
  enzymes 153–154, 321–322, 324, 326–329, 403
  fermentation 279, 290–291, 462
  immobilization techniques 329*f*
  industrial sectors 324–325
  laccases 325
  mechanisms 167–168, 168*f*
  metal extraction 163–168
  pesticides 52
  processes 280–282, 323–324
  prokaryotic cell 10*f*
  recycling 171–173
  screening 150
  strains 167, 173
  treatment 447
  xylanases 325–328
Microbial bioremediation 58–59, 80, 386, 401*f*
  advancements in 405–406
  bioaccumulation 402
  bioaugmentation 400
  bioleaching 401
  biosorption 401
  biostimulation 400
  enzymatic oxidation/reduction 399–400
  factors affecting 402–403
  future of 406–407
  precipitation 402
  techniques & mechanisms 7–26
Microbial cell immobilization 106–120
  adhesion & adsorption 108*f*, 108–109
  advantages & disadvantages 115–116, 116*t*
  applications 115–119, 119*t*
  cross-linking/covalent binding 107–108, 110–111, 111*f*
  encapsulation 111–115, 112*f*
  entrapment 109*f*, 109–110
Microbial consortia (MC) 57, 148–149, 171, 287–288
  bioremediation 62–64
  designing 145
  and mixed cultures 61–70
Microbial desalination cell (MDC) 312
Microbial electrochemical systems (MESs) 81–83, 96–97
  construction & configuration 82
  principles & mechanisms 81–82
  waste management 82–83
Microbial electrolysis cells (MECs) 280, 284, 289–290, 312, 331
Microbial fuel cells (MFCs) 2–3, 22–27, 23*f*, 280–283, 299–302, 304, 306, 310, 314, 323–324, 329–331, 330*f*, 333–334, 352–353, 392
  air cathode 22
  animal wastes 305–306
  BPEC 313–314
  capacitor-based power management 309–311
  challenges & perspectives 314–316
  constructed wetland 23
  current density 301
  direct power utilization 311–314
  energy storage 308–311
  energy-generating capacity 301–302
  hybrid systems coupling 331–333
  photoelectrocatalytic system 313–314
  photosynthetic 331–333
  plant wastes 306–308
  power density 302
  scale-up 331–333
  sediment 24–25
  soil-based 23
  techno-economic barriers 333–334
  tubular 331–333
  upflow 331–333
  wastes by 300*f*, 302–308, 308*t*
Microbial hybrid systems 81–97
  challenges 95–96

ion exchange 93*f*, 93–94
pollutant removal 94*t*
Microbial photoelectrochemical cell(MPEC) 313–314
Microbial remediation strategies 2–3, 18, 27
advances & technologies 22–26
bioaugmentation process 8–9
chlorinated solvents 20–22
dye remediation 18–20
HMs & metalloids 9–12, 11*t*
synthetic biology & omics 25
Microbial technologies 331–335
advantages 325–326
chemical processes 326*t*
fundamentals of 323–326
Microbial technologies in e-waste 170, 174
BES 170
bioaccumulation 166
bioleaching 163*f*, 163–166, 165*t*
biosorption 166
enzymatic degradation 166–167
genetic engineering 168–169
microbial consortia 169
nanotechnology integration 169
Microbial waste management, AI 342–347, 356
application 347–350
artificial neural networks 344*t*
bin detection 349–350
decision trees 347
random forest 346–347
support vector machine 346
vehicle & route optimization 350
Microencapsulation 112*f*, 112
Micro/nanofabrication techniques 378–379
Mineralization 152
Mint innovation 171–173
Miscanthus cultivation 307*f*
ML. *see* Machine learning (ML)
MNPs. *see* Magnetic nanoparticles (MNPs)
MOFs. *see* Metal-organic frameworks (MOFs)
Moisture content (MC) 246
Molecular weight determination 150
Mother Earth Foundation (MEF) 434–435
MPEC. *see* Microbial photoelectrochemical cell (MPEC)
MRFs. *see* Material recovery facilities (MRFs)
MSW management, India 419–420
current practices 419–420
MSW. *see* Municipal solid waste (MSW)
MTX. *see* Methoxychlor (MTX)
Municipal solid waste (MSW) 341, 385
Mycorrhizal fungi 397–398


## N


NA. *see* Natural attenuation (NA)
NAMASTE scheme 429–430
Nanobiosensors 184, 187, 190–192
Nanocarrier materials 368–369, 369*f*
Nanoencapsulation microbial cells
bacteria performance 123–124
pollutant removal mechanism 125*f*, 125–126, 128*t*
Nanofertilizers 187–189
nutrient use efficiency 191
Nanoherbicides 187
Nanomaterials 186–193
agricultural use 186–187
benefits 121
coated microbial cells 120–126
Nanoparticles (NPs) 107, 120–123, 184, 186, 188–191
classification 184–185
encapsulated bacteria 123–126
metal 121–122
Nanopesticides 187–188
Nanosilica & polymeric nanospheres 369
Nanotechnology 25, 120, 183–184, 187, 192–193
in agriculture 183–184, 190–192
managing agricultural waste 189
nanoremediation 191–192
plant cultivation 187–189
resilience 192
soil health 189–192
National Green Tribunal (NGT) 426–427
National Mission on Sustainable Habitat (NMSH) 429
Natural attenuation (NA) 55–56
Natural carrier materials 368
NER. *see* Normalized energy recovery (NER)
NGT. *see* National Green Tribunal (NGT)
NMSH. *see* National Mission on Sustainable Habitat (NMSH)
Normalized energy recovery (NER) 306–307
NPs. *see* Nanoparticles (NPs)


## O


ODF. *see* Open Defecation Free (ODF)
OLR. *see* Organic loading rate (OLR)
OMSW. *see* Organic municipal solid waste (OMSW)
Open Defecation Free (ODF) 427–428
OPP. *see* Organophosphorus pesticides (OPP)
OPs. *see* Organochlorine pesticides (OPs)
Organic loading rate (OLR) 263

Organic municipal solid waste (OMSW) 393–394
Organic NPs 184–185
Organic waste 433–434
  anaerobic digestion 431
  composting 430–431
  vermicomposting 431
Organochlorine pesticides (OPs) 52, 65–66
Organophosphorus pesticides (OPP) 68–69
Oxidative treatments 397
Oxygen 265

## P

PAC-MBR hybrid system 86*f*, 86–87
PAHs. *see* Polycyclic aromatic hydrocarbons (PAHs)
Partial least square (PLS) method 349
Particle swarm optimization (PSO) 352
PCBs. *see* Polychlorinated biphenyls (PCBs)
PCR. *see* Polymerase chain reaction (PCR)
PEG. *see* Polyethylene glycol (PEG)
Persistent organic pollutants (POPs) 1–2, 4–5, 12
  microorganisms & metabolic pathways 13*t*
Pesticides 51–52
  *Actinomycetota* for bioremediation 64–65
  biochemical 52
  classification 52
  conventional treatments 54
  pathways 54
  remediation 15–17
  risk level 52–53
  runoff 394–395
Pesticides pollution 51
  eco-friendly solutions 54–55
  primary pathways 53
Petrochemical (PHC) residues 1–2, 6, 12–14, 24
Petroleum hydrocarbons 394–395
PFR. *see* Plug-flow reactor (PFR)
PGP. *see* Plant growth-promoting (PGP)
PHA. *see* Polyhydroxyalkanoates (PHA)
Phenol degradation 378
*Phormidium autumnale* 19
Photocatalytic oxidation 313
Phylum *Actinomycetota*
  bioremediation potential 60–61
  characteristics 59–60
Physicochemical methods/process 54, 105, 448
Plant
  cell culture systems 289
  pathogens 186–187
  wastes 306–308
Plant growth-promoting (PGP) 15–16
Plastics 447
  degradation 386–387
PLS method. *see* Partial least square (PLS) method
Plug-flow reactor (PFR) 264
PMC. *see* Pune Municipal Corporation (PMC)
Pollutants 105, 115–116, 123–124, 126, 128*t*
Polychlorinated biphenyls (PCBs) 3–4, 12
  antibiotics 12
  biodegradation 63
Polycyclic aromatic hydrocarbons (PAHs) 6–8, 12, 14–15, 22–24
Polyethylene glycol (PEG) 110, 114
Polyhydroxyalkanoates (PHA) 286–287
Polylactic acid 286–287
Polymerase chain reaction (PCR) 404
Polymeric NPs 185
Polysaccharides 247
Polyurethane foam 117
Polyvinyl alcohol (PVA) 113–114, 117
Polyvinyl chloride (PVC) 143–144
  biodegradation process 147*f*
  bioremediation in environment 153–154
  plastic waste occurrence 144*f*
  polymers 145
POPs. *see* Persistent organic pollutants (POPs)
Porous NPs 185
Proteomics 405
Protoplast fusion 146
*Pseudomonas putida* 387
*Pseudomonas* sp. strain PS6 15–16
PSO. *see* Particle swarm optimization (PSO)
Pune Municipal Corporation (PMC) 436–437
*Purpureocillium lilacinum* 400
PVA. *see* Polyvinyl alcohol (PVA)
PVC degradation
  analytical methods 150–152
  biodegradation approaches 145–146
  biofilms in marine 148
  bioprospecting microbes 146–150
  fungal species 148
  genetic engineering 146
  marine bacteria 148
  microbes 147–148, 150*t*
  microbial enzyme application 146
  multienvironment pretreatment 146
  potential bioprospecting environments 149*f*
  pure culture isolation & identification 145
  thermophiles & halophiles 149
PVC. *see* Polyvinyl chloride (PVC)

PWM Rules, 2016 421–422
  and amendment 422*t*
Pyrolysis 268, 431–432

## Q

Quantitative PCR (qPCR) 404

## R

Radioactive pollutants 396
Radioactive waste 5
  remediation 17*t*, 17–18
Random forest (RF) 346–347, 349, 351–352, 355–356
  and ANN 352
  microbial biodegradation 354–355
RBC-UF hybrid system 88–89
RCRA. *see* Resource Conservation & Recovery Act (RCRA)
RDF. *see* Refuse-derived fuels (RDF)
Reactive oxygen species (ROS) 122
Recycling & circular economy 387–389, 388*f*
Redox reactions 81–82
Reductive dechlorination 20
Refuse-derived fuels (RDF) 429
Resource Conservation & Recovery Act (RCRA) 3, 6
Resource recovery 280, 282–283, 289–290
  bioenergy production 284–285, 285*f*
  bioproduct synthesis 286–287
  bioremediation 287–288
  microbial conversion 288–289
  nutrient recovery 285–286, 286*f*
RF. *see* Random forest (RF)
ROS. *see* Reactive oxygen species (ROS)
Rotterdam Convention 1

## S

*Saccharomyces cerevisiae* 391
Sanitary landfilling 385
SBM. *see* Swachh Bharat Mission (SBM)
SDGs. *see* Sustainable Development Goals (SDGs)
Siderophores 10–11, 15–16
SIP. *see* Stable isotope probing (SIP)
Sludge solubilization 251–252
Smart cities mission 428–429
SMES. *see* Superconducting magnetic energy storage (SMES)
SmF. *see* Submerged fermentation (SmF)
Soft-rot fungi 214
Soil
  contamination 53
  ecosystems 147–148
  fungi 148
  harbors 147
  pollutants 395
  replacement/washing 396–397
  waste 348, 388
  waste generated in India 417, 419
Soil vapor extraction (SVE) 396–397
Solid waste management (SWM) 419–421, 426–429, 433–434, 436–438
  sustainability challenges 432–433
Solid-state fermentation (SSF) 217
SRB. *see* Sulfate-reducing bacteria (SRB)
SSF. *see* Solid-state fermentation (SSF)
Stabilization/solidification methods 397
Stable isotope probing (SIP) 405
Stockholm Convention 1
*Streptomyces* sp. 61, 65–69
Submerged fermentation (SmF) 217–219
Sulfate-reducing bacteria (SRB) 259–260, 265
Sulfide 265
Superconducting magnetic energy storage (SMES) 309
Support vector machines (SVMs) 342, 346, 349, 355–356
Surface topography 152
Sustainable Development Goals (SDGs) 418
  waste management 418*t*
SVE. *see* Soil vapor extraction (SVE)
SVMs. *see* Support vector machines (SVMs)
SWaCH cooperative 436–437
Swachh Bharat Mission (SBM) 419
  urban & rural 427–428
SWM Rules, 2016 420–421, 426–427
SWM. *see* Solid waste management (SWM)
Synthetic
  and engineered consortia 379
  biology techniques/tools 2–3, 406–407
  pesticide residues 395

## T

Thermal desorption 396–397
Total solids (TS) 246
T-RFLP & ARISA techniques 405
Tricarboxylic acids (TCAs) cycle 12, 14–15
Trichloroethylene (TCE) 8, 17–18, 20–21
Tubular MFCs 331–333

## U

Upflow anaerobic sludge blanket (UASB) 264
Upflow MFCs 331–333

## V

Valorization methods 460–462
Vermicomposting 452–454, 453*f*
Volatile fatty acids (VFAs) 246–247, 258*f*
Volatile organic compounds (VOCs) 6, 20–22, 26, 394

## W

Waste
  based carriers 378
  biomass sources 370
  classification & sorting systems 348–349
  collection 348–350
  derived feedstocks 391
  generation prediction 347–348
  valorization 280–281, 287–288, 461*t*
Waste management 106, 341, 385–388
  bioadsorption systems 86–87
  biological membrane filtration 88–89
  biological oxidation 90–92
  biophotocatalysis 84–85
  digital technologies & IoT 432
  immobilization approaches 106*f*
  India 436–437
  Japan 435–436
  MESs 82–83
  Mexico 433–434
  microbial 342–347, 344*t*
  Philippines 434–435
  smart 348*f*
  sustainable 417–418
  traditional 341–342
Waste-to-energy (WtE) technologies 385–387, 389*f*, 389–393, 431–432
  BES 392
  biological conversion 391–392
  electroactive systems 390–391
  hybrid/integrated systems 392–393
  thermal conversion 390–391
Wastewater 79–80
  purification methods 80
Wastewater treatment 85–86, 90, 93–96, 364, 446
  advancements 378–379
  biological nitrogen removal 375–378
  metal-organic frameworks 370
Water pollutants 394–395
Water treatment 80–85
Waterborne polyurethane (WPU) 113
White-rot fungi 215
Wind/water erosion 53
Wood–Ljungdahl pathway (WLP) 258
Woody biomass 202–203
WPU. *see* Waterborne polyurethane (WPU)
WtE technologies. *see* Waste-to-energy (WtE) technologies

## X

Xylanases 321–326, 328–329
β-D-xylosidase 208–209

## Y

Yeasts 58

## Z

Zero-waste agriculture 447, 463
Zinc oxide NPs 187–188, 191–192
Zwietering–Gompertz model 243

Printed and bound by CPI Group (UK) Ltd, Croydon, CR0 4YY
21/09/2026
02220078-0006